Mathematics Education in Different Cultural Traditions- A Comparative Study of East Asia and the West

New ICMI Study Series

VOLUME 9

Published under the auspices of the International Commission on Mathematical Instruction under the general editorship of

Hyman Bass, President Bernard R. Hodgson, Secretary-General

The titles published in this series are listed at the end of this volume.

Frederick K.S. Leung
Klaus-D. Graf
Francis J. Lopez-Real
(Editors)

Mathematics Education in Different Cultural Traditions- A Comparative Study of East Asia and the West

The 13th ICMI Study

Springer

Frederick K.S. Leung
The University of Hong Kong
China

Klaus-D. Graf
Freie Universitaet
Germany

Francis J. Lopez-Real
The University of Hong Kong
China

Library of Congress Control Number: 2005935746

ISBN-10: 0-387-29722-7 (HB)
ISBN-13: 978-0387-29722-4 (HB)
ISBN-10: 0-387-29723-5 (e-book)
ISBN-13: 978-0387-29723-1 (e-book)

Printed in the United States of America.

9 8 7 6 5 4 3 2 1

springer.com

Contents

Preface

1. BACKGROUND

This volume is the final outcome from the 13th ICMI Study Conference held at the University of Hong Kong in October 2002. Such Study Conferences have been organized by ICMI on a regular basis since the 1980's and have focused on topics deemed to be of significance, in terms of both theory and practice, for the international mathematics education community. Recent years have seen an increasing interest in comparative studies across nations, particularly in terms of students' achievement and performance, as witnessed by such large-scale projects as TIMSS and PISA. Reports from such studies have generated much discussion and not a little controversy. Whilst they undoubtedly provide valuable and rich sources of data, many educators have felt the need to interpret such findings within the subtler context of the underlying cultural traditions of the participants. The perspective of different cultural traditions is also an important one when we consider the broader aspects of mathematics education, not simply the measurement of performance, in an international context.

The ambitious idea of a comparative study conference organized through ICMI, and a subsequent Study Volume, was first suggested by Professor Dianzhou Zhang to the Executive Committee of ICMI in 1997. The main premise of the proposal was that the study should use the lens of different cultural traditions through which to make analyses of mathematics education in an international context. An International Programme Committee (IPC) was formed in early 2000 and its first task was to produce a Discussion Document about the proposed study. Since this document formed the framework for the Study Conference, and indeed was the initial framework for the present Volume, we have included the full text at the beginning of this volume. For logistical reasons it was realized at the outset that it would be impossible to cover a wide range of different cultures and the decision was made to limit the study to East Asia and the West.

2. THE STUDY CONFERENCE

After dissemination of the Discussion Document to the international community, a large number of papers were received by the IPC as possible contributions to the study. Eventually 42 papers were selected and their authors were invited to attend the Study Conference in Hong Kong. These 42 papers were published by the Faculty of Education of the University of Hong Kong in a 'Pre-Conference Proceedings' in October 2002. As with previous ICMI Study Conferences, it was not intended that these papers should necessarily form the subsequent Study Volume arising out of the conference. Rather, they would form the *starting points* for discussion, out of which further elaboration of the papers could be undertaken, or new collaborations for other papers could be established. Based on the range and emphasis of the papers, the conference was organized around five themes. These themes are shown below together with the group leaders for each theme:

Context (FAN Liang-huo, Walther FISCHER)
Curriculum (Margaret WU, ZHANG Dian-zhou)
Teaching & Learning (Colette LABORDE, LIN Fou-lai)
Values & Beliefs (Alan BISHOP, Katsuhiko SHIMIZU)
Textbooks (PARK Kyungmee, LEUNG Koon Shing Frederick).

However, in addition to this organized framework, certain sessions were timetabled as 'Ad-hoc Discussion' sessions. The intention was to provide the opportunity for discussions to take place between participants from different groups, as a kind of cross-fertilisation process. These ad-hoc discussions also proved to be very fruitful.

3. STRUCTURE OF THE STUDY VOLUME

By its nature, a Study Conference has a fluid and organic structure that is likely to produce on-going change throughout the study. This comparative study conference was no exception and the final structure of the present Volume is the result of the changing emphases and concerns that took place during the study. The main themes identified above have been adhered to, but the chapters focusing specifically on textbooks have been incorporated into the broader theme of Curriculum. In the same way, the original subsection on Assessment (in the Discussion Document) has also been subsumed into this theme. Hence this volume consists of four major sections:

Context of Mathematical Education
Curriculum
Teaching and Learning
Values and Beliefs

There is also an 'Introduction' section that includes the Discussion Document and a chapter by Frederick Leung that 'sets the scene' for the whole Volume. Finally there is a last section that we have called 'Outlook and Conclusion' in which, apart from the editors' attempts to provide some overall conclusions arising from the previous sections, we include a chapter that considers, albeit briefly, methodological issues, a chapter on some other cultures, and a final chapter by Alan Bishop that asks the reader "what comes next?"

4. OTHER PERSPECTIVES AND FUTURE RESEARCH

It is inevitable in a volume such as this, that one's initial wide-ranging and ambitious intentions can never be entirely fulfilled. However, this can also be looked at in a positive light in the sense that it can provide some signposts for the directions that further work in this area could focus on. Here we identify just three possibilities.

4.1 Technology

Under the section "Aspects of the study" in the Discussion Document, the use of technology in the teaching and learning of mathematics was not specifically highlighted. However, the sub-section 'Methodology and media' gave clear indications that this topic was an important one to consider. And in the wider context, discussions on the impact of globalisation are often linked to advances in technology. Within the international mathematics education community, increasing attention has been paid to the impact of technology in conferences and journal papers. Indeed, recent years have seen the establishment of new international journals with a specific focus on this area of mathematics education. It was therefore rather a surprise to us that we did not receive more papers on this topic after the initial call for papers and that this area was not explored more fully in the Study Conference. This volume does include three chapters in Section 3 (Teaching and Learning) that specifically focus on technology and distance learning and other chapters do touch on technology but not as the main focus of interest. Whatever the reasons for this, we feel that this should be a fruitful area for further comparative research in mathematics education

4.2 Teacher education

By contrast with the technology aspect above, the Discussion Document did propose a specific subsection on teacher education. In fact, the full heading for the subsection was 'Teachers, teacher education, values and beliefs'. While there were many papers in the original submissions for the Study Conference that focused on values and beliefs and on the cultural influences on teachers' instructional practices, we were again surprised by the lack of comparative studies on the preparation of teachers. That is, what differences and similarities exist in the preparation and training of mathematics teachers between East Asia and the West? For example, in terms of both the emphases in the respective teacher education curricula and the expected roles of the teacher that are inherent in such curricula. Questions such as these are touched on implicitly in some of the chapters but we suggest that the specific area of teacher education is one that could be another fruitful area for further research.

4.3 Comparisons of other cultural traditions

As we have described earlier, this volume is essentially limited to the comparison of just two cultural traditions, East Asia and the West. Even within this limitation it is clear, as discussed in some of the chapters, that no simplistic homogeneity of cultural tradition exists under these two broad categories. We certainly hope that this volume will be considered as a valuable contribution to the comparative study of the cultural influences on mathematics education across these two particular traditions. However, we hope it may also be seen as a starting point for further research into similarities and differences *within* these broad cultural traditions and also as a stimulus for work that looks at *other* cultural traditions across the global context. We have attempted to give at least a flavour of this broader perspective in Chapter 2-1 and Chapter 2-8 and in Section 5 of the volume.

MATHEMATICS EDUCATION IN DIFFERENT CULTURAL TRADITIONS: A COMPARATIVE STUDY OF EAST ASIA AND THE WEST

Discussion Document

PREAMBLE

Education in any social environment is influenced in many ways by the traditions of these environments. As a consequence the results of such education will naturally differ with different traditions in different environments. Indeed, this is necessary since one of the intentions of education is to support the traditional continuity of structure and function of a special environment.

On the other hand, today we are observing a growing interdependence between environments like regions, states, countries, and different cultural areas of the world. In many respects they have to rely on corresponding or equivalent standards of education, and differences can cause irritations.

In mathematics education also, taking an international and intercultural point of view, we face this split phenomenon of difference and correspondence, linked with the perpetual challenge to improve the quality of mathematics education. A study attempting a comparison between mathematics education in different traditions will be helpful to understand this phenomenon in detail and to exploit it for the sake of mathematics education. From this, paths will be discovered leading to adequate and effective applications of differences, as well as correspondences, in national and international environments.

Due to the size of an ICMI Study, in manpower and in time, this enterprise must be limited to only a selection of cultural traditions. Those based in East Asia and the West seem particularly promising for a comparison, since similar interests in differences and correspondences have existed for a long time and experiences in equivalent research have been gathered.

A rich variety of aspects of mathematics education is to be considered in this comparative study, ranging from the host of social, economic and other contexts, curricula, teachers, students, goals, contents, methodology, media etc. to the nature of mathematics and the future of mathematics as well as mathematics education. Traditions of teaching and learning that are deeply embedded in history and culture will have to be compared, with a consideration of the rich experience growing out of them as well as their resistance to change.

At the same time, this comparative study must consider present developments in society, science and technology as well as ethics. Changing attitudes between generations are influencing the teacher-student relationship, as are the new information and communication technologies. In addition, these technologies define new roles for both the teachers and learners and the reaction is different in different traditions.

What kind of subjects will there be in schools of the future and how much planning is going on? In what ways will mathematics education of the future be comparable to that of today and how will it differ? What forces are competing in this field?

Exchanges of experiences and expectations will be an important part of the study and critical considerations will be inevitable.

Previous ICMI studies normally proceeded in three steps: Discussion Document, Study Seminar and Study Volume. In our case we will insert an Electronic Discussion Forum before and possibly even after the seminar.

First, the IPC offers a Discussion Document to the mathematics education community and people from interested contexts. We will welcome applications for a study seminar by invitation which we expect to take place in Hong Kong in October 2002. Contributions can come from individuals as well as jointly from colleagues who are already engaged in comparative activities about different traditions in mathematics education. This will allow an operationalising of the study by referring to case studies, for example.

Second, the Electronic Discussion Forum will allow statements about the theme of the study in general and corresponding comments and questions from any colleagues interested in the study.

Among other intentions the Forum should especially enable colleagues interested in the same or similar field of comparison to meet and to cooperate in preparing a contribution in general or a case study in particular.

Third, the Study Seminar will consist of presentations identifying and interpreting consequences from different traditions to a variety of aspects of mathematics education. Moreover, a great deal of work has to go into the comparing of observations and findings, for example in focus groups. In this way the seminar will arrive at recommendations for the applications mentioned above, serving to make differences and correspondences fruitful for national and international education.

Fourth, a Study Volume will be published for the mathematics education community and the interested public, containing the results from the communications and comparisons at the seminar.

1. WHAT IS IN THIS STUDY?

1.1 Scope of the study

In this study, we confine ourselves to mathematics education of school age children (although we are not confining ourselves to mathematics education in regular schools) and related areas such as teacher education, and will not specifically study issues such as vocational education or education at tertiary level. Problems in mathematics education to be investigated may include issues such as curriculum, assessment, policy, influences of information and communication technology (ICT) and multi-media, community and family background etc. These will be further elaborated in Section 4.

1.2 What do we mean by "cultural tradition" in this study?

"Tradition" is an equivocal word which can refer to many different things. As this study is on mathematics education, we obviously would like to study "traditions" which have a bearing on mathematics education. Two obvious choices are the *education* tradition and the *mathematics* tradition in different countries or regions. Indeed it is reported in the literature that mathematics education in a given country or system is very much influenced by the underlying education tradition and mathematics tradition. However, in this study we would argue that both the education tradition and the mathematics tradition are related to a deeper level of tradition, that of the culture. And it is this deeper level of *cultural tradition* with reference to which we would like to compare the mathematics education in different regions.

Culture is "one of the two or three most complicated words in the English language". It may refer to "the fabric of ideas, ideals, beliefs, skills, tools, aesthetic objects, methods of thinking, customs and tradition", or the "*configuration* or *generalization* of the 'spirit' which inform the 'whole way of life' of a distinct people". For this study, culture refers essentially to values and beliefs, especially those values and beliefs which are related to education, mathematics or mathematics education. An example of a value that pertains to education is the importance attached to education in different cultures. An example concerning mathematics is the view of the nature of mathematics (e.g. whether or not it is essentially a pragmatic discipline). An example of a belief that affects mathematics education is the attribution of success and failure in mathematics (e.g. attribution to effort versus innate ability).

Although "culture" is a crucial concept in this study, it is not the intention of this paper to offer a comprehensive definition of the term. Rather, the brief discussion and examples above are meant to indicate the level of depth we are referring to when the term "tradition" is used and to stimulate discussion on what cultural values and beliefs are relevant to a discussion on mathematics education. Indeed, one important purpose of this study is exactly to identify the aspects in our cultural values which have impact, directly or indirectly, on mathematics education.

1.3 What are "East Asia" and "the West"?

This study is on the cultural traditions in East Asia and the West. In using the terms "East Asia" and "the West", we do not merely refer to geographic areas. Our contention is that cultural divisions are much more meaningful than political or geographic divisions in explaining differences of educational practices in mathematics. East Asia and the West in this study are therefore cultural demarcations rather than geographic divisions, roughly identified as the Chinese/Confucian tradition on one side, and the Greek/Latin/Christian tradition on the other. We acknowledge that neither of these "poles" is well defined, as with any label given to any culture. But we use the two terms to point to the scope that we want to confine ourselves to in this study.

In identifying these two "poles", we are not claiming that the two cover all major cultural traditions in the world. Nor are we implying that these two are the most important human traditions. For example, it has been pointed out that there is a distinctive East European tradition in mathematics education which is definitely worth studying. Equally worth studying are traditions in South Asia (in particular that of India) and Africa. However, it would not be possible for a single ICMI study to cover all important traditions worldwide. What we hope to achieve, by choosing these two poles for study, is a balance between using pertinent examples to study the relationship between cultural traditions and mathematics education on the one hand, and choosing two major traditions that have attracted attention in the field of mathematics education on the other. More justifications on the choice of the two traditions will be discussed in section 2.

1.4 What do we mean by a comparative study?

To compare means to identify similarities and differences, and to interpret and explain the similarities and differences identified. It may not be as easy as it is conceived. Given two things or concepts, there may exist infinitely many aspects of similarities and differences, and hence in a comparative study, we are always confining the comparison to a particular theme or some particular themes. For our study therefore, we are comparing practices in mathematics education (as defined in 1.1 above) along the theme of *cultural traditions* (as described in 1.2 above). Reminding ourselves of this obvious point is important. In studying mathematics education in different countries, we will definitely come across important aspects of mathematics education that are of interest to us. But in deciding which of those aspects should be included in our study, we have to ask ourselves whether or not those aspects are related to this theme of cultural traditions. Some aspects of mathematics education which we deem to be related to different cultural traditions will be discussed in Section 4.

After we have identified similarities and differences within a certain theme, the next question is what to do with them. A simple juxtaposition of similarities and differences does not in itself explain. There is a need for analysis based on certain theoretical frameworks, or in the absence of a suitable theoretical framework, a need for the establishment of one, based on the differences and similarities observed. More about how this comparative study is operationalised will be discussed in Section 3.

2. THE RATIONALE FOR THIS STUDY

2.1 Why is this comparative study important?

2.1.1 Pressures and needs from outside mathematics education

Rapidly developing information and communication technologies have an enormous influence on mathematics, science, production, society, politics, education and even lifestyle. Increasing globalisation is encouraging the assumption of universalism in mathematics education. The increase in journals and books about mathematics teaching, the multitude of conferences in every part of the world, the availability of materials via the World Wide Web, and the activities of multinational computer companies all increase the pressure for adopting similar practices in mathematics teaching around the world.

At the same time, however, the globalisation processes are producing reactions from mathematics educators in many countries who are concerned that regional and local differences in educational approach are being eradicated. This is not just a mathematical ecology argument, about being concerned that the rich global environment of mathematical practices is becoming quickly impoverished. It is also an argument about education, which recognises the crucial significance of any society's cultural and religious values, socio-historical background and goals for the future, in determining the character of that society's mathematics education.

Policy makers also recognise the importance of adjusting to the changing world and mathematics education reform movements can be found in many countries at this moment. A number of international studies have also taken place in the last decade to provide policy makers with information on the relative standing and effectiveness of their education systems.

2.1.2 Pressures and needs from within mathematics education

Mathematics educators from all around the world are continuing to make efforts to improve the quality of mathematics education. One way to achieve this is through learning from different countries. For instance, many schools in the US are importing the mathematics text books of Singapore, while Asian countries are mimicking the approach taken by US textbooks that have been developed as a result of many specific educational projects. While these exchanges do not yet show clear evidence about which approach is the most appropriate for each country, they stress the need for more awareness of the cultural traditions in which the respective mathematics teaching is embedded. More in-depth studies on the relative merits of the different approaches within different education systems are needed.

2.2 Why is this study focused on a comparison between East Asia and the West?

A complete comparison of the background, perspectives and practice of mathematics education around the world is far beyond the scope of an ICMI study. Realistically we need to limit the range of a comparative study while still ensuring that clearly different traditions are being examined. Recently, comparative studies such as SIMS and TIMSS have produced data indicating that there may be some systematic reasons for differences in achievement and practice between some regions. East Asian countries such as Japan, Korea, Taiwan, China, Singapore and Hong Kong consistently outperform western countries in North America, Europe and Australia in these international tests. These results have brought about a growing interest for policy makers and educators to find out the factors behind Asian students' high performance in mathematics.

Surprisingly, a superficial look at mathematics teaching in Asian countries indicated that teaching methods in these countries are not perceived as advanced as in Western countries. For instance, mathematics education in the East Asian countries mentioned above is typically characterised by the following: They are very often content oriented and examination driven. Large class sizes are the norm and classroom teaching is usually conducted in a whole class setting. Memorization of mathematical facts is stressed and students feel that their mathematics learning is mainly learning by rote. Teachers feel guilty for not teaching enough problem-solving during their classes. Students and teachers are subjected to excessive pressure from highly competitive examinations, and the students do not seem to enjoy their mathematics learning.

This parity between the high mathematics performance in Asian countries and a lack of modern teaching methods is puzzling. It prompts for a call for in-depth studies about mathematics learning that goes beyond classroom teaching. Cultural traditions, for example, could be highly pertinent to students' learning. In particular, there are marked differences in cultural traditions between the East and the West. These cultural differences provide us a unique opportunity to gain a deeper understanding about students' learning and achievement.

This study presupposes that the impact of cultural tradition is highly relevant to mathematics learning. Cultural traditions encompass a broad range of topics. It includes the perceived values of the individual and the society, as well as social structures such as the relationship between parents and children, or the relationship between teachers and students. There are clear differences in all these areas between Asian and Western traditions. For instance, some scholars have identified features of East Asian mathematics education, together with their underlying values, by contrasting them to the West.

Apart from scientific interests in comparing mathematics achievements in the East and the West, there is also a strong political impetus behind such studies. After World War II, Western countries have dominated world economic development for a long time, but in the last ten years, we saw the emergence of East Asia as strong competing economic powers. The United

States government, for example, has given very strong support in international comparative studies of mathematics and science, with particular emphasis in the comparison between the U.S. and Japan. The Australian government, on the other hand, recognising the importance of its geography proximity to Asian countries, has funded special studies aiming to get a better understanding of mathematics learning in Asian countries.

The culmination of all of the above issues has made a study between Asian and Western traditions in mathematics education not only interesting but also important. There is also some urgency for such a study to take place now, because with the current advances of technology and communication, the differences between cultural traditions are diminishing. Before long, we will not have such contrasting traditions to provide us with rich information to carry out such as a study.

2.3 What can be achieved from this study?

The focus of this study is on differences in educational traditions which have cultural and social implications for the environment and practice of mathematics education. This focus does not mean a narrow interest on how to improve scores in international tests. The study is about gaining a deeper understanding of all aspects of mathematics learning and teaching, and about what each tradition can learn from the other.

To develop this point further, the study should not be a 'horse-race-type' of study resulting in a linear ranking list. Instead, in this study we want to start with a mutual understanding of the mathematics education systems and processes in different traditions that lead to more or less satisfactory success in the learning and application of mathematics. Given this degree of 'success', according to some agreed criteria, then we need to identify the conditions that contribute to this success (e.g. learning techniques, external influences, natural abilities of the children, language etc). Only then can any kind of transfer of methods take place, if desired, from one tradition into another. It is not a simple import-export business. As an old Japanese proverb puts it: *We cannot easily plant a good seed of another field into our own field.*

In summary, we are hopeful that this study can achieve the following:

1. By contrasting the different traditions, we gain a deeper understanding of various aspects of mathematics learning and teaching. For example, we may gain an insight into the cognitive processes of "doing mathematics", such as learning about place values, or grasping the concepts of abstract representations in algebra.

2. By contrasting different traditions, we develop a process of self-reflection on our traditional ways that we often take for granted. There is the opportunity to take a fresh look at our usual practices and beliefs, and, in the process, we gain a better understanding about our own traditions.
3. By contrasting the different traditions, we share between us the latest educational development and research. We learn from each other's successes and failures, and develop a common goal for improvement for the coming years.

3. HOW WILL THE STUDY BE OPERATIONALISED?

This study is rather different from the earlier ICMI studies in that it is specifically concerned with comparing practices in different settings and with trying to interpret these different practices in terms of cultural traditions. It is also the intention of the IPC to ensure that the study will result in various products and outcomes including, of course, a book for the ICMI series, but equally important is the process by which the study will proceed. This means that there is a need to create certain kinds of activities for engaging the participants in operationalising the study, and through these activities to develop specific kinds of contributions to the study. The IPC has identified the following as being some of the most important activities and contributions for this study.

3.1 Identification and analysis of previous studies

As has been pointed out in the previous section, there is clearly a need to build on the several previous studies which have been carried out on East/ West differences, and we wish to encourage contributors to the conference to be aware of the available background literature. In particular we note that there have been various achievement-based studies such as SIMS and TIMSS, and other studies such as PISA etc.

Contributions and proposals will therefore be useful which give some synopses and critical analyses of these studies. In particular, as well as focusing on the various ideas and results that have come from these studies, we will have an interest in the methodologies used in them. Most of the analyses performed on the data of these international comparison projects have been quantitative in nature, and we are more interested in a qualitative aspect which seems to have been ignored so far. The development of productive cooperative researching is one of the outcomes we seek and awareness of previous cooperations will be very helpful in this development.

3.2 Joint contributions

In the same spirit we wish to encourage joint contributions from colleagues who are already engaged in East/West cooperative or comparative activities, as well as from those who wish to use the study as an opportunity to begin to engage in such activities. Previous international conferences such as PME, HPM, ICME have already provided some contexts for cooperative international activities and it will be important to build on those activities. These conferences have already demonstrated the value of building trust between cooperating colleagues, as well as the importance of taking advantage of the different backgrounds and knowledge of the participants.

The nature of this ICMI comparative study is such that it will be highly dependent on trust and productive cooperation between the participants, and we realise that this kind of cooperation can take time to develop. There is much to learn from each cultural tradition before one can begin to seek meaningful contrasts and to develop productive explanations. In some senses the IPC sees this study as being an opportunity both to take stock of the progress made so far in this area, and also to sow the seeds for future cooperative research and development. Developing joint contributions is one way in which study participants can begin to engage in the spirit of this ICMI study, and seeking these is one way in which the IPC and ICMI can help to foster future collaborative ventures.

3.3 Case studies

In a sense this ICMI comparative study is one large case study, as has already been mentioned, so the IPC sees the need for there to be several case studies, with a comparative flavour at the heart of the study, which focus on specific aspects of mathematics education (see Section 4). Thus it is important that participants are aware of the nature and value of case-study activity in order to develop the study to its fullest potential. It is only through such case-studies that the richness of educational and cultural interactions can be presented and interpreted.

The case studies hopefully will demonstrate the observation and understanding of a range and variety of typical phenomena in the different traditions, together with their importance in mathematics education. Cases should not only focus on demonstrating and analysing differences but will also explore aspects of similarity between the traditions. The depth of analysis made possible by such case studies will help to challenge the naive policy and practice of attempting to merely copy and transport specific practices from one tradition to another.

As well as welcoming contributions regarding previous case studies and the presentation of current comparative case studies, the IPC sees this study as creating the opportunity for the development of future comparative case study activity. It will welcome proposals for such work.

3.4 Variety of documentation

The IPC is aware of the value and importance of including a variety of documentation in this study. Mathematics education reveals itself through various media and materials. Government publications and documents demonstrate intentions, values and contextual features of a system's policies and practices, but in order to study the influences of the different traditions to the depth desired it is necessary to seek further documentation. The IPC will therefore also welcome contributions which are based on other kinds of documentation.

Textbooks give more information about intended practices as well as about the roles of the teacher and the students but other teaching materials are also revealing. Videotapes of classrooms are increasingly being used as research data in cross-cultural research studies. Aspects of student achievement are well used in comparative studies, but can give misleading and shallow information if not adequately interpreted through deeper cultural and social perspectives. Data on students' and teachers' attitudes, beliefs and values will be particularly interesting for this study as these often reveal more about the significant aspects of difference between cultural traditions.

3.5 Different perspectives on mathematics

Differences in mathematics education intentions and practices are often stimulated by the different influences coming from various groups of professional working in and with the area of mathematics education. In addition different cultural traditions view mathematics itself in different ways, as has already been pointed out in an earlier section.

In this study therefore the IPC wishes to seek contributions from people who work in different parts of the mathematics education community. In a rich comparative study such as this, different perspectives are crucial, and colleagues who work in areas such as mathematical applications, informatics, and the history and philosophy of mathematics are encouraged to participate.

3.6 Different perspectives on the study of mathematics education

In the same sense as in the previous point, there are several different approaches to the study of mathematics education which need to be recognised in this study. Although the IPC sees case studies as being important kinds of contributions, it also does not seek to restrict the methodology of the studies presented. It recognises value for example in psychological, sociological, and anthropological approaches as well as in contributions from other areas of the educational sciences.

The IPC is also aware that as this comparative study develops, particular differences in methodology between cultural traditions may become revealed, for example concerning the position and role of researchers. It is conscious of the dangers of applying certain methodologies from one cultural tradition inappropriately in another cultural tradition. The IPC therefore hopes that one outcome of this study is an increased awareness in the international mathematics education community of the need for cultural sensitivity in carrying out future comparative studies.

4. ASPECTS OF THE STUDY

4.1 Context

Mathematics education does not take place in a vacuum, but there is always a host of different contexts within which the practice of mathematics education takes place. These contexts may be social, political, economic, philosophical or ethical, but they are of course all related one way or another to the underlying cultural values. What are the elements within these contexts which are relevant to mathematics education? What are the "givens" from which we organize our mathematics education, and what are the constraints within which we carry out the education?

4.2 Determinants of the curriculum

In the literature on sociology of education, an important characterisation of an education system is to identify the significant figures or interested parties that determine education policy in general and the curriculum in particular. In this study, however, we are interested not in these general sociological issues, but in the determinants of the mathematics curriculum. Who are the major players (e.g. mathematicians, mathematics educators, school teachers, bureaucrats etc.) in determining a particular mathematics curriculum? Are there significant cultural differences in the constellation of determinants of the mathematics curriculum? In addition to the issue of "who", we are also interested in the mechanism through which the mathematics curriculum is determined. Are there cultural differences on how the mathematics curriculum is determined (e.g. centralized versus decentralized)?

4.3 The role and place of mathematics in the overall curriculum

Mathematics occupies a central place in the curriculum of nearly all countries in the world, yet there may be subtle differences in the importance attached to mathematics as a school subject in different countries or cultures. As far as the education system is concerned, what role is mathematics playing in terms of sifting or filtering students through the education ladder in different countries? Is mathematics a highly aspired subject? What place does mathematics occupy within the overall curriculum (this may be reflected, for example, in the number of hours devoted to mathematics)? Is mathematics viewed as a service or instrumental subject, a subject for mental training, or a subject essential for the development of a cultured citizen? The different perceived roles of mathematics may affect the way mathematics education is conceived and organized in different countries and cultures. These are related to points (6) and (11) below.

4.4 Teachers, teacher education, values and beliefs

The teacher is one of the most crucial elements in the implementation of the mathematics curriculum, and hence a study of mathematics education in different traditions should definitely focus on the teacher. Are there cultural differences on the image or role of teachers in the education process? Are there differences in the way mathematics teachers are educated or expected to be educated? What are the relative emphases on the subject matter (mathematics), pedagogy, and pedagogical content knowledge in the teacher education curriculum?

The literature shows that there is a high correlation between the attitudes and beliefs of teachers on the one hand, and their instructional practices on the other, and it will be both important and interesting to see whether there are cultural differences in teacher attitudes and beliefs. For example, how do teachers in different cultures perceive the nature of mathematical knowledge? The literature shows that mathematics is viewed by teachers as more or less a subject of absolute truth or as a fallible subject. Are there any cultural differences in these attitudes? Also, are there cultural differences in the patterns of attribution of success and failure in mathematics? Does the strength of the relationship between attitudes and practices differ in different cultures?

4.5 Students, learning styles, attitudes, role of students in the teaching/learning process

Just as teachers in different cultures differ in their attitudes towards mathematics and mathematics teaching and learning, students in different cultures are reported to differ in their attitudes towards mathematics and mathematics learning. Do these differences in attitudes lead to differences in learning styles? For example, are Asian students more passive in mathematics lessons? Does physical passivity necessarily imply mental passivity? Can we account for the different achievements of students by differences in attitudes and learning styles?

Also, does the pattern of gender differences in attitudes differ in different cultures? Are boys and girls treated differently in different cultures? How are students of different abilities (e.g. gifted children, disadvantaged students) treated in different cultures? Is the role of student in the teaching and learning process perceived differently in different cultures? These differences, if they exist, will definitely affect both the way teachers organize their teaching as well as the way students participate in the teaching and learning activities.

4.6 Intentions and goals

Studies of different cultural traditions often reveal differences in the goals of education, and it is likely that there are also important differences in the goals and aims for mathematics education. This does not just concern the formal and published goals and aims in, for example, government publications, but concerns also the informal, understood but not stated intentions which people from a cultural tradition take as natural, assumed, and unimportant to discuss. So, are there formally described differences in the goals and aims of mathematics education between different cultural traditions, and what are they?

Intentions however are more general phenomena than goals and aims, and concern personal values as well as societal expectations. In this study it is hoped to reveal more of the informal intentions that the different cultural traditions hold for mathematics education as well as the more formal goals. Is it possible through this study to identify these informal intentions and to explore the sytematic differences between them?

4.7 Content

The differences between the content taught in various countries would appear to be important in this study, particularly in explaining variation in student's performance. School mathematics syllabuses might ostensibly be similar in different societies, but the arrangement of the content and the approach to particular topics could be very different. For example, the content of geometry taught in school might have different emphases on coordinate geometry, vector geometry, Euclidean geometry, transformation geometry or indeed on the integration of these geometries in different countries. Whether algebra is or is not considered as a means for mathematical argument can result in very different approaches to algebra.

So, the question is, how do the different aspects of content chosen and the approaches taken interact with, and how are they influenced by, particular cultural values?

School mathematics is developing. For example, the modeling of situateions through programming and computer algorithms has developed a certain appropriate content for school mathematics, such as testing the correctness of algorithms. How do different cultural traditions react to such new content?

Comparison of content can also focus on some specific issues. For example, how are ideas of proving and testing in mathematics introduced and developed? How is discrete mathematics implemented in school? What has been the legacy of the New Math that was exported from the West to many countries in the 50s? Such issues may also have significant cultural roots.

4.8 Methodology and media

Methodology of teaching and learning is one of the master keys opening up differences in mathematics education, both in theory and practice. Researchers have shown rising interest in comparisons in different countries for the last two decades and they generally agree that differences are both substantial and striking.

Major and coherent questions are:

1. What are the real differences of teaching and learning in the classroom?
2. Why are there differences?
3. How can studies of differences help us to improve teaching and learning?

Section 2.3 has already been dealing with question 3) and we will not go into any more details here.

As to question 1) differences have been observed and studied in many aspects like classroom organisation and routines, teaching sequences, instructional expectations from students, teacher-student interactions, representations for mathematical concepts and procedures. Are these the most influential aspects of differences and which others should be considered? For example, what about the process of planning, analysing and evaluating lessons? Or a deeper insight into the changing teacher-student interaction influenced by general changes in the attitudes of different generations towards each other and the growing interaction of students with each other about learning and with informal teaching and learning resources, such as the internet?

There have been attempts by researchers to characterise differences in mathematics lessons in Japan (e.g. as built around a consideration of multiple approaches to carefully chosen practical examples or activities). In lessons in American schools some researchers see teachers presenting information and directing student activities and exercises with a unique feature of the multiplicity and diversity of both topics and activities.

How can these issues be investigated further and understood more deeply?

As to question 2) reasons for differences observed have been found in different theories about teaching and learning. Education in general and teaching methods in particular do rely on very different fundamentals e.g. philosophy-oriented, science-oriented or application-oriented (or dialectical, hermeneutical, empirical-analytical). Schemes for the structuring of a lesson also can follow a dialectically oriented model of the process of education (Bildung), a process of development as in nature or technology, or a model oriented at the problems to be handled.

Theories about learning are often related to psychological considerations, resulting in a behaviourist or action-oriented view, e.g. Which concepts are dominating in different traditions and what are the consequences, good and bad?

In practice, teaching and learning is strongly dependent on the formal organisation of lessons and strings of lessons, following principles like education (Bildung) oriented, subject matter oriented or student oriented for the structuring of lessons. So to answer the question why there are differences we have to find out in detail about a mixture of philosophical, social, psychological, cultural, political, economical and even ecological intentions and principles for teaching and learning.

Strong changes in teachers' and students' activities in class are reported from many countries. Control of teachers on the activities of students is diminishing. Students rely on outside sources or on information from their peers. They become more active in class and they want to have influence on contents and goals. Models like "students teach students" or "learning by teaching" are practised. How do mathematics educators in different traditions judge these activities? How do they try to exploit them for mathematics education purposes or how do they try to avoid them happening?

Some of the observations just mentioned originate from the new role of the didactic component "media". Compared to classical media like blackboard and chalk or the OHP, modern information technology based media like multimedia computers have a stronger touch of educational intelligence. They can furnish students with information as well as with some advice for learning and about correctness of findings. Communication technology can open up the classroom for geographically and – this is more important – culturally distant information, giving rise to activities like "distance learning". These experiences have a very strong impact not only on cognitive but also on emotional intentions in class (motivation, creativity, ambition). Our study will not only have to evaluate the quality of these attempts but also compare reactions to this kind of challenge in different traditions.

Do we see modern media as a new component in the conservative teacher-student structure of the classroom? What chances and what risks do we see in the new media and what are we doing about it? How can we explain the unchanged importance of teachers for the learning process to society, and how the new role? Even if we see the risks of the new media, can we avoid them? Many societies have seen a "metacognitive shift" going on at all levels of information and communication processes. There is less interest in the contents of these processes and growing interest in the medium carrying the contents. How do we uphold the intentions of mathematics education as far as contents and methods are concerned?

4.9 Assessment of students' achievement

There have been different views and approaches to the assessment of students' achievement in mathematics. For example, in East Asian societies, students, teachers, and parents view written tests and examinations as one of the most important things in a students' school life and as a key to the success of their future life. There are also high-stakes college entrance examinations. This does not seem to be the case in many Western societies. Moreover, there have been great differences in the approaches to assessment of students' achievement in different societies with different traditions. For example, in Western countries (e.g. the United States), standardized tests and multiple choice questions have been used for a long time in the assessment of students' achievement. Recently, alternative assessment methods have been increasingly used in Western schools. Again, this is not the case in Asian countries, though we can see some influence of the standardized test and alternative assessment in some Asian educational systems. The question is, what are the exact differences in assessment of students' achievement in different societies with different traditions? Why are there such differences? Will there eventually be a universal view and approach to the assessment of students' achievement because of the growth of globalization and information technology?

4.10 Different aspects of achievement

Many available comparative studies, especially large-scale achievement assessments, have shown that East Asian students overall outperform their Western counterparts in mathematics. However, some recent researches have suggested that the superior performance of East Asian students is more evident in certain aspects of mathematics achievement, such as using basic skills of computation and algebraic manipulation, solving routine problems and school mathematics problems by applying algorithms, and it is less so in some other aspects. In some studies, researchers found that Western students performed as well as, or sometimes even better than, their Asian counterparts in aspects such as using visual and graphical representation and solving open-ended problems.

Some people have also argued that East Asian students might be better at abstract thinking in mathematics, but Western students might be better at intuitive thinking. Is this really true? In other words, do Asian students perform better in some aspects of mathematics achievement, and worse in some other aspects than Western students? If so, do the differences in these different aspects of achievement reveal certain differences in cultural traditions?

Furthermore, it has been shown in some international studies that students in East Asia, such as in Japan, do not like mathematics as much as, but they performed academically better than, students in Western countries. The question is, to what extent are students' attitudes towards mathematics and mathematics learning affected by their cultural traditions, by their social environment, or by their school learning experiences?

4.11 Views on the nature of mathematics

Researchers have documented the fact that different kinds of mathematics have developed through different cultural traditions. For example, there are significant differences between ancient Chinese mathematics and ancient Greek mathematics. The former is more algorithm-oriented and application-oriented, and the latter is more oriented towards deductive reasoning and not towards the application of mathematics.

The question then is, how do different cultural perspectives on the nature of mathematics influence mathematics education in different traditions? Moreover, what is the relationship between the cultural traditions and the societal views on mathematics and mathematics education in different traditions. In other words, how do they influence or interact with each other?

4.12 Uses, misuses, and abuses of mathematics

There are different views on the relationship between mathematics and society. It has been claimed that mathematics is essential to a modern society, and mathematics can help people become informed citizens and make wise decisions in their individual and social life. This seems particularly true in the modern information age. However, it is not uncommon to see that mathematics, including statistics, is misused or even abused intentionally or unintentionally on different occasions and for different purposes. The question in this connection is, how is mathematics used, misused, and abused in different societies with different traditions? Why is mathematics used, misused, and abused in this way or that way in different societies? What can we do in mathematics education to help students as well as the general public to understand mathematics and its usefulness appropriately so they will not mathematically mislead others or be misled by others with numbers, data, graphs, statistics, or other forms of mathematical information.

4.13 Non-formal mathematics education

In addition to formal school mathematics education, there are other forms of mathematics education, e.g., Juku schools in Japan, Bu-Xi-Ban in Hong Kong and Taiwan, private tuition in Singapore, mathematics clubs within schools, internet-based and home-based learning, to name a few. In this regard, we are interested in the following questions: what are the other forms of mathematics education existing in different societies with different traditions? Why do they exist in different societies? What role do they play in the whole picture of mathematics education? How do they interact with formal mathematics education?

4.14 Evolution of mathematics education

Mathematics education has a long history in human civilization. The question of interest is, how has mathematics education been developing and changing in different countries or societies in different traditions? As the world is increasingly globalised and information technology helps people in different places communicate and share their questions, ideas, and information more freely and conveniently, will there be a universal approach to mathematics education? What can we learn about mathematics education from the past in order to ensure a better future for mathematics education?

The IPC, as well as ICMI, is interested to have approximately equal number of participants from East Asia and the West, like the composition of the IPC. English, however, will be the language of the conference. We are well aware that this may mean a handicap for many individuals whose first language is not English, but we would nevertheless like to encourage such people to participate. We would also like to encourage the native English speakers to take special care of this situation. We will have little chance to succeed in a real comparative study if we do not succeed in managing the language problem in the Study Conference.

It is expected that every participant be active in discussion and other modes of activity during the conference.

INTERNATIONAL PROGRAMME COMMITTEE

The members of the International Programme Committee (IPC) were:

- Alan BISHOP (Australia)
- FAN Lianghuo (Singapore)
- Walther FISCHER (Germany)
- Klaus-Dieter GRAF (Germany, Co-chair)
- Bernard HODGSON (Canada, ICMI Secretary)
- Colette LABORDE (France)
- LEUNG Koon Shing Frederick (Hong Kong, Co-chair)
- LIN Fou Lai (Taiwan)
- Francis LOPEZ-REAL (Hong Kong, Chair of the Local Organising Committee)
- PARK Kyungmee (Korea)
- Katsuhiko SHIMIZU (Japan)
- Jim STIGLER (USA)
- Margaret WU (Australia)
- ZHANG Dianzhou (China)

MATHEMATICS EDUCATION IN EAST ASIA AND THE WEST[1]: DOES CULTURE MATTER?

LEUNG Koon Shing Frederick
The University of Hong Kong

1. INTRODUCTION

Students in East Asia outperformed their counterparts in the West in large-scale international studies of mathematics achievement. For example, the top-ranking countries[2] in the Third International Mathematics and Science Study (TIMSS) in 1995 (Beaton *et al*, 1996; Mullis *et al*, 1997), its repeated study in 1999 (TIMSS-R or TIMSS 1999) (Mullis *et al*, 2000), and the succeeding trend study in 2003 (TIMSS 2003) (Mullis *et al*, 2004) as well as the OECD Programme for International Student Assessment (PISA) studies in 2000, 2001 and 2003 are all East Asian ones (OECD, 2001, 2003, 2004). These results are consistent with those of earlier studies (for example, Second International Mathematics Study (SIMS) (Robitaille and Garden, 1989), International Assessment of Educational Progress (IAEP) (Lapointe

[1] In this paper, East Asia refers to countries or education systems such as China, Hong Kong, Japan, Korea, Taiwan and Singapore, and the "West" refers to countries in North America, Europe and Australia. Here "East" and "West" are cultural rather than geographic demarcations, with the Chinese or Confucian tradition in the East and the Greek/Latin/ Christian tradition in the West. We acknowledge that neither of these "poles" is well defined, as with any label given to any culture. But we use the two terms to point to the scope that we want to confine ourselves to in this chapter.

[2] While some education systems (e.g. Hong Kong) participating in these international studies are not countries, the generic term "countries" will be used as a convenient way to refer to all participants.

et al, 1992), and a study by Stevenson and Stigler (Stevenson *et al*, 1990, 1993), and it seems that the superior performance of East Asian students is rather stable over time.

How do we account for the superior performance of East Asian students? Student achievement is enmeshed in a host of variables at different levels, and international comparative studies in achievement should prompt us to identify similarities and differences in educational practices, understand the different traditions underlying the different practices, and seek to explain the similarities and differences. In explaining differences, the crucial question to ask is "what matters?" That is, of the variables construed to be related to student achievement, which of them actually show a pattern that matches with that of student achievement in international comparative studies?

In this paper, we will look at a number of variables including societal resources, characteristics of the education system, and teacher attitudes and attributes in order to identify whether there are any patterns that can be matched with the pattern of student achievement. Finally, we will examine the underlying cultural values in East Asia to see whether they can be used to account for the superior performance of the East Asian students.

2. EAST ASIAN STUDENTS' MATHEMATICS ACHIEVEMENT IN TIMSS-R

The high achievement of East Asian students is best illustrated by their results in TIMSS-R (Mullis *et al*, 2000). There were only five East Asian countries (namely Chinese Taipei, Hong Kong, Japan, Korea and Singapore) in TIMSS-R, but they topped the 38 countries in grade eight mathematics (see Figure 1). Furthermore, the difference in the level of achievement between these high performing countries and many of the other TIMSS-R countries was rather substantial. All the five countries were more than three standard deviations above the lowest scoring country, and more than one standard deviation above 15 of the countries.

3. STUDENTS' ATTITUDES TOWARDS MATHEMATICS

Students' achievement in mathematics is usually highly correlated with a positive attitude towards mathematics (Hammouri, 2004; Ma and Kishor, 1997; McLeod, 1992). However, in international studies of mathematics achievement, the superior performance of the East Asian students did not

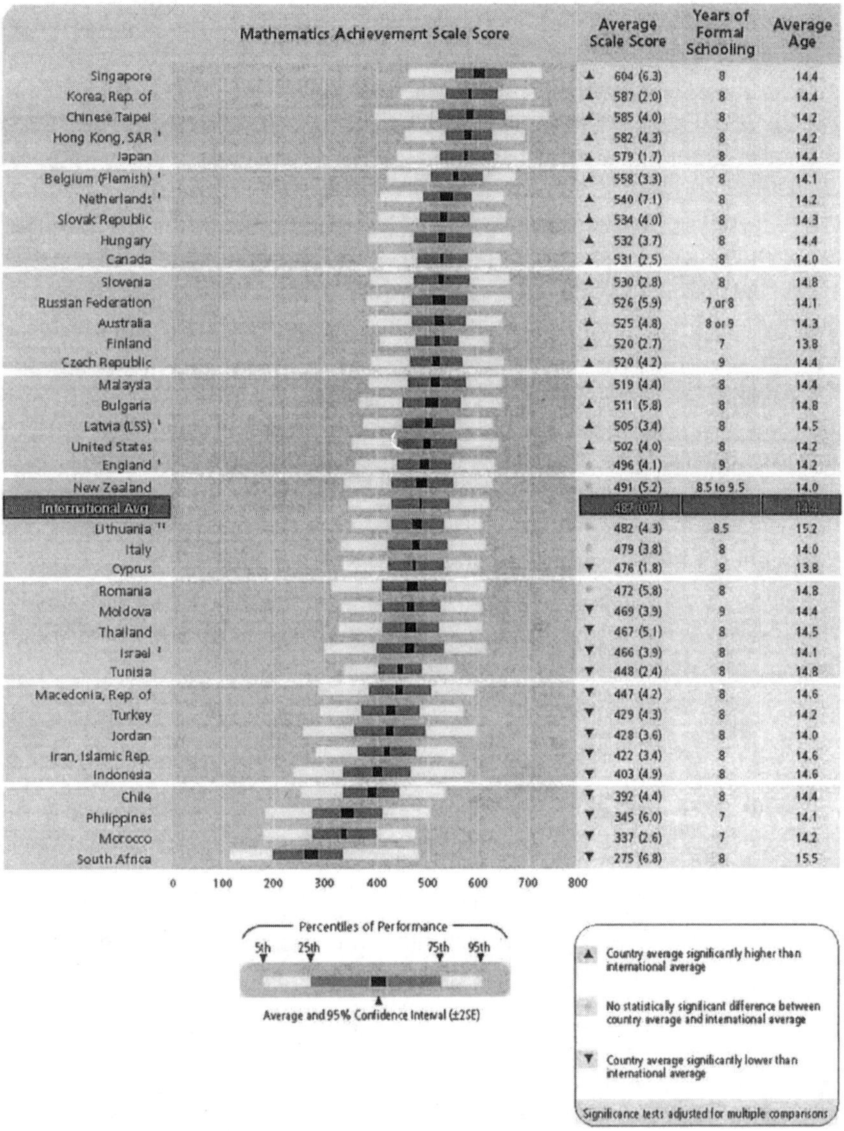

† Met guidelines for sample participation rates only after replacement schools were included (see Exhibit A.8).

1 National Desired Population does not cover all of International Desired Population (see Exhibit A.5). Because coverage falls below 65%, Latvia is annotated LSS for Latvian-Speaking Schools only.

2 National Defined Population covers less than 90 percent of National Desired Population (see Exhibit A.5).

‡ Lithuania tested the same cohort of students as other countries, but later in 1999, at the beginning of the next school year.

( ) Standard errors appear in parentheses. Because results are rounded to the nearest whole number, some totals may appear inconsistent.

Figure -1. Distribution of TIMSS-R Mathematics Achievement (Adopted from TIMSS 1999 International Report by Mullis et al, 2000) seem to be accompanied by correspondingly positive student attitudes towards mathematics and mathematics learning. In the PISA study, for example, Korean students' mean score in the "index of interest in mathematics" is -0.27, which is the lowest among all the PISA 2000 countries[3] (OECD, 2001: 266). This is consistent with the findings from the TIMSS-R student questionnaire, as discussed below.

3.1 The importance of mathematics

The TIMSS-R results show that students all over the world attached a lot of importance to mathematics. In no country was there less than 88% of the students who thought that it was important to do well in mathematics, and the average percentage of students across all the 38 TIMSS-R countries who agreed that it was important to do well in mathematics was 96%. Figure 2 shows students' report on "whether it is important to do well in mathematics" for a selected number of countries[4]. Relatively speaking, we can see that students in East Asia attached less importance to mathematics. Four of the five East Asian countries (with the exception being Singapore) were below the international average of 96%, and Japan, Chinese Taipei and Korea were actually the three countries with the least number of students who thought that mathematics was important.

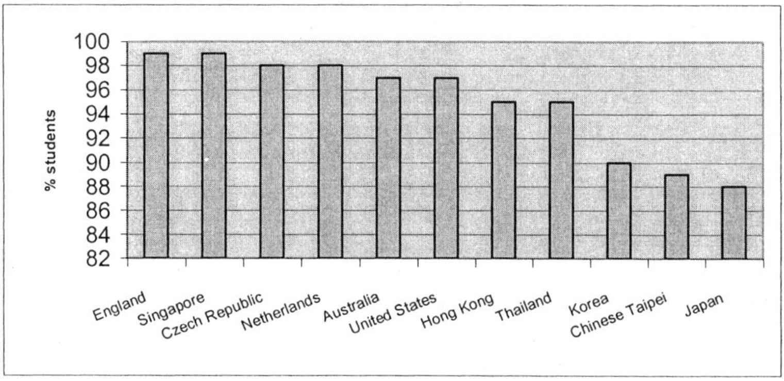

Figure -2. Students' Report on Whether It is Important to do Well in Mathematics (International Avg.=96% of Students)

[3] Korea and Japan are the only two East Asian countries which participated in PISA 2000, but data on the index of interest in mathematics for Japan was not available.

[4] Since there are so many TIMSS-R countries, this selected group of countries will be used for data presentation in this chapter. The international averages cited, however, are the averages for all the 38 TIMSS-R countries.

3.2 Positive attitudes towards mathematics

To bring up students with a positive attitude towards mathematics is a common goal in the mathematics curriculum of many countries, and a positive attitude towards the subject is usually highly correlated with achievement. TIMSS-R developed an index of positive attitudes towards mathematics based on a number of questions in the student questionnaire, and the results for some of the TIMSS-R countries are shown in Figure 3. As can be seen in Figure 3, students in East Asian countries (again with the exception of Singapore) were below the international norm in terms of positive attitudes towards mathematics.

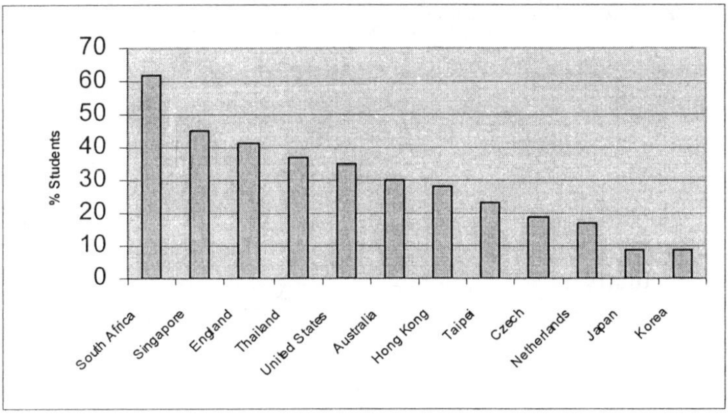

Figure -3. (International Avg.=37% of Students)

3.3 Self-concept in mathematics

In addition to the index of positive attitudes towards mathematics, TIMSS-R also developed an index of students' self-concept in mathematics. The index reflects students' confidence in their ability in mathematics. From Figure 4, it can be seen that students in the East Asian countries, including Singapore, all had self-concept in mathematics lower than the international average.

From the results presented above, it is not difficult to conclude that East Asian students' superior achievements were not accompanied by correspondingly positive attitudes towards mathematics. How do we account for such superior achievement and yet negative attitudes? In the sections below, we will look at variables of different levels from the results of the TIMSS-R questionnaire and other studies in order to seek some understanding of the phenomenon.

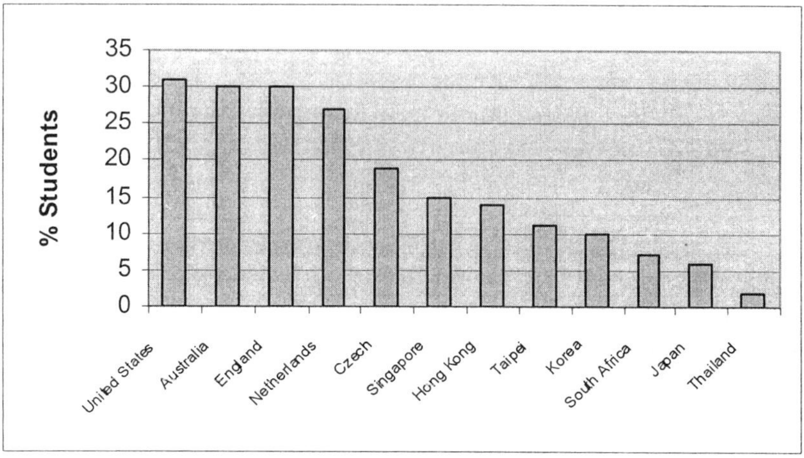

Figure -4. Students' Self-Concept in Mathematics
(International Avg.=18% of Students)

4. SOCIETAL RESOURCES

Whenever a country wants to improve student achievement, a common strategy is to put in more resources. This is based on an assumption or belief that somehow student achievement should be positively related to the resources put into education. Indeed, within a country, academic achievement is usually found to be highly correlated with the social and economic status (SES) of the student. At the international level, can the superior performance of East Asian countries be explained by their wealth or by the resources they put into education?

4.1 The wealth of a country

As can be seen from Table 1, the East Asian countries in TIMSS-R were indeed relatively affluent countries among the TIMSS-R countries. But it should be noted that there were many other affluent countries in TIMSS-R which did not do well in mathematics. So the wealth of a country alone does not guarantee that its students will perform well.

Table -1. GNP per Capita (1999)

Country	GNP per Capita
Japan	US$38,160
Singapore	US$32,810
United States	US$29,080
Hong Kong	US$25,200
Australia	US$20,650
Canada	US$19,640
Chinese Taipei	US$13,235
Korea	US$10,550
TIMSS-R countries average	**US$10,584**

4.2 Education expenditure

Although the East Asian countries were relatively affluent, the amount of money they put into basic education was not particularly high. As can be seen from Table 2, all of these East Asian countries were bclow the international average in public expenditure on basic education spending as measured by percentage of GNP. Of course the expenditure of East Asian countries in absolute terms might not be low compared to those countries with low GNP. But note that there were wealthy Western countries with high GNP, which also invested a higher percentage of their GNP to basic education, and yet their achievement in mathematics was low compared to the East Asian countries.

Table -2. Education Expenditure

Country	% of GNP
Canada	6.9%
Australia	5.5%
United States	5.4%
Chinese Taipei	4.9%
Korea	3.7%
Japan	3.6%
Singapore	3.0%
Hong Kong	2.9%
TIMSS-R countries average	**5.1%**

4.3 Educational resources

Education expenditure only reflects the amount of money governments in different countries allocated to education. Whether the allocated money results in adequate resources available for mathematics instruction in the schools is another matter. Furthermore, in many countries, private resources are additionally provided at home to support the education of children. Since East Asian parents are known for their emphasis on the education of their children, it will be of interest to see whether East Asian students have better educational resources compared to students in other parts of the world.

The TIMSS-R questionnaire results show that school resources for mathematics instruction varied in these East Asian countries (Mullis *et al*, 2000: 229-233), but surprisingly, home educational resources for East Asian students were in general unfavourable (see Figure 5).

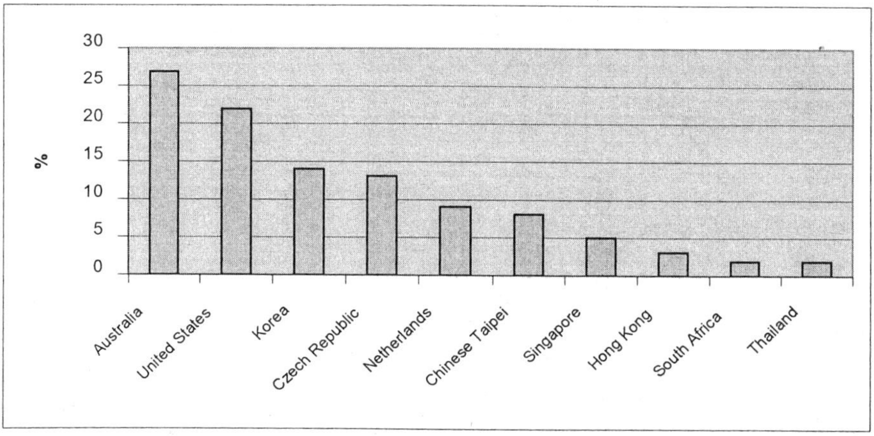

Figure -5. Home Educational Resources (International Avg.= 9%)

From the discussions above, we can see that variables due to societal resources which usually explain within-country differences in achievement fail to explain across-country differences in achievement as far as these East Asian countries are concerned. In fact, from the findings shown above, we can say that societal resources for education are relatively unfavourable in the East Asian countries.

5. CHARACTERISTICS OF THE EDUCATION SYSTEM

If East Asian countries are not better resourced in education, is the high achievement of their students due to a more efficient education system or a better designed curriculum?

5.1 The education system and the curriculum

A common characteristic of all East Asian education systems is that they are highly centralized, and one may argue that centralization increases efficiency. But the TIMSS-R results show that actually many Western systems were also highly centralized (Mullis *et al*, 2000: 148-9), and hence it is unlikely that it is the efficiency of a centralized system that produces high achieving students.

Nor is it likely that the high achievement is due to differences in the content of the curriculum. Actually, the mathematics curriculum is very similar worldwide (Howson and Wilson, 1986), although it is reported that the curriculum content is more demanding in East Asian countries (Silver, 1998; Stigler and Hiebert, 1999).

5.2 Time spent on mathematics instruction

A more demanding curriculum, however, does not necessarily imply that more curricular time is spent on mathematics instruction. The TIMSS-R results show that East Asian countries did not devote a lot of time to mathematics instruction (Figure 6). Except for Hong Kong, all the East Asian countries spent less time on mathematics instruction per year than the international average of 129 hours.

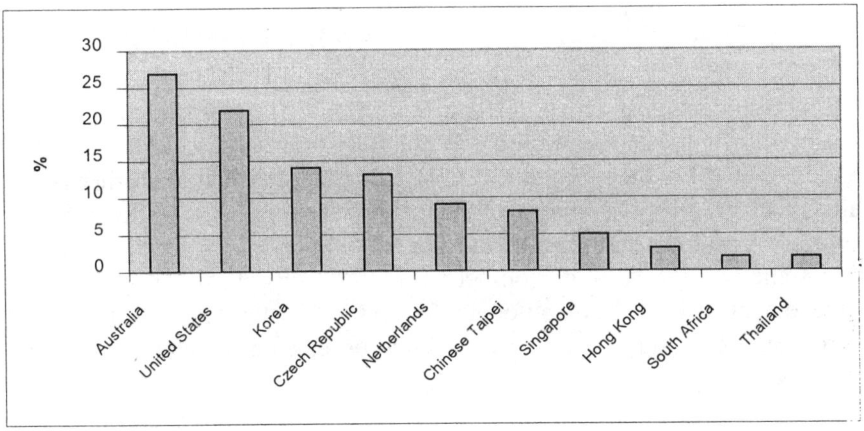

Figure -6. Hours of Mathematics Instruction per Year (International Avg.=129 Hours)

5.3 Out-of-school time studying mathematics

Although East Asian countries did not devote particularly more time to mathematics instruction in schools, there is a common perception that their students spend a lot of time outside regular schooling studying mathematics. This however is not supported by the TIMSS-R data, which show that East Asian students did not spend more time than their Western counterparts out of school studying mathematics or doing mathematics homework[5] (Figure 7).

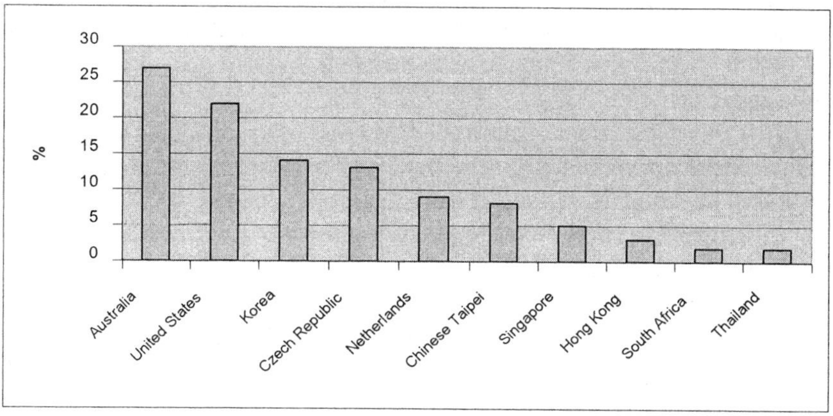

Figure -7. Out-of-School Time Studying or Doing Mathematics Homework (International Avg.=1.1 Hours per Day)

To sum up, we find that the variables at the level of the education system cited above fail to explain the high achievement of East Asian students as well.

6. TEACHER ATTITUDES AND ATTRIBUTES

Since students learn most of their mathematics from their teachers in the classroom, it is reasonable to conjecture that student achievement is related to the quality of the teachers and their teaching. An effective teacher is often thought to be one who is confident in her teaching, who holds a positive attitude towards the subject that she teaches, who employs appropriate teaching methods, and who is competent in the subject matter. Is the high achievement of East Asian students due to the fact that they are taught by a more confident and competent teaching force, employing better teaching methods?

[5] There may be different interpretations of what "out of school" means in different countries, and so the results on time spent out of school should be interpreted with care.

6.1 Teacher attitudes towards mathematics and mathematics education

In a comparative study conducted by the author (Leung, 1992), a questionnaire on attitudes and beliefs in mathematics and mathematics education was administered to about 900 junior secondary school mathematics teachers from Beijing, Hong Kong and London. Results and issues pertaining to the theme of this paper will be discussed below.

6.1.1 Attitudes towards mathematics

In the questionnaire, the SIMS *Mathematics as a Process* scale (Robitaille and Garden, 1989:199) was used to measure "the extent to which (teachers) view the discipline (of mathematics) as rule-oriented or heuristic, as fixed or changing, and as a good field for creative endeavour or not" (Robitaille and Garden, 1989, p.199). The attitudes of teachers from the three cities are shown in Figure 8.

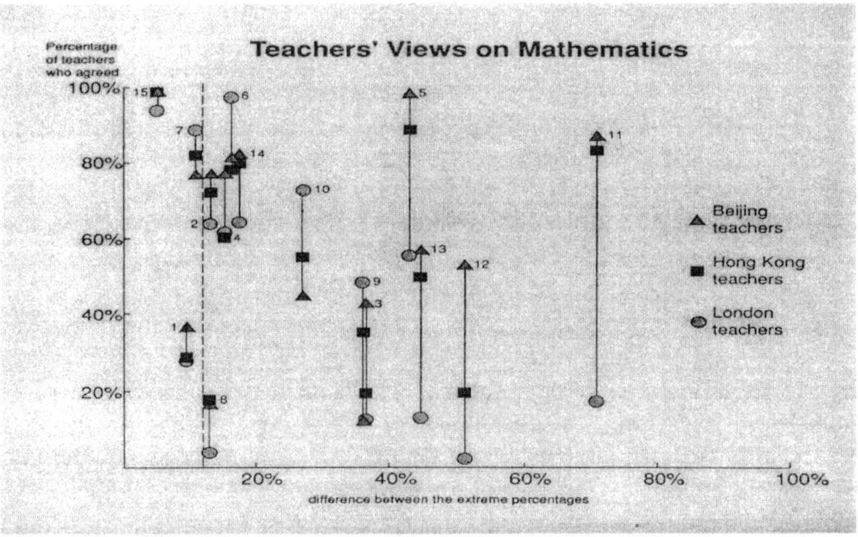

Figure -8. Views of Teachers from Beijing, Hong Kong and London on Mathematics

From Figure 8, we can see that teachers from the three cities differed tremendously in their attitudes towards mathematics. ANOVA tests showed that the responses of teachers from the three cities differed significantly in 12 of the 15 items on the scale. The item that attracted the greatest discrepancy of responses is item 11: "There is always a rule to follow in solving a mathematics problem". 87.6% and 83.1% of the teachers in Beijing and Hong

Kong respectively agreed with this statement (only 2.3% and 7.2% respectively disagreed) while only 17.0% of the London teachers agreed (64.7% of them disagreed). This statement perhaps best describes the difference in perception of mathematics among the teachers in the three cities. The Beijing teachers tended to view mathematics as a rule-oriented and fixed discipline, while teachers in London perceived mathematics as more heuristic and changing, and the attitudes of the Hong Kong teachers lay between these two extremes.

The more significant and interesting finding, however, is that a pattern of responses emerges across the three cities. All except three items follow a trend - with the response of the Hong Kong teachers lying between those of the Beijing and London teachers. (The three exceptional items, 4, 6 and 8, are near the middle of the list above, showing that they are items that attracted neither strong agreement nor strong disagreement among teachers in the three cities.) This pattern persists in the other items of the questionnaire, as we will see in later parts of this chapter.

6.1.2 Attitudes towards mathematics education

Three items in the questionnaire that measure teachers' attitudes towards mathematics teaching and learning will be discussed here.

(i) Aims of mathematics education

In one of the items in the questionnaire, teachers were asked what they thought the most important aims of mathematics education were for junior secondary school students, and their views are shown in Figure 9.

From Figure 9, we can see that teachers in all three cities did not think appreciating mathematics for its own worth an important aim in mathematics education. For the four other aims supplied in the question, teachers from different cities had different responses. Teachers in Beijing stressed mathematics as a tool for other subjects (52% thought this was the most important aim) and teachers in Hong Kong stressed the training of the mind (52%), while teachers in London stressed the ability to communicate logically and concisely (33%) and the applications of mathematics (39%).

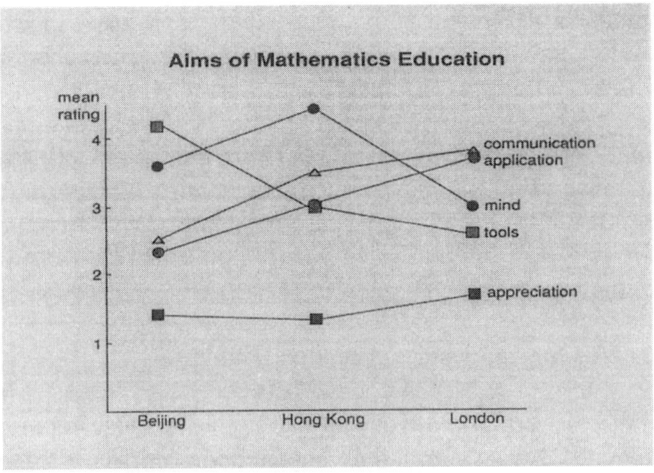

Figure -9. Teachers' Views on the Aims of Mathematics Education

(ii) Influences on mathematics content taught in the classroom

When teachers were asked what should have the greatest influence on the mathematics content taught in the classroom, some interesting contrasts between the views of teachers in the three cities emerged (see Figure 10).

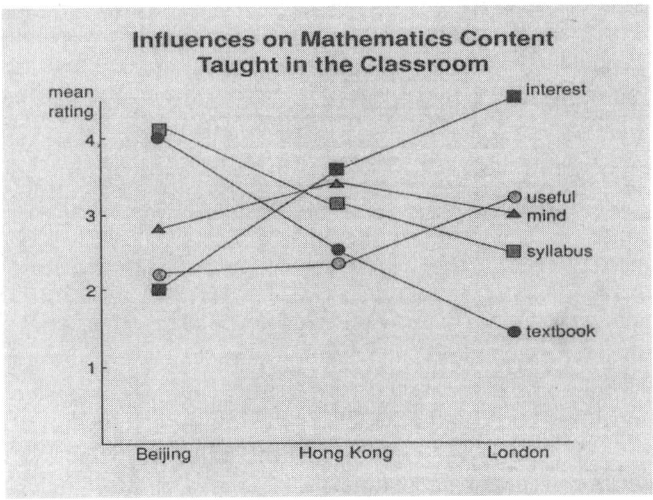

Figure -10. Teachers' Views on the Influence on Mathematics Content Taught

It can be seen from Figure 10 that in determining the mathematics content taught in the classroom, the views of the Beijing and London teachers were nearly diametrically opposed to each other. The most important consideration for the London (and Hong Kong) teachers seemed to be whether the content was interesting and meaningful for students, while for the Beijing teachers, interest did not seem to be a consideration at all.[6]

In Beijing, and to a lesser extent in Hong Kong as well, the syllabus seemed to dominate the content of mathematics taught in the classroom. The textbook was also thought to be an extremely important determinant of the content in Beijing, and in contrast, it was not an important consideration in the minds of the London teachers at all.

(iii) Factors affecting students' success and failure

When teachers were asked what they thought the most important factors affecting students' success or failure in mathematics were, teachers in all three cities did not consider luck or family support important factors (see Figure 11). But for the three other options, again an interesting pattern is obtained.

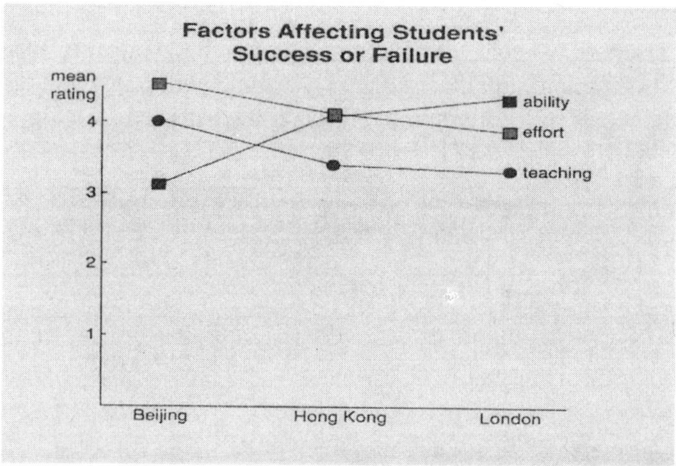

Figure -11. Teachers' Views on Factors Affecting Students' Success or failure

It seems from Figure 11 that teachers in Beijing believed in effort (put in by students and teachers) very much, and thought that the ability of students

[6] This view of the Beijing teachers may be accounted for by the traditional Chinese attitude towards studying and learning. For the Chinese, studying or learning is a serious endeavour and a hardship, and one is not supposed to "enjoy" the studying (Garvey and Jackson, 1975).

was not as important in determining success or failure in mathematics[7]. This contrasts sharply with the attitudes of the London teachers, who in general thought that the ability of students was the most important factor. The attitudes of the Hong Kong teachers again lay between those of their Beijing and London counterparts.

6.2 Confidence in preparation to teach mathematics

In TIMSS-R, an index of "teachers' confidence in preparation to teach mathematics" was developed from the teacher questionnaire to gauge teachers' confidence in teaching the different mathematics topics in the TIMSS curriculum framework. The results for some of the TIMSS-R countries are shown in Figure 12, and it can be seen that the East Asian teachers were in general not very confident in their preparedness to teach mathematics.

From the discussions above, we can see that the East Asian students are not taught by teachers who hold very positive attitudes towards mathematics and mathematics education, or are very confident in their teaching. It is hard to conclude from these findings that East Asian students' superior achievement is related to the attitudes of their teachers.

6.3 Reported teaching style

In past literature, it was reported that the teaching style in East Asia was rather traditional. Teaching was predominantly content oriented and examination driven. Instruction was very much teacher dominated, and student involvement was minimal. Teaching was usually conducted in whole class settings, with relatively large class size. There was virtually no group work or activities, and memorization of mathematics facts was stressed. Students were required to engage in ample practice of mathematical skills, mostly without thorough understanding (Brimer and Griffin, 1985; Biggs, 1996; Leung, 1995, 2001; Wong and Cheung, 1997; Wong, 1998).

In a comparative study of primary school mathematics teachers in China and the US by Ma (1999), teachers were interviewed on their approaches to teaching selected primary school mathematics topics. They were presented with some classroom scenarios and were asked how they would react to those scenarios. In an attempt to relate Ma's findings for the US and

[7] The Chinese conception of ability and effort is quite different from that in the West. The Chinese believe that ability is not "internal and uncontrollable" (Good and Weinstein, 1986), but something that one can "develop", as shown by the Chinese proverb "Diligence can compensate for stupidity".

Shanghai teachers to other East Asian teachers, a small-scale replication of Ma's study was carried out in Hong Kong and Korea (Leung and Park, 2002). Nine teachers from each of Hong Kong and Korea were interviewed using the four Teacher Education and Learning to Teach Study (TELT) (Ball, 1988) tasks that Ma used in her study, and some results of this replication study are reported below.

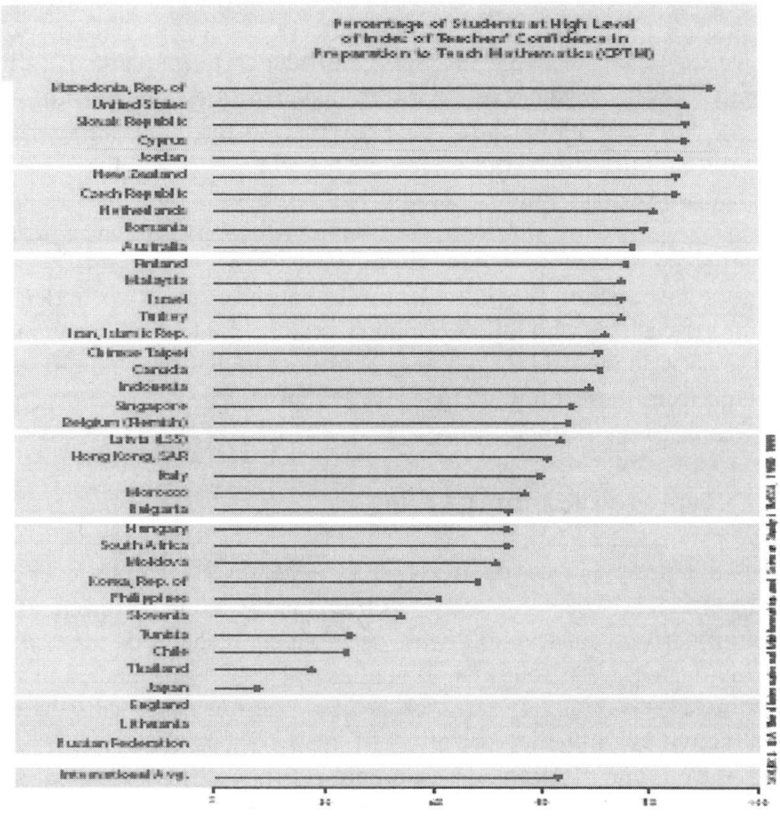

Figure -12. Index of Teachers' Confidence in Preparation to Teach Mathematics (adopted from TIMSS 1999 International Mathematics Report by Mullis et al, 2000)

Teachers' reaction to the following scenario reveals their teaching strategies in dealing with the topic of multi-digit number multiplication:

Task: Multi-digit Number Multiplication

Some sixth-grade teachers noticed that several of their students were making the same mistake in multiplying large numbers. In trying to calculate

```
   123
 x 645
```

the students seemed to be forgetting to "move the numbers" (i.e., the partial products) over on each line. They were doing this:

```
   123
 x 645
 -----
   615
   492
   738
  ----
  1845
```

instead of this:

```
   123
 x 645
 -----
   615
  492
 738
 -----
 79335
```

While these teachers agreed that this was a problem, they did not agree on what to do about it. What would you do if you were teaching sixth grade and you noticed that several of your students were doing this?

In contrast to the Shanghai counterparts in Ma's study, the majority of the teaching strategies reported by Hong Kong and Korean teachers were procedurally rather than conceptually directed (see Table 3). The following excerpt of an interview with a Hong Kong teacher is a typical example of the reactions of the Hong Kong and Korean teachers in the replication study:

I: What would you do if you find your students are making this kind of mistake?

T: I'll teach them that if you use the hundreds place times the unit place, your answer should be written under the "6", align with this "6". Now the textbook leaves two empty spaces there, and I'll tell them if you have empty spaces, it is easy for you to align wrongly. So I'll teach them after you get the product, say six times three equals eighteen, you should immediately put down two zeroes in the two empty spaces. If you have something there, you won't align wrongly. So the first thing is to align, ... (It is) the same with the second number. It is the tens place, so you should put your answer under the tens place, and you put a zero at the unit place ...

How do the reported teaching strategies of Hong Kong and Korean teachers in the replication study compare with those of Shanghai and US teachers in Ma's study? Table 3 below combines the results of Ma's study and Leung and Park's replication study:

Table -3. Teaching Strategies

	US	Shanghai	Hong Kong	Korea
Procedurally directed	16 (70%)	9 (13%)	7 (78%)	6 (67%)
Conceptually directed	7 (30%)	63 (88%)	2 (22%)	3 (33%)

From Table 3, we can see that as far as reported teaching strategies are concerned, Hong Kong and Korean teachers were more akin to the US teachers rather than the Shanghai teachers in Ma's study.

6.4 Teachers' mathematics competence

Is the procedural teaching reported by the Hong Kong and Korean teachers in the replication study a reflection of their lack of competence in mathematics? In the interview, it was found that although the reported teaching style of East Asian teachers was rather traditional and procedural, these teachers actually possessed conceptual understanding of the procedures. In the multi-digit number multiplication discussed above, when teachers were probed further in the interview on the reasons for the procedures of "aligning" and "putting in zeroes", it was found that these teachers fully understood the rationale for "moving the numbers" (i.e. the partial products). Nearly all of them said that the students' problem was a lack of understanding of "place value", and they would break the multiplication into three partial multiplications to point out students' mistakes:

I: Why are students making this kind of mistake?

T: It's a problem with place value. In fact the 6 here stands for 600. I can change 645 into 600 + 40 + 5, and then do it step by step, dividing (the multiplication) into 123 × 600 + 123 × 40 + 123 × 5. Then I will address the problem of aligning the numbers (i.e. the partial products). I will tell them that you need to add in zeroes ...

So we can see that, unlike the US teachers in Ma's study, the East Asian teachers in the replication study possessed conceptual as well as procedural understanding of the algorithm behind multi-digit number multiplication (see Table 4), and it seems that they had the mathematics competence needed to teach the topic. However, as pointed out above, the majority of the teaching strategies reported by Hong Kong and Korean teachers were procedurally rather than conceptually directed (Table 3), despite their conceptual understanding of the mathematics involved.

Table -4. Teachers' Knowledge of Multi-digit Multiplication Algorithm

	US	**Shanghai**	**Hong Kong**	**Korea**
Procedural	14 (61%)	6 (8%)	0 (0%)	1 (11%)
Conceptual + procedural	9 (39%)	66 (92%)	9 (100%)	8 (89%)

Teachers' mathematics competence is also reflected in their response towards the following task in the study.

Task: Division by Fractions

People seem to have different approaches to solving problems involving division with fractions. How do you solve a problem like this one?

$$1\ 3/4 \div 1/2 =$$

Imagine that you are teaching division with fractions. To make this meaningful for kids, something that many teachers try to do is to relate mathematics to other things. Sometimes they try to come up with real-world situations or story-problems to show the application of some particular piece of content. What would you say would be a good story or model for

$$1\ 3/4 \div 1/2\ ?$$

All the teachers in Hong Kong, Korea and Shanghai were able to calculate the division correctly (see Table 5). Also, at least half of the teachers in each of these East Asian countries understood the division by fractions concept well enough to be able to generate a representation of the problem by providing a story.

Table -5. Teachers' Knowledge of Division by Fractions

	US	**Shanghai**	**Hong Kong**	**Korea**
Correct answer	9 (43%)	72 (100%)	9 (100%)	9 (100%)
Correct story	1 (4%)	65 (90%)	6 (67%)	5 (50%)

The results above show that most East Asian teachers were rather competent mathematically. They had a good grasp of the underlying concepts of school mathematics, and they were particularly proficient in the calculation procedures.

7. EXPLANATION AT THE LEVEL OF CULTURE

From the discussions in the preceding sections of this paper, we can see that other than the possible difference in competence of the teachers, none of the variables we reviewed at the levels of society, education system and teachers seemed to be able to explain the high achievement of East Asian students. However, a number of findings mentioned above prompt us to look at explanations at the level of culture. This will be further discussed below.

7.1 The low self-concept in mathematics

One of the striking findings of TIMSS-R discussed above is the negative attitudes towards mathematics and the lack of confidence in doing mathematics for students in the East Asian countries. One possible reason for this negative finding may be due to the stress in the cultures of these countries on the virtue of humility or modesty. As the author pointed out elsewhere (Leung, 2002: 106):

> Children from these countries are taught from when they are young that one should not be boastful. This may inhibit students from rating themselves too highly on the question of whether they think they do well in mathematics, and so the scores may represent less than what students are really thinking about themselves. On the other hand, one's confidence and self image are something that is reinforced by one's learned values, and if students are constantly taught to rate themselves low, they may internalize the idea to result in really low confidence. Furthermore, the competitive examinations systems coupled with the high expectations for student achievement in these countries have left a large number of students classified as failures in their system, and these repeated experiences of a sense of failure may have further reinforced this lack of confidence.

So this low self-concept in mathematics, and perhaps the negative attitude towards mathematics in general as well, may have a deeper cultural root. It may be something that students acquired while being brought up in the East Asian culture.

7.2 Teacher attitudes

The same argument may hold for the low confidence of teachers in preparation to teach mathematics mentioned earlier in this paper. But as far as teacher attitudes are concerned, the more important finding cited above is not that the negative attitudes of the East Asian teachers are due to the underlying cultural values, but that the attitudes of teachers from Beijing, Hong Kong and London, as reflected in their response to the questionnaire, followed a trend according to the cultural location of the place they come from. As was pointed out earlier in this paper, the attitudes of teachers in Beijing, Hong Kong and London nearly always form a pattern, with the attitudes of Hong Kong teachers falling between those of Beijing and London teachers.

The significance of this Beijing-Hong Kong-London pattern is attributable to the fact that Hong Kong was a British colony for more than one and a half centuries before reverting back to China in 1997. But because of her origin from and proximity with China, Hong Kong has never lost her cultural link with the motherland. On the contrary, being the most homogeneously ethnic Chinese community outside Mainland China and Taiwan, most of the Chinese traditions and values are still retained. Hong Kong's dramatic prosperity under British rule and the consequent development into one of the world's major communication and financial centres has however made it a city prone to influence by Western cultures. So culturally, Hong Kong may be considered to be located somewhere between China and England, and the findings discussed above do seem to confirm this influence of culture on teacher attitudes. This may be taken as an evidence of the effect of the cultural values on the attitudes of teachers (and students).

7.3 Teaching style

Another interesting finding discussed in this chapter is the predominantly procedural manner of teaching reported by Hong Kong and Korean teachers despite their good grasp of the underlying mathematical concepts. Is there a cultural explanation to this phenomenon?

The interview results of the replication study cited above show that Hong Kong and Korean teachers seemed to believe that, for elementary school, there is no need to teach in a conceptually rich manner. It would be inefficient or even confusing for elementary school children to be exposed to rich concepts instead of clear and simple procedures. The following view of a Korean teacher expressed during the interview illustrates this mentality well.

> Teacher E:When I teach multiplication of three fractions in the sixth grade, I usually emphasize that multiplication should be done from left to right. For example, when I teach 4/7 ×5/3 ×1/2 , I always ask my students to multiply 4/7 ×5/3 first. The associative law holds for multiplication, so in fact there is no need to highlight this fact. You can say that I have taught a wrong method to students. However, I stress that fact because mentioning the order is helpful when students perform mixed operations with multiplication and division. Consider the operation 4/7 ÷5/3 ×1/2. Here, the order is important. If students do the multiplication first, they will get the wrong answer. I ask students to follow the rule of order, which is not strictly correct mathematically, because I consider the operation which will be dealt with in the next chapter.

From the excerpt above we can see that this Korean teacher understood the mathematics concepts behind the multiplication of fractions, but she deliberately taught students a rule which is "not strictly correct mathematically" in order not to confuse the students. This stress on the procedure instead of concepts for the sake of efficiency illustrates very well the pragmatic philosophy in the East Asian culture. Such a pragmatic mentality compels East Asian teachers to deliberately teach in a procedural manner for pedagogical reasons.

7.4 Why do East Asian students excel?

Given that East Asian students possess such negative attitudes towards mathematics and hold such low self-concept in mathematics, and are taught by teachers using teaching methods that stress procedures rather than concepts, why do the East Asian students perform so well in international studies of mathematics achievement?

Paradoxically, one may argue that this negative correlation between students' confidence in mathematics and their achievement is something to be expected:

> Over-confidence may lower students' incentive to learn further and cause them to put very little effort into their studying, and hence result in low achievement. This is exactly the kind of justification for the stress on humility or modesty in the East Asian culture. The Chinese saying "contentedness leads to loss, humility leads to gain" illustrates the point well.
>
> (Leung, 2002:16)

Also, it should be noted that procedural teaching does not necessarily imply rote learning or learning without understanding. Understanding is "not a yes or no matter, but a continuous process or a continuum" (Leung, 2001). The process of learning often starts with gaining competence in the procedure, and then through repeated practice, students gain understanding. Much of the mathematics in the school curriculum may need to be practiced without thorough understanding first. With a set of practicing exercises that vary systematically, repeated practice may become an important "route to understanding" (Hess and Azuma, 1991). As Marton pointed out (1997), in the East Asian culture, repetitive learning is "continuous practice with increasing variation". This is perhaps the way both teachers and students in East Asia acquire their mathematics competence.

7.5 The competence cycle

In the replication study cited above, Hong Kong and Korean teachers were asked when and where they acquired their mathematics competence. They indicated that their mathematics and pedagogical competence were mainly acquired while they were still students in school. So it seems that mathematics competence is "inherited". East Asian students, taught by their competent teachers, acquire competence in mathematics. When they graduate and join the teacher force, they in turn become competent teachers. Once a good cycle starts, the positive effects cumulate and increasingly reproduce themselves. Unfortunately, this holds true for a vicious cycle as well.

But how did the good cycle start in the first place? One possible root for the start of the good cycle is the underlying Confucian cultural values common to the East Asian countries. It is well known that in the Confucian culture, there is strong emphasis on the importance of education. Under the influence of this philosophy, East Asians consider learning or studying a serious endeavour, and students are expected to put in hard work and perseverance in their study. This is reinforced by a long and strong tradition of public examination, which acts as a further source of motivation for learning. This high expectation on the student to achieve provides an important source of motivation both for teachers to teach conscientiously and for students to learn well.

Related to this strong emphasis on the importance of education, there is a tacit expectation and strong tradition in the East Asian culture that the teacher should be an expert or a learned figure in the subject matter. This expectation provides incentives for East Asian teachers to strive to attain competence in mathematics. Probably it is this prevailing cultural value of the emphasis on education and the scholar-teacher that starts and keeps the good cycle of competent teachers-competent students in East Asia.

8. CONCLUSION

This paper is prompted by the superior performance of East Asian students over their Western counterparts in international studies of mathematics achievement, and it seeks to explore whether differences in other aspects of education or the system may be used to explain the differences in achievement. A number of factors at the levels of societal resources, the education system, and the teachers, thought to be related to student achievement, were examined. The findings indicate that despite a more or less common legacy of Greek mathematics and the recent globalization in education, mathematics education in East Asia and the West retains important differences. There are differences in terms of student achievement, student attitudes, teacher attitudes, teaching style, and teacher competence. The various institutional and societal variables examined fail to explain these differences, and possibly it is factors beyond those falling in these levels that are at work in having an effect on student achievement.

Since these high achieving countries share a common culture, roughly referred to as the Chinese or Confucian culture, a number of characteristics of the culture were examined to see whether they may be used to explain the differences in student achievement and other variables in mathematics education. As the discussion in this chapter shows, there are indeed different cultural values pertinent to education that may explain the differences. This is of course no proof that differences in student achievement are caused by cultural differences. But in the absence of clues from variables at other levels, it is probable that culture does matters.

If culture does matter, what are the implications of the cultural differences for mathematics education in both East Asian countries and those from the West? These are issues that will be discussed in the rest of the chapters in this book.

REFERENCES

Ball, D., 1988, *Knowledge and Reasoning in Mathematical Pedagogy: Examining What Prospective Teachers Bring to Teacher Education*", unpublished doctoral dissertation, Michigan State University, East Lansing.

Biggs, J.B. 1996, Western Misconceptions of the Confucian-Heritage Learning Culture, in: *The Chinese Learner: Cultural, Psychological and Contextual Influences*, Watkins D.A. and Biggs, J.B, eds., Comparative Education Research Centre, Hong Kong, pp.45-68.

Beaton, A.E., Mullis, I.V.S., Martin, M.O., Gonzalez, E.J., Kelly, D.L. and Smith, T.A., 1996, *Mathematics Achievement in the Middle School Years*, Centre for the Study of Testing, Evaluation and Educational Policy, Boston College, Boston.

Brimer,A. and Griffin,P., 1985, *Mathematics Achievement in Hong Kong Secondary Schools*, Centre of Asian Studies, University of Hong Kong, Hong Kong.

Dahlin, B. and Watkins, D., 2000, The Role of Repetition in the Process of Memorisation and Understanding: A Comparison of Views of German and Chinese Secondary School Students in Hong Kong, *British Journal of Educational Psychology*, **70**:65-84.

Garvey, A. and Jackson, B., 1975, *Chinese Children. Research and Action Project into the Needs of Chinese Children*, National Education Research Development Trust, Cambridge.

Good, T. and Weinstein, R., 1986, Teacher Expectations, in: *Improving Teaching: 1986 ASCD Yearbook*, Zumwalt, K., ed., Association for Supervision and Curriculum Development, Alexandria, VA.

Hammouri, H.A.M, 2004, Attitudinal and Motivational Variables Related to Mathematics Achievement in Jordan: Findings from the Third International Mathematics and Science Study (TIMSS), *Educational Research*, **46**(3):241-257.

Hess, R.D. and Azuma, M., 1991, Cultural Support for Schooling: Contrasts between Japan and the United States, *Education Researcher*, **20**(9):2-8.

Howson, G. and Wilson, B., 1986, *School Mathematics in the 1990s*, Cambridge University Press, Cambridge.

Lapointe, A.E., Mead, N.A., and Askwe, J.M., 1992, *The International Assessment of Educational Progress Report No. 22-CAEP-01: Learning Mathematics*, The Centre for the Assessment of Educational Progress, Educational Testing Service, New Jersey.

Leung, F.K.S., 1992, *A Comparison of the Intended Mathematics Curriculum in China, Hong Kong and England and the Implementation in Beijing, Hong Kong and London*, Ph.D. thesis, the University of London.

Leung, F.K.S., 1995, The Mathematics Classroom in Beijing, Hong Kong and London, *Educational Studies in Mathematics*, **29**:197-325.

Leung, F.K.S., 2001, In Search of an East Asian Identity in Mathematics Education, *Educational Studies in Mathematics*, **47**:35-51.

Leung, F.K.S., 2002, Behind the High Achievement of East Asian Students, *Educational Research and Evaluation*, **8**(1):87-108.

Leung, F.K.S. and Park, K.M., 2002, Competent Students, Competent Teachers?, *International Journal of Educational Research*, **37**(2):113-129.

Ma, L., 1999, *Knowing and Teaching Elementary Mathematics*, Lawrence Erlbaum Associates, Publishers, Mahwah, N.J.

Ma, X. and Kishor, N., 1997, Assessing the Relationship between Attitude towards Mathematics and Achievement in Mathematics: A Meta-Analysis, *Journal for Research in Mathematics Education*, **28**(1):26-47.

Marton, F., 1997, Student Learning: East and West, public lecture delivered at the Chinese University of Hong Kong, March 18.

McLeod, D.B., 1992, Research on Affect in Mathematics Education: A Reconceptualisation, in: *Handbook of Research on Mathematics Teaching and Learning,* D.A. Grouws, ed., MacMillan, New York, pp. 575-596.

Muijs, R.D., 1997, Predictors of Academic Achievement and Academic Self-Concept: A Longitudinal Perspective, *British Journal of Educational Psychology*, **67**:263-277.

Mullis, I.V.S., Martin, M.O., Beaton, A.E., Gonzalez, E.J., Kelly, D.L. and Smith, T.A., 1997, *Mathematics Achievement in the Primary School Years*. Centre for the Study of Testing, Evaluation and Educational Policy, Boston College, Boston.

Mullis, I.V.S., Martin, M.O, Gonzalez, E.J., Gregory, K.D., Garden, R.A., O'Connor, K.M., Chrostowski, S.J. and Smith, T.A., 2000, *TIMSS 1999 International Mathematics Report*, International Study Center, Lynch School of Education, Boston College, Boston.

Organisation for Economic Co-operation and Development, 2001, *Knowledge and Skills for Life: First Results from PISA 2000*, OECD Publications, Paris.

Organisation for Economic Co-operation and Development, 2003, *Literacy Skills for the World of Tomorrow - Further Results from PISA 2000*, OECD Publications, Paris.

Robitaille, D.F. and Garden, R.A. ed., 1989, *The IEA Study of Mathematics II: Contexts and Outcomes of School Mathematics*, Pergamon Press, Oxford.

Stevenson, H.W., Lummis,M., Lee,S. and Stigler,J., 1990, *Making the Grade in Mathematics: Chinese, Japanese and American Children*, National Council of Teachers of Mathematics, Reston, VA.

Stevenson, H.W., Chen, C. and Lee, S.Y., 1993, Mathematics Achievement of Chinese, Japanese, and American Children: Ten Years Later, *Science*, **25**(9):53-58.

Stigler, J.W. and Hiebert, J., 1999, *The Teaching Gap*. Free Press, New York.

Wong, N.Y., 1998, In Search of the 'CHC' Learner: Smarter, Works Harder or Something More? Plenary lecture, in: *Proceedings of the ICMI-East Asia Regional Conference on Mathematical Education*, H.S. Park, Y.H. Choe, H. Shin, and S.H. Kim, eds., **1**:85-98.

Wong, K.M. and Cheung, W.W., 1997, A Survey of the Current State of Primary Mathematics Teaching in Hong Kong, *EduMath*, **4**:3-15.

Section 1

CONTEXT OF MATHEMATICAL EDUCATION
Introduction

FAN Lianghuo[1] and Walther FISCHER[2]
[1] *Nanyang Technological University, Singapore;*
[2] *University Erlangen-Nürnberg, Germany*

Among the 14 aspects of the study identified by the International Programme Committee, the first is "context". As one can see in the discussion document, it says that "mathematics education does not take place in a vacuum, but there is always a host of different contexts within which the practice of mathematics education takes place"; moreover, these contexts may be social, political, economic, philosophical, or ethical, and they are related one way or another to the underlying cultural values.

Focusing on the aspect of "context", the first section of this study volume presents what we have achieved in this ICMI study. In general, there are three related questions to the topic of the section: (1) what are the elements within these contexts that are relevant to mathematics education? (2) What are the "givens" from which we organize and practice mathematical education? And (3) what are the "constraints" within which we carry out mathematics education?

There are seven chapters in the section. They are arranged into the following three subsections: (a) Eastern and Western Perspectives; (b) Differences and Similarities of Educational Mathematics based on Cultural History; (c) Givens and Constraints of Mathematical Education in different Cultural Traditions.

Three chapters are under the subsection of "Eastern and Western perspectives". Hirabayashi's contribution, entitled *A Traditional Aspect of Mathematics Education in Japan – Mathematics as Gei (art), its Jutsu (technique) and Do (way)*, deals with a commonly seen phenomenon not just in Japan but, in our view, also in many other parts of the world, that is, many students do not enjoy learning mathematics. The chapter examines the concept and characteristics of Gei as a culture in Japanese tradition, including both Gei-Do and Gei-Jutsu, and their implication on mathematics teaching and learning in the Japanese educational settings. Hirabayashi argues that if mathematics education has some degree of the spirit of GEI, mathematics will be joyfully learned by all young pupils, or at least, will not be so much hated as now.

One can see that, interestingly, Hirabayashi's view is to a degree echoed by Ueno in the following chapter: *From Wasan to Yozan: Comparison between Mathematical Education in the Edo Period and the One after the Meiji Restoration.* In this chapter, Ueno discusses how Japanese accepted western mathematics (yozan) in the Meiji period, although in the Edo period Japanese mathematics (wasan) had bloomed. According to Ueno, in the late Edo period all over Japan people enjoyed solving problems and making new problems, but today, this tradition has almost disappeared since mathematical educators have forgotten to prepare interesting mathematical problems. Instead, students in Japan learn mathematics because it is useful for entrance examinations, while logical thinking is still missing in Japanese society and this makes mathematical education more difficult.

It has been well known since the 1980s that Japanese mathematics education has been widely acclaimed in international comparisons, especially in the West. Nevertheless, it seems to us that because of many constraints and limitations one can easily see concerning these comparisons, it is not unusual that the conclusions are often made based on limited observations, quick snapshots, and decontextualized figures and statistics, which often, though understandably, lead to misinterpretation, over-simplification, and over-generalisation. For example, we have seen that the practices about Chinese and Japanese education have been described and viewed very differently by researchers in these two countries and in the West. Therefore, more exchanges and collaborations are clearly needed between "insiders" and "outsiders" in this connection. We hope and believe that the pictures of Japanese mathematics education described in these two chapters by the two "insiders", who show a strong sense of "self-criticism" and "self-reflection" on Japanese mathematics education, will bring interested readers an enlightening perspective to think about and learn from the experiences of Japanese mathematics education, and moreover, about relevant issues concerning comparative study.

In a large sense, Keitel's contribution: *Perceptions of Mathematics and Mathematics Education in the Course of History: A Review of Western Perspectives*, is a response and dialogue with regard to the first two chapters from a Western perspective, as the subtitle suggests. Keitel looks at mathematics and mathematics education mainly as social phenomena, and thus tries to establish a more direct link to mathematics education as a social task. For example, she argues that "rigid examinations and tests have been developed as assessment instruments for military and economic purposes first". This perspective provides us with a broad viewpoint of issues concerning examinations or testing in mathematics education. The discussion in the chapter suggests that in many aspects social structures and social needs are main determinants for mathematics education.

Fischer's chapter compares four classical ancient mathematics textbooks from different times, from different cultures, written in different languages – China's *Jiuzhang Suanshu* (about 150 B.C.) and *Suanfa Tongzong* (1592), Japan's *Jinko-ki* (1627), and a German one, *Rechnen auf den Linien und mit der Feder* (1522). The chapter aims to use historical topics as indicators for the existence of fundamentals in educational mathematics and highlight the similarities and differences in these four textbooks from a perspective of intercultural comparison. Fischer concludes that conformities between the East and the West enable us to understand each other, while differences present us with the chance to enrich and to complete each other.

The last subsection, "Givens and constraints of mathematical education in different cultural traditions", contains the last three chapters in this section. In Wong's chapter *From "Entering the Way" to "Exiting the Way": In Search of a Bridge to Span "Basic Kkills" and "Process Abilities"*, the author tries to explain and understand the phenomenon of Chinese mathematics education by examining the practice of mentoring in various Chinese traditions such as calligraphy, martial arts (Kung Fu), acupuncture, and Buddhism, and thus rejuvenating the use of practices ("entering the way") in the gradual deve-lopment of understanding and higher-order learning ("leaving the way"). According to Wong, meritable practices play a key role in enhancing students' learning of mathematics. In particular, mere repetitive practice may lead to learning by rote, but repetition with variations may lead to under-standing.

In the following chapter, Li argues that a key belief in China about teaching and learning of mathematics is that *Practice Makes Perfect*. The chapter identifies and discusses four sources of this belief: mathematics, learning, teaching and examination. The chapter also discusses two levels of meaning of manipulation or practice in practical teaching in China.

In the last chapter, Leu & Wu present an empirical study, which explored students' awareness of their teachers' mathematics pedagogical values in two Taiwanese elementary classes, with the teacher in one class being mainly influenced by Confucianism and the other by Buddhism. According to the study, Leu and Wu argue that Confucianism and Buddhism did have an influence on the teachers' mathematics pedagogical values, and students could be aware of these values as well. The chapter highlights the influence of culture on teachers' beliefs about mathematics teaching and learning and correspondingly their teaching behaviours. The chapter also briefly discusses the implication of the findings.

We hope that readers will find the chapters contained in the section informative and enlightening.

Chapter 1-1

A TRADITIONAL ASPECT OF MATHEMATICS EDUCATION IN JAPAN

Mathematics as GEI (art), its JUTSU (technique) and DO (way)

Ichiei HIRABAYASHI
Hiroshima University, Japan

Maybe we were born to play,
Maybe we were born to amuse,
When I hear children singing in play,
I feel myself merrily shaking.
(Monk-Emperor Goshirakawa Ed.: *Ryojinhisho* 1192, translated By The Author)

1. SOME INTRODUCTORY REFLECTIONS

1.1 Present status of mathematics education in Japan

If we summarize, according to the recent IEA report 1999, the present status of Japanese pupils' performance in mathematics, we can say that they do fairly well as compared with other countries but they have a strong aversion to mathematics. If I say in a rather cynical way, it seems in our country that the aim of mathematics education is to make pupils hate mathematics, then in this point we may have very much succeeded.

But, which of the next two cases is considered to be better: one is 'to be not so able but like mathematics very much' and the other is 'to be able but not to like it'? I would prefer the former because if they like mathematics, even if they cannot do it so well at present, they may be expected to re-learn again when it is needed in the future, but in the latter case they will never return to mathematics throughout their later lives. In fact, a Japanese proverb says "What one likes one will do well."

From this point of view, I cannot be so optimistic as to appreciate the present status of mathematics education in our country. I often ask my colleagues or intimate school teachers: what is the residue of our effort in mathematics education? In reality, we could send to college or university many students among whom splendid mathematicians or skillful users of mathematics in technology and science may be born, but most of our graduates leave high school with a bad impression toward mathematics and even if they were able in mathematics they will forget it very soon after leaving school, only retaining a big prejudice that mathematics was a very difficult and uninteresting subject.

To support this conjecture, I would like to introduce two episodes from my teaching experience. These episodes are something like comical stories and you may read them with some laughter but I wish you to know that they suggest a serious problem for us, mathematics educators, which cannot allow us to remain with laughing. I also hope that these will be used as some documentation of a qualitative aspect in this ICMI-study which was sought by the co-chairs in their 'Discussion Document'.

1.1.1 Episode (1)

The first is an episode of a new student in the college where I was working after my retirement from a national university. In this college, one unit of mathematics was a compulsory subject in the first year even for literature-course students. After the first day of my lecture he came to me with an irritating attitude and said: "I preferred this college to others because there was no mathematics among the entrance examination subjects, so why should such a *math-hater* as me have to learn mathematics again after the entrance?"

Recently in our country many universities or colleges, especially in their literature departments, tend to exclude mathematics from the entrance examination subjects aiming to gather as many students as possible under the rapid decrease of the young population. As a result, in the secondary school, students will not learn mathematics and other exact-sciences because they are useless in the entrance examination. We refer to this phenomenon in the secondary school as *mathematics-detachment* or science-detachment and now some people concerned in education have begun to worry seriously about it.

1.1.2 Episode (2)

The second episode was one which happened when I was a young professor at a women's college. At the beginning of my first lecture I wished to know how much these students possessed of the secondary school mathematics and I gave them a small test, in which, among others, I asked them to state the Pythagorean theorem (which is called the "three squares theorem" in our country and I do so). To my surprise there were very few who could do it (to my memory it was less than 20%), even though I emphasized it was not required to prove the theorem but only to write down the proposition. Some of them wrote only the formula: $a^2 + b^2 = c^2$. Then I asked one of them what did a, b and c mean, but the answer was very shocking: she said, "I don't know, but isn't it trifling to know it?" Among other papers, the most impressive that I still remember was something like a letter to me, on which I could read the following: "I am very sorry I cannot do this. Though I remembered until the graduation of the secondary school, it elapsed long time after leaving mathematics, and I've forgotten almost all of the mathematics that I learned." Only several weeks had elapsed since they left the secondary school, so the amount of mathematics they had learned was decreasing rapidly, as suggested above, and when they left college to go into society, it would rapidly tend to zero, I believed.

In our country, the Ministry of Education examines the level of pupils' ability in mathematics through an achievement-test almost every ten years, and it has been usual to say that the result was almost satisfactory. But I think this result cannot be considered as reliable for knowing how much mathematics they have acquired that will last long in their lives, because the test was held immediately after their learning; for instance in this test, knowledge of the 5th grade mathematics is examined at the beginning of the 6th grade. In order to know a person's real acquirement of mathematics we should examine adult people in the society to determine what mathematics learned by them at school still remains with them.

Generally speaking, what is really said to be the achievement of education, should be examined by a test like this to adult people in the society. But in reality it will be impossible, or at least very difficult, to hold such an examination. In the case of our country, however, I believe in my heart that the average mathematical ability of all adults may be equal to that of the 4th grade pupils. Most knowledge of mathematics higher than the 5th grade, if it was taught in schools, would be forgotten very rapidly when they begin their usual social lives. Mathematics education – is it such a vain human performance as this? If it were an economical enterprise, it would go bankrupt very soon under such a poor result as this.

Someone might say that: certainly they forget much of the mathematics they have learned, but it happens only in such domains as knowledge and skills. However, mathematical attitudes, ways of thinking and abilities to solve problems, if they are well fostered through a good education, will still remain and continue to be maintained through their lives. I myself hope so, but we have not yet succeeded in showing this fact enough to satisfactorily persuade the public and political authorities, and this is also the *status quo* of mathematics education as a scientific discipline in Japan and perhaps in many other countries.

1.2 Motivations to learn mathematics

With all these defects, why are people so eager to make their children learn mathematics? It's only because mathematics is the key subject of entrance examination to higher schools and a high schooling is believed to continue to a high status in the future society. Then, may we say, mathematics is learned only for the examination not for its educational qualities?

But I think there are some other facts that suggest different motivations or reasons to learn mathematics and in some degree encourage us, educators and researchers in mathematics education. In the following I will refer to some of them.

The first fact that I wish to show is that most teachers in our country, especially in Primary Schools, believe that they teach mathematics not merely for the entrance examination. They work to give all children not only fundamental knowledge and skills but also habits and attitudes, which are expected as essential to develop their sound intelligence to think reasonably in their daily work and treat their personal problems logically. Some parents may be dissatisfied with such teachers because their work does not seem to teach mathematics for the entrance examination, and they send their children to a special *Juku* (a small private school) after school to train them for the examination, but teachers in ordinary schools will not wish to be entirely involved in preparing children only for the entrance examination.

The second is the fact that there still exist many *Soroban Juku* (small private schools for abacus). In our country, 'abacus' meant 'arithmetic' and was one of the 'Three Rs' until the Meiji Restoration (1868) and, after the introduction of western mathematics, it has gradually fallen and today it only remains as a small topic in the 3rd and 4th grades mathematics. But while we can have an electronic calculator very cheaply, why do some parents send their children to this Juku which has no concern with the entrance examination? The director of the "National Association of Soroban Education" says there are about 30 thousand Soroban Juku and 15 thousand teachers of it in the country. I think, in Juku like these, we can see a very interesting traditional atmosphere of mathematics education still existing and attracting pupils' minds. It is a spirit which can be seen in most of the traditional ways of training in cultures.

Perhaps, some of you know that there are many traditional cultures which are taught not in schools but in a small Juku: for instance, that of the tea-ceremony, flower-arrangement, traditional dancing, song, calligraphy, drawing, poetry (俳句 *haiku*, 短歌 *tanka*), etc. We refer to these traditional cultures as "GEI (芸, 藝)" and perhaps the abacus is also still learned as one of these GEI. And in fact in our country there is evidence that the traditional mathematics of our country which is called *wasan* (和算) in itself was considered to be one of these GEI and it is my intention in this paper to argue that even today's mathematics in schools may not be possessed by children if it does not take a form of GEI. I wish to call this traditional mode of culture as "GEI-esprit" and to discuss what it is and in what form it is maintained, or should be maintained, in mathematics education in Japan.

2. THE CONCEPT OF GEI

It was a historical necessity that many domains of Japanese culture have been influenced by the "CHC (Confucian Heritage Culture)" which was described in Professor Wong's thesis in the Proceedings of ICMI EARCOME 1 in Korea in 1998. However, this is not to say that there are not many other unique traditional features which are proper to our cultural field, even in today's mathematics education. As one of them, I wish to notice in this paper the GEI-esprit mentioned above in studying mathematics in schools.

What is GEI? Though it is very difficult to explain it, especially to foreign peoples, it may be permitted for the present to be taken as a kind of self-culture or hobby which has no concern with earning money or making a living except in the case of some professionals of it. Most students or learners of GEI of any kind, wish to learn it only because it's fun and they may gain from it some good effects in living happily or sometimes it could give them a good reputation in their society.

To understand the concept of GEI, it would be important to know that its substance is manifested in the personality of its master (師匠, *Shishou*) and the master is the living model for disciples (弟子, *Deshi*) to follow in all respects. GEI cannot be studied through only books or manuals but through a direct personal guidance of the master, or seniors who are certificated as the teacher by the master. For this reason, even today, in some GEI there are master-families (家元, *Iemoto*) who have inherited from their ancestors not only GEI itself but even the monopolistic right of certification to teach the GEI, for instance in tea-ceremony, flower-arrangement, traditional dancing etc. And it would be natural that they formed an association or sometimes a closed fraternity under the master and often competed with other ones of the same GEI field. This would be a weak aspect of the system of this society in developing their GEI but would have a merit to maintain its uniqueness or purity for a long time.

The fact that this peculiar aspect of GEI was existing even in the traditional mathematics, was for the first time indicated by a historian of mathematics **Dr. Yoshio Mikami** (1875-1950) in the twenties of the last century in his study of *wasan* (Japanese mathematics) which had developed since the first half of the 17th century and was ruined rapidly after the Meiji-Restoration (1868) with Westernization of our country.

Until around the end of the War II we had two eminent historians of mathematics, one of whom was the above mentioned Dr. Mikami and the other was **Dr. Kin-nosuke Ogura** (1885-1962). They were colleagues in the same university and had almost the same opinion about the causes of the decline of *wasan*; these were the shortage of logic, little connection to philosophy and natural sciences, defects of symbolism etc. But with one point they were not in accord; Dr. Ogura said from his materialistic historical viewpoint that *wasan* had been ruined by its 'guild'-like nature as we saw in the medieval ages in Europe, but Dr. Mikami insisted that there was not such a system as guild in this country: *wasan* was nothing but one kind of GEI and *wasan*-mathematicians had only enthusiastically enjoyed it without regard to any other things. I myself heard this fact personally from him soon after the War when I was a young teacher in a high school.

I think that in schools of our country, this atmosphere, so to speak GEI-esprit, may be implicitly existing in the mode of teaching and learning mathematics and if it is true, as someone says, that a decline of mathematics education is now beginning in schools of our country, it would be due to the shortage of GEI-esprit in mathematics education in recent years. Indeed, teachers in schools may be apt to become a mere living teaching machine with little attractive effect produced from his/her own personality and humanity. To see the circumstances of today's mathematics education in Japan from this aspect, we should analyze the GEI-esprit more closely in its substance. In this regard I wish to mention two characteristics of GEI and its training.

(1) **The Mind is aimed at**: The first is that what should be learned in GEI is not only the technique but the mind which sustains the GEI from inside, and because of the possibility to be possessed of this mind, GEI has a good educational quality above the mere acquisition of the technique.

(2) **The Teacher is crucial**: The second is that GEI is believed to be learned only through the personality of the teacher and cannot be taught without the direct guidance of the teacher.

In the following I would like to show these two unique characteristics of GEI-education by referring to some examples which are also quoted from domains other than mathematics education.

3. JUTSU (TECHNIQUE) AND DO (WAY) IN GEI

In GEI-training we may discern two features: one is JUTSU(術, technique) and the other is KOKORO(心, mind) which supports the technique from inside. To acquire the technique with a good mind is the ultimate aim of students and it would be given under the personal guidance of their master who him/herself had already attained this aim to a higher degree after the severe training process. This training is often compared to the walking of a long way and in this sense GEI-training, emphasizing the feature of having a sound mind, is often called GEI-DO (芸道), where "DO (道)" means "way" and it implies the training process to follow in order to acquire the GEI in the most proper sense. GEI which aims to have the technique exclusively may be called GEI-JUTSU (芸術, artistry), but if it has little support of such a mind as above, it is looked down as merely GEI-TO(芸当, feat) with a little change of spelling from GEI-DO. In most of GEI in our country, most people, except someone who wishes to be a professional in their GEI, learn GEI with a focus on the mind through the GEI-training besides becoming an expert of its technique, because such a mind would be very valuable for living soundly in any other field.

About the first characteristic mentioned above in (1), I wish to allude to JUDO (柔道) which is our traditional sport and has now spread worldwide. But in former time JU-DO was called JU-JUTSU (柔術) and was one of the martial arts, that is, a warrior's technique to fight against the enemy in the battlefield. So, when and why did JU-JUTSU change to JU-DO? It was in 1882 in Tokyo when **Jigoro Kano** (1860-1936), who is called "the father of JUDO", opened his training-hall KODOKAN (講道館).

Since his younger days, he had a strong intention to revive the decayed JUJUTSU as a practice to train young men's bodies and minds and this plan had developed to open his private training hall. On the occasion of the inaugural meeting of this hall he made a speech which was worthy of attention to understand the distinction between JUTSU and DO even from the viewpoint of mathematics education. In the following, some lines are extracted from this speech. Though they are a little long, I wish to quote them from an English edition of his biography.

> "It's no reflection on any of you, but nowadays few men of good character would pursue an interest in jujutsu for long. Those who do are

generally roughnecks, men who are fond of fighting or who don't have enough mental discipline to get an education. My own belief is that jujutsu training should improve a man's character as well as his physical powers. I hope you agree."

"In my opinion the ideal should be to prevent fights, to promote education, and to cultivate good manners and civilized behavior."

"From today we will no longer practice jujutsu. We will practice something new, which we call judo."

"As you know, the word *jujutsu* is composed of two parts: *ju* means "gentle or flexible" and *jutsu* means "technique". The *ju* of judo is the same as in jujutsu – we will preserve gentleness and flexibility – while *do* means "path or way". In judo, we will focus above all on the way – the path itself. Technique will be secondary to achieving an understanding of the way. To train men of good character for life, judo is the ideal way." (Watson, *Judo*, 51)

Mathematics is not a sport and not all the assertions quoted above may be applied to mathematics education. But most essential parts of today's school mathematics are not the techniques, which should not be forced on all children to be acquired completely. As a subject of common education, mathematics should be considered to be material to train the intelligent part of pupils' personality and should be organized as such in its curriculum as well as in its teaching. What is to be learned is not only the *technique* but the *way* to develop their personality – it is the fundamental recognition in learning GEI for common (not professional) peoples and it should also be the primary motivation to learn mathematics for common pupils.

We can see the distinction between JUTSU and DO even in today's school mathematics in our country. In our current study of mathematics education, we often distinguish the aims of this education into two parts; one is called 'substantial aim' and the other 'formal aim'. The former implies the acquisition of mathematical knowledge and skills, while the latter means to be equipped with good intelligent attitude and habit, which perhaps corresponds to the mind in GEI-DO. In the 'Mathematics Program' issued by the Ministry of Education of Japan, which is the current national curriculum of mathematics, we can read in its outset the 'Overall Objectives' which is the aim of mathematics education through *all* grades, as follows: (I mention here, as an example, only that of the Lower Secondary School which is compulsory for all pupils.)

> "For the students to understand deeply the fundamental concepts, principles and rules relating to numbers, quantities, figures and so forth. For students to acquire methods of mathematical expressions and strategies, and to improve their abilities to relate phenomena mathematically. For students to enjoy mathematical activities, to appreciate the importance of mathematical approach and ways of thinking and to inculcate in them the right attitudes necessary to make use of mathematics." (Japan Society of Mathematics Education, *PROGRAM*, 21)

While the two aims, 'substantial' and 'formal', referred to above are so much blended or even mingled in this quotation that we cannot easily separate the two clearly, the overall stress here is on formal aims rather than substantial ones. But in the 'Objectives' (aims) of *each* grade mathematics, this Program alludes only to substantial aims, for instance, to understand equations, to deepen understandings on properties of figures etc. and it is natural that teachers are inclined to stress the substantial aims only under today's severe competition in the entrance examination. It is in this inclination that we can see a clear sign of the decline of GEI-esprit in our country.

4. TEACHER AS A MODEL OF LEARNING

Here we will be concerned about the second characteristic of GEI which I mentioned in (2) of Section 2.

In all countries which have been under a Confucianist educational influence, it may be a common tradition of education for a pupil to pay high respect to his teacher, and in the training of GEI-DO of our country it is the essential attitude for disciples to entertain a high respect for their master in order to be well possessed of GEI. Without such a respect it is believed to be impossible to have good success, because the teacher (master, 師匠, *shisho*) him/herself is regarded as the incarnation of GEI or the model for students (disciples, 弟子, *deshi*) so as to be in accord with his/her personality under his/her direct guidance.

Today this aspect of teaching and learning might be seen in the most typical form in SUMO-wrestling which is still the favorite traditional professional sport and still maintains the habits of its original appearance. In SUMO, the authority of master is almost absolute for disciples, because the master himself walked the severe way of training during his younger ages and is believed to manifest all JUTSU (technique) and KOKORO (mind) in his whole personality.

Here I do not wish to insist that the teacher of mathematics should be like the master of SUMO, but I only wish to suggest that there is an important factor of teaching which is inevitably concerned with the personality of the teacher, especially if we wish to regard mathematics as an activity proper to the human mind instead of seeing it as a mere technique pragmatically useful in life. I wish to say that the teacher should not be satisfied with being a mere transmitter of technique like a teaching machine but should be a good personal model for pupils in enjoying and using mathematics actively and effectively, and in this regard we can well refer to the mode of teaching and learning in GEI-DO.

In fact, there may be two modes of teaching and learning in anything; one can be done using just books or manuals, while the other intrinsically needs to be done under the teacher's personal guidance. This distinction of education was well known from early ages in our country. A typical example can be read in the recent historical novel of **Ryotaro Shiba**: *Kukai no Fukei* (Scenery of Kukai, 1975). I would like to introduce one of its scenes:

Kukai (778-835) was a monk, the founder of a Buddhist sect. When he returned from China, another monk **Saicho** (767-822) was already a top leader of Buddhist society of our country. Saicho heard that Kukai had returned with a new sutra of Buddhism, and asked to borrow it from him to know the new doctrine of Buddhism through it. But, in spite of repeated requests from this elder senior in this religion, Kukai will not lend it and at last he refuses with a letter in writing roughly as follows:

> "The essence of this new doctrine could not be understood only by reading a sutra, and if you really wish to know it, come to me by yourself, I will teach it personally."

Here we should remember that Kukai was a mere young junior and Saicho was the greatest elder senior at that time.

Since very olden times, 'being taught personally from teacher' (師承, *shishou* in Shiba's terminology) instead of 'learning only through books' (筆授, *hitsuju*) was a common mode of teaching and learning not only in Buddhist society but in many traditional cultural regions of our country. Though the latter is a common mode of learning explicitly pervasive in today's classroom, the former mode should be implicitly embedded in the teaching of mathematics especially at primary school level, and the shortage of this mode may be a cause of decline of mathematics education and produce many math-haters and math-dropouts.

If mathematics can be seen as an activity of the human mind, it could not be taught or learned merely through a mechanical procedure, because it would be a very complex and delicate activity. Certainly mathematics is a cognitive subject, but to learn it we need to have a good emotion toward it.

Emotions like this may be too delicate to express in words or letter. It may be possible only through the whole personality and humanity of the teacher. For instance, can a teacher really help their pupils to like mathematics if he/she hates mathematics? Can a teacher teach mathematics successfully if he/she is neither loved nor respected by pupils?

To my great regret, recently in our country, we sometimes hear that pupils did violence to teachers. Especially the teacher of mathematics may not be respected but be mostly hated by pupils for his/her transcendental attitude.

If a teacher is a model of mathematics learning for pupils, teacher education is the most crucial to improve mathematics education, and in our country it would be the responsibility of universities, especially its department of education. But in our country, mathematics professors are not eager to take up teacher education but rather mathematics education.

5. FASCINATION OF GEI

Among trends which have effected the recent reformation of the school curriculum in our country, there are two great ones: *globalization* and *equalization*. The former is clearly indicated in the Discussion Document of this ICMI Study, but I wish to be concerned with the latter.

First of all we should notice the fact that in its historical origin there was not a direct relation between primary education and secondary education and they developed almost independently; the former was for children of common peoples and the latter was for a few intelligent elites who were expected to be future scholars or social leaders. But after World War II, especially in the secondary level the curriculum made for these few elites has been equally imposed to all children under the name of 'popularization of secondary education' and in responding to the strong public claim for equalization; all children were forced to learn the mathematics that was required only by a very few intelligent elite children in former times.

It will be suitable to introduce here another episode which I saw in a TV drama of some years ago:

5.1 Episode (3)

It happened in a mathematics classroom of a secondary school that a student suddenly stood up and asked his teacher saying: "My father keeps a Chinese restaurant and makes and sells Chinese noodles everyday, but I never saw that he used mathematics like this factorization and others in his work. I am asked to succeed him in his work in the future. Why should such a man like me learn mathematics like this?"

The teacher persuaded him one way or another, but after returning to his room he began to wonder seriously why he as a teacher should teach such mathematics to such a pupil; he might be able to persuade his pupil but could not persuade himself!

Since then, it has become my custom to tell this story to my students of future teachers and ask: "If you were this teacher, how would you persuade this pupil?" Answers were of many varieties and were very interesting, however, I will not talk about it in further detail. But I only wish to say that it is not good in the first place that a pupil should ask such a question: if this pupil is really enjoying mathematics, he would not ask such a 'philosophical' question. In most GEI-training learners will not ask such a question, perhaps because GEI has a unique attraction even if it has much pain to bear. Indeed in *juku* of mathematics they seem to attend joyfully even though being tired after school. Perhaps education in *juku* has something similar to GEI-training in its nature.

GEI-DO may be well defined as a systematized hobby so as to be understood by foreigners. In our country this systematization of GEI as a culture had begun since very old times; for example, in the tea-ceremony (茶道, *sado*) and flower-arrangement (華道, *kado*), it began during the 14th-15th century. But long since before then, GEI existed with a close relation to human nature.

In this respect it is interesting to know that at the early ages of the 12th century the **Emperor Goshirakawa** (1127-1192) indulged himself deeply in the learning of a kind of folk song, *Saibara*, and collected them and edited a book on it, from where I have quoted the poem at the beginning of this paper. It was said that he devoted himself too much in learning, almost forgetting food and sleep. GEI has such a kind of fascination and induces the whole human mind to learn and enjoy it.

GEI-DO is a long way to be possessed of the GEI; it resembles very much a journey with its various pains as well as many enjoyments. It has a strong attraction to human nature in its nature, and even if it happens that its training gives some pains or troubles with it, they will be overcome by this irresistible attraction. If mathematics and its learning has the form of GEI, it will become a great effective material as a school subject for *all common pupils* as well as for some elites.

6. CONCLUDING REMARKS

As I said before, the product of GEI-training is not only the technique but rather the mind which accompanies the training, and to have this mind is the real aim of the training, especially for common peoples other than those who wish to be a professional of this GEI. This mind would be demonstrated, for instance in the case of the tea-ceremony, in making his/her daily behavior elegant or reasonable. If mathematics learning is reformed in a way comparable to GEI-training, the effect of learning would be seen in the learners' way of thinking or activities in many domains of their future lives and because of this effect mathematics would be able to occupy its paramount place among school subjects for all pupils.

I will close this paper with one more episode. If it has any implications, I hope, they will be left to the readers' own consideration.

In our country, equalization of education appears to be misunderstood, even among educational authorities, as if it can be realized merely by a levelling-down of content or reduction of difficulties; especially in mathematics, they seem to believe that all pupils can learn the same mathematics by doing this. It may also be a reaction against the education of the past, which was much too academic, as if all pupils would be mathematicians in the future.

6.1 Episode (4)

After the recent reformation of the national curriculum in which difficult topics have been greatly reduced, I happened to meet one of my intimate principals of the Lower Secondary School and asked him if the population of pupils who like mathematics has increased more than before. He decisively replied:

"No, absolutely no! Incapable pupils are still incapable and dislike mathematics even though it became easy, and more than that, able pupils have left mathematics because it became easy and trifling for them. Then, the population of math-haters has increased more than before."

REFERENCES

Beaton, A. et al., 1997, The Mathematics Achievement in the Primary School Year, *IEA's Third International Mathematics and Science Report*.

Beaton, A. et al., 1997, The Mathematics Achievement in the Middle School Year, *IEA's Third International Mathematics and Science Report*.

Japan Society of Mathematics Education, 2000, *Mathematics Program in Japan*.

Mikami, Yoshio, 1922, Charactor of Japanese Mathematician and Nation, *Studies of Psychology* **125**:311-333. (only in Japanese)

Watson, Brian N., 2000, *The Father of JUDO, A Biography of Jigoro Kano*, Kodansha International, Tokyo.

Shiba, Ryotaro, 2002, *Scenery of Kukai*. Chukou-bunko, Chuou-kouron-shinsha, 152. (only in Japanese)

Wong, Ngai-Ying, 1998, In Research of the "CHC" Learner: Smarter, Work Harder or Something More?, *Proceedings of ICME-EARCOME 1*, **1**:85-98.

Chapter 1-2

FROM WASAN TO YOZAN

Comparison between Mathematical Education in the Edo Period and the One after the Meiji Restoration

Kenji UENO
Kyoto University

1. INTRODUCTION

From the middle of the seventeenth century Japanese mathematics developed in its own way and at the end of the Edo period even in a small village there were many farmers interested in mathematics. Development of wasan is different from that of western mathematics. Wasan mathematicians were interested in finding interesting problems of geometry rather than developing a general theory and mathematical concepts. This is partly because in the Edo period Japanese society was stable, and serious mathematics was only used for the calendar and surveying. On the other hand, in the Edo period the soroban (Japanese abacus) was popular. The Soroban was a powerful tool for calculation and was an indispensable tool for commerce.

In the Edo period elementary education was supported by the private sector and many children went to private schools called Terakoya. Literacy and numeracy were very high in the education of Terakoya in the late Edo period. In the first section of the present paper we shall quote the first paragraph of "Teaching of Yoshishige" written by an old farmer, which will tell of the education system of the Edo period and the popularity of mathematics.

In Section two we shall discuss briefly about "Jinkoki", the most successful text book for basic mathematics. The first edition was published in 1627. In the East Asian tradition a textbook of mathematics is basically a collection of problems with answers, and also a method of how to solve a problem is given. Usually it does not explain the theory itself behind the problems. A student is supposed to learn the theory implicitly by solving problems. On the other hand "Jinkoki" contains not only useful problems for daily life but also contains many mathematically interesting problems. Many people were interested in solving such problems and found fun with mathematics. Wasan was born and developed in such an atmosphere.

In Section three we shall discuss Confucianism in the Edo period to understand the philosophical background of wasan and education. In Section four we shall briefly describe the development of wasan. We also discuss characteristics of wasan, part of which has been mentioned above.

After the Meiji restoration the government tried to import western technology and culture eagerly. The government introduced the public education system and for mathematics the government decided to use western mathematics. But soroban was so popular and useful it remained a useful tool for mathematical education. In Section five we shall discuss how western mathematics has dominated Japan. Under the strong influence of Chinese culture, Japan developed an educational method based on solving problems to understand mathematical theory. Wasan followed the system. Even after the Meiji government introduced western mathematics for education, this tradition strongly survived. The Greek tradition of mathematical reasoning has been respected in modern Japanese education but mathematical education has failed to convince students of its importance. This is partly because Japanese society does not respect logical reasoning but rather intuition and neighbours' opinions. Even today many mathematicians and mathematical educators follow the habit. Also many problems which attracted students become old fashioned because of the rapid change of our society. We do not have text books like "Jinkoki". This makes the reform of mathematical education in Japan difficult. This also shows that education heavily depends on the society and its tradition.

2. TEACHING OF YOSHISHIGE

The Edo period dated from 1600 to 1868, the year of the Meiji restoration. In 1600 TOKUGAWA Ieyasu[1] achieved military supremacy in the battle of Sekigahara and in 1603 he received the title of shogun from the emperor. Ieyasu created the political "bakuhan" system under which government functioned through two political mechanisms: bakufu or shogunate and han or daimyo (local lord) domain. There were about 250 han's. The

[1] In the following we follow the East Asian tradition that we first write a family name then a given name.

Tokugawa shogunate introduced a four-class concept: the warrior (samurai), the farmer (hyakusho), the artisan (shokunin) and the merchant (shonin). Most samurai lived in towns and most of them were bureaucrats. Artisans and merchants lived mainly in towns. Farmers lived in villages. Farmers had a long tradition of self-governing to use water properly for their rice fields. Local lords used the system and administration was done by documents so that rich farmers were supposed to read and write well. Moreover, farmers were proud of producing rice, which was a basis of the economy in the Edo period. In the Edo period many agrarian risings broke out. There were secret rules for agrarian risings and they usually asked a local lord to replace his bureaucrats who ill-treated farmers so that they would have better treatment (Hosaka 2002). They used the social consensus of Confucianism in which a local lord should rule well according to Confucian ideals. Such agrarian risings often succeeded. But these facts also tell us that even in villages they had a system of education. In the Edo period education was done mostly privately. Many private elementary schools called Terakoya were built. Teachers were often priests in the villages or jobless samurai in towns. Also there were many women teachers. Many Terakoya accepted boys and girls but they had different desks often separated by a screen.

Figure 1-2-1. Terakoya[2], EISHOUSA Chouki "Onna Imagawa"

[2] Boys and girls were studying at a Terakoya. Usually they were separated by a screen. But in this picture there is no screen. (Kumon Institute of Children, 2000), p.171.

In Terakoya they taught mainly how to write (calligraphy) and sometimes the soroban (Japanese abacus). Also a writing master was chosen in such a way that children could learn the basics of their future lives as townsmen or farmers.

In the text "Teaching of Yoshishige" (Tamura, 1981) written by TAMURA Yoshishige (1790-1877) in 1873, we can find a vivid description of how education of the Edo period was done in a village. The writing begins as follows.

> I was born on 10th of October 1790. When I became a boy my parents advised me to go to Terakoya to learn how to write. But at that time I hated to practice calligraphy, I did not follow their advice. So my parents tried to teach me at home but I did not practice at all. One day my mother told me that a boy like you who hates learning would find his way into the gutter in the end. My grandmother was next to me and she helped me saying that the boy liked handiwork, so he would be suitable to be a carpenter. But my father was disgusted with the conversation and said that even a carpenter had to write down numbers on lumber but the boy could not do it. Listening to their conversation I felt awkward but I could do nothing. In this way I lived in my youth. I only wrote labels of seeds and agricultural diaries in bad handwriting if it was necessary. But I worked very hard in farming from early in the morning to late in the evening. When I became eighteen years old, a mathematician visited our village to teach mathematics to young men in my village. My grandfather and my uncle advised me to take the mathematical lessons and they kindly offered me all expenses for the lessons. But I rejected their kind offer. I said to them. "I am thankful for your offer but until now I have learned nothing. I am afraid that the study of mathematics for forty days would not give me a good understanding of mathematics. It is shameful for me not to understand mathematics even if I would have a good teacher and worked hard. It is difficult for me not to follow your advice but I hope you will understand me." Then they agreed with me. Thus I only worked very hard in farming without learning calligraphy and mathematics.

Though Yoshishige did not learn calligraphy and mathematics in his youth, he wrote several important books on agriculture. There remain also several beautiful calligraphies of Yoshishige. The above writing shows how the three R's education was done in the Edo period. In this writing Yoshishige also wrote a mental attitude for farmers which is based on Confucianism. There is a phrase describing how a good farmer should keep away from learning mathematics by halves. This also shows that in the late

Edo period mathematics was popular even in villages and many farmers spent much time on learning mathematics.

There remains a record of education of HAGIWARA Nobuyoshi (1828-1909), who belongs to the last generation of wasan mathematicians. At eight years old he went to Terakoya in which a priest of a temple of his village taught him calligraphy until he was thirteen years old. From fourteen to seventeen years old he took private lessons on the Japanese abacus under a wasan mathematician living in his village. From eighteen to twenty years old he studied Chinese classics. Then from twenty three to thirty three years old he studied wasan under SATO Yoshinori and got three licenses of the Seki school out of five. Note that Hagiwara was a middle class farmer and almost all his life he stayed as a farmer. He found the time to study mathematics only at night. He attended Sato's lessons on foot. He had to walk more than fifteen kilometers.

3. JINKOKI

Many young farmers and townsmen were interested in wasan in the Edo period. This is because they had the good textbook "Jinkoki" by YOSHIDA Mitsuyoshi (1598-1672). Its first edition was published in 1627 and widely read though the Edo period. There is an English translation of Jinkoki (Yoshida 2000). A detailed description of Jinkoki can be found in this English translation. Here I only point out two facts. The first is that Jinkoki showed how to use the soroban for multiplication and division and described mathematical subjects necessary for merchants and artisans but also it contains many interesting mathematical problems, which interested many people. The second is that Mitsuyoshi was troubled with an unauthorized publication of Jinkoki and he published several revised versions. With careful study of such revised editions we can find that Mitsuyoshi always tried to improve and rearrange the contents of the book according to his experience of teaching mathematics. Jinkoki covers almost all important mathematics useful for daily professional life. Since in the Edo period the society was stable, there were not so many subjects to change according to the change of the society. Jinkoki was widely read and was one of the best sellers in the Edo period.

Many books named after Jinkoki were published in the Edo period and even in the Meiji period. There is a list of more than 300 books in (Yoshida, 2000). Here I copy two pages of "Dinkoki", which is an introductory book for multiplication and division by soroban but in the front page (back page of the cover) we can find two interesting problems copied from Jinkoki. In this way many people had chances to consider certain interesting problems and many people were interested in solving mathematical problems. And in the late Edo period they even made up problems by themselves. History of Jinkoki shows that careful choice of mathematical problems is very important in mathematical education, at least in Japan. Today our society is changing rapidly and we need to be aware of the fact that certain problems are universally interesting but other interesting problems depend on the society.

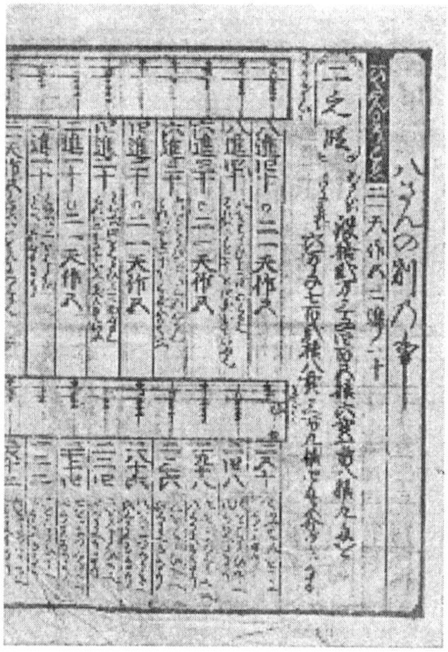

Figure 1-2-2. The first page of Dinkoki. Division by soroban is explained.[3]

4. A FEW WORDS ABOUT CONFUCIANISM IN THE EDO PERIOD

Tokugawa shogunate supported Zhu Xi orthodox (Shushigaku) and also many local lords supported it by inviting teachers to their official school for samurai class. But important Confucian philosophers in the Edo period founded their own interpretation of Confucian classics. Here I mention only two philosophers: ITO Jinsai (1627-1705) and OGYU Sorai (1666-1728).

[3] No textbook of arithmetic of the Edo period explained addition and subtraction by soroban. All the children learned addition and subtraction by soroban from their parents or neighbours.

Jinsai studied Zhu Xi orthodoxy and also Zen Buddhism both of which could not satisfy him. Later he proposed a doctoring "Kogaku" that Confucian classics should be read according to the old meaning when the classics were written. He built a private school "Kogi-do" in Kyoto and taught his doctrine. But his teaching was far advanced. He discussed with his students about interpretations of a phrase in the classics and he chose the best interpretation even if proposed by his student. In the Edo period a teacher's opinion was absolutely correct.

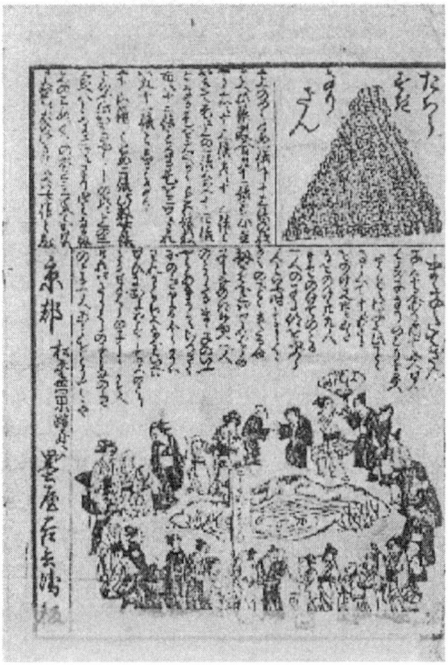

Figure 1-2-3. Dinkoki, The back page of the cover. Mamakodate[4] and Sugizan[5] are explained.

Ogyu Sorai was opposed not only to Zhu Xi orthodoxy but also Jinsai's Kogaku. He pointed out the importance of phonetic study in the study of Confucian classics and he himself studied Chinese. His doctrine was called "Kobunjigaku" and it became popular. He educated many Confucian scho-

4 In the western literature this problem is known as Josephus' problem.

5 Sugi is the Japanese name of a cedar. The original problem is counting the numbers of logs of cedar piled up. In almost all the mathematical textbooks of the Edo period they used straw rice bags instead of logs of cedar as in the figure. The problem is to calculate $1+2+\cdots+n$.

lars. Sorai was interested in not only philosophy but also almost everything. For example he criticized wasan. He wrote:

> About mathematics, mathematicians today are only interested in technical matters just searching complicated things and it is useless.

Later we will come back to his statement about wasan. Wasan mathematicians opposed this statement but some of them, for example an important mathematician of the Seki school MATSUNAGA Yoshisuke (1692-1744), agreed partially with Sorai in his letter to KURUSHIMA Yoshihiro (?-1757) and asked him to do serious mathematics. Sorai was an exception and other Confucian philosophers were not interested in wasan.

5. WASAN

Wasan is a mathematics developed in the Edo period under influence of Chinese mathematics in the Song and Yuan dynasties. The mathematic books "Suanxue quimeng" written by Zhu Shijie in 1299 and "Suanfa tongzong" written by Cheng Dawei in 1592 were brought into Japan when TOYOTOMI Hideyoshi invaded Korea. It is shame for Japan that we could not import such books in a peaceful way. (Early modern porcelain makers were also brought to Japan from Korea forcefully and they founded early modern porcelain factories in Japan.)

In the sixteenth century, Japanese merchants had trade with China and they brought the Chinese abacus. Soon, the abacus became popular in Japan and many people started to learn. "Suanfa tongzong" was a textbook for abacus in China. "Suanxue quimeng" is a book on algebra, mainly the theory of equations: Tianyuang Shu. In this book, for the first time Japanese mathematicians learned Tianyuang Shu. But the book is written in such a way that the reader can easily understand Tianyuang Shu. The first Japanese book correctly using Tianyuang Shu is SAWAGUCHI Kazuyuki's "Kokon sanpo-ki" (Old and New Mathematical Methods) published in 1637. In this book he had correctly describe how to express unknowns by letters as was described in "Suanxue quimeng". This method was later completed by SEKI Takakazu (1640?-1708). Seki named his method Boshoho (side writing method) and in wasan any number of unknowns could be used to write down equations. Nevertheless SEKI and all wasan mathematicians had no concept of polynomials. Seki's Boshoho lacked the symbol of equality as with Chinese Tianyuang Shu in which one unknown could be used for higher degree equations.

Zhu Shijie's book "Siyuan Yujian" was not brought to Japan at that time so that wasan mathematicians did not know Zhu Shijie's method to use four variables. But still there remains a question how SEKI invented or completed his Boshoho and what kind of relation Seki had with Taniguchi's teacher HASHIMOTO Masakazu (?-1683?) and Taniguchi's student TANAKA Yoshizane (1651-1719) who lived in Osaka and Kyoto. Tanaka used Boshoho and there remains evidence that Seki and Tanaka had some contact. Also there is a natural guess that Jesuit missionary gave some influence to wasan. Carlo Spinora (1564-1622), who came to Japan in 1602 and was killed in Japan, taught western mathematics in Kyoto and gave certain influence to the earlier development of wasan. Since Christianity was strictly forbidden in the Edo period, no evidence is left. We found vague evidence that certain mathematicians in the early Edo period seemed Christian. At the moment we cannot find evidence that a mathematician like Hashimoto or Taniguchi had contact with Korean mathematicians and learned Tianyuang Shu from them. Since in the early Edo period many Japanese went to Asian countries for trade, we need to study such a possibility. After Seki's invention of Boshoho, wasan developed rapidly. At the time of Seki's student TAKEBE Katahiro (1664-739) wasan reached its highest peak methodologically speaking. For example, the calculation of π interested many wasan mathematicians. Seki used a regular 2^{17}-gon to calculate π, then he applied the Atkin acceleration, a method found in the twentieth century in numerical analysis, to obtain a correct value of π of 15 places of decimals. Takebe studied it from a slightly different viewpoint and he applied the Richardson acceleration, which was also found in the twentieth century in numerical analysis, to calculate π. From a regular 2^{10}-gon he found a very precise value of π, correct up to 42 places of decimals. But both Seki and Takebe did not explain the reasons why their acceleration methods gave correct answers. This fact already tells us a fundamental character of wasan. For example, wasan mathematicians were ingenious to calculate coefficients of power series expansions, but had very little interest in creating a general theory of power series expansion. So after Seki and Takebe there were many interesting discoveries in wasan, but these results were essentially refinements of earlier investigations. In other words, for wasan mathematics is not a theory but a collection of many interesting results.

Wasan lacked the concept of functions. Though many wasan mathematicians used power series expansion they did not realize that such power series define functions. For wasan, a power series expansion is an algorithm to calculate an area of a configuration or a definite integral. WADA Yasushi (1787-1840) spent much of his life to calculate and make tables of definite integrals $\int_0^1 x^m(1-x)^n\,dx$. But he never discovered the possibility of changing the interval of the integrations. Thus wasan mathematician could not find the fundamental theorem of calculus.

A lack of strict logical reasoning is another characteristic of wasan. Indifference to logical reasoning is still common in Japan today so that we could not blame wasan mathematicians for lacking serious logical thinking. Of course they had a vague notion of logical thinking. For example in the history of π in wasan we can find a lack of logical reasoning. Jinkoki used 3.16 which is roughly equal to $\sqrt{10}$. Though many mathematicians in the Edo period found that π is 3.14 ···, the value 3.14 was not used in the textbook of elementary mathematics written after Jinkoki (Itakura, Nakamura and Itakura, 1990).

Ogyu Sorai did not accept the calculation of π by using regular polygons. He argued that polygons are polygons even if they have a huge number of sides and they are different from a circle. Sorai argued that calculations by wasan mathematicians were just technical ones and they never gave the real reasoning. Sorai was partially correct in a sense that no wasan mathematician at that time had correct logical reasoning so that they could not show that their results for π were close enough to the real value of π.

On the other hand what Sorai expected was an explanation by a philosophical principle which was also not based on logical reasoning (Itakura, Nakamura and Itakura, 1990). Even today many Japanese do not respect logical reasoning.

But wasan had a different aspect which is important from the view point of education. As I wrote above, in the late Edo period many people were interested in solving problems and making difficult problems. If they could solve difficult problems they often made Sangaku, dedicated it to the Temple or Shrine and hung it so that many people read it (for an example see Figure 4). This became very popular and many mathematicians and amateur mathematicians followed the custom. In this way they could exchange their ideas for solving problems.

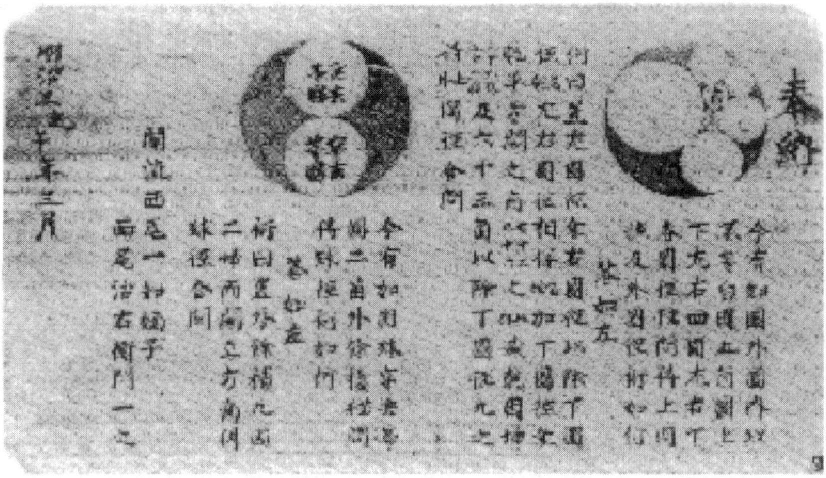

Figure 1-2-4. Sangaku, Yamashita Shrine, Ishikawa, 1870.

Often they say that wasan was a kind of art (Gei), and each wasan school formed a closed society. This is partially correct but partially wrong. Even in the early stage of wasan mathematicians in Edo, Osaka and Kyoto exchanged their ideas. After ARIMA Yoriyuki (1714-1783), who was a local lord and mathematician, published secret results of the Seki school, many advanced mathematical books were published. For example, wasan mathematician NAKASONE Muneyoshi (1824-1906) had copied mathematics books of different schools (Nakasone, 2001). Each school had its own secret results but mathematically speaking they were not so important. Also there were many amateur mathematicians all over Japan and they just enjoyed making and solving mathematical problems. Thus situation of wasan was different from serious Gei like No play or Bugei (martial arts) and wasan schools functioned as an education system. I emphasize the fact that in the Edo period many people enjoyed mathematics and there were many professional mathematicians who lived by teaching mathematics.

6. YOZAN (WESTERN MATHEMATICS)

In the Edo period western mathematics came through Chinese translations. For example, a Chinese translation of Euclid's book on geometry came to Japan in the eighteenth century but no wasan mathematicians accepted its importance. They thought that western mathematics was at a low level, since it only treated simple configurations. But through Chinese translations wasan mathematicians learned the tables of trigonometric functions and logarithms. INOU Tadataka (1745-1818) surveyed the whole of Japan to make a precise map of Japan. He used the table of logarithms to check his measurements by using trigonometry. The table he used still survives. Surprisingly he corrected certain values of the tables, which tell us that he or his pupil understood how to make such tables. Thus in the Edo period they could use freely the table of logarithms but they did not find the notion of function behind such tables.

Western mathematics also came through Dutch books. Some Japanese people started to learn western mathematics seriously when Tokugawa shogunate opened the naval academy in Nagasaki in 1855. When Tokyo Sugaku Gaisha, which later became the Mathematical Society of Japan, was founded in 1878 certain numbers of former students of the naval academy joined it. The Japanese accepted western mathematics because it was useful to military science. Wasan mathematicians were not interested in military science and they continued to study wasan after the Meiji restoration. Western mathematics became popular through education in the elementary schools. In 1872 the Meiji government introduced the public education system so that in every elementary school western mathematics should be taught. This means that they tried to introduce calculation on paper and abandon calculation by soroban. But immediately they found that it was impossible, since soroban was so common and for daily calculation soroban was very powerful. The education of soroban still remains in school.

Also in the early Meiji period the government invited many western scholars for higher education, in which western mathematics was taught. Gradually wasan mathematicians started to learn western mathematics through Chinese translations. Such an example can be found in the writings of SATO Noriyoshi (1820-1896). Sato Noriyoshi was a mathematician teaching wasan at Seishi-kan, the official school of the domain of Fukuyama from 1846 to 1872. After Seishi-kan was closed he opened a private school of wasan at home. He left many handwritten books and notes on wasan; many of them are copies of books of old masters. Some of his books were donated to the Kyoto University by his grand child. Figure 5 is a copy of a page of his notes.

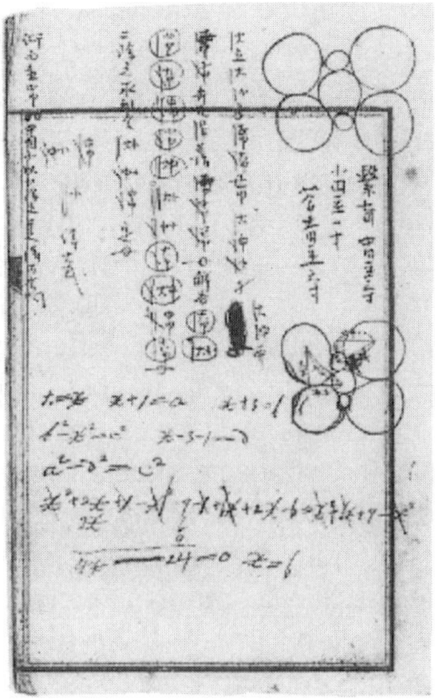

Figure 1-2-5. Sato Yoshinori: Sanpo Senmonshokai, In the upper part the problem is solved by the method of wasan, and the same equations of western mathematics were given in the lower part.

In this page a problem is given and solved by a typical method of wasan and usual western mathematics. The problem is as follows. Three kinds of circles are given. The diameter of the small circle is 1, the one of the middle circle is 3. Two small circles, one middle circle and four big circles are circumscribing as in the picture. What is the diameter of the biggest circle?

In this way, for wasan mathematicians it was not difficult to learn western mathematics up to a certain stage. Moreover, for mathematics in daily life soroban and wasan were enough. When the Meiji government started to survey lands for new tax systems, some wasan mathematicians collaborated in surveying lands. Their knowledge was enough for surveying (see for example [Nakasone, 2001]). In the provinces wasan mathematicians undertook these activities until the end of the Meiji period. Almost all of them believed that wasan was superior to western mathematics, since wasan treated more complicated configurations. Hence wasan mathematicians could not realize the fact that western mathematics had many applications to other sciences and technology. Wasan was never taught in public education. Wasan gradually disappeared and western mathematics took its place.

Also the Meiji government sent young men to Europe to study western sciences and technology. They came back to Japan and started to teach in higher education. In mathematics the most important figure is KIKUCHI

Dairoku (1855-1917) who studied mathematics in England. He became Professor of the Imperial University and later became the Minister of Education. He was a big influence on mathematical education in Japan.

In education in the elementary schools, the first Japanese translations of western textbooks were used. Gradually many Japanese mastered elementary parts of western mathematics and they wrote many textbooks. At the highest peak, all over Japan each province had different textbooks. To control quality of textbooks the government introduced an official examination system. In 1902 a bribery scandal about the official examination was made public and the Minister of Education Kikuchi Dairoku was forced to introduce the state textbooks. This was the end of the tradition of Jinkoki, in which the private sector could choose subjects to teach in mathematics. Of course the basic subjects were determined by the government already in the early Meiji period but still there remained certain freedom. The textbook writers lost their jobs. They tried to publish reference books for examinations in which they could try to convey their ideas of mathematics. This succeeded partly. For example, for reference books to the entrance examinations to universities we can find SEIMIYA Toshio' book (Seimiya, 1968) in which the author explains the methods for finding new theorems in elementary geometry to high school students.

On the other hand mathematical educators gradually lost their interest in having a wide scope of mathematics in education. They just discussed the arrangement of subjects but not the contents of the subjects. Also they have not been sensitive to the rapid change of our society and they forgot to make interesting problems for their students. Also, the old tradition of understanding mathematics by solving many problems has produced the widely spread misunderstanding that mathematics was a subject to memorize formulas and apply them to solve problems. These facts made the Ministry of Education take much control over the content of mathematics in elementary education, which has brought about the present miserable situation of mathematical education in Japan.

7. CONCLUSION

The transition from wasan to yozan (western mathematics) was smoothly done after the Meiji government chose western mathematics as a subject of elementary education. But soroban was very popular until recently. For wasan mathematicians it was not difficult to learn the elementary parts of western mathematics but they missed the importance of logical reasoning of mathematics and they could not develop a general theory. For them mathematics was not a science, but was a collection of many interesting results.

On the other hand, wasan attracted many people and in the late Edo period all over Japan people enjoyed solving problems and making up new problems.

Today in Japan students learn mathematics since it is useful for entrance examinations. But many students find fun in solving mathematical problems. Today, this tradition has almost disappeared, since mathematical educators have forgotten to prepare interesting mathematical problems that fit the modern society. Logical thinking is still missing in Japanese society and this makes mathematical education more difficult.

REFERENCES

Hosaka Satoru, 2002, *Hyakushou-Ikki to sono Saho* (Agrarian Rising and its Rules), Yoshikawa Kobunkan (in Japanese).

Itakura Kiyonobu, Nakamura Kunimitsu and Itakura Reiko, 1990, *Nihon ni okeru kagakukennkyu no houga to zasetu* (Germination and Breakdown of Scientific Research in Japan), Kasetsusha (in Japanese).

Kumon Institute of Children, 2000, *Ukiyoe ni miru Edo no Kodomo-tachi* (Children of the Edo period in Ukiyoe), Shougakukan.

Nakasone Muneyoshi, 2001, *Nakasone's Collection of Books and Notes on Wasan* (CD-ROM), Okumura Hiroshi, ed., Iwanami.

Seimiya Kiyoshi, 1968, *Kikagaku–Hakkenteki Kenkyouhou–* (Geometry, –Study method to find new theorems–), Kagaku-Shinkosha (in Japanese).

Tamura Yoshishige, 1981, Yoshishige Ikun (Teaching of Yoshishige), Nihon Nogyo Zenshu (Complete Collection of Writings on Agriculture in Japan), vol.21, Nobunkyo.

Yoshida Mitsuyoshi, 2000, *Jinkoki* (English translation), Wasan-Kenkyujo.

Chapter 1-3

PERCEPTIONS OF MATHEMATICS AND MATHEMATICS EDUCATION IN THE COURSE OF HISTORY - A REVIEW OF WESTERN PERSPECTIVES

Christine KEITEL
Freie Universitaet Berlin

1. INTRODUCTION

Giving a summarizing survey of perceptions of Western mathematics and mathematics education and relating it to an outlook on Japanese mathematics and mathematics education, as provided by Hirabayashi and Ueno (this section), is a somewhat big task, which would require a few more pages than available here. Moreover, for somebody not closely familiar with Japanese mathematics and mathematics education, past and present, it is intriguing to witness the tremendous modernity of science and technology in Japan on the one hand, and on the other to understand a statement like "Japanese society does not respect logical reasoning but intuition and neighbour's opinion" (Ueno). Is that ironical self-criticism? Understatement? A play with slight nostalgia? Overall, Hirabayashi's paper gives, for the specific field of mathematics and mathematics education, a substantial example of how the fabulous leap of modernisation that Japanese society made since the middle of the 19th century, had – as every revolution does – to be paid for by painful losses of precious traditions. It is also intriguing that the turbulance of this deep change in Japanese life– whether its most important phase is rather placed in the time after 1858 or after 1945 – apparently have not yet completely settled.

For me, the problem is not alleviated by the fact that Ueno and Hirabayashi exhibit rather contrasting positions. So I apologize for concentrating on a review of the historical development of mathematics and mathematics education in Europe as an attempt to represent a Western perspective. However, at the end, I shall try to give some comments on the ideas of my Japanese colleagues, which hopefully can be understood in the light of my preceding considerations.

Taking into account that mathematics has been determined since early on by social practices and developed in relation to social and technological challenges and changes (Keitel, Kotzmann & Skovsmose 1993), I shall look at my subjects mainly as social phenomena, and thus try to establish a more direct link to mathematics education as a social task. The close relationship of mathematics and social practices implies that in distinct cultural traditions different mathematical practices and representations may be identified. Looking for the origin of such differences permits one to identify underlying commonalities and original approaches as well.

2. HISTORICAL ACCOUNTS OF MATHEMATICS: ART OR TECHNOLOGY?

2.1 Mathematics as a distinctive tool for problem solving in social practices

Since the beginning of social organisation, social knowledge of exposing, exchanging, storing and controlling information in an either ritualised or symbolized (formalised) way was needed, developed and used, in particular information that is closely related to the production, distribution and exchange of goods and organisation of labour. This is assumed to be one of the origins of mathematics, which developed in response to social needs and changes of social organisations, which demanded more control and means for government or administration. Thus early concepts of number and number operations, of time and space, came into being to fulfil social demands (Nissen et al. 1990, Damerow & Lefèvre 1981). In early social organisations, control of social practices and the transmission of necessary knowledge were mostly secured by direct participation in social activities and direct oral communication among the members. Ritualised procedures of storing and using information have been developed since the Neolithic revolution, during the transition to agriculture and permanent living sites, which demanded planning the cycles of the year.

The urban revolution and the existence of stratified societies with a strong division of labour induced symbolical storage and control of social practices by information systems based on mathematics, which were bound to domain-specific systems of symbols with conventional meaning. The earliest documents available are the clay tablets from Uruk (3000 BC), in which mathematics appears as a necessary and useful tool for solving problems of agriculture and economic administration – "bookkeeping" of production and distribution of goods in a highly hierarchically structured slavery-based society (Nissen et al.1990). To secure transmission of the necessary knowledge, to be able to use existing mathematics or to develop it further, in early Babylonian texts, collection of tasks have been discovered, which describe various common social problems and their solution, partly detached from real contexts as typical didactical exercises or teaching tasks. We witness the mathematics of that time being employed as a technique and a useful and necessary tool. The scribe who disposed of the appropriate knowledge to handle this tool, becomes an important man. More generally, the governing class or group disposes of an additional instrument of securing and extending its power and authority.

2.2 Mathematics as theoretical system and a worldview

A new, and eventually most consequential perspective of mathematics, emerged in Ancient Greece: Mathematics (more correctly: geometry) as a theoretical system, as a philosophy, as the queen of sciences, and as a universal divine mental force for mankind. Greek societies were differentiated into two classes with two distinct social practices: the Non-Greek or slaves for all practical and technical manual labour necessary for the maintenance and practical life of the society, and the Greek citizens for warfare, physical activities such as leisure (sports) and spiritual activities in politics, philosophy, rhetoric and the other of the seven "liberal arts", mathematics among them. For the Greek therefore, mathematics was detached from the needs of managing ordinary daily life and from the necessity of gaining their living. Instead, by the scientific search for funda-mental, clearly hierarchically ordered bases, creating connections and elements of a systemic characterisation of existing formal mathematical problem solving techniques and devices, independent of any specific practical intention, they reformulated mathematics as a scientific system and philosophy, a (Platonic) ideal theory to be further discovered and con-structed by human theoretical thinking and reasoning, not by doing or solving practical problems.

It is this distinction between *mathematics as the queen*, as a science of formal symbols, notations, definitions, concepts, rules, as elements of a formal universal language with an unambiguous grammar, providing algorithms, reasoning procedures and logical argumentation, hierarchies as elements of formal routines, as an ideal system of connections of concepts in theorems, networks, models and holistic theories, mathematics as a science of formal systems, which claims to be used for description of all kind of scientific or social relationships, that is opposite to mathematics as a simple technique or real problem solving tool that was not disregarded, but only and simply used: *mathematics as the servant.*

Prominent names of Greek scholars as the "discoverer of the mind" (Snell 1948) and as universally known, are for example, Euclid and Pythagoras. Euclid collected and unified the geometrical knowledge developed and available until his time and gave it a clear theoretical representation in his handbook. He introduced a structure defining mathematical thinking as logical reasoning from axioms to concepts and theorems to proofs. His handbook eventually became a manual for instruction in use in Europe up to the 19th century. Pythagoras is celebrated as one of the greatest mathematicians. As a philosopher, he assigned mathematics a quasi-religious aura, in viewing mathematics as the structure underlying the construction of the cosmos, and *number* as the basis of the universe. He became the founder of a circle of adepts with somewhat cabbalistic tendencies. Pythagoras was most influential in the later history of European mathematics for first emphasizing a hermetical character of the mathematical community, and for laying the ground for the high esteem of mathematics by the Christian church in the middle ages.

2.3 Mathematics as a human endeavour for scientific and social development from the Middle Ages to the 17th century

In the Middle Ages the feature of two quite distinct spheres of mathematics persisted, although according to the cultural changes, their appearance and orientation partly changed.

The idea of mathematics as a sublime, leisurely pursued occupation of free citizens (as in Greek society) got lost. Instead, the high spiritual rank of mathematics seemed essentially worthy of serving to praise God. By means of cloisters spread over the Christian countries, the church for centuries treasured the knowledge available at the time, including surviving classical (i.e. Greek and Roman) manuscripts. Mathematics was adopted in three major fields. Astronomy, including the establishment of reliable, universally applied calendars; a better understanding of the Creation – where however the holy Scriptures gave the guidelines for which mathematics was supposed to furnish substantiation; and finally both geometry and numbers for enriching the pervasive use of symbolism in architecture and book-illumination. None of these activities required or inspired an extension or advancement in the mathematical domain itself. So, with respect to mathematics, there is a shrinking rather than a progress in this time.

The same applies to the tracks of mathematics for ordinary use. There was not much of it. Schools, which also were administered by the church, first offered no mathematics at all for daily life use, although later on a little arithmetic was included.

However, over the centuries, the traces of structuring the world by human rational activity became more numerous, thus imposing a need for dealing with it appropriately. There were several fields in which the mathematisation of the real world and of social life advanced more remarkably, among these notably architecture, military development in both fortification and armament, mining industry, milling and water-regulation, surveying, and above all, manufacturing and trade. The extension of trade from local business to exchanges over far distances prompted the emergence of banking, and for the functioning of this an unambiguous, clear and universal regulation was needed: the system of book-keeping was invented. This was the first consistent, comprehensive mathematical structuring of a whole field of a social practice.

All these endeavours culminated in the period of the European Renaissance, which we regard as a unique confluence of a wide range of contributions: inventions, discoveries, and human genius. The rediscovery of Greek culture incited a revolutionary change of perceptions, secularisation, the idea of man as an autonomous individual, and a merging of all of his various capacities and powers in this one notion of the individual. A prototype of this new man is Leonardo da Vinci, painter, architect, mathematician, engineer, inventor, scientist, writer, cartographer etc. A key-achievement in renaissance mathematics is linear perspective, and interestingly, it is in this point that renaissance mathematics and art converge. It is not surprising if we identify Leonardo's activity in both art and mathematics, as visual research.

Mathematics was seen as the queen *and* the servant of sciences, as a practical *and* theoretical tool, as artful theory and general philosophy and a base for the development of technology and natural sciences, to discover the "world" and "tame" or "dominate nature".

2.4 Mathematics as rationality and common sense

Universities, founded since the 13th century, helped to emancipate knowledge from clerical purposes. Renaissance interest in antique culture also contributed to rediscovering and re-editing classical texts and old knowledge, and book-printing made them available to a wider public. Knowledge from non-European cultures, such as Indian, Arab, Chinese (indirectly) had partly been accumulated in former times, and now was partly complemented by books brought back from commercial travellers, embassy personnel, explorers and others. So the later decades of the 16th and 17th centuries possessed a rich fund to continue and complement building of cultures as newly conceived in the Renaissance. The sciences were emancipated from religious and philosophical restrictions, as mathematics was from religious and philosophical bonds. The abundance of knowledge in the recent past became itself a subject of analytical reflection and of new philosophical approaches.

Descartes conceived the idea of the "rational man". He developed algebra as a general method for mathematics, and moreover for rational thinking in general. Descartes believed that mathematics itself could become so 'easy' or easily understandable, accessible and acceptable by all people that it could be considered as part of "common sense", "le bon sens pour tout le monde"[1] (Davis & Hersh 1986, Glucksmann 1987, Habermas 1971, Lenoir 1979, Kline 1985, Restivo et al. 1993). Leibniz, one of the inventors of calculus, shared this perspective of rational mankind. In evaluating the discovery of calculus, he believed that since rational discourse and strict mathematical reasoning was unlimited it would solve all problems in the world. The call "Calculemus (let us compute!)" sums up Leibniz's approach to arguments. It encourages those engaged in a dispute to turn it into computing. By his famous contention he argues that whenever and wherever a dispute arose, calculation should solve it, and finally save the whole world from controversies, from hostile actions and even from war! The application of a mathematical, rational argumentation and calculation was considered as the universal remedy for any personal or social problem, as it solved problems in a way understandable and acceptable for everybody accessible to rational processes. Like Descartes, Leibniz perceived mathematics and thinking mathematically as the fundamental basis of 'sane' mind building as a general

[1] Descartes referred to the *bon sens* everyone had to understand his revolutionary ideas in mathematics and philosophy (Glucksmann, 1987), but this trust in peoples' natural common sense has been abandoned by scientists over subsequent centuries. Voltaire's (1764) ironic reply to Descartes' assumption that "*Le bon sens est la chose la mieux partagée du monde*" ["Common sense is the best shared thing in the world"] did not share Descartes' optimism: "Common sense is not so common". But the ideal to achieve a "unified universalisation", as the general means for systematic theorising was a programmatic taken up by educationalists later who wanted to build the general reasoning competencies in everybody's mind, but for a long time restricted to upper social classes.

reasoning competency, the facilitator and creator of rationality and a rational mankind.

Mathematics continued to be the queen and servant at the same time. In the newly appearing natural sciences the services of mathematics on a high level were essential, and mathematical research even stimulated solutions to new problems. On the lower level, mathematics could offer to master daily life problems in various fields, and this development went towards conveying more mathematics to more people. Mathematics had become a necessary craft knowledge or the professional knowledge for certain practitioners. The increasing importance of trade and commerce demanded extensive computation skills in trade, commercial and banking companies. Also manufacturing, quality control of production, and distribution necessitated new mathematical tools. The availability of Arab-Indian mathematics and their connotation system allowed for written computation with ciphers, decimal fractions, formal solutions for practical problems of craft, trade and commerce in terms of calculation rules, and hence appropriate schooling was demanded and propagated. For computation schools, textbooks were needed to directly address the problems of the practitioners: problems of commerce, of production of goods in various crafts. One of the most famous new textbooks was widely used and distributed on the European continent, written in German by Adam Ries (Ries 1522, Winter 1991). The explicit use of vernacular as the common language for instruction in computation schools contrasted with the teaching and learning in academic oriented Latin-schools and universities. Computation schools served as a secular complement and a necessary element in various kinds of vocational or professional training. Practical mathematics was extended to a popular activity and more and more widely accepted, and the early textbooks like the one of Ries[2] can be considered as the typical pattern for similar textbooks of those times, and moreover as an often copied model for teaching in later common and public schools offering elementary mathematics and the respective schoolbooks.

[2] Winter evaluates Ries' textbook as a revolutionary book: Ries did not only use vernacular and gave access to knowledge to ordinary people, but also provided a guided tour into arithmetic by combining theoretical explanations for concepts and systematic connections of rules and algorithms; he also provided a collection of well-chosen problems using as context the specific professional practices of his addressees, he always tried to show the general rules in special cases, but turned back to the general. The formulation of the "regula de tri" as underlying most of the proposed calculations as a kind of general theorem is not very different to those school books for elementary mathematics for many years up to the 20th century.

3. GROWING SOCIAL NEEDS FOR MATHEMATICS EDUCATION

3.1 Professionalisation and specialisation of knowledge and the need for education

The achievements of the 15th to 17th century entail an explosion of trades, crafts, manufacturing and industrial activities. From 1751-80 Diderot and d'Alembert publish their monumental Encyclopedia, which tells a lot about the diversity, ingenuity, and craftsmanship (mostly mechanical) developed and required in numerous professions. The ability of a greater part of the population to appropriately deal with fundamental systems of symbols like writing and calculating becomes a condition for the functioning of societies: Elementary (mathematics) education and training is established as a reaction to social demands and needs, either prior to or during vocational training and various professional practices[3]. Parallel to upcoming educational institutions and in concert with them, mass production for unlimited reproduction of knowledge *enables* and *needs standardisation* and *canonical bodies and representations of knowledge.* A reflection and restructuring of existing knowledge on a higher level is asked for: Meta-knowledge has to be developed that offers standards of knowledge and their canonical representations for *educational purposes*; at the same time it becomes an implicit condition for developing *new systems of knowledge*, in particular for sciences like mathematics that are perceived to a greater part as independent of immediate practical purposes.

Later on in the 19th century, the competition between the bigger European states, inspired by a strong and fateful ideology of national superiority and ambition, draws attention to "knowledge as power", making school education a central interest of governments.

Industrialisation is accompanied by an increasing autonomy of systems of scientific and practical knowledge. To be a mathematician, somebody who does mathematics and nothing else, is a new profession. As their work is conceived as autonomous, hence without immediate practical use in other domains, mathematicians are employed often as either university or secondary school teachers.

[3] Until today the most widely used models for book-keeping in economic and business administration, and for investment and planning can be related back to those invented in the16th century.

As in the sciences, specialisation and professionalisation of experts become requirements in all branches of the economy, as in social services and administration, and constructing and creating new knowledge become a *precondition* for the material reproduction of society instead of a *conses-quence*. Specialisation is a condition for creating new knowledge, but at the same time it entails a risk because it does not quasi-automatically integrate the newly developed knowledge into a system and therefore makes integration in a wider context difficult. Partial knowledge must be generalised and incorporated into a level of meta-knowledge.

3.2 Education as a public task

In the 19th century in many countries, public and state controlled bipartite school systems are created: higher education as mind-forming for an elite, elementary education to transmit skills and working behaviour for the majority, the future working class.

Mathematics becomes a subject in higher education institutions because of its formal educational qualities, e.g. educating the mind independent of a direct utilitarian perspective, and fostering general attitudes to support the scientific and science-driven technological development.

In the elementary or general school for workers and farmers, only arithmetic teaching in a utilitarian sense is offered: to secure the necessary skills for the labour force, to secure acceptance of formal rules and formal procedures set up by others. Mathematics education for the few is strictly separated from the skill training for the majority. This corresponds to a separation of mathematics education as an art and science in contrast to mathematics education as a technique; conceptual thinking versus algorithmic, machine-like acting.

For educational institutions under state control, criteria and measures to control access to higher education are also set up: entrance examinations. It was a revolutionary idea to define access to higher education for the best achievers in exams, instead of only for the noble or rich. This measure contributed to the realisation of the idea of the enlightenment that knowledge is (political) power and access to knowledge will empower people. However, examinations later turned into one of the major constraints and obstacles for learning and teaching mathematics.

3.3 Mathematics education for all

In the 19th and 20th century, mathematics became the driving force for almost all scientific and technological developments: mathematical and scientific models and their transformation into technology impact not only on natural and social sciences and economics but also on all activities in the social, professional and daily life. This impact increased rapidly by the development of the new information and communication technologies based on mathematics, which radically changed the social organisation of labour and our perceptions of knowledge or technique to an extent that is not yet fully explored.

On the one hand, mathematics as a human activity in a social environment is determined by social structures, hence it is not interest-free or politically neutral. On the other hand, the continuous application of mathematical models, viewed as universal problem solving procedures, provide not only descriptions and predictions of social actions, but also prescriptions. The increasing social use of mathematics makes mathematical methods and ways of argumentation to quasi-natural social rules and constraints, and creates a mathematised social order effective in social organisations and hierarchical institutions like bureaucracy, administration, management of production and distribution, institutions of law and the military etc. Social and political decisions are turned into facts, constraints or prescriptions that individual and collective human behaviour has to follow. How to cope with these demands? What is the necessary knowledge to be provided and an appropriate way of education to be offered, if the individual is meant to act according to an understanding of what is in his/her interest?

Since the middle of the 20th century, a perspective of general education gained more and more acceptance and political support: Mathematics Education for All. The New Math movement started to introduce mathematics for all by a formally unified, universally applicable body of theoretical knowledge of modern mathematics exposed to all, but had to be revisited and discarded as a solution. In the last 50 years, intensive work in curriculum development created a wide range of different and more and more comprehensive approaches combining new research results in related disciplines like psychology, sociology, and education and developed this vision further (Howson, Keitel, Kilpatrick 1981). Today, a variety of conceptions have been developed, that promise to integrate scientific mathematical practices and common vocational or professional practices and their craft knowledge, or conceptual and procedural knowledge, or mathematical modelling and application. The Dutch conception of "realistic mathematics" as starting from mathematical practices to explicate and theorize the embedded, even advanced mathematics, might be considered the most balanced conception of "Mathematics for All".

But new notions entered the arena: "Mathematical literacy", "Educational standards" and "Benchmarks" are key issues of the recent political debates and disputes in mathematics education that started after the release of international comparative studies like TIMSS and PISA, and their ranking results of tests, and might act as a counter to a movement like "Mathematics for all". Proclaiming that the PISA tests are based on "definitions of mathematical literacy" that are underpinned by fundamental and widely accepted research results, and that it is absolutely unproblematic to test such kinds of competencies or proficiencies on a global scale to rank countries' performances, produced strange and urgent political measures taken in some of the countries that did not perform well, called for by the alarmed public and the media, although it is still far from being clear what are the actually necessary competencies and who are the right experts to decide this.

Jablonka (2003) excellently unfolds what research on "Mathematical literacy" can do and what not. She investigates different perspectives on mathematical literacy that shows clearly that these considerably vary with the values and rationales of the stakeholders who promote them. The central argument underlying each investigation is that it is not possible to promote a conception of mathematical literacy without at the same time – implicitly or explicitly – promoting a particular social practice of mathematics, be it the practice of mathematicians, of scientists, of economists, of professional practices outside science and mathematics etc. She argues that mathematical literacy focussing on citizenship in particular refers to the possibility or need of critically evaluating aspects of the surrounding culture of the students – a culture that is very much shaped by practices that involve mathematics. Her conclusion emphasises that the ability to understand and to evaluate these different practices of mathematics, and the values behind them, should form a component of Mathematical Literacy. This analysis of various conceptualisations of Mathematical Literacy and the related research rationale (or its lack) is followed by her development of a theoretically based categorisation of approaches to Mathematical Literacy that provides criteria for selection and construction of teaching and learning materials and for curriculum development oriented towards Mathematical Literacy.

4. CONCLUDING REMARKS

Viewed from outside, the Japanese debate on actual problems of mathematics education, as presented in Hirabayashi's and Ueno's chapters, has a somewhat peculiar ring. Because – and that is what I have tried to show in the previous paragraphs – the problems with mathematics education are basically the same in all industrialised modern societies. Hence, reconsidering advantages and failures of WASAN and YOZAN will not really match them. Let me put one simple question: Would WASAN have been – be – a beloved occupation if it was a compulsory school discipline and achievement in it controlled by heavy and most consequential and rigid exams that directly and decisively impact on future life expectations?

In fact, these problems directly spring from the complementary character of fundamental properties and perceptions of mathematics: mathematics is art *and* technology, queen *and* servant. Unfortunately, in the mathematics education of our modern societies these complementary properties inevitably turn into antagonisms: Art or technology? Equity or elite-formation? In any of these cases both would be needed, but we are told that we cannot have both. Society claims equity and equal chances for all, and then society demands selection. There is no solution for this dilemma inside mathematics education. The best that can be done, and in fact what many of us have been concerned with for a long time, is to find the best balance of contradictory demands. This eventually includes a fair balance between the societal interest and interests of the student; it may be skill training on the one side, and intellectual formation on the other.

Decisions have to be made on all three levels: on the level of the organisation of the educational system, in the construction of the curriculum including pedagogical and methodical approaches, and on the mode of assessment. The choices are influenced by many determinants, partly more specific to particular societies, partly more general: educational traditions, traditions of values, ideologies, new insights or achievements in the sciences that form the school subjects and their epistemologies, and political and economic interests.

The pervasiveness of economic thinking and interests have successively created so high a pressure of economic orientation that educational aims and the subject matter are marginalized unless they prove justification in terms of economic interest.

Rigid examinations and tests have been developed as assessment instruments for military and economic purposes first. Rigorous testing and exams in order to best control school achievement is a constant threat for the student and the most secure way to prevent or to kill affection for learning a subject. Examination or testing is not in the interest of the educational process; it is the economic interest to obtain best human capital, and to select it by appropriate quality control, but quality of what and for what?

Is it completely wrong to read Hirabayashi's paper as how, in a dream, he returned to the times of WASAN, when exams did not suffocate real learning, understood in a more comprehensive sense, and when spare time was available for leisure and GEI, recreation and pleasure with mathematics, allowing one to indulge in most elegant and artful ways of doing mathematics?

This idea is appealing to me, for one could have similar thoughts with respect to our old German public education. In the 19th century Humboldt reformed the education system in Prussia. He created the notion of "Bildung" (in English mostly translated as "formation", but that does not really cover the meaning). I imagine a relationship to GEI. "Bildung" comprises learning as universal as ever possible with strong emphasis on humanities: philosophy, history, literature, art, music, but also mathematics and sciences. The ideal was the completely cultivated, best educated man, and again similar to GEI, "Bildung" was not a process ending at the end of one's studies, but just the base laid in the youth to be enlarged and enriched during the whole life; "Bildung" as an attitude and a path as much as an accomplishment. And that was conveyed, or tried to be conveyed, by ways and means of a public education, at secondary schools and at universities.

After World War II "Bildung" – or more specifically at that time "Allgemein-Bildung" (i.e. general education or formation) still was the officially declared overall objective of public education, and in theory, it still is. But over the years and the reforms succeeding each other, its practical impact faded away, and there is not much left but the notion, attacked and replaced by more modern slogans like Mathematical Literacy or "Grundbildung" (basic education) or basic competencies and standards.

But nostalgia may rather immobilize. As an outsider, one may be tempted to view the fact that Japan actually has come to know two quite different approaches to mathematics and mathematics education, as a privilege rather than a problem. And one tends to imagine, how both approaches, WASAN and YOZAN, would be brought to a fruitful dialogue in school discussions, illuminating the specific characteristics of each other.

REFRENCES

Burton, L., 1999, Mathematicians and their epistemologies – and the learning of mathematics, in: *European Research in Mathematics Education*, Schwank, I. ed., Forschungsinstitut für Mathematikdidaktik, Osnabrück, pp.87-102.

d'Ambrosio, U., 1985, Ethnomathematics and its Place in the History and Pedagogy of Mathematics, *For the Learning of Mathematics*, **5**(1):44-48.

Damerow, P., Lefévre, W., eds., 1981, *Rechenstein, Experiment, Sprache. Historische Fallstudien zur Entstehung der exakten Wissenschaften*, Klett-Cotta, Stuttgart.

Davis, P., 1989, Applied Mathematics as Social Contract, in: C. Keitel et. al., eds., *Mathematics, Education and Society*, UNESCO Document Series No.35, UNESCO, Paris, pp.24-28.

Davis, P. J., and Hersh, R., 1985, *The Mathematical Experience*. Birkhäuser, Boston.

Davis, P. J., and Hersh, R., 1986, *Descartes' Dream*. Harcourt, San Diego.

Devlin, K., 1992, *Sternstunden der modernen Mathematik*, (Mathematics – the Golden Age), DTV, München.

Ernest, P., 1999, Forms of Knowledge in Mathematics and Mathematics Education: Philosophical and Rhetorical Perspectives, *Educational Studies in Mathematics*, **38**(1-3): 67-83.

Gellert, U., Jablonka, E., and Keitel, C., 2001, Mathematical Literacy and Common Sense in Mathematics Education, in: *Sociocultural Aspects of Mathematics Education*, B. Atweh et al., eds., Lawrence Erlbaum, New York, pp.57-73.

Glucksmann, A., 1987, *Descartes c'est la France* [Descartes is France], Flammarion, Paris.

Habermas, J., 1971, *Toward a Rational Society*, Heinemann, London.

Hammond, A.L., 1978, Mathematics – Our Invisible Culture, in: *Mathematics Today*, Steen, L.A., ed., Springer, Berlin.

Howson, G.A., Keitel, C., and Kilpatrick, J., 1981, *Curriculum Development in Mathematics*, Cambridge University Press, Cambridge.

Jablonka, E., 2003, Mathematical Literacy, in: *Second International Handbook of Mathematics Education*, Bishop, A., Clements, K., Keitel, C., Kilpatrick, J., and Leung, F., eds., Kluwer, Dordrecht, pp.77-104.

Joseph, G.G., 1992, *The Crest of the Peacock: Non-European Roots of Mathematics*, Penguin Books, London.

Keitel, C., Kotzmann, E., and Skovsmose, O., 1993, Beyond the Tunnel Vision: Analysing theRelationship between Mathematics Education, Society and Technology, in: *Learning from Computers: Mathematics Education and Technology*, C. Keitel and K. Ruthven, eds., Springer, Berlin, pp.242-279.

Keitel, C., 2000, Cultural Diversity, Internationalisation and Globalisation: Challenges and Perils for Mathematics Education, in: *Cultural Diversity in Mathematics Education*, A. Ahmed, J.M. Kraemer, and H. Williams, eds., Ellis& Horwood, Chichester, pp.41-60.

Kline, M., 1985, *Mathematics and the Search for Knowledge*. Oxford University Press, Oxford.

Lenoir, T., 1979, Descartes and the Geometrization of Thought, *Historia Mathematica*, **23**: 355-379.

Nissen, H., Damerow, P., and Englund, R.K., 1990, *Frühe Schrift und Techniken der Wirtschaftsverwaltung im Vorderen Orient: Informationsspeicherung und -verarbeitung vor 5000 Jahren* [Early writing and technologies of the management and administration of economics in the Near East: Information storing and processing 5000 years ago], Bad Salzdetfurth, Franzbecker, Germany.

Renn, J., 2002, Wissenschaft als Lebensorientierung – eine Erfolgsgeschichte? Preprint 224, Max-Planck-Institut für Wissenschaftsgeschichte, Berlin.

Restivo, S., van Bendegem, J. P., and Fischer, R., eds., 1993, *Math Worlds: Philosophical and Social Studies of Mathematics and MathematicsEeducation*. SUNY Press, Albany.

Ries, A., 1522, *Rechnen auf der Linien und mit Federn*. Annaberg Reprint 1999.

Snell, B., 1948, *Die Entdeckung des Geistes bei den Griechen*. Claasen & Goverts, Hamburg.

Winter, H., 1989, *Entdeckendes Lernen im Mathematikunterricht. Einblicke in die Ideengeschichte und ihre Bedeutung für die Pädagogik*, Westermann, Braunschweig.

Woodrow, D., 2003, Mathematics, Mathematics Education and Economic Conditions, in: *Second International Handbook of Mathematics Education*, Bishop, A., Clements, K., Keitel, C., Kilpatrick, J., and Leung, F., eds., Kluwer, Dordrecht, pp.11-32.

Chapter 1-4

HISTORICAL TOPICS AS INDICATORS FOR THE EXISTENCE OF FUNDAMENTALS IN EDUCATIONAL MATHEMATICS

An Intercultural Comparison

Walther L. FISCHER
University Erlangen-Nürnberg, Germany

1. PRELIMINARIES – METHODOLOGICAL PRESUPPOSITIONS FOR A COMPARATIVE STUDY

In a *comparative study* it is not sufficient just to state differences and correspondences between the compared topics. Foremost it is necessary to establish *criteria and norms* under which the comparison is carried out and which in the end enable an evaluation of the differences and correspondences of the stated facts to be made, especially with respect to the realization of certain aims. The constitution of criteria and norms presupposes the existence of a kind of neutral *tertium comparationis*.

2. EVENTS IN HISTORY OF MATHEMATICS AS A TERTIUM COMPARATIONIS – THESES

With respect to the aims of the ICMI-Study-13 our General Thesis is:

> Comparisons of events in the history of mathematics and of mathematics education in different cultures supply a tertium comparationis for the ICMI-Study-13.

In times of rapid developments in all fields of our human life, we need, on the one hand, power enough and ideas to follow up the (dynamical) social and economic changes - and on the other hand, we also need fixed points, in the literary sense of the phrase, or more closely to the reality of the historical process of ideas: we have to determine *cognitive and conceptual invariants* within the flow of time to master the situation of life in the future.

Our Special *Theses* are:

(1) With respect to the ICMI-Study such cognitive and conceptual invariants constitute a tertium comparationis for the Study.

They can be explored by observing the interrelations between history of mathematics and educational mathematics in different cultures, for:

There are anthropological constants and intercultural invariants in the flow of time, and in consequence there are fundamentals in mathematics education. (Fischer 1990; 1991)

(2) Indicators of such invariants and fundamentals in mathematics and mathematics education are to be found in events in the history of mathematics and in the history of mathematics education, for:

The different stages of history of mathematics represent a paradigm of the development of mathematical thinking for essentially they are indicators of universal schemes of the cognitive development in human beings.

Moreover, certain events in the history of mathematics indicate that there are everlasting fundamental educational principles, and that these invariants deliver guidelines for mathematics education in the time to come too.

Our theses are based on the presumption that in the course of time not all of the phenomena are changing. If everything would change and change again, an accumulation of knowledge would not be possible.

3. COMPARISON OF TEXTBOOKS FROM DIFFERENT TIMES AND CULTURES

3.1 Four textbooks from the Far East and the West

In evidence of the assertion of our theses we will refer to certain events in the history of mathematics. *In detail, our theses are substantiated by a comparison of four famous mathematical textbooks* from different times, from different cultures, written in different languages: From ancient China *Jiuzhang Suanshu* (ca 150 B.C.) and CHENG Dawei, *Suanfa Tongzong* (1592), from Japan Mitsuyoshi YOSHIDA, *Jinko-ki* (1627) and a German texbook from Adam RIES , *Rechnen auff der Linihen und Federn* (1522).

3.2 China: Jiuzhang Suanshu (about 150 B.C.)

From the Far East we choose from China the *Jiuzhang Suanshu*, the "Nine Chapters on Arithmetical Art", compiled in the early Han-Period (about 150 BC), a work representative of the development of ancient Chinese mathematics from *Zhou* and *Qin* to the Han dynasties. The *Jiuzhang Suanshu* consists of a collection of 246 problems closely concerned with situations of practical life. The problems are worked out as far as they are accompanied by rules which display the method of solution.

3.3 China: CHENG Dawei, Suanfa Tongzong (1592)

Next we draw the attention to CHENG Dawei (1533-1606) *Suanfa Tongzong*, "Systematic Treatise on Arithmetics" from 1592. This book was of great importance for the further development of mathematical ideas and techniques in China and also because of its influence and impact on the development of mathematics in Korea and Japan. We explicitly point to the fact that, although published about 1700 years later, in 9 of the 17 chapters CHENG Dawei almost literally repeats the topics of the 9 chapters of the *Jiuzhang Suanshu*.

3.4 Japan: Mitsuyoshi YOSHIDA, Jinko-ki (1627)

With respect to Japan we refer to the most famous textbook in the Edo-Period (1603-1867), to Mitsuyoshu YOSHIDA (1598-1672) *Jinko-ki*, "The Book of the Large and Small Numbers" from 1627. Though we know that YOSHIDA got lots of his knowledge from CHENG Dawei's book, it would be wrong and unjustified to state that he just repeated CHENG's book in a shortened form. YOSHIDA took in the contents and methods of the *Suanfa Tongzong* and made it his own and also dropped some of the more difficult topics of the Chinese work. However, he rearranged the remaining problems and self-reliantly adapted the questions to the contemporary situations and to the bare necessities and requirements of life in Japan and added new problems. Of course, via CHENG Dawei there is a historical string leading back even to the *Jiuzhang Suanshu*. (Also see Ueno, 2004)

3.5 Germany: Adam RIES, Rechnen auf den Linien und mit der Feder (1522)

As a typical example of a Western textbook we select a book from the renowned German Adam RIES (1492-1559), the second edition of his "*Rechnen auf den Linien und mit der Feder*" from 1522, ("Calculation on the Lines and with the Quill"). Adam RIES is known as the "German Master of Calculation". In this book calculation using a kind of abacus ("Lines") and using the "new" Indian-Arabian script (by drawing the numerals with a "Quill") was taught.

4. EARLY MATHEMATICAL TEXTBOOKS I: CONFORMITIES, CORRESPONDENCES – FUNDAMENTALS

4.1 List of Contents – the early texts used in instruction

Already a glimpse at the lists of contents, at the titles of the chapters and the main topics handled in the quoted books shows *many correspondences, conformities and concordances in topics of application and likewise with respect to the problem's reference to mathematical conceptions, to concepts and methods.*

Comparing in detail and side by side just the lists of contents of the chosen exemplary books, we may state that all of the textbooks are *collections and compilations of particular (singular) problems*. They evidently were used as textbooks to support the oral instruction in applied elementary mathematics. The topics refer to *the domains of problems relevant in everyday life in the respective times. Some problems of recreational mathematics complete the works.*

4.2 Style of presentation

Moreover, similarities are obvious with respect to the style of the books, displaying the different problem-fields in the form of *sequences of problems* and to the *presentation of the particular problems* in detail.

The *style of the problem's presentation* in the chosen texts is in question-answer-form and the arrangement of the problems shows that a kind of inductive method is practised in teaching the different types of problems proceeding from simply to more complex problem situations.

Even the *formulation of the problems* within the texts of the quoted books is rather stereotyped. It follows a 5-step scheme: *Problem Situation, Statement of the Question, Answer, Solution-Rule, Probation*. The *wording of the problem-settings* mirrors and represents thus the basic elements and the basic type of mastering of any problem situation whatever and wherever. In this respect, it is *a fundamental element* of mathematical thinking and mathematical instruction.

4.3 The solution-rules as fundamentals

The (solution-) rules in the Eastern and Western textbooks describe step sequences of solution-procedures to establish in the end the wanted result of the particular problem, the answer to the question. And again a glimpse at the problems, or just at the list of contents, shows that the '*solution-methods*' are largely the same in all of the textbooks. Formally the '*rules*' consist of a sequence of steps in the form of algorithmic routines, and this in most cases in colloquial prose. They can be considered as fundamentals as well.

4.4 Mathematical conceptions and concepts

In consequence of the above, the contents of the textbooks are largely concordant with respect to the *mathematical concepts* and to the tools constituting the *algorithmic solution-methods*.

Seen from the point of mathematical conceptions (subject-matters), these time-honoured textbooks deal with *integers* (natural numbers - the Chinese books already with negative integers), with *fractions*, with *order structures* of the corresponding numbers and *quantities* (*measuring numbers*), with the *linguistic and symbolic representation* of the elements of these number domains – especially already using a decimal system of notation, with the *elementary arithmetical operations* addition, subtraction, multiplication, division, with powers and roots.

A central part in the foundation of the solution-procedures is played by *proportions*, *proportionalities* or as we call it today *linear functions*. In consequence of that, the most frequently used method is the *rule of three* (Regeldetri). The *rule of double false* (Regula falsi) is explained and applied. Problems of measuring and calculating areas and volumes refer to *basic figures and solids of elementary geometry*. They call for methods to extract *square* and *cubic roots*.

There are problems concerning the solution of *systems of linear equations* and of *diophantine equations*.

In some respects the *Jiuzhang Suanshu* is already more advanced than RIES' book in dealing with systems of linear equations, using a kind of *matrix representation* and calculation and already a scheme known in modern mathematics as HORNER's scheme. – A particular chapter is devoted to applications of PYTHAGORAS' theorem.

4.5 Number aspects

With respect to the different aspects of the number concept we state that all of the different aspects occur in the textbooks: the number aspect of *cardinal number*, of *ordinal number*, of *measuring number*, numbers in the function of an *operator* and, of course, numbers as *elements in algorithmic processes*.

As most of the problems deal with situations of daily life, *magnitudes* play the most important role in all the early textbooks, i.e. measuring and the aspect of *measuring number* are foremost. In consequence of that, *transformations of different measuring units into one another using tables of measuring units* are found in all early textbooks.

4.6 Topics and domains of subject-matters of the textbooks in detail – strings of topics

With respect to the *subject-matters*, the *application-modes* and *–topics*, already the list of contents and the titles of the chapters manifests that many, many themes and subjects and methods coincide across and over times and cultures too – in our wording: they are fundamentals. According to the vocational necessities of the clientele of the teachers using the quoted textbooks the problems refer to *problem-fields in the domain of applied mathematics*, i.e. they refer to situations of everyday life, of craft and trade, problems in civil community, of goods exchange, of currency, of conversion of measuring units, of land surveying and engineering. Problems of recreational mathematics complete the stock of subject-matters. Lots of the subject-matters are fundamentals.

A comparison of the textbooks in West and East discloses not only on the large scale an extensive conformity in problem- and application-domains. The same is valid on the small scale. Certain topics almost literally are to be found in different times in West and East. *Strings of special (subject or formal) topics pass through the texts, through times and culture*; they are fundamentals.

In the following we display just one example of such an everlasting problem-type to deliver another evidence of the existence of fundamentals which were and will be of a central importance with respect to mathematics education. The selected example is an evidence for our *thesis why today, and in the time to come, certain classical contents are fundamentals and will be justified as topics in mathematics instruction. (Fischer, 1990)*

As a classical example for such a string of topics we choose the *problem field of motions of bodies* ('Pursuit and Meeting'). The general question reads: "When and where do they meet?"

Some of the first problems of this type occur in China. E.g. Problem 14 in Chapter VI of the *Jiuzhang Suanshu* concerns the '*Pursuit of Dog and Hare*'. The solution is done by a proportion or respectively by using the rule of three. The same problem and solution method is known from India and Arabia. In the West it occurs in different variations under the Latin title "*De cursu canis et fuga leporis*" (ALCUIN, 9th cent.). It is found in the works of FIBONACCI (1202), GEORGIUS VON HUNGARIA (15th cent.), WOLLACK (1467), Ulrich WAGNER (1483), Petrus APIANUS (1527), etc..

Those problems of motion which our students still have to solve today are everlasting problems. For they are and function as

(1) Paradigms for certain Fundamental Mathematical Aspects: Rule of Three, Proportionality, Proportions in Arithmetic and Geometry, Linear Equations, Linear Functions.

(2) They are Expansive Problems: Beyond the particular cases they induce more generalized questions, they point beyond the particular case towards generalizations of certain mathematical structures and to different domains of applications.

(3) They are the very beginning of Physical Kinematics: – They establish a pre-stage of a part of theoretical physics.

Many problems in everyday life, in sciences and techniques even today, are variations of such problems: We think of *traffic situations*, the graphical *time-tables of railways*, the *world-lines* in EINSTEIN-MINKOWSKI Geometry, think of the board-computers and the navigation-systems in our cars; in astronomy there are the motions of the *planets*, of *satellites*; in physics there are problems of motions of sub-atomic particles in accelerators, etc..(Fischer, 2001(1))

Considered from this point of view it is not surprising, that lots of particular problems from the ancient textbooks were treated in the course of times all over the world. Some of them were transmitted almost literally e.g. from China to Europe, are repeated nearly in the same wording in the textbooks of the German masters of calculation and still are found in our mathematical school-books today.[1]

Even problems from the domain of recreational mathematics pass through times and cultures. Just for curiosity we mention:

(1) Counting-off Problems: in Europe the "Josephus Problem"; in Japan's Jinko-ki "Mamako.Date".
(2) Problems of Transfusion: in Europe known from TARTAGLIA (1499 – 1547), in Japan again from YOSHIDA in Jinko-ki III/11).
(3) Magic Squares: known from ancient China as the Luo Shu and the He Tu. Later on they are to be found in CHENG Dawei's book, found in YOSHIDA's Jinko-ki. The Luo Shu is put as a problem in Adam RIES' books and is present even in the textbooks for Primary Schools of today in Germany.
(4) We could continue in our collection: "The problem of the 30 birds", the "Chinese Residue Theorem" and many, many others.

4.7 A summary – the reasons for all such conformities

The comparison of historical mathematical textbooks elucidates that certain events in history of mathematics indicate *the existence of fundamentals in mathematics and at the same time of fundamentals in mathematics education* with respect to the topics and to the *basic fundamental mathematical concept structures, the fundamental algorithmic structures.*

If we ask for the *reasons for such conformities* we may state: All these exemplar historic events refer to hitherto unchanged problem situations in practical life. And, as we may complete and state in some keywords: *Similar life situations induce similar tasks, similar problems, similar questions. Similar problem structures induce – culture and language over-lapping – similar conceptual (mathematical) modelling structures (models) and similar solution-techniques (methods) in answering the questions.*

Many topics and ideas and methods remained unchanged, for they refer *to universal basic physical and human situations of everyday life* – yesterday in the ancient times and still today with us – and they depend on the *anthropological constants of the cognitive apparatus of the human mind.* (Fischer, 1997; 2001(2))

[1] The history of such everlasting singular themes was followed up by many authors. We just point to SMITH's "History of Mathematics (1951,1953), to ROUSE-BALL's Recreations (1931), to the books of Martin GARDNER and to many books dedicated to problems of recreational mathematics.

5. EARLY MATHEMATICAL TEXTBOOKS II: ON THE ROUTE TO MATHEMATICS

5.1 The early authors on the route to mathematics

Mathematical texts and especially textbooks in the history of mathematics for a long time and in all cultures[2] were on the whole nothing other than collections of singular (particular) problems completed by rules of solutions or by the solution of particular problems carried out in detail. The rules were nothing other than descriptions of particular algorithmic solution methods adapted to the particular problems. The rules were neither formulas nor theorems but recipes. So, on the one hand *the early textbooks cannot be considered as mathematic books in the narrower sense of the meaning.*

On the other hand, such an assessment of the textbooks and of the achievements of their authors and compilers is just one half of the truth. We should bear in mind that *the conception of the essentials, the objectives and the tasks of mathematics and of mathematical cogitation cannot be characterised uniquely. These conceptions have changed several times in the course of the history of mathematics. More than once the self-conception of mathematics has changed too*[3].

The self-conception of mathematics and of mathematical cognition worldwide developed starting in a *pre-mathematical phase* progressing to *proto-mathematical* and to mathematical phases, stepping onward from *naïve*, to critical (axiomatic) and to *formalistic (non-standard) stages*. In other words, mathematical cogitation firstly started with an *enactive pre-conceptual thinking (reflecting)* by acting on the basis of sense-perceptions, next arrived at the phase of *thinking (reflecting) using iconic proto-concepts* and ended up in the phase of *conceptual thinking* – increasingly on the route *from inductive probabilistic conclusions to deductive inferences and syllogisms*. These steps were accompanied by the formation of certain *types of language* being increasingly "precise" (formalistic structured) and apt to mathematical ideas and procedures. (Fischer, 1997)

If we want to do justice to those authors from long ago, after a more detailed analysis of the books and their history we have to state: *The early texts and their authors were truly already on the route to mathematics, were*

[2] We think of the texts of the ancient Babylonians as well as those of the Egyptians too.

[3] This is true even of the situation of mathematics in the19th and 20th century. We just mention for the last 60 years the name Bourbaki and with respect to educational mathematics the keyword "New Math".

on the route from mere problem-solving to structure- and system-cognition. Therefore, the quoted texts only from our point of view represent the very *beginning of mathematics*. However, from the true historical point of view they are already *certain final points of the scientific development*. The extant texts are compilations of the results and works of many unknown authors before. The solution-rules – before being selected and compiled together with the related problems from already existing materials – had to be elaborated by human beings of an extraordinary kind of mathematical intellect (Fischer, 2001(2), 242). And of course, this development necessarily involved pure mathematical aspects too.

5.2 The route from particular problems to paradigms, from heuristics to theorem and proof

To sum up: The *old-aged texts* are not only compilations of problems concerning the practical needs of everyday life, they are really filled in with mathematical ideas and thinking. They introduced and used in a non-formal language many fundamental elements of mathematical thinking (see 4), moreover they already used, for example, the idea of variables, a kind of inductive way of argumentation, being on the way to theorem and proof by heuristic considerations and comments. Especially the grouping of the problems into chapters shows, that they were *on the route from particular problems to idealized problems and to problem-types, from problem-types to paradigms, from heuristics to theorems and proofs*.[4] This general schematic sequence in mathematical cogitation[5] can be partly shown in LIU Hui's (ca 260 AD) commentaries to the *Jiuzhang Suanshu*. (Bai, 1983; Wu, 1982; Li, 1982; Zhong, 1982; Siu 1993; Fischer, 2001(2), 244)

[4] As an Example: Following SWETZ/KAO (1977) we have to distinguish between the person of the Greek philosopher PYTHAGORAS (ca 580-500 BC), the person who first detected the relation between the areas of the squares of the lengths of the legs of special right triangles, the person who first got an insight into the generality of the relation and who formulated the so-called Pythagorean Theorem, and the person who for the first time presented a formal correct proof of the theorem.

[5] in the course of history as well as in mathematics education as well as in the cogitative development of the singular individual.

6. DIFFERENCES BETWEEN EASTERN AND WESTERN TRADITIONS

6.1 Differences within the frame of invariants – differences in educational mathematics

Invariants within a dynamical system constitute a kind of frame within which the phenomena are able to vary their shapes and attributes. This means that *intercultural historical studies at the same time do not only disclose correspondences* via times, languages and cultures, but *also disclose differences with respect to the assessment and self-conception of mathematics and mathematics education.* Of course, these differences are dependent on the state of development in the single historical periods and on the cultural background.

With respect to our theme we may state that *today* on the whole, *mathematics curricula* worldwide are in a certain sense rather uniform concerning the mathematical contents. Differences are to be seen especially in *educational principles, methods, and in instructional practice* which closely depend on the cultural context.

The differences in *educational and instructional methods* derive from the background of the human environment, of social life and its necessities on the one hand, and from the self-comprehension of the human beings within the world, from their self-conception and the assessment of the value placed on their cogitation and way of thinking on the other hand. Especially with respect to mathematics education, differences in the different countries derive from the different assessment of the value placed on mathematics in everyday life. This includes the different accentuation and evaluation of the two poles: *Relation of mathematics and mathematics education to application and intention to an elucidation of concrete and abstract structures.* And all this is closely related to the different estimation and conception of what is meant by *'being', and is closely related to the worldview of the special language of the respective cultural region.*

6.2 Different accentuations of logic in the West and East – different language structures

Every language mirrors a special view of the world and is co-existentially bound with a special type of logic. Again, the concept and role of logic cannot be uniquely determined in the course of time. Roughly speaking Western logic is a 2-valued formal logic coined by truth-values and – since ARISTOTELES (384 – 322 BC) - by syllogisms governed by the principle of contradiction. Eastern logic is a 'logic of ontological polarities' governed by the principle of complementary, harmonizing the different opposite poles.[6] ZHANG Dongsun (CHANG Tungsun) (1952) called this type of logic a "*logic of analogy*", a "*logic of correlations*".[7] The same difference causes distinctions in the use of logical operations (connectives) here and there. While Western logic stresses the logical operator of the exclusive-or (either-or, lat. *aut-aut*), Eastern logic stresses the operator of a non-exclusive-or (lat. *vel*).

Therefore, in spite of certain conformities and concordances between the early Eastern and Western textbooks – caused by the reference of the problems to applications in everyday life – and in spite of the fact that the Eastern texts distinguish between true and false solutions too (i.e. that they followed up in their argumentations implicitly a *2-valued truth-scheme*) the *accentuation of thinking in the Occident was later different from that in the Orient*. Even the application-oriented Western textbooks are thoroughly founded on the basis of ARISTOTELEAN logic, i.e. on the Greek attitude of thinking – let us say: on EUCLID's conception of mathematics. Thinking and reasoning even in inductive stages of oral instruction was (and is) in the end implicitly based on the deductive methodology of EUCLID's "Elements". Especially in the Occident the development of calculation, of calculation techniques and that of pure mathematical ideas and structures in number theory and geometry occurred in front of a background of conceptions and methods oriented in EUCLID's "Elements".

There are further differences, for example between the *character and role of definitions* depending on the different structure of sentences of the Greek and the Chinese languages. In scientific cogitation in the West the definition of a concept is done by abstraction and consists in a formation of an (equivalence-)class combining all attributes (intension) and all objects (extension) being covered by (comprising) the concept. Definitions of this type depend on the linguistic fact that in the Greek language (in contrast to the structure of Chinese language) there is a definite article (Snell, 1948). In the East, definitions are first and foremost 'ostensive', consisting in a presentation (naming) of some typical elements from the set of objects subsumed under (comprised by) the concept.

[6] See the principles of Yang and Yin and the logic of the Yi Jing.

[7] In the archaic period of Greece (before Aristoteles) a fundamental principle in the explanation of nature and the world was the formation of analogies too. See MEHL, 2003.

6.3 The role of the EUCLIDEAN scheme

To specify these general aspects from another view we point to the famous EUCLIDEAN Scheme with its steps: *Analysis, Construction, Proof, Determination* (Fischer 1989). While (even) in the instructional texts in the West the solution procedure largely follows this scheme, in the Eastern texts only some of the steps are realized or realized just in modifications. One early exception in China is LIU Hui (ca 260 AD) in his commentaries to the *Jiuhang Suanshu* (Bai, 1983; Wu, 1982; Fischer, 2001(2)).

In Western texts there is a kind of "*proof*" after having established the solution. The "*answer*" is followed by the "*Probe*" (probation). Reversing the sequence of steps in the solution procedure it is shown that the solution is not only a necessary condition but a sufficient condition too. In this way we "make sure", that the solution really is a solution to the question. The EUCLIDEAN step to "*determination*" (how many solutions are possible? – uniqueness?) is to be found in the old Eastern texts only in a quite unsystematic manner by the presentation of several solution procedures.

The original difference of the logical attitude in the East and West can be specified in detail, for example by the different presentation of the conception and method of what is called today the "*EUCLIDEAN Algorithm*" to determine the greatest common divisor of two numbers. In Jiuzhang Suanshu the solution-rule to Problem I.6. (*yo fen* – cancellation of fractions) just names the step sequence of the procedure 'subsequent mutual subtraction' without any commentaries, explanations or hints on its foundation while in EUCLID, "Elements" Book VII Proposition 1 and 2, the algorithm is deductively proved.

6.4 Notation of numbers

The development and formation of a special mathematical language using special shapes of the numerals, special symbols (token) to designate mathematical relations, operations and functions took place only rather late in the course of history. Thus solution–rules, the step sequences of algorithms in the early texts were represented in terms of colloquial language, i.e. in a non-formalistic manner. The introduction of an adequate mathematical script influenced and accelerated the development of mathematics and mathematics education enormously. This effect may be demonstrated following the introduction of the HINDU-ARABIC shapes of numerals in the West in the 12th century (in some countries like Germany in the 15th century), with a phase transition in Japan after the *Meiji*-Reform at the end of the 19th century.

7. CONCLUSION AND CONSEQUENCES

7.1 The value of classical contents in school mathematics

The comparison of the old textbooks demonstrates that in the course of history there are *common mathematical-formal and subject-oriented strings of themes and topics, which as fundamentals determine the basic knowledge of mathematics education still today*. Those fundamentals, as a kind of invariants, will be valid and have to be observed and realized in educational mathematics of the future too. Especially, *certain "classical" contents, classical formal and applicatory subject-matters of mathematics education will still have a value in the curricula in the time to come.*

7.2 Differences

Besides such conformities there are *differences*, in the past and of course today in educational mathematics and in the daily practice of instruction. Those differences in characterization and estimation of educational principles in the course of the daily classroom work depend on the respective cultural context. The differences vary not only between East and West, they vary sometimes even between single parts of the countries here and there.

7.3 The problem of "problem-solving" – applied and pure aspects

Worldwide in school-mathematics there are *differences in the estimation of applied and pure aspects* of mathematics instruction too – variations quite often from teacher to teacher.

To master the future the younger generation needs a kind of flexibility. Therefore one problem of educational mathematics today is *the realisation of the balance between applied and pure aspects*. Especially with respect to the keyword "*problem-solving*" we may observe that already the ancient texts tended to combine problem-solving and structural-thinking. And that means: within educational mathematics *applied and pure aspects have to be observed and realized likewise.*

7.4 Chances

Conformities between the East and the West enable us to understand each other – differences present us with the chance to enrich and to complete each other.

REFERENCES

Bai Shangshu, 1988, *Annotation of Jiuzhang Suanshu and its Commentaries*. Science Press, Beijing. (in Chinese)

Chang Tungsun, 1952, A Chinese Philosopher's Theory of Knowledge, *ETC. A Review of General Semantics*, IX, **3**:203-226.

Cheng Dawei, 1990, *Suanfa Tongzong*. Jiaoyu chubanshe, Anhui. (in Chinese)

Deschauer, Stefan, 1992. Das zweite Rechenbuch von Adam Ries. Vieweg, Braunschweig.

Fischer, Walther L., 1982, Educational Mathematics on the Road to Mathematics. Bull. Math.Education Study, Math. Education Soc. Japan, **23**(3/4):56-83.

Fischer, Walther L., 1989, The Euclidean Scheme as a Universal Principle in Mathematics and in School-Mathematics. Bull. Western Japanese Academic Society of Mathematics Education, Hiroshima University/Japan, Nr.15:1-22. – And in: FISCHER (1996): Mathematikdidaktik: 204-232.

Fischer, Walther L., 1990, Sind klassische Inhalte im modernen Mathematikunterricht überholt?, in: *Herausforderung der Didaktik*, Beckmann, H.K./Fischer, W.L., ed., *Bad Heilbrunnn*, Klinkhardt Verlag, pp.163-182.

Fischer, Walther L., 1991, Some Fundamentals in Mathematical Instruction, *Proceedings*, 5, Southeast Asian Conference on Mathematical Education (ICMI-SEAMS 5 1990), University Brunei, Darussalam, Bandar Seri Begawan/Brunei, pp.133-137.

Fischer, Walther L., 1996, *Mathematikdidaktik zwischen Forschung und Lehre*, Klinkhardt Verlag, Bad Heilbrunn.

Fischer, Walther L., 1997, On Some Fundamentals in Mathematical Education and their Origins–From Premathematics to Protomathematics to Mathematics. Lecture, Tokyo Science University (Japan Society of Math. Education), Tokyo/Japan.

Fischer, Walther L., 2001, De Cursu Canis et Fuga Leporis – Bewegungsaufgaben in alter und neuer Zeit, Eichstätter Kolloquium zur Didaktik der Mathematik. Eichstätt: Universität Eichstätt, 1:1-10 .

Fischer, Walther L., 2001, Mathematica Perennis – Historical Topics as Indicators of Fundamentals in Mathematics Education. Keynote Address at the '4th Intern. Symposium on the History of Mathematics and Mathematical Education using Chinese Characters', 20.Aug.1999, Maebashi/Japan. Proceedings, Maebashi/Japan, **2**:231-250.

Jiuzhang Suanshu, 1990, Guji chubanshe, Shanghai. (in Chinese)

Li Di, 1982, The Mathematical Inferences of Liu Hui, in: Wu Wenjun, ed., pp.95-104.

Mehl, Andreas, 2003, Zwei folgenreiche Prinzipien in Natur und Welterklärung durch Griechen archaischer und frühklassischer Zeit, in: *Naturrezeption. Graz*, Liedtke, M., ed., Austria Medienservice.

Rouse Ball, W.W., 1931, *Mathematical Recreations and Essays*, Mac Millan and Co, London.

Siu Man-Keung, 1993, Proof and Pedagogy in Ancient China: Examples From Liu Hui's Commentary on Jiuzhang Suanshu, *Educationasl Studies in Mathematics*, **24**:345-357.

Smith, David Eugene, 1951, 1953, *History of Mathematics*, Vo.l.I, Vol.II. Dover Publications, New York.

Snell, Bruno, 1948, *Die Entdeckung des Geistes bei den Griechen*. Claasen & Goverts, Hamburg.

Swetz, Frank and Kao, T.I., 1977, *Was Pythagoras Chinese?* Pennsylvania State University Press, University Park, Pennsylvania.

Ueno, Kenji, 2004, From Wasan to Yozan. This volume.

Wu Wenjun, ed., 1982, *Jiuzhang Suanshu and Liu Hui*. Shifan daxue chubanshe, Bejing. (in Chinese)

Yoshida, Mitsuyoshi, 1977, *Jinko-ki*, Oya; Shinichi, eds., Iwanami-shoten Co, Tokyo. (in Japanese)

Yoshida, Mitsuyoshi, 2000, *Jinko-ki*, Wasan Institute, Tokyo. (in Japanese/English)

Zhong, Shanji, 1982, Arithmetic in Nine Sections and its Commentarist Liu Hui, in: Wu Wenjun, cd., pp.2-11.

Chapter 1-5

FROM "ENTERING THE WAY" TO "EXITING THE WAY"[1]: IN SEARCH OF A BRIDGE TO SPAN "BASIC SKILLS" AND "PROCESS ABILITIES"

WONG Ngai-Ying
The Chinese University of Hong Kong

1. THE PHENOMENON OF THE CHC LEARNER

For centuries, the Chinese have been fleeing from wars and famines, fighting for survival from North to South, and from East to West. Through it all, they have tried to build a brighter future for their *next generation* through *education*. This mentality has made the children of the CHC (Confucius Heritage Culture) the hardest working learners on earth.

During the last twenty years, the outstanding performance of Far Eastern Asian students, as demonstrated in international comparisons, especially in the area of mathematics, has attracted the attention of sociologists, educationalists and psychologists. Numerous studies have been conducted to account for this "phenomenon of the CHC learner." Social, cultural and pedagogical factors have been used to explain the apparent excellence of these students (Bond, 1996; Watkins & Biggs, 1996, 2001; Wong, 1998b). On the one hand, it was perceived that the general characteristics of the CHC learning environment (which emphasised recitation and memorisation, large classes with passive learners, and was teacher-centred with teacher-as-

[1] The author wishes to pay tribute to his Ph.D. thesis supervisor, Dr. David Watkins, for initiating the author into the fruitful research fields of the phenomenon of the CHC learner.

authority) were quite opposite to what was found to be conducive to learning (Biggs & Moore, 1993). On the other, research has repeatedly shown that CHC students were more strongly inclined to deep approaches to learning – the opposite to rote learning – than were Western students (Biggs, 1994; Watkins & Biggs, 1996).

Social-achievement orientation (as opposed to individual-achievement orientation; see Yu, 1996), collective thinking, diligence and attributing success to effort (rather than talent)[i] are often thought to be the factors constituting the CHC learner's phenomenon (Bond, 1996; Bond & Hwang, 1986; Hau & Salili, 1996; Ho, 1986; Lee, 1996). In brief, the CHC learner is envisioned as one who works harder instead of one who is smarter. The factor contributing to such outstanding performance among students of the Confucian tradition seems to be their orientation towards social achievement, the origin of which could be traced to the Confucian ideology and culture of de-emphasis of individual non-mundane pursuits.[2ii] Moreover, the degree of a person's success in life is judged not only by whether his or her achievements are passed on to the next generation[iii], but also by a person's worldly career and his or her contribution to the welfare of society[iv]. These factors are often seen as the origin of the achievement orientation of CHC societies (Qian, 1945, pp. 7-10).

Along this line of thought, an individual is valued not as an individual but for his or her role[v] in a vast network of kinship. Thus, it is of utter importance that a person plays precisely the role he or she inherits when born. In fact, rites[vi] and social norms (cardinal relations[vii]) are central themes of Confucianism. One of the major functions of education then, is to cast youngsters into a certain role that they will take up in society and act and behave accordingly in the light of their family background and socio-economic status. This predisposition, however, may be reversed when the examination system comes into play – because this system could bring about social mobility. Social mobility, in turn, sets new rules for a new game. Thus, education, in conjunction with the examination system, has the function of enabling an individual to strive for the best role s/he could attain[viii]. It is understandable, therefore, that the CHC learner (albeit an adapted one) would be strongly motivated towards high academic achievement when it is measured against conventional tests and examinations (Wong, 1998b).

However, there have been counter-arguments to the effect that the above associations could be an "over-Confucianisation" and there is also doubt as to whether a "causal relationship" between cultures and phenomena

[2] When Confucius was asked about life after death, he replied: "We know so little about this life, how can we know about life after death!"

("Culture X → Behaviour Y") exists (see Wong & Wong, 2002). Furthermore, the "examination culture," which is often used to describe the CHC learning environment (Lee, Zhang, & Zheng, 1997; Leung, 1995; Zhang, 1993; Zhang & Lee, 1991), can only be regarded at most as a bi-product of Confucianism after it is institutionalised (Wong, 1998b, 2000; see also Biggs, 1996).

Attention is thus turned to the identification of meritable practices (or even just practices) in various regions irrespective of their cultural origins (whether it is "Confucian" or "Daoist", "Eastern" or "Western"). In the study of meritable practices, we shall confine our area to cultural potentials and in what follows, we shall inspect some Chinese views of pedagogy as exemplified by two old traditions, namely calligraphy and martial art, to see what lessons we can learn from them.

2. THE ROLE OF PRACTICES: ENTERING THE "WAY"

As remarked by Biggs (1994), we have to distinguish rote learning with repetitive learning in our discussions. Marton also pointed out in a public lecture[3] that continuous practice with increasing variations could deepen understanding (see also Watkins, 1996). Confucius' words, "Learn the new when revising the old" (Analects, 2:11)[ix], were also quoted in the lecture. In fact, scholars have pointed out that the first stance of the Analects of Confucius on learning, "Learn and practice frequently" (Analects, 1:1)[x], should be interpreted as "Learn and put your knowledge into practice frequently." Confucius did not particularly advocate rote learning and over-drilling (see also Lee, 1996). Research studies do support the hypothesis that excellent academic performance of Asian learners may be due to a synthesis of memorising and understanding, a practice that is uncommon among Western students (Marton, Tse, & dall'Alba's, 1996; Marton, Watkins, & Tang, 1997; Watkins, 1994). It was also found that recitation is a common practice among CHC learners to bring about sharp focus for better understanding (Dahlin & Watkins, 2000).

That there is always a right way ("rite") to do anything is one of the central beliefs in Confucianism. This is the "rite" that we ought to follow no matter whether the issue is big or small[xi]. When Confucius entered the great ritual hall, he asked about the details (of the proper way) in each step.

[3] *Student learning: East and West.* Public lecture presented at The Chinese University of Hong Kong, Shatin, Hong Kong, 18 March 1997.

Someone called him ignorant but he replied, "This is the rite."[xii] This is also true in learning. As pointed out in Biggs (1994): "Chinese educators … believe that art should not only be beautiful but morally good; the idea of one right way pervades teaching." (p.28). So, there is a standard path, a routine that the learner can and should follow. In mathematics, though there are sometimes different ways to solve a problem, the one posed in the textbook is often the "best" (i.e., most elegant and simplest) way to solve it. Along this line of thought, it may not be advisable for beginners to deliberately create solutions different from the standard ones (Wu, 1994).

Following stringently the standard way and practice until one acquires "fluency" could be the first step in learning. The learning of Chinese calligraphy is a very good example, i.e. the conventional way of learning is to start working with "copy books"[xiii]. Various basic skills (e.g., the application of brush to effect different kinds of strokes) are also performed during the teaching lesson. Then, the master will choose for the disciple an exemplary calligrapher, say Yan Zhenqing[xiv], so that the disciple may imitate[xv] the calligraphy of Yan to a state that one cannot easily distinguish the calligraphy of Yan from that of the disciple. That is called "entering the Way[xvi]." The emphasis is on doing each thing (brush-painting in this case) in the prescribed (or right) way.

Chinese martial art (Kung Fu) shares the same approach. Practising of such footwork as "horse-riding postures[xvii]" (see Figure 1), as well as other kinds of basic physical training like sandbag punching[xviii] comprise the basics of Chinese martial art (Figure 2). Then there are standard sequences[xix] for one to practise (see Figure 3). In simple terms, these sequences are "fictitious" courses of fighting (terms like "shadow boxing" in Xi et al, 1984). In other words, this is another kind of imitation of the "right" way to attack and defend. For some traditions of kung fu like the Northern Mentis style[xx], there are special kinds of "sequences with partners," one acting as the attack side while the other acts as the defense. In most traditions, "paired practices" are common (see Figure 4). There are also other kinds of combat skill training. Of course the final stage is free fight[xxi]. In the Form-and-Will[xxii] tradition, the complete programme of learning was depicted as "standing postures – force training – paired practices – actual fighting."[xxiii] In other words, the right way of learning is imitating the "right" way of fighting, accompanied by training of basic skills. This can only be done with a lot of drill and practice.

3. INITIATION: EXITING THE "WAY"

3.1 Initiation by the teacher

However similar the imitation is in respect of the original, even to a state that the works of the imitator and the imitated are indistinguishable, it is still considered as the first step in learning. Using once again the example of calligraphy, one enters the "gate", i.e. to get to know the basics of a certain specialised knowledge, by way of painting in exactitude the calligraphy style of Yan. Ultimately, a "personalised Yan-style" should emerge so that people well-versed in calligraphy can see that the character of the calligrapher is revealed in the calligraphy just as clearly as the style which originated from the Yan style. This "looking similar but being different"[xxiv] phenomenon shall be called "exiting the Way"[xxv], i.e. transcending the original cast. However, there is no obvious way to tell how a learner can arrive at this stage. Some believe that through incessant practice and a long period of "indulgence" and "hatching," then, mystical though it may sound, at a certain point, insight can be obtained. This is clear from the maxim "familiarity breeds sophistication" (commonly translated as "practice makes perfect"[xxvi]). Familiarity is a necessary but in itself is not a sufficient condition for sophistication. The case is similar for martial art and it has been said that "when one has practised for a long time, 'it' [the skills] will naturally be acquired"[xxvii] (Tang, 1986, p. 38). Eventually, the learner might come to a stage where he would exclaim 'aha' when the posture comes about naturally without his making the effort "[xxviii] (p. 126).

The master's guidance and initiation are of vital importance. It is not easy to show a clear path of instruction, because initiation strategies are situation-dependent. Skilful teachers can grasp the right moment and use appropriate means to trigger students' "sudden enlightenment" when he sees that the disciple is ready. So it is said that, "I let you see the mandarin duck when it is sewn, but I will not let you have the golden needle (for sewing)."[xxix] This indicates that the teaching method cannot be disclosed. That is why the great Chan (Zen)[xxx] master Huang Bo[xxxi] exclaimed, "I did not say that there is no Chan, I only said that we don't have masters!"[4][xxxii] The following story (Zhang, 1979) illustrates such an initiation process.

[4] Abbot Huang Bo told the mass, "You guys, ... do you know that there are no Chan masters in the whole of the Tang Empire?" At that time, a monk went out and said, "There are so many (people) in various places leading disciples, how do you explain that?" Bo said, "I did not say that there is no Chan, I only said that there is no master!"

> When the great zitherist Bai-Ya[xxxiii] learned to play the zither under the Master Cheng-Nian[xxxiv], there came a time when the Master told him that he had taught him all he knew and that he was going to introduce his own master to him. Thus, Master Cheng-Nian led Bai-Ya to the seashore and asked him to wait for him while he rowed away in a boat to meet his master. After some days, the Master did not return so Bai-Ya passed the time practising the compositions that his Master had taught him. After a while, he realised that his music had become inseparable from the voice of Nature. At that instant, he realised that Nature was his master's teacher.

Obviously, this concept is in line with the thinking of the ancient masters that Nature is the great teacher[xxxv], and that lay people make use of Nature's wisdom without noticing[xxxvi].

3.2 Doubt and reflection

Many believe that the process of initiation also depends on the calibre of disciples. This is without doubt why Hui-Neng, the Sixth Patriarch of Chan Buddhism[xxxvii], said that his Way is meant for those with great wisdom and of high calibre[xxxviii]. However, we can still pick out some commonalties in such ways of doing.

By analysing the Chan classic The Lineage Transmission in Jing-de[xxxix], Cai (1986) identified three categories of "non-routine" modes of teaching used by Chan masters: use of symbols, use of words, and use of gestures (including the raising of the eyebrow, erection of the fingers, snapping of the fingers, slapping the face, waving of the hand, beating and shouting) and silence (pause, remaining in solitude, rising from the meditation seat, dismissal and returning to the Master's room). The provision of experience transcending words, bringing about a threat of discomfort, and initiating reflections and mindfulness were also identified in Wong (1998a).

In brief, reflection forms the core of Confucian and Chan's way of "pedagogy" (bringing about realisation). Thus the major task of the master is to arouse within the disciple a state of discomfort and perplexity. To quote from a popular citation of Confucius: "Enlightenment comes when one is stunned and understanding when one is aroused" (*Analects*, 7:8)[xl]. Doubt and realisation form the central theme of learning in Confucianism. Zhu Xi[xli], the central figure of neo-Confucianism in the 10th century, pointed out that "Reading books [learning] is to arouse doubt when one does not doubt and let those in doubt settle in the state of no doubt. This is how one grows"[xlii] (Li, 1270, p.296).

According to Cronbach (1955), the threat of discomfort is a prerequisite to the problem-solving process. The arousal of doubt in a disciple is also a main theme of Chan initiation, as it was said, "A little doubt leads to small understanding while strong doubt leads to deep understanding."[xliii] It was also said, "To make effort, one is only to have great doubt." "Great faith, great doubt and great diligence"[xliv] are also identified as the "three pillars of Zen (Chan) Buddhism" (Kapleau, 1980) and it seems that doubt – reflection – realisation is a formula for "exiting the Way." This principle can clearly be seen from the words of the Sixth Patriarch of Chan Buddhism, Hui-Neng in Chapter 10 of his *Platform Sutra*[xlv] where he teaches his disciples how to transmit the doctrine:

> If someone asks you for the meaning [of something], use "emptiness" in response to questions on "reality," use "reality" in response to questions on "emptiness"; use "saint" in response to questions on "mundane," use "mundane" in response to "saint." When two phenomena are contrasted together, the middle way will emerge. So we have questions and answers. All other questioning will follow this method and the truth will not be lost. Suppose someone asks, "What is darkness?" The answer could be, "Brightness is the cause, darkness is the condition; where there is no brightness, there is darkness." We use brightness to illustrate darkness and use darkness to show brightness. Our minds go to and fro between these two notions, and the middle way will emerge. Other questioning will follow this. When you transmit the dharma [doctrine] in the future, you should adopt this kind of teaching so that the spirit of our school will be maintained.[xlvi] (see Wong, 1998a)

3.3 Repetition and variation

Repetition until understanding is internalised, could be a general strategy employed to bring about reflection and hence deeper understanding. This is well explained in the following words of Zhu Xi:

> Generally speaking, in reading, we must first become intimately familiar with the text so that its words seem to come from our own mouths. We should then continue to reflect on it so that its ideas seem to come from our own minds. Only then can there be real understanding. Still, once our intimate reading of it and careful reflection on it have led to a clear understanding of it, we must continue to question. Then there might be further progress. If we cease questioning, in the end there'll be no more progress.[xlvii] (see also Lee's chapter in Watkins & Biggs, 1996)

In addition, recitation, when undertaken with reflection, is a means of bringing about repetitive learning. The following words of Zhu Xi give a clear explanation: "The method of reading books is, read once then reflect once; reflect once then read once again. Recitation is a means of enhancing reflection ... if reading is just done in the mouth and not reflected in the mind, you cannot remember the passage well "[xlviii] (see Li, 1270, p. 170).

In fact, reinterpreting earlier findings in phenomenography (e.g., Bowden & Marton, 1998; Marton & Booth, 1997) leads to the conclusion that one way of experiencing a phenomenon can be characterised in terms of those aspects of the phenomenon that are discerned and kept in focal awareness by the learner (see also Runesson, 1999). Since discernment is an essential element to learning and variation is crucial in bringing about discernment, repetition by systematically introducing variations could be the key to bringing about learning and understanding.

In this light, repetition and practice of basic skills (entering the "Way") form the basis for developing process abilities (exiting the "Way") and thus should not be overlooked. The heart of the matter seems to lie in the quality rather than quantity of these practices, including the systematic introduction of variations.

Let us project the case on to mathematics and look into the example of introducing the formula for factorising quadratic polynomials. After students have learned how to factorise quadratic polynomials for completing squares, a series of related problems could be set to let students experience the generalisation of the completing square method into the quadratic polynomial formula. For a first trial, one can easily factorise polynomials like x^2+4x+4 and x^2+6x-7 by completing square. Gradually, we can introduce some variations like $2x^2+8x-8$, $2x^2+4x-7$ and $3x^2+5x-4$. Later on, some parameters can be further introduced, such as: x^2+4x+k, $x^2+2bx+4$, $x^2+2px+q$, x^2+px+q. Finally, the factorisation of ax^2+bx+c would come about naturally. Huang (2002; see also Gu, Marton & Huang, in preparation; Huang & Leung, 2002) also concludes that learning through exercises with variations is one of the characteristics of Chinese mathematics teaching (p. 236).

4. VARIATIONS: A BRIDGE OVER "BASIC SKILLS" AND "PROCESS ABILITIES"?

Some of the critical aspects in the "middle zone" between the "East" and the "West" have already been identified (Gu, 2000; Leung, 2000; Wong, 2000). The two extremes of "product" (content, basic skills, drills, etc.) and "process" (higher-order abilities, creativity, discoveries, etc.) is one such aspect, for which the debate can be dated back at least to the 1960s when there was the "New Math Reform" and the "Back to Basics" movement afterwards (Hill, 1976; NACOME, 1975; Wong & Wong, 2001). This may suggest a similarity with the notion of *Gei* (art)[xlix] in Japanese education, which comprises both *Jutsu* (technique)[l] and *Do* (way)[li] (Hirabayashi, 2003). In particular, in the ICME-Study conducted in 1986, the "process-based curriculum" was queried (Howson & Wilson, 1986, pp.25-26) and the conclusion of striking a balance between "content" and "process" was arrived at (Howson & Wilson, 1986, p. 35 & p.51). It is clear from the above discussion that CHC pedagogy may recognise the continuum between "content" and "process", not just recognising that the acquisition of the former is a foundation on which the latter could be developed. It has been further pointed out that "these abilities cannot be developed out of the mathematics context. The problem does not lie in striking a balance but letting the introduction of mathematics knowledge be the foundation of the development of higher-order abilities" (Wong, 1995, p.71). This can be echoed by a common saying in Chinese martial art that "If you only practise combat skills but do not develop the inner energy, your efforts will be in vain when you become old; but if you only practise the inner energy and not the combat skills, you would be like a boat that has no rudder"[lii].

As society becomes more and more technologically advanced and information-based, more emphasis will be placed on higher-order thinking and basic skills will be de-emphasised. In spite of such a shift of emphasis, the base of knowledge should not be ignored. We may find some inspirations in Peddiwell's (1939) famous narration of the "Saber-tooth curriculum":

> In Chellean times, "Mr. New-Fist" began to construct the curriculum of "fish-grabbing-with-the-bare-hands," "woolly-horse-clubbing" and "saber-tooth-tiger-scaring-with-fire." It has been very successful. However, after some time, the Ice Age was approaching. The world has changed. The dirt and gravel carried by the glacier made the water too muddy to catch the fish. The woolly horse ran away from the chilly weather and the place was occupied by antelopes which were so shy that no one could get close enough to club them. Furthermore the new dampness in the air spreads a disease among the saber-tooth tigers and most of them died. The bear that came recently were not afraid of fire. There came an urge among the community to upgrade the curriculum into "fishnet-making and using," "antelope-snare construction and operation" and "bear-catching and killing." That initiated heated debate between the conservatives and the radicals. Finally came the wise old man who said, "Don't be foolish. We don't teach fish-grabbing to grab fish; we teach it to develop a generalized agility which can never be developed by mere training. We don't teach horse-clubbing to club horses; we teach it to develop a generalized strength in the learner which he can never get from so

> prosaic and specialized a thing as antelope-snare-setting. We don't teach tiger-scaring to scare tigers; we teach it for the purpose of giving that noble courage which carries over into all the affairs of life and which can never come from so base an activity as bear-killing." "If you had any education yourself," he continued to say severely, "you would know that the essence of true education is timelessness. It is something that endures through changing conditions like solid rock standing squarely and firmly in the middle of a raging torrent. You must know that there are some eternal verities, and the saber-tooth curriculum is one of them!" *(pp. 42–44)*

This analysis details the "Chinese" way of learning, that is, repeated learning could form the basis of understanding. It is not the quantity of practice that matters but the quality, in particular, whether variation is systematically introduced. A master may, at certain stages in the course of learning, point out the critical aspects, which is crucial to students when they try to transcend from the stage of "Entering the Way" to that of "Exiting the Way". Genuine understanding emerges because of and not in spite of the fluency of skills.

5. EPILOGUE: OTHER ROLES OF PRACTICE

Certainly, there are other goals of repetition. Automation, which is necessary in some disciplines, is one. Undoubtedly, reflex response is needed in martial arts. Wong (1949) sees this as particularly difficult when compared with the learning of literature (pp.3-4)[liii]. The Sixth Patriarch of Chan Buddhism also stressed that responses should be done "without thought" (without hesitation) since Chan Buddhism always emphasises that realisation is something which is beyond words[liv]. Thus technical fluency, at times, is considered a means but at other times, as an end in itself . Practices have these dual purposes too and hence the Chinese stress that "combat skills should not be separated from the fist nor songs from the mouth[lv]" (practice should be performed frequently). In the classic "Treatise of Martial Arts," it was said that:

> Combat skills concern bodily movements; this is closely related to day-to-day practice, until these movements become fluent. As it is said: "Having practised a movement a thousand times, [only then,] can the bodily movements become natural."[lvi] *(Wan, 1964, p. 9)*

In mathematics, we often solve problems by applying the "fastest strategies" before going for "backup strategies" (Kerkman & Siegel, 1997).

The notion of "techniques", being a bridge between tasks and conceptual reflection, has also been put forth; and application of any such techniques should go through the process of naturalisation and internalization (Artigue, 2001). We may conclude in the following figure the role of practice:

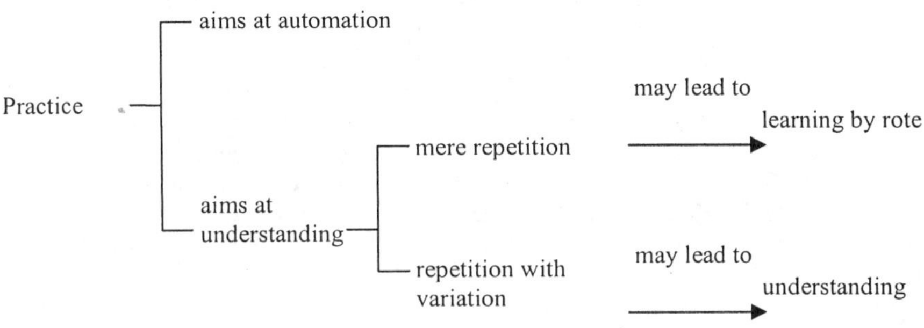

(see also Wong, 2001)

REFERENCES

Artigue, M., 2001, Learning Mathematics in a CAS Environment: The Genesis of a Reflection about Instrumentation and the Dialectics between Technical and Conceptual Work, Paper presented at the CAME meeting, Utrecht, July 2001.

Biggs, J., 1994, What are Effective Schools? Lessons from East and West [The Radford Memorial Lecture], *Australian Educational Researcher*, **21**:19–39.

Biggs, J.B., 1996, Western Misconceptions of the Confucian-heritage learning Culture, in: *The Chinese learner: Cultural, Psychological and Contextual Influences*, D.A. Watkins & J.B. Biggs, eds., Comparative Education Research Centre, Hong Kong and The Australian Council for the Educational Research, Victoria, Australia, pp.45–67.

Biggs, J.B., and Moore, P.J., 1993, *The Process of Learning*. Prentice Hall, New York.

Bond, M.H. ed., 1996, *The Handbook of Chinese Psychology*, Oxford University Press, Hong Kong.

Bond, M.H., and Hwang, K.K., 1986, The Social Psychology of Chinese People, in: *The Psychology of the Chinese People*, M.H. Bond, ed., Oxford University Press, Hong Kong, pp.213–266.

Bowden, J., and Marton, F., 1998, *The University of Learning*. Kogan Page, London.

Cai, R.T., 1986, *The Enlightening Methods of Chan Masters*, Wen Shu Publications, Taipei. [lvii] [in Chinese]

Cronbach, I.F., 1955, The Meaning of Problems, in: *Readings in Educational Psychology*, J.M. Seidman, ed., Houghton Mifflin, Boston, pp.193–201.

Dahlin, B., and Watkins, D.A., 2000, The Role of Repetition in the Processes of Memorising and Understanding: A Comparison of the Views of Western and Chinese School Students in Hong Kong, *British Journal of Educational Psychology*, **70**: 65-84.

Gu, L., 2000, *Exploring the middle zone*. Paper presented at the 9th International Congress of Mathematics Education (Gathering of Chinese scholars), Tokyo/Makuhari, Japan, 31 July–6 August.

Gu, L., Marton, F., and Huang, R., in preparation, Two Theories of Variation in Chinese Mathematics Learning, in: *How Chinese Learn Mathematics: Perspectives from Insiders*, L. Fan, N.Y. Wong, and J. Cai, eds., World Scientific, Singapore.

Hau, K.T., and Salili, F., 1996, Achievement Goals and Causal Attributions of Chinese students, in: *Growing up the Chinese Way*, S. Lau, ed., The Chinese University Press, Hong Kong, pp.121–145.

Hill, S., 1976, Issues from the NACOME Report, *Mathematics Teacher*, *October*, 440–446.

Hirabayashi, I., 2003, A Traditional Aspect of Mathematics Education in Japan: Mathematics as *Gei* (art), its *Jutsu* (technique) and *Do* (Way). Chapter // of this volume.

Ho, D.Y.F., 1986, Chinese Patterns of Socialization: A Critical Review, in: *The psychology of the Chinese people*, M.H. Bond, ed., Oxford University Press, Hong Kong, pp.1–37.

Howson, G., and Wilson, B., eds., 1986, *School Mathematics in the 1990s*. Cambridge University Press, Cambridge.

Huang, R. and Leung, F.K.S., 2002, Is there a Chinese Approach? – A Comparison on the Ways of Teaching the Pythagoras Theorem among Australia, Czech Republic, Hong Kong and Shanghai, in: *Proceedings Second East Asia Regional Conference on Mathematics Education and Ninth Southeast Asian conference on Mathematics Education*, D. Edge & Y.B. Har, eds., **2**:247-252. Association of Mathematics Educators & National Institute of Education, NTU, Singapore.

Huang, R., 2002, *Mathematics Teaching in Hong Kong and Shanghai – A Classroom Analysis from the Perspective of Variation*. Unpublished Ph.D. dissertation. The University of Hong Kong, Hong Kong,

Kapleau, P., 1980, *The Three Pillars of Zen: Teachings, Practice, Enlightenment*, Anchor Books, Garden City, NY.

Kerkman, D.D., and Siegel, R.S., 1997, Measuring Individual Differences in Children's Addition Strategy Choices, *Learning and Individual Differences*, **9**(1):1–18.

Lee, W.O., 1996, The Cultural Context for Chinese Learners: Conceptions of Learning in the Confucian Tradition, in: *The Chinese Learner: Cultural, Psychological and Contextual Influences*, D.A. Watkins and J.B. Biggs, eds., Comparative Education Research Centre, Hong Kong, The Australian Council for the Educational Research, Victoria, Australia, pp.25–41.

Lee, P.Y., Zhang, D., and Zheng, Z., 1997, Examination Culture and the Teaching of Mathematics, *EduMath*, 4:96–103.[lviii] [in Chinese]

Leung, F.K.S, 2000, In Search of an East Asian Identity in Mathematics Education — The Legacy of an Old Culture and the Impact of Modern Technology. Regular lecture presented at the 9th International Congress of Mathematics Education, Tokyo/Makuhari, Japan, 31 July–6 August.

Li, J.D., 1270, *Selections from the conversations of Master Zhu, arranged topically*. Ching Chung Bookstore, Taipei, reprint 1961, English translation: D.K. Gardner, Trans., 1990, University of California Press, Berkeley, CA.[lix]

Marton, F., and Booth, S., 1997, *Learning and Awareness*. Lawrence Erlbaum Associates, Mahwah, NJ.

Marton, F., Tse, L.K., and dall'Alba, G., 1996, Memorizing and Understanding: The Keys to the Paradox?, in: *The Chinese Learner: Cultural, Psychological and Contextual Influences*, D.A. Watkins and J.B. Biggs, eds., Comparative Education Research Centre, Hong Kong, The Australian Council for the Educational Research, Victoria, Australia, pp.69–83.

Marton, F., Watkins, D., and Tang, C., 1997, Discontinuities and Continuities in the Experience of Learning: An Interview Study of High-school Students in Hong Kong, *Learning and Instruction*, **7**:21-48.

National Advisory Committee on Mathematics Education [NACOME], 1975, *Overview and Analysis of School Mathematics, Grades K-12*, Conference Board of the Mathematical Sciences, Washington, DC.

Peddiwell, J.A., 1939, *The Saber-tooth Curriculum*.McGraw-Hill, New York.

Qian, M., 1945, *Soul and Mind*, reprinted in M. Qian, 1976, *Soul and mind* [in Chinese]. Jing Lian Publications, Taipei.[lx]

Runesson, U., 1999, *The Pedagogy of Variation: Different Ways of Handling a Mathematics topic* [in Swedish]. Unpublished doctoral dissertation, University of Göteborg, Sweden. English summary available Internet: http://www.ped.gu.se/biorn/phgraph/civial/graphica/diss.su/runesson.html

Tang, Y., 1986, *Brief Introduction to Form-and-Will Style Chinese Martial Art*, Form-and-Will Study Group, Hong Kong.[lxi] [in Chinese]

Wan, L.S., 1964, *A Treatise of Martial Art*, Ping Ping Publications, Taipei.[lxii] [in Chinese]

Watkins, D., 1994, Memorising and Understanding: The Keys to Solving the Mysteries of the Chinese Learner? Paper presented at the 23rd Congress of the International Association of Applied Psychology, Madrid.

Watkins, D.A., 1996, Hong Kong Secondary School Learners: A Developmental Perspective, in: *The Chinese Learner: Cultural, Psychological and Contextual Influences*, D.A. Watkins and J.B. Biggs, eds., Comparative Education Research Centre, Hong Kong and The Australian Council for the Educational Research, Victoria, Australia, pp.107–119.

Watkins, D.A., and Biggs, J.B., eds., 1996, *The Chinese learner: Cultural, Psychological and Contextual Influences*, Comparative Education Research Centre, Hong Kong and The Australian Council for the Educational Research, Victoria, Australia.

Watkins, D.A., and Biggs, J.B., eds., 2001, *Teaching the Chinese Learner: Psychological and Pedagogical Perspectives*. Comparative Education Research Centre, The University of Hong Kong, Hong Kong.

Wong, H.F., 1949, *Writings on the Mentis Style Chinese Martial Art*, Hong Kong Jing-Wu Athletic Association Chinese Martial Art Section, Hong Kong.[lxiii] [in Chinese]

Wong, H.F., 1955, *The Paired Sequence "Beng Bu"*. Hon Fan Gymnasium, Hong Kong.[lxiv]

Wong, N.Y., 1995, Hong Kong Mathematics Education in the Universal Education and Post-universal Education Periods, in: *Retrospect and Outlook on Mathematics Education in Hong Kong*, M.K. Siu, ed., Hong Kong University Press, Hong Kong, pp.69–87.[lxv] [in Chinese]

Wong, N.Y., 1998a, The Gradual and Sudden Paths of Tibetan and Chan Buddhism: A Pedagogical Perspective, *Journal of Thought*, **33**(2):9-23.

Wong, N.Y., 1998b, In Search of the "CHC" Learner: Smarter, Works Harder or Something More? Plenary lecture, in: *Proceedings of ICMI – East Asia Regional Conference on Mathematical Education 1*, H.S. Park, Y.H. Choe, H. Shin, and S.H. Kim, eds., pp.85-98, Korean Sub-Commission of ICMI; Korea Society of Mathematical Education; Korea National University of Education, Seoul, Korea.

Wong, N.Y., 2000, Mathematics Education and Culture: The "CHC" Learner Phenomenon. Paper presented at the 9th International Congress of Mathematics Education (Topic Study Group 22), Tokyo/Makuhari, Japan, 31 July–6 August.

Wong, N.Y., 2001, Practice, Rote Learning and Backwash Effect – From Examination to Learning [in Chinese], in: *Holistic Review of the Mathematics Curriculum: What's Next?*, S.K. Leung, ed., Department of Curriculum and Instruction, The Chinese University of

Hong Kong and the Hong Kong Association for Mathematics Education, Hong Kong. In press.[lxvi]

Wong, N.Y., and Wong, K.L., 2001, The Process of the 'Modern Mathematics' Reform and the Lessons We Learn on Contemporary Mathematics Education [in Chinese], in: *The Long Path of Mathematics Education in Hong Kong: 'Modern Mathematics' as the Opening Chapter*, N.Y. Wong, ed., Hong Kong Association for Mathematics Education, Hong Kong, pp.9-111.[lxvii]

Wong, N.Y. and Wong, W.Y., 2002, The "Confucian Heritage Culture" Learner's Phenomenon, *Asian Psychologist*, **3**(1):78-82.

Wu, H., 1994, The Role of Open-ended Problems in Mathematics Education, *Journal of Mathematical Behavior*, **13**(1):115–128.

Yu, A., 1996, Ultimate Life Concerns, Self and Chinese Achievement Motivation, in: *The Handbook of Chinese Psychology*, M.H. Bond, ed., Oxford University Press, Hong Kong, pp.227-246.

Zhang, D., 1993, Success and Inadequacies of Mathematics Education in Chinese Communities [in Chinese], in *Proceedings of the International Symposium on Curriculum Changes for Chinese Communities in Southeast Asia: Challenges of the 21st century*, C.C. Lam, H.W. Wong, and Y.W. Fung, eds., Department of Curriculum and Instruction, The Chinese University of Hong Kong, Hong Kong, pp.93–95.[lxviii]

Zhang, D., and Lee, P.Y., 1991, Examination Culture and Mathematics Teaching, *Proceedings of ICMI – China Regional Conference on Mathematical Education at Beijing, 1*, 8, Beijing Normal University, Beijing.

Zhang, X.B., 1979, Story of the water fairy [in Chinese], *China Man Monthly*, **1**:84–85.

Figure 1-5-1. Horse riding postures

Figure 1-5-2. Basic physical training

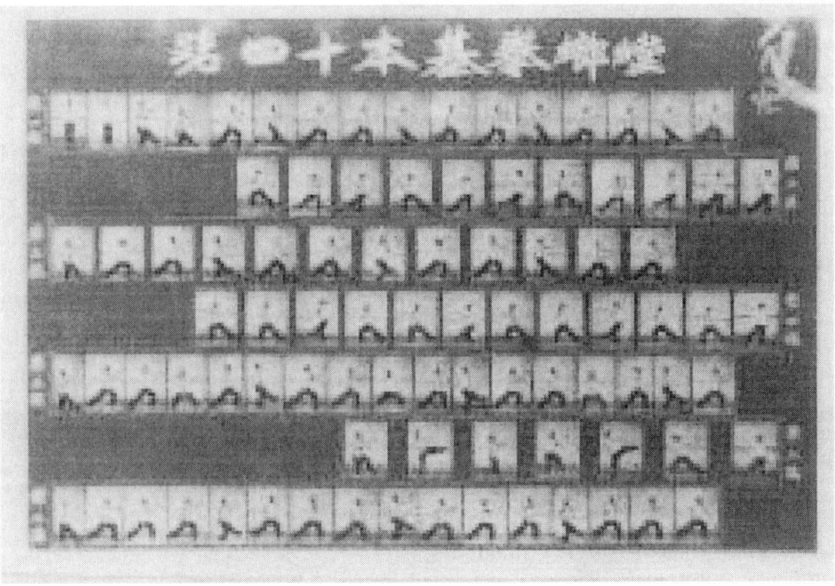

Figure 1-5-3. An example of sequence

Figure 1-5-4. A paired sequence (taken from Wong, 1955)

[i] 勤能補拙
[ii] 《論語・先進第十一》：「未知生、焉知死？」
[iii] 子嗣
[iv] 三不朽
[v] 本份
[vi] 禮
[vii] 五倫
[viii] 力爭上游，出人頭地
[ix] 《論語・爲政第二》：「溫故而知新。」
[x] 《論語・學而第一》：「學而時習之。」
[xi] 《論語・學而第一》：「有子曰：『禮之用……小大由之』。」
[xii] 《論語・八佾第三》：「孔子入太廟，每事問。」
[xiii] 描紅

xiv 顏真卿
xv 臨摹
xvi 入法
xvii 紮馬
xviii 打沙包
xix 套路
xx 北螳螂
xxi 「紮馬→套路→散手對練→搏擊」
xxii 形意
xxiii 「站樁、試力、推手、實擊」
xxiv 求其似、求其不似
xxv 出法
xxvi 熟能生巧
xxvii 「習之若恆久，不期自然至。」
xxviii 「技到無心始見奇。」
xxix 行機禪師：「鴛鴦繡出從君看，不以金針度與人。」—《續傳燈錄》（明・圓極居頂，?–1404）。
xxx 禪
xxxi 黃檗
xxxii 「黃檗示眾云：『汝等諸人，……還知大唐國裏無禪師麼？』時有僧出云：『只如諸方匡徒領眾，又作麼生？』檗云：『不道無禪，只是無師』！」—《碧巖錄》（宋・圜悟，1125）。
xxxiii 伯牙
xxxiv 成連
xxxv 《道德經》：「人法天、天法地、地法自然」；《論語・陽貨第十七》：「天何言哉？四時行焉，百物生焉，天何言哉？」
xxxvi 《易繫辭上傳》：「百姓日用而不自知。」
xxxvii 六祖慧能
xxxviii 《六祖法寶壇經・般若品第二》：「此法門是爲大智人說、爲上根人說。」
xxxix 《景德傳燈錄》
xl 《論語・述而第七》：「不悱不啓，不憤不發；舉一隅不以三隅反者，則不復也。」
xli 朱熹
xlii 《朱子語類・卷十一讀書法下》：「讀書無疑者須教有疑、有疑者卻要無疑。到這裏方是長進。」
xliii 「參禪須是起疑情，小疑小悟，大疑大悟。」— 《禪關策進題解：袁州雪巖欽禪師普說》（雲棲袾宏，頁 436–438）；《佛光大藏經：禪藏，禪關策進卷》（高雄：佛光出版社，1994，頁 437）；又見《雪巖祖欽禪師（宋）語錄：袁州仰山禪寺語錄》（昭如、希陵編，頁 11635–11733）；《禪宗集成》17 冊（台北：藝文印書館，1968，頁 11661）。
xliv 明州剛宗軟禪師示眾：「大凡做工夫，只要起大疑情。」—《續指目錄》卷十二（清・聶先。台灣：新文豐出版公司，頁 193）。
xlv 法寶壇經
xlvi 「若有人問汝義，問有將無對，問無將有對。問凡以聖對，問聖以凡對。二道相因，生中道義。如一問一對，餘問一依此作，即不失理也。設有人問何名爲暗。答曰：明是因，暗是緣，明沒則暗。以明顯暗，以暗顯明。來去相因，成中道義。餘問悉皆如此。汝等於後傳法，依此轉相教授，勿失宗旨。」

xlvii 《朱子語類・卷十讀書法上》：「大抵觀書先須熟讀，使其言皆若出於吾之口。繼而精思，使其意皆若出於吾之心。然後可以有得爾。然熟讀精思既曉得後，又須疑不止如此。庶幾有進。若以爲止如此矣，則不復有進也。」

xlviii 《朱子語類・卷十讀書法上》：「讀書之法，讀一遍又思量一遍；思量一遍又讀一遍。讀誦者，所以助其思量……若只是口裏讀、心裏不思量，看如何也記不仔細。」

xlix 藝

l 術

li 道

lii 「練拳不練功，到老一場空；練功不練拳，猶如無帽船。」

liii 「……武術之運用，動輒危及生命，拳腳交攻之中，刀槍齊舉之際，俄頃之間，生死勝敗立見，此又非七步倚馬可望其萬一也。」

liv 《六祖法寶壇經・行由品第一》：「火急速去，不得遲滯，思量即不中用。見性之人，言下須見，若如此者，輪刀上陣亦得見之。」

lv 「拳不離手，曲不離口。」

lvi 「身功乃講身法，此則在素日練拳之精熟矣。所謂：『拳打千遍，身法自然』，故拳須打過千遍，始可研求自然之境，如不於拳上努力，欲求身法進退躲閃之自然是不可能也！」

lvii 蔡榮婷(1986)。《禪師的啓悟法》。台北：文殊出版社。

lviii 李秉彝、張奠宙、鄭正亞(1997)。考試文化與數學教學。《數學教育》，第 4 期，96–102。

lix 宋・黎靖德(1270)。《朱子語類》。台北：正中書局。

lx 錢穆(1945)。靈魂與心。載錢穆(1976)。《靈魂與心》。台北：經聯出版社。

lxi 湯又覺(1986)。《意拳淺釋》。香港：意拳研習會。

lxii 萬籟聲(1964)。《武術匯宗》。台北：平平出版社。

lxiii 黃漢勛(1949)。《螳螂拳術隨筆．讀書與練拳》。香港：香港精武會國術部。

lxiv 黃漢勛(1955)。《領崩步》。香港：漢勛健身院。

lxv 黃毅英(1995)。普及教育期與後普及教育期的香港數學教育。載蕭文強(編)，《香港數學教育的回顧與前瞻》(頁 69–87)。香港：香港大學出版社。

lxvi 黃毅英(2001)。操練、不求甚解、倒流效應——從考試到學習。載梁淑坤(編)。《評核與數學教育—「數學課程全面檢討：之後又如何？」研討會跟進論文集》，122-127。香港：香港中文大學課程與教學學系及香港數學教育學會。

lxvii 黃毅英、黃家樂(2001)。「新數學」運動的過程及對當代數學教育之啓示。載黃毅英(編)《香港近半世紀漫漫「數教路」：從「新數學」談起》，9-111。香港：香港數學教育學會。

lxviii 張奠宙(1993)。華人地區數學教育的成功與不足。載林智中、黃顯華、馮以浤(編)，《東南亞華人社會的課程改革：二十一世紀的挑戰國際研討會論文集》(頁 93 - 95)。香港：香港中文大學課程與教學學系。

Chapter 1-6

PRACTICE MAKES PERFECT: A KEY BELIEF IN CHINA

LI Shiqi
East China Normal University

1. DIFFERENT VIEWS ABOUT MANIPULATIVE PRACTICE

Mathematics educators in the West usually put emphasis on understanding in mathematics teaching and learning (Commission on Standards for School Mathematics 1989; National Council of Teachers of Mathematics 1980). They consider the creative aspects of mathematics as the most important goal and regard drill and practice as imitatively behavioral manipulation. As Hiebert and Carpenter reviewed and commented comprehensively (1992), when the importance of both conceptual and procedural knowledge is confirmed, the question of whether to be concerned with conceptual relationships or procedural proficiency first is still left. Western educators often stress the need to build meaning for written mathematical symbols and rules before practicing the rules for efficient execution. The inherent reason is that when a procedure is practiced it can become fixed, making it difficult to think of the problem situation in another way. So well-practiced learners are reluctant to connect the rules with other representtations that might give them meaning. For practical teaching their suggestion is that the environment should help students build internal representations of procedures and conceptual networks before encouraging the repeated practice of procedure (cf. Hiebert & T. Carpenter, 1992, p. 78-79).

In China, as well as East Asian countries, routine or manipulative practice is an important mathematics learning style. *Practice Makes Perfect* is the underlying belief. Many mathematics teachers and also students believe it and consider it a general principle for mathematics teaching and learning. Through imitation and practice again and again, people will become highly skilled. Students from East Asia often top the list in international assessments of mathematics education and mathematics competitions, for example, IAEP 1992, TIMSS (Mullis, et al. 2000) and IMO. From the perspective of insiders their achievements would be attributed to a large amount of routine practice, problem solving and frequent tests. Based on the point of view of the dual nature of mathematics conception (Sfard, 1991), the analysis shows that the mechanism of routine practice is not simply interpreted as a way in which students only mechanically imitate and memorize rules and skills. Manipulation is the genetic place of mathematical thinking and the foundation of concept formation. It provides students with a necessary condition of concept formation and is the first step of mathematics comprehension (Li, 2000).

2. CULTURAL SOURCES OF THE BELIEF

Many scholars have probed the cultural sources of Chinese beliefs in mathematics education (e.g. Zhang, 1998; Leung, 1998; Wong, 1998). Generally the cultural roots of the belief *Practice Makes Perfect* could be summed up in the following aspects:

- the tradition of mathematics
- the traditional belief about study
- the examination culture
- the basic ideas in the syllabus (or curriculum standard)

China has a long historical tradition of mathematics, for example the masterpiece of mathematics《九章算术》(Nine Chapters of Arithmetic). It differs greatly from ancient Greek mathematics in ideas and principles. In *Nine Chapters of Arithmetic* there are basically no theorems concerning the topics of number theory, arithmetic or geometry. It tends towards solving a collection of different practical problems while Euclidean geometry from Greece is a way of deduction and generalization, which is more similar to rational thinking in contemporary mathematics. 算学 ("Subject of Computation") has been the synonym of mathematics for a long time in China. Under the influence of this tradition even today not a few Chinese adults still call mathematics "arithmetic". Although more and more teachers think that rules and formulas should be considered as useful tools or paths to the formation of important mathematical ideas and methods, there are many mathematics teachers who see mathematics as nothing but a set of rules and skills for solving all kinds of problems. Mastering the rules and skills is the first important task in general.

In accordance with Chinese culture, study itself is not an easy and light work but an arduous struggle. This idea could even be traced to Chinese pictographic characters of 教育 (education). 子 in the lower-left part of 教 indicates young people. The meaning of upper-left part is hard burden. The whole character of 教 means take hard burden to the shoulder of young people. The character of "育" means development. So 教育(education) presents the idea: young people grow and develop under the condition in which they make every endeavor to tackle tough tasks. However Confucius said in his 《论语》 (*Analects*): 学而时习之，不亦悦乎. Translated into English: "It is pleasant to learn and practice time and again". The pleasantness of study must come from best endeavor. As in the West, mathematics is commonly regarded by Chinese as an abstract and difficult course in school. But most Chinese students, teachers and even parents believe that mathematics is a subject every one could learn well since it is not intelligence but diligence that is essential for success. Although not every student is so clever to learn mathematics easily, most of them could learn it by working hard for success. One of the reasons people believe *Practice makes perfect* is that *diligence could remedy mediocracy* (Wong, N.Y., in press). Students are often encouraged to dedicate greater efforts to mathematics than to other courses. The ordinary way to engage them in learning is doing mathematics – solving problems according to rules and examples. As the Chinese idiom *Slow bird should fly earlier* points out, the greater efforts students devote to mathematics, the greater progress they will make. Diligent practice makes them grasp basic abilities to solve routine problems successfully.

A tradition about teaching in the syllabus of school mathematics in China is that *basic knowledge and basic skill* are emphasized as one of the goals. This phrase began from the beginning of the 1950s after the People's Republic of China was born in 1949. The syllabus has been revised several times in about 50 years, but the *two basics* idea about teaching has been kept until now. Since Chinese mathematics educators strongly believe that a solid base is the prerequisite of a stable skyscraper, it is really a distinctive characteristic of mathematics education in China. In school, the first thing for students is to build a strong, well-knitted mathematics foundation. All aspects and steps of mathematics education including the curriculum, teaching, review, exercise, homework and evaluation serve towards this goal. To put *two basics* into effect, the teaching process is usually deliberately organized to guarantee that teachers and students concentrate on concepts, theories, rules, skills and techniques. For further study, students grasp "two basics" proficiently through practical exercises in classroom and at home.

Frequent quizzes and tests are a driving force too to push students' work for "two basics" harder in manipulation and understanding.

Examination is a cultural tradition too in the long history of China. Tracing back to about 1400 years ago, the Emperor began to choose his officers by examination and a fixed system of officer selection was gradually created from then. All people who could be in the top of the examination list would be appointed as officers in some important positions. It appears equitable in a feudal age with a hereditary system. The principle of equity of the examination is confirmed today though we talk about it now in the area of schooling. Because of the large population in China students are confronted with serious competition for the chance of university admission. Under the pressure of examinations students hope to get higher scores and outperform others. Diligent practice of all learned knowledge and skills is an effective means for this purpose. "Education for exam" has been almost a directional convention for years. Undoubtedly students are driven by exams to engage in learning activities and this is reinforced by a large amount of practice so that their performance could be improved.

Summarizing this as "hardworking + rigid exam", the cultural traditions of Chinese mathematics education lead people to believe that routine practice is the efficient way for mathematics learning.

3. PRACTICE: TWO MEANINGS IN ACTION

Penetrating the behaviors in teaching and learning, *Practice makes perfect,* in Chinese 熟能生巧, is usually understood and interpreted in two ways or is even considered at two levels. The meaning of the character "熟" goes beyond "practice" or "do". It means both *familiarize with* and *be proficient at.* This subtle distinction is differentiated and leads to potentially different ways or even different levels in action. At the first level, it appears as repeating manipulation or real routine. Teachers merely let their students follow some rule mechanically to do many routine exercises such as $3 \times (-8) = -24$, $(-3) \times 8 = -24$ time and again.

It is worth noticing that experienced teachers always arrange their manipulation in a different way. They deliberately work out a set of systematic problems with hierarchy. For example, when we investigated in a school, to grasp the formula: $x^2 - y^2 = (x + y)(x - y)$, the whole domain of the symbols x and y is under consideration and to be exercised steadily by students, for example, $4a^2 - 9b^2$, $1/16s^2 - 25/49t^2$, $0.81x^2 - 0.0036y^2$, $a^2x^4 - b^6y^8$ etc., varied here in coefficients, letters and indexes. So they have different meanings in content.

Another example is in a case study of mathematics teaching. A lesson videotaped in a classroom in Shanghai was analyzed for how the teacher taught the concept of midpoint connector (a segment connecting the midpoints of two sides) of a triangle. We saw that after the concept was introduced, it is natural for the teacher to give students four questions with a system of more and more complicated diagrams (Figure 1): if the thick segment can be seen as the third side of all possible triangles respectively in the four figures, draw out all possible midpoint connectors in these triangles which have the thick segment as the third side. The left figure is the simplest case. Then the next two figures are more complicated. Students may be misled, for example, to draw a segment that is not a midpoint connector of any triangle in the figure. In the right figure three overlapped triangles shear the thick segment as the third side. Often students overlook the small triangle in the middle position and do not find out all possible midpoint connectors. The idea of designing this group of questions is to provide students with variant cases to imagine the midpoint connectors of triangles so that they are prepared to deal with complicated situations when they further their studies.

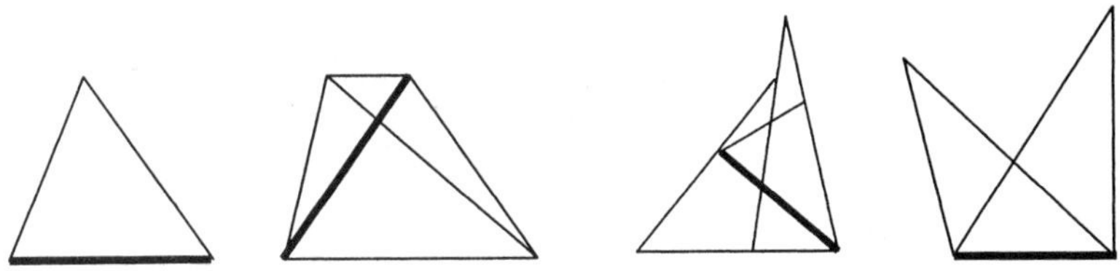

Figure 1

The other kind of change is about the method or strategy. For finding the power of a number, say 98^2, 998^2, 9998^2 ... , we saw experienced teachers begin with the usual way as 98x98 according to the definition of square. It is exactly routine. Then they would turn to the second stage to analyze these numbers and let students recognize the characteristics in these groups. Students were encouraged not to use direct calculation but to explore other ways to cope with these problems of finding the value of a square. Some would find the way to apply the formula $(x - y)^2$ to this context. $98^2 = (100 - 2)^2$, then $98^2 = 100^2 - 2\times100\times2 + 2^2 = 10004 - 400 = 9604$. Furthermore, more ingenious ways might be probed by some students:

$98^2 = 9604$

$998^2 = 996004.$

Then the following answers are conjectured:

$9998^2 = 99960004$

$99998^2 = 9999600004,$

and for the generalization

$999\ldots998^2 = 999\ldots996000\ldots004$

From here it clues students to see a similar way for computing powers such as 103^2, 999^3, 102^3, 104^4 etc.. This is a typical way of manipulation in Chinese classrooms called "一题多解"：solving one problem in variant ways.

Such a way of manipulation is usually called 变式训练, *variant manipulation* or *variation* in English. We call it meaningful manipulation – a transitional stage that lies halfway from simply repeated manipulation to understanding and has its distinctive function. It is *meaningful* because students are provided with a period of time and space to broaden their views to a rule or principle as much as possible. The mechanism within variant manipulation is usually called 举一反三，which means literally "to draw inference from reflection on one example". Students can experience themselves thoroughly in variant (just purely mathematical but not real) situations step by step so as to reflect on and to gain insight into what they practice. It therefore opens a door leading them from proficiency to understanding, i.e. lays a foundation for weaving a concept framework.

4. RESEARCH ON THE CONTROL OF OVERDONE

Undoubtedly manipulation is a sword with two edges since it may bring us a negative by-product too, even when it has a positive function. In the long run, being familiar with processes of mathematics (i.e., can compute, can use formulas and rules) is no more than the initial step. Immoderate practice may delay students' development of object formation (Li, 2000). Unfortunately some teachers may be afraid that their students could not use rules and formulas fluently, so they let them have too much assignment and make a long stay at the process stage. In contrast to the case in the West, a serious problem confronted with in China is that many students are often burdened with too much manipulation and their understanding and creativity are hindered. There is some research that has explored and reflected on the result of this kind of practice. One of the astonishing cases is an investigation into 79 college students who majored in mathematics in a small college. Subjects were asked to answer the following four questions (two items in every one):

1. Among all plane figures with same circumference, which one has biggest area? Why?
2. What is the relation between the hypotenuse and two arms in a right triangle? Why?

3. Can you compute the value of a determinant? Do you know what a determinant means?
4. Can you calculate $d(\sin x) / dx$? Why?

The finding is that all subjects know "how" but few know "why" (Table 1).

Table 1-6-1. The result of the investigation

	Question 1	Question 2	Question 3	Question 4
Know the first	100%	100%	100%	100%
Know both	0%	7.6%	1.2%	6.3%

To treat this phenomenon a project was undertaken by some mathematics educators in Jiangsu Province (Tian, et al, in press). Using both quantitative and qualitative approaches they have developed special assessment scales as guidance for manipulative skills, including polynomial operation, mathematical inference and geometric visualization and construction so far. The idea is that in practical teaching after a period of practical training, teachers could use these scales to evaluate the proficiency of students' performance. If a student or a class of students meets the corresponding criterion, it reminds teachers to stop training for this sort of practice. It means keeping within proper bounds and binding the sack before it is full. We hope such kind of research will be beneficial to the control of overdone manipulation in Chinese classrooms.

As an example, the scale for integral expression skill has 38 items and at least 75 steps (See appendix). There are different requirements in different grades (Table 2).

Table 1-6-2. Norm-referenced table for Integral expression skill

	Average Level		Pass Level		Excellent Level	
	Correct Items	Correct Steps	Correct Items	Correct Steps	Correct Items	Correct Steps
Junior 2	17.90	32.02	≥13	≥22	≥27	≥50
Junior 3	21.84	38.93	≥16	≥26	≥32	≥62
Senior 1	27.67	50.06	≥24	≥41	≥34	≥68
Senior 2	30.06	55.59	≥26	≥45	≥36	≥71

According to the Norm-referenced Table, teachers could choose appropriate level (pass, average or excellent) as a reference to evaluate their students in some grade. For instance, one in junior 3 (grade 9) could pass it when s/he gets more than 16 correct answers of items or 26 correct steps, while one in senior 3 (grade 12) should be correct in 28 items or 49 steps.

This presents a picture of the actual proficiency level of students in Jiangsu Province.

We would note that the test is timed and all items of the scale should be finished independently by students in 10 minutes. This reflects an aspect of point of view on skill proficiency from Chinese mathematics educators.

APPENDIX

Scale for Polynomial Skill

I. Simplifying (1 - 5):

1. $-3x^2y + 5x^2y =$
2. $(1/4)ab^2 - 2ab^2 =$
3. $(1/3)xy^2 \cdot (-6)x^2y =$
4. $6ab^2c \div (-9ac) =$
5. $(-3xy^2)^3 =$

Factorizing (6 – 8):

6. $x^{2m} - 9 =$
7. $x^2 - 3x + 2 =$
8. $y^2 + y + (1/4) =$

Making perfect square:

9. $x^2 - 3x + 1 =$

II. Simplifying (10 - 14):

10. $-(2/3)ab + (3/4)ab + ab$
 =
 =
11. $-y^2 - 2x^2 - (-3y^2)$
 =
 =
12. $3x^2y \cdot (1/2)x \cdot (-2xy^2)$
 =
 =
13. $3x^2y^3 \div (1/3)xy \div (-xy)$
 =
 =
14. $[(-2n)^2]^3$
 =
 =

Factorizing (15 - 17):

15. $(a+b)^2 - (x-y)^2 =$
16. $(x+y)^2 + 5(x+y) + 6 =$
17. $(m-n)^2 + 4(m-n) + 4 =$

Making perfect square:

18. $1/2\,x^2 + x + 3/2 =$

III. Simplifying (19 - 20):

19. $4x^3 - (-6x^3) + 9x^3$
 =
 =
20. $2a^2by \cdot (-1/3\,by) \div a^2by$
 =
 =

Factorizing (21 - 23):

21. $a^2 - ab + ac - bc$
 =
 =
22. $m^2 - n^2 + am - an$
 =
 =

23. $x^2 + 2xy + y^2 - z^2$
=
=

Making perfect square:
24. $x^2 + px + q =$

IV. Simplifying (25 - 30):

25. $-1/2\ ab^2 (b^2 + 3a^2b)$
=
=

26. $(x - 2y^2)\cdot(-2x^2y)$
=
=

27. $(m^3n + mn^2) \div 1/3\ mn$
=
=

28. $(a^3b^4c - 2ab^3c) \div (-2ab^2)$
=
=

29. $(-2ab^2 + a^2b + 3ab^2)^2$
=
=

30. $(-a^2b)^5 \div a^6b^2$
=
=

V. Simplifying (31 - 32):

31. $(4x^2y - 5xy^2) - (3x^2y - 4xy^2)$
=
=

32. $6ab^2\cdot(-1/3)\ ab^4 \div 2a\cdot(-ab^2)$
=
=

VI. Simplifying (33 - 38):

33. $2s^2t + 1/2\ s^3t^2 \div 3/2\ st + 2/3\ s^2t$
=
=
=

34. $ab^2c^2 - ab\cdot 1/2bc^2 - ab^2c^2$
=
=
=

35. $x^4y^2z - xy\cdot x^2y\cdot(-xy)$
=
=
=

36. $-m^5n^3 \div 1/3\ m^2n^2 \div mn + 4m^2$
=
=
=

37. $(21a^2b^2 - 35a^3b^3) \div 7a^2b^2\cdot 2ab$
=
=
=

38. $2x^5y^4 \div 1/2xy^2 + (-3x^2y^2)\cdot 2x^2$
=
=
=

REFERENCES

Commission on Standards for School Mathematics, 1989, *Curriculum and Evaluation Standards for School Mathematics*. National Council of Teachers of Mathematics, Reston, VA.

Hiebert, J. and Carpenter, T, 1992, Learning and Teaching with Understanding, *Handbook of Research on Mathematics Teaching and Learning*, D. Grouws, ed., Macmillan Publishing Company, pp.65-97.

Leung, F.K.S., 1998, The Traditional Chinese Views on Mathematics and Education – Implications for Mathematics Education in the New Millennium, *Proceedings of ICMI-EARCOME 1*, **1**:69-76.

Li, S., 2000, Does Practice Make Perfect, paper presented at the 9th International Congress of Mathematics Education (Regular lecture).

Mullis, I. et al., 2000, *TIMSS 1999 International Mathematics Report*, International Study Center, Lynch School of Education, Boston College.

National Council of Teachers of Mathematics, 1980, *An Agenda for Action*. The Council, Reston, VA.

Sfard, A., 1991, On the Dual Nature of Mathematical Conception: Reflections on Processes and Objects as Different Sides of the Same Coin. *Educational Studies in Mathematics*, **22**:1-36.

Tian, Z. et al., 2003, *Explorative Researches on Basic Knowledge and Basic Skill of Mathematics*. ECNU Press, Shanghai. (In Chinese)

Wong, N.Y., 1998, In Search of "CHC" Learner: Smart, Works Harder or Something More, *Proceedings of ICMI-EARCOME 1*, **1**:85-98.

Wong, N.Y., in press, *Conceptions of Doing and Learning Mathematics among Chinese*.

Zhang, D., 1998, Some Characteristics of Mathematics Education in East Asia – An Overview from China, *Proceedings of ICMI-EARCOME 1*, **1**:49-59.

Chapter 1-7

THE ORIGINS OF PUPILS' AWARENESS OF TEACHERS' MATHEMATICS PEDAGOGICAL VALUES: CONFUCIANISM AND BUDDHISM - DRIVEN

LEU Yuh-Chyn and WU Chao-Jung
National Taipei Teachers College, Taiwan, R.O.C.

1. INTRODUCTION

The teaching of mathematics involves mathematics, teachers, and students. These three elements are not value-free; in fact, they are value-carriers. For example, some students may carry the opinion that mathematics is only useful for the purpose of examination and they lack the motivation of learning mathematics. These kind of opinions or values are usually produced by the mathematics teaching at school. And, unfortunately, some values from students are not appropriate (Bishop, 1991).

Research on the values involved in the teaching of mathematics is in its infancy (e.g. Leu, 1999; Leu & Wu, 2002; FitzSimon, Seah, Bishop & Clarkson, 2000; Chin & Lin, 2000.). However there is little research on whether these mathematics pedagogical values from the teachers would influence the students. If the teachers' mathematics pedagogical values were to transfer to their students, and consequently to influence the students' values and behaviour, then an investigation on teachers' pedagogical values and whether these values are received by the students is highly worthwhile.

This paper explores students' awareness of their teachers' mathematics pedagogical values in two classes. One teacher is mainly influenced by Confucianism and the other by Buddhism. Therefore, we shall first introduce both the ideas of Confucianism and Buddhism so that the reader can understand the teaching behaviours of those two teachers. The influence of culture difference on mathematics teaching is demonstrated as well.

1.1 Confucianism

The Chinese people believe that the most important thing to be a human being is to seek the inner spirit of one's heart. As a result, the Chinese people emphasize inner benevolence and kindness as a way to achieve the completeness of morality.

In Confucianism, the seeking of the inner spirit can be realized intuitively mainly by experiencing and cultivating. It cannot be understood or proven by means of debate and logical inference. The practice of introspection and self-perception encourages the pursuit of perfection and the fear of making mistakes, which inhibits the development of rationality of human beings (Wu, 1995).

In Confucianism, the ethical system of benevolence, loyalty, and decorum was the system that kept stratum relationships during feudal times. For example, family members must obey the family elders and civilians must obey the monarchy emperor and his courtiers (Hwang, 1994). The emphasis on stratum of nobility and common people causes the lack of a spirit of equality and democracy.

Confucius, the founder of Confucianism, believes that everyone is educable. The Confucianists believe that education and learning are always associated with effort (Lee, 1999).

Success in one's life is highly valued in Confucianism. During the Suei Dynasty (581A.D.-604A.D.), the imperial examination system was established. Since its establishment, intellectuals studied day and night to memorize various Chinese Classics in order to pass the imperial examination. Those who passed the examination would attain positions or jobs in the imperial government as well as bring honours to their family (Hwang, 1994). Therefore, scholars would place examination scores as their first priority and would do everything to yield to the examination (Chang, 2001).

The imperial examination system was abolished soon after the founding of the Republic of China in 1912. Yet, the National Examination still remains as the main selection standard for employment in government jobs. This forces teachers and students to pursue good scores on examinations. The projection on mathematics teaching is test-oriented teaching and doing simulation problems is the major classroom activity. Everything is done to obtain high scores. Thus, the purpose of mathematics education is narrowed down to achieving high scores on examinations (Chang, 2001).

1.2 Buddhism

Originally inherited from India during the Han dynasty (25A.D.-220A.D.), Buddhism and its development had been influenced by Chinese culture along with a combination of certain aspects of Confucianism and Taoism. Buddhism got localized in China. Buddhism has one similar aspect with Confucianism, that is, there is no omnipotent supernatural creator existing beyond the universe. However, Buddhism encourages individuals to seek for emancipation, and to free oneself from the restraints of Confucianist dogma in order to reach self-actualization. Buddhism also emphasizes that self-improvement relies on oneself and enlightenment is controlled by oneself. It also stresses individual meditation. The goal is to become an enlightened being like Buddha and not like a god in the Judea-Christian tradition. On the other hand, Confucianism encourages self-realization for developing a standard of morals and ideals (Hwang, 1994).

As most of the residents of Taiwan are immigrants from the southeast of Mainland China, about a quarter of Taiwanese are followers of Buddhism. The other sixty-percent of the population in Taiwan is associated with multiple folk beliefs of Confucianism, Buddhism, Taoism and other various spiritual faiths. For example, the most well known Buddhist organization, Tzu-chi, has more than one million followers. The Tzu-chi Teachers Association has more than ten thousand members, representing about 5% of the teacher population. Buddhism thus has a huge hidden impact on Taiwanese society in general, and on education in particular (Chin, Leu & Lin, 2001).

2. TEACHERS' MATHEMATICS PEDAGOGICAL VALUES

2.1 The research methods and process

We work within a framework based on the theory of Raths, Harmin and Simon (1987), in which the value is defined as any beliefs or attitudes, activities or feelings that satisfy the criteria of choosing, prizing and acting. What satisfies the criterion of choosing is the belief or attitude chosen under free will, among several different options or after thoughtful consideration. What satisfies the criterion of prizing is the belief or attitude of cherishing, showing pride, or willingly making public. The belief or attitude that satisfies the criterion of acting is a performance acted out repeatedly.

The research subjects are two teachers, Ms. Chen and Ms. Lin, and their students. Before looking into students' awareness, we need to first introduce the two teachers' mathematics pedagogical values. Ms. Chen and Ms. Lin have been teaching at elementary school for 21 years and 9 years respect-tively. Ms. Chen is a devout Buddhist and is currently a mentor of an intern teacher. On the other hand, Ms. Lin considers herself as a Confucianist.

We defined the teachers' pedagogical values through classroom observations and follow-up interviews. Regarding Ms. Chen, we observed a total of 12 lessons of her mathematics teaching during a period of five months, and we interviewed her 12 times, with about 1.5 hours per interview. As for Ms. Lin, we observed a total of 10 lessons of mathematics teaching and we interviewed her 18 times. Classroom observations were aimed to find out the repetitive behaviours or the crucial events. The purpose of the interviews was to realize the reasons why these behaviours and crucial events happened and to formulate some assumed values that could be research targets. Through the interviews, we examined if the assumed value satisfies the criteria of "choosing" and "prizing".

2.2 The research results

In the following paragraphs that discuss the teachers' mathematics teaching behaviours and their related interviews, if they meet the criteria of Raths et al., they are explained in the parentheses. The teachers' repetitive behaviours (acting) are described first, and then the evidence of choosing and prizing is presented.

In Taiwan, more than 70% of the elementary school teachers don't know about the concept of constructivism or how to implement it in their mathematics teaching (Hue, 1997). This implies that most teachers in the elementary schools still conduct instruction by traditional lecturing. However, Ms. Chen conducts her lesson by raising questions for students to solve by themselves (acting).

Ms. Chen said that, "Traditional lecturing can only teach students how to memorize mathematical knowledge. The new teaching method, however, encourages students to provide an explanation for their own problem-solving strategy. By doing so, the students have better grasp of the concept but the drawback is that their calculation ability is weaker. Nevertheless, I am more for the new teaching method." From this above-mentioned quote, it showed that Ms. Chen chose her teaching method after understanding and evaluating the advantages and disadvantages on emphasizing the understanding of concepts or the practice of rote calculation (choosing).

Ms. Chen's teaching had two purposes. One was to teach students the content of mathematics. She once told her students' parents, "When solving problems, to get the correct answer is important. But what's more important is to get the reasoning right." From the way that Ms. Chen would publicly state her idea of emphasizing the understanding of mathematics, it demonstrates that she upholds this value (prizing). The other more important purpose was to edify students' attitude toward mathematics problems and problem-solving. She said, "I encourage students to apply the spirit of solving mathematics problems to daily life. That is, when confronted with real life problems, they should not give up, even if no one ever taught them how to solve the problems." From the way Ms. Chen placed much more emphasis on the attitudes of problem-solving than the learning of mathematics knowledge, it is evident that she cherishes this value (prizing).

Ms. Chen always asked her students to concentrate during the class time (acting). The purpose of her acting is to make her students to have tranquil minds and steadiness so as to unfold wisdom and reinstate students' enlightenment. Besides, when students solved mathematics problems incorrectly, Ms. Chen would ask them to perform self-reflection (acting). With this practice, Ms. Chen offered her students the opportunity to think further and to reinstate their enlightenment. Ms. Chen stated that "Confucianism only teaches things in this present life, while Buddhism elaborates a person's past, present, and future. It's all clear... I want to reinstate students' awareness. The informative knowledge learned daily can't keep us from misery and mishap." (choosing and prizing). Ms. Chen set her goal of education to reinstate students' original enlightenment because she was a devout Buddhist.

Ms. Chen seldom let the class answer as a whole or discuss in groups (acting). If other students told the appointed student the answer, or if the appointed student peeped at other students' answers, he/she would be stopped by Ms. Chen (acting). She said, "I scarcely do group discussion because I couldn't distinguish whether individual students learned or not."(choosing). Ms. Chen once told the intern teacher publicly: "Don't ask students the questions that are supposed to be answered by the class as a whole or you won't be able to distinguish who gets it from who doesn't."(prizing). Ms. Chen's value, that is, mathematics learning depends on individual efforts and apprehension, is consistent with one of the concepts of Buddhism, "self-improvement depends on oneself."

The other teacher, Ms. Lin, had her own teaching behavioural patterns in mathematics including systematically asking students to preview, asking students to answer testing questions, neglecting wrong solutions, and emphasizing review (acting).

As Ms. Lin was asked to compare the advantages and disadvantages of the teacher-guided and the student-centered teaching approach, she answered that the merit of the former one is that "The leaning speed and effect is faster as a whole"; the drawback is that "some students become less concentrated sometimes." As for the latter teaching approach, Ms. Lin thought that it's good because "...the easily-made mistakes were usually presented by students and other students could learn that that kind of problem-solving strategy was wrong."; and it's deficient because "...the schedule may be delayed. And some students were more impressed by the wrong solving strategy." Finally, Ms. Lin chose to teach in the former way (choosing).

Ms. Lin said, "Going over textbooks three times is better than going over them twice; twice is better than once." These words revealed that she stressed the idea that students must gain much knowledge from textbooks (prizing).

Therefore, "teaching students the mathematics knowledge in the textbook in order to obtain good scores" is defined as one of Ms. Lin's mathematics pedagogical values. This value was profoundly influenced by Taiwan's examinations culture. From a survey result, 87.7% of elementary school teachers asked students to do mathematics problems as skillfully as they can, and 96.8% of elementary school teachers make up questions for mid-term examination according to textbooks and workbooks (Ku, 2000). We believe that this value is likely to be a popular mathematics pedagogical value in Taiwan.

Why does Ms. Lin neglect student's wrong solutions? She said, "It's because I corrected some wrong writing of Chinese characters on the blackboard but many students copy the wrong writing instead of the correct one. Later on, I learned to encourage students to find right answers. Then they won't copy the mistakes. There won't be any negative effect." Another reason is the consideration of students' self-esteem. Some students might feel ashamed if their mistakes are shown to the class. In Taiwan, the percentages of elementary school teachers who hold some similar opinions are 30.37% and 21.48%, respectively (Hue, 1997).

3. PUPILS' AWARENESS OF TEACHERS' VALUES

3.1 Research methods and process

The research is aimed to understand the pupils' awareness of Ms. Chen's mathematics pedagogical values and behavioural execution of these values, through a questionnaire and interviews. The items in the questionnaire were designed from a video-taped teaching section of Ms. Chen. All of the 36 students in Ms. Chen's class completed the questionnaire and 8 of them were interviewed. We edited Ms. Chen's recorded teaching and selected sections that were best representatives of her pedagogical values. After playing each short section, the pupils were asked to answer questions about the section. Here is a sample item in the questionnaire: "For what reason do you think Ms. Chen gives quizzes in class?" The questionnaire included six questions and it took about half an hour to complete. The questions in the interviews were mainly aimed to investigate students' awareness of Ms. Chen's mathematics teaching behaviour based on her values (such as the sample question 1 stated below), to examine which values her students received from Ms. Chen were more important: to acquire knowledge, to deal with people and life, or to reinstate enlightenment (sample question 2), etc.

Sample question 1: How does Ms. Chen teach the mathematics lesson?
Sample question 2: During the two-year period when Ms. Chen is your teacher, what is the most impressive or memorable things/words?

The researchers would ask further questions according to the contents and reasons of students' answers. Every student was interviewed for about half an hour.

However, we did not adapt the video taping of Ms. Lin's mathematics teaching because it consisted of more traditional lecturing that is less characteristic. After interviewing Ms. Chen's students, we tried to transform the interview questions into a questionnaire for the 25 students of Ms. Lin, in the effort to understand how they became aware of Ms. Lin's pedagogical values. This questionnaire was composed of 8 items and it took about 40 minutes to complete. Some sample questions are the following: "What do you think is the purpose of Ms. Lin's mathematics teaching, to learn how to do the math questions in the textbook, to learn how to solve the math questions in the real life, or other aims?"; "What do you think Ms. Lin values more, your grades of mathematics, your learning attitudes of mathematics, or other aspects?"

Due to the situation that some data collected from the questionnaire were quite different from the result we had observed, we then decided to further examine those data by the means of interviewing Ms. Lin's students. We also collected new data at the same time. Fourteen students were interviewed. Some sample questions of the interview were, "What learning attitude does Ms. Lin emphasize? Why do you think so?"; "How does Ms. Lin teach in mathematics lesson?". Every student was interviewed for about half an hour.

3.2 Research results

The research results showed that these two classes of students were quite aware of their teachers' pedagogical values or behavioural execution of these values. For example, when students were asked the question, "Which does your teacher emphasize more, grades or learning attitudes?", most students thought their teacher emphasized learning attitudes more than grades. But the contents of each teacher's emphases were different.

The eight Ms. Chen's students interviewed all thought that she heavily emphasized learning attitudes. The learning attitudes Ms. Chen emphasized included the aspect of examination culture, such as "work hard on homework" and "think more on one's own." Students would say the attitudes toward life and self-correction were also involved. For example, some students commented "Don't stop solving problems just because you think you can't." and "Don't be afraid of making mistakes, just correct the mistakes and try not to repeat them." These were the attitudes Ms. Chen emphasized about self-improvement.

Ms. Chen once said, "I want to help students reinstate their enlightenment. That's because the knowledge learned during youth cannot protect from sorrow and misery. I think scores are important but attitudes toward life are much more important." These statements not only declared Ms. Chen's belief in Buddhism but also revealed that Ms. Chen couldn't exclude the mainstream tide of emphasizing examination scores in traditional social culture. That is why Ms. Chen's students were aware that she emphasized good attitudes toward examination culture and Buddhism.

In the questionnaire, when asked if Ms. Lin emphasized their learning attitudes, 24 out of 25 students answered "Yes". Only one student left the question unanswered. When the students were asked the question, "What attitudes does Ms. Lin emphasize?" Ms. Lin's students' answers were quite consistent. The attitudes Ms. Lin emphasized were concerned with examination culture. Nine students talked about the attitude: "We, as students, must preview, review and correct the workbook." Five students mentioned the attitude of "being serious and concentrated in class." Four students commented that "Students are encouraged to turn to the classmates or the teacher with questions." When her students were asked the question "Why did Ms. Lin emphasize these attitudes?", Ms. Lin's students answered, "Good learning attitudes lead to good grades," and "The teacher doesn't want us to get good scores by mere memorizing." From the above information, we concluded that Ms. Lin's students were aware of her pedagogical value that she heavily emphasized mathematics scores. However, the reason that some answers in the questionnaire were "Ms. Lin emphasized learning attitudes" instead of "Ms. Lin emphasized mathematics scores" was that Ms. Lin often reminded her students that she was about to grade their learning attitudes at the end of each semester.

Let us now see how students became aware of the way the two teachers taught. Ms. Chen's students had similar answers, such as "Ms. Chen would give us questions and we must write down the procedures of solution and answers on our own. We cannot allow others to see our answers. Then Ms. Chen would ask two classmates to demonstrate their solutions and answers on the blackboard. The whole class would check the answers on the blackboard and compare with our own answers." Another student answered, "Ms. Chen always asks us to do the problems on our own instead of discussing with the classmates." The students' description of Ms. Chen's teaching was consistent with our classroom observation. When researchers asked, "What happens if you discuss with classmates before writing down the solution?" the students answered firmly, "Of course not, then the teacher won't know if you get it or not." Another answer was "The teacher doesn't want us to be influenced by other's thinking." It is apparent that her students had received the information that Ms. Chen emphasized individual learning and having their own problem solving strategies. Through the questionnaire, we found that 72% of the students were aware of Ms. Chen's value that she prized students to learn mathematics on their own.

Ms. Lin's students answered differently than Ms. Chen's students but their answers were congruent. For example, one student said, "Ms. Lin wants us to read out loud the textbook problems and she would explain the meaning and solving strategies of the problems. Then we solve the problems on our own and check the answers. If we have any questions, we raise our hands and then Ms. Lin would explain." Only one lesson out of the ten observed lessons proceeded through group discussion. Only two students out of the fourteen students we had interviewed mentioned that Ms. Lin would do group discussion in mathematics class. The way students described Ms. Lin's teaching method revealed that her students' were aware of Ms. Lin's expectation that students must learn the mathematics knowledge from textbooks.

Using the same questions for two different classes of students, we could see most students were aware of the characteristics of their teacher's teaching and provided different answers.

When the researcher asked, "What happened if a student's solution on the blackboard was wrong?" Ms. Chen's students answered, "Ms. Chen would ask the demonstrator whether he/she knows where they had gone wrong. If he or she didn't know, then the teacher would have a class discussion, and the demonstrator should correct the error." The answer was consistent with Ms. Chen's own description that "I won't tell what's wrong; otherwise students might lose the sense of awareness." Ms. Lin's students said, "If any student said anything wrong, the teacher would clarify again."

One question in the interview was "If the teacher finds mistakes when supervising in the classroom or when correcting students' homework, would the teacher discuss the errors on the blackboard?" Ms. Chen's students answered, "Ms. Chen would discuss some odd or wrong problem solving strategies on the blackboard." Ms. Lin's students said, "Rarely. That's because telling us the correct solutions are more helpful to us. If we discuss about wrong ones, we could absorb nothing good." Other students said, "Not really. If so, isn't it very embarrassing?" Another student's answer was "No! The student who made the error would be laughed at by other kids. Then he/ she might lose confidence in mathematics." The answers showed those students of two classes were aware of their teacher's way of dealing with students' errors respectively.

4. CONCLUSION AND DISCUSSION

4.1 Conclusion

This research found that culture deeply influences the mathematics teaching behaviours and pedagogical values for the elementary school teachers and what is more important was that their students were aware of these behaviours and values. Even though the students from both Ms. Chen's and Ms. Lin's class thought that their teachers valued learning attitudes more importantly than grades, the contents and aspects of learning attitudes were different. Ms. Chen's students were mostly aware of the attitudes of self-reflection and self-correction, the attitudes related to reinstate enlightenment. On the other hand, Ms. Lin's students were mostly aware of the attitudes of previewing and reviewing the lesson and being concentrated in the class, the attitudes related to getting good grades.

There were 72% of Ms. Chen's students who were aware that their teacher would like them to solve the problems by themselves, not by group discussion or asking other students. Some of her students were also aware that Ms. Chen would like the students to find out the mistakes in their problem-solving strategy and correct them by themselves. As for Ms. Lin's students, 86% of them were aware that their teacher helped students learn through lecturing. And her students were aware that the reasons that Ms. Chen didn't like to present the wrong solutions were the concern of students learning wrong solutions and hurting students' self-confidence.

4.2 Discussion

4.2.1 The aspect of cultural influence on teacher's mathematics pedagogical values

In teaching mathematics, Ms. Chen emphasized that students should solve problems on their own, while Ms. Lin emphasized that she should present the solution to students. The difference would be that Ms. Chen believed in the concepts of Buddhism that self-improvement relies on oneself. Ms. Lin adopted much lecturing because of the influence of Confucianism.

The stratum relationship stressed by Confucianism is presented in the "teacher-centred" lecturing method in school education. The relationship between teachers and students is stratified. So the communication is one-way. Modeling is one important approach of learning in Confucianism. In Confucianism, people are encouraged to observe others' behaviours. Then they should imitate the good behaviours of others but avoid and correct bad behaviours (Lin, 1999). Therefore, Ms. Lin explains the fast, correct solutions to the students and lets them learn and imitate. That is the exact presentation of modeling.

Ms. Lin and Ms. Chen had one thing in common: they seldom allowed students to discuss in groups. But mathematics education in the United States has emphasized that instruction should vary to include opportunities for discussion between teacher and students, and among students themselves (NCTM, 1989). In England, they have stressed "opportunity for greater cooperation and mutual support among those who teach mathematics in school" (Cockcroft, 1981). Raymond (1997) found that among her research subjects, 4 out of 6 beginning elementary teachers used 'primarily non-traditional' mathematics teaching or 'all nontraditional' mathematics teaching. It is 2/3 of all subjects. Non-traditional mathematics teaching includes students learning mathematics only through problem-solving activities and students learning mathematics through cooperative group interactions. The difference in teaching method possibly resulted from the influence of western culture.

Western culture upholds the values of equality and democracy. Under this kind of cultural influence, student-centred teaching, including group discussion and cooperative learning, is more likely to dominate.

When the mistakes occurred as students wrote their solutions on the blackboard, Ms. Chen expected her students to find the errors by themselves. But Ms. Lin would explain the correct solution herself. When mistakes were found while supervising in the classroom or correcting students' homework, Ms. Chen would discuss the mistakes on the blackboard, while Ms. Lin neglected students' mistakes. The different ways of dealing with students' errors were due to the different values each teacher upholds. Ms. Chen believes that she should provide the opportunity for self-reflection which would reinstate students' enlightenment. The reasons that Ms. Lin didn't discuss the students' errors were not only to avoid the negative impact but also to avoid hurting students' self-confidence. That's because she was influenced by Confucianism which encourages the pursuit of perfection.

How about the western teachers? Some teachers like to demonstrate the correct solutions. Negative instances are not allowed to appear either on the blackboard or on the overhead screen because "...poor learners could pick them up and store them!"(Bauersfeld, 1992). The fact that Ms. Lin doesn't want her students to copy the wrong solution shares this common point of view. Some teachers discuss the wrong solution with their students because "...the discussion of what is missing or wrong with a negative example can contribute to the student's constructive rendering and contouring of the desired result." (Bauersfeld, 1992). Whether or not the teachers discuss students' mistakes, western teachers focus on the learning of knowledge. But Ms. Lin takes student's self-esteem and losing face into consideration. What concerns Ms. Lin more is the seeking the inner spirit.

4.2.2 The aspect of pupil's awareness of teachers' values

This research found that students could be aware of their teachers' mathematics pedagogical values. Nevertheless, there is one premise of this research, that is, the teachers and students in this research are living under similar cultural influences. What would the result be if the teachers and the students are living under different cultures, in the cases such as the teacher or the student migrates from one culture to another culture? Would the students still be capable of becoming sensitive to their teacher's mathematics pedagogical values? Would there be any conflicts in values between teachers and students? These are worth further investigation.

This research demonstrates that the methods of questionnaire and interview are suitable for investigating sixth-grade students' awareness of their teachers' mathematics pedagogical values. Are these two specific methods of questionnaire and interview applicable to younger school children? This is another question that needs to be further addressed.

5. IMPLICATIONS

Recently, elementary mathematics education in Taiwan has been actively reformed. The purpose of mathematics education has shifted from knowledge retention to the improving of students' ability in solving problems, communicating and reasoning. The method of teaching has changed from traditional lecturing to group discussions. Will the reform be successful eventually? Tye (2000) proposed five main reasons for failures in education reforms and three of them are related to culture and tradition. They are "the social context: conventional wisdom about schooling", "parent and community expectations" and "the complexity of teachers' lives."

This research has found that Confucianism and Buddhism would influence elementary teachers' mathematics pedagogical values. Students could be aware of these values as well. If the purpose of mathematics education in Taiwan is to change or cultivate values, then it would take more than just passing down the traditional values. Therefore, how to change or cultivate teachers' mathematics pedagogical values through teacher education is becoming more important. Meanwhile, it is a profound subject for further research.

Acknowledgements: Thanks is given to the teachers Ms. Chen and Ms. Lin, who were kind and open in sharing their ideas with us. We thank our assistant Ms. I. H. Hong, who did a great job on the audio-video tape transcriptions. We are grateful to the National Science Council (NSC) for funding the projects (NSC88-2511-S-152-002; NSC89-2511-S-152-001). The opinions expressed do not necessarily reflect the views of NSC.

REFERENCES

Bauersfeld, H., 1992, Classroom Cultures from a Social Constructivist's Perspective, *Educational Studies in Mathematics* **23**:467-481.

Bishop, A.J., 1991, Mathematical Values in The Teaching Process, in: *Mathematical Knowledge: Its Growth Through Teaching*, Bishop et al., eds., Dordrecht: Kluwer.

Chang, D.C., 2001, *Mathematics Education with Fine Quality: Set a Firm Base and Enhance creativity* (July 12, 2001); http://www.schoolnet.edu.mo/msnvc/vc013/b16.htm.

Chi, M., 1993, Joy-oriented Culture and Sin-oriented Culture—the Discussion on the Difference Between Christianity and Chinese Confucianism, *Philosophy magazine*, **3**: 130-140.

Chin, C., and Lin, F.L., 2000, A Case Study of a Mathematics Teacher's Pedagogical Values: Use of a Methodological Framework of Interpretation and Reflection, in: *Proceedings of the National Science Council Part: Mathematics, Science, and Technology Education*, **10**(2):90-101.

Chin, C., Leu, Y.C., and Lin, F.L., 2001, Pedagogical Values, Mathematics Teaching and Teacher Education: A Case Study of Two Experienced Teachers, in: *Making Sense of Mathematics Teacher Education*, F.L. Lin and T.J. Cooney, eds., Dordrecht: Kluwer Academic Publishers, pp.247-269.

Cockcroft, W.H., 1981, *Mathematics Counts*, London: Commercial Colour Press.

Elliott, D., 1999, Learning to Think: Asians are Trying to Prepare Kids for the Information Age. Can Creativity be Taught? *Newsweek*, September, **6**:38-41.

FitzSimon, G., Seah, W.T., Bishop, A.J., and Clarkson, P., 2000, Conceptions of Values and Mathematics Education held by Australian Primary Teachers': Preliminary Findings from VAMP, in: *Proceedings of the History and Pedagogy of Mathematics Conference*, National Taiwan Normal University, Taipei.

Hue, C.W., 1997, Elementary Teachers' Attitude toward a Constructivist Approach of Teaching. Nat. Chengchi University, *Journal of Education and Psychology*, **20**:55-70.

Hwang, K.K., 1994, Self-realization and Value Change in Chinese Society, in: *Chinese Values—from the View of Social Science*, K.S. Yang, ed., Laureate Publishers.

Ku, M.F., 2000, Research on Elementary School Teachers' Attitude Toward Mathematics Assessment, *Research Report of National Hsin-chu Teachers College*, **13**:1-38.

Lee, W.O., 1996, The Cultural Context for Chinese Learners: Conceptions of Learning in the Confucian Tradition, in: *The Chinese Learner: Cultural, Psychological and Contextual Influences*, D.A. Watkins and J.B. Biggs, eds., Hong Kong: Comparative Education Research Centre & Australia: Australian Council for Educational Research Ltd, pp.25-41.

Leu, Y.C., 1999, The Influences of Buddhism on an Elementary Mathematics Teacher's Professional Development, in: *Proceedings of the National Science Council: Part D, R.O.C.*, **9**(3):65-75.

Leu, Y.C. and Wu, C.J., 2002, *Mathematics Pedagogical Value System Oriented Toward the Acquirement of Knowledge in Elementary School*, Paper presented in the 26th Conference of the International Group for the Psychology of Mathematics Education, Norwich.

Lin, H.Y., 1999, Mothers' Beliefs, Goals and Child-Rearing Behaviors (I): An Analysis of the Themes, *Research in Applied Psychology*, **2**:143-180.

National Council of Teachers of Mathematics (NCTM), 1989, *Curriculum and Evaluation Standards for School Mathematics*, Reston, Va.: The Council.

Raths, L.E., Harmin, M., and Simon, S.B., 1987, Selections from Values and Teaching, in: *Value Theory and Education*, P. F. Carbone, ed., Malabar: Krieger.

Raymond, A.M., 1997, Inconsistency Between a Beginning Elementary School Teacher's Mathematics Beliefs and Teaching Practice, *Journal for Research in Mathematics Education*, **28**(5):550-576.

Tye, B.B., 2000, *Hard Truths: Uncovering the Deep Structure of Schooling*, New York: Teachers College Press.

Wu, K.J., 1995, How the Difference between Eastern and Western Humanism Influence the Attitude of Education, *Chinese Culture Monthly*, pp.43-64.

Section 2

CURRICULUM
Introduction

Margaret WU[1], PARK Kyungmee[2] and LEUNG Koon Shing Frederick[3]
[1]The University of Melbourne; [2]Hongik University ; [3]The University of Hong Kong

In comparing mathematics education in different cultural traditions, one important source of information is the comparison of curricula in different countries. The mathematics curriculum of a country may be viewed as a centrepiece in the whole picture of mathematics education. At one end, political and cultural traditions have an influence on the establishment of a curriculum. At the other end, the curriculum has a direct impact on student learning and performance. In this study, the comparison of curricula goes well beyond a superficial mapping of similarities and differences of curricula across countries. A closer examination of the determinants of curriculum and the consequential impact of curriculum is the focus of this section, particularly in relation to East and West cultural traditions. Furthermore, an examination of the curriculum of a country is not only carried out through scrutinizing curriculum documents of a country. The curriculum of a country is also reflected in the textbooks used, in the approach to the assessment of mathematics, and in the performance of students in international studies.

1. DETERMINANTS OF CURRICULUM

The mathematics curriculum of a system is shaped by many factors, including political impetus, cultural traditions, and influential individuals working in the field of mathematics. The first paper (Chapter 2-1, Bessot and Comiti) in this section examines the historical impact of mathematics education on current day curricula in France and Vietnam. These two countries had a link in history for about one hundred years from the second half of the 19th century. Yet there are some differences in the approach to teaching mathematics in the two countries. Bessot and Comiti used the topic of vectorial geometry as an example to illustrate the differences in mathematics education in the two countries, and traced the differences to political and historical factors. Bessot and Comiti further linked student performances to the various emphases in the respective curricula.

The paper by Wu and Zhang (Chapter 2-2) takes a critical look at the determinants of mathematics curriculum. Not only are the determinants identified, but questions are also raised about relative merits, problems, and the impact of the determinants on mathematics education. For example, a surface classification of whether the curriculum is centrally determined or decentralised is not a useful one. One needs to look into the implemented curriculum in the classrooms to assess the degree to which the curriculum differs within a country. Further, irrespective of whether the curriculum is determined by policy makers or educators, the tension between the two groups is always present. How does one resolve the competing demands of stakeholders of education? Through a survey of curriculum development in a number of countries, a deeper understanding is gained about the determinants of curriculum. In particular, cultural traditions can be seen as playing an important role in shaping the curriculum development.

2. MATHEMATICS TEXTBOOKS

A great deal can be learned about the intended and implemented mathematics curriculum from the way textbooks are organised, presented and used in the classrooms in a country. Three papers in this section examine and compare mathematics textbooks used in the East and the West.

Li and Ginsburg (Chapter 2-3) compared the classification of mathematical knowledge, and the degree to which content areas are segregated, in textbooks used in Hong Kong, Mainland China, Singapore and the United States. They found that in Asian countries, textbooks contain clearly separated topics, while in the United States, the mathematics content within each chapter of a textbook is more heterogeneous. Further, teachers in China, Singapore and Hong Kong are required to teach according to the textbooks, while teachers in the United States are encouraged to bring in knowledge from outside the classroom in presenting their lessons. Li and Ginsburg linked these findings to cultural traditions of the East and the West, in that a positional compliance ideology dominates the East (where everyone has his/her own place in the society/home), while a contractual compliance ideology is the norm for the United States (where there are negotiating processes between the individual and the authority). Consequently, the curriculum differences between the East and the West is not simply a matter of choice or preference, but they relate to deeper cultural issues and therefore would be difficulty to change overnight.

Yeap, Ferrucci and Carter (Chapter 2-4) compared the portrayal of arithmetic problem solving in textbooks used in Singapore and the United States. Mathematical problem solving has gained a great deal of attention in these two countries in the past two decades. This paper examines three aspects of problem solving: whether problem solving processes are emphasized; whether the role of problem solving is merely a vehicle for teaching mathematics, or whether the role of problem solving is for the sake of teaching problem solving. Yeap et al found that in American textbooks, problem solving processes are more explicitly presented than in Singapore textbooks. However, in the United States, problem solving tends to be used as a tool for teaching mathematics, while in Singapore, problem solving is taught for the sake of teaching problem solving. Further, problem solving material is more structured in Singapore in that isomorphic problems are presented and grouped semantically (by task demand), while in American textbooks, problems tend to be grouped by contextual theme rather than by mathematical content. Interestingly, Yeap et al found that, while there are marked differences between the presentation of problem solving in text-books, there are no distinctive differences between curriculum statements on problem solving in Singapore and the United States. They concluded that the authors of the Singapore textbooks have followed the traditions of the East in implementing the formal curriculum statements into classroom practice.

The third paper (Chapter 2-5, Park and Leung) on the comparison of textbooks examines various features of mathematics textbooks used in China, England, Japan, Korea and the United States. Five aspects of the textbooks are considered: uniformity versus diversity in terms of publication policies, essential versus discretionary in terms of the choice of content, absolute versus relative in terms of the role of textbooks in teaching and learning, plain versus colourful in terms of physical appearance, and lastly, the organization, style and approach in terms of the content. Park and Leung found that textbooks in the United States are more diverse, contain more elective topics, and cater for the individual by including high and low-level problems. On the other hand, textbooks in Asian countries are used as the teaching material, and look less attractive in appearance. In terms of content, American textbooks tend to be thematic and spirally structured, while East Asian textbooks tend to be linearly structured by mathematics strands. Park and Leung related the observed differences in textbooks to dissimilar social and cultural values, and suggested that the textbooks may in turn reinforce underlying cultural differences. From this point of view, textbook developers have an important task of finding the best balance to benefit students in ways that are consistent with the cultural values of the country.

3. ASSESSMENT OF MATHEMATICS

Two papers in this section look into the assessment of mathematics, and how the process of assessment can reflect cultural differences between the East and the West. Wu (Chapter 2-6) compared the results of two international surveys: the Programme for International Student Assessment (PISA 2000) and the Trends in International Mathematics and Science Study (TIMSS 1999). The two surveys differ in the emphasis of curricular content in the test instruments. PISA has a "literacy" orientation and is not curriculum based, while TIMSS uses the curricula of participating countries as the foundation of the mathematics framework. A comparison of results showed that the group of East Asian countries performed relatively better in TIMSS than in PISA. To gain a better understanding of the results, an item-by-item examination was carried out. Some distinctive differences were observed in the performance of East and West countries at the item level. In particular, Western countries performed better on items involving less formal mathematics, items with verbal response and items with interpretations of graphs. In contrast, Asian countries performed better on items with numerical computation and items involving algebra. These results reflect different emphases of mathematics education in the East and the West.

The second paper on assessment (Chapter 2-7, Zhao, Mulligan and Mitchelmore) reported a study carried out to compare mathematics assessment practices in Australian and Chinese primary schools. The study found that assessment practices in these two countries are quite different, where Australian schools adopt an outcome-based approach while Chinese schools adopt a content-based approach. In addition, the assessment practice in Australian schools is based on portfolios of work samples, while Chinese schools use formal tests and examinations for assessment. However, in both China and Australia, the implemented assessment practices do not conform to what is intended. The intended assessment encompasses a much wider scope of different assessment methods. Through interviews with parents and teachers, Zhao et al linked the respective assessment practices to different cultural values in East and West societies. For example, even in Australia, parents from China still hold the same view as Chinese parents in China. This is an indication that cultural background has a deep root and has a profound influence on the expectations of assessment practices.

4. PROCESS OF CURRICULUM REFORM

While it is helpful to identify curriculum differences between the East and the West, a critical review of the process of curriculum reform is also of great importance. The last paper in this section (Chapter 2-8, Bienvenido) provides us with a valuable lesson about the trials and tribulations of curriculum reform through several decades of curriculum changes in the Philippines. Owing to the lack of resources and expertise in the Philippines, mathematics curriculum reform has been carried out by foreign consultants and advisers through projects funded by World Bank. Typically, Western curricula were brought into the Philippines, with insufficient time for trialling and evaluation before a full implementation was carried out. The end result was that little improvement was seen in students' mathematics achievement over the decades. Bienvenido pointed out that the main reason for failure was that the reforms did not originate at the school level where teacher input was paramount in the successful implementation of a curriculum. Bievenido concluded that a country should not blindly copy another country's curriculum. Rather, foreign curricula should be used as "mirrors" to reflect strengths and weaknesses of a country's own curriculum. The best curriculum for a country is one that takes the cultural values of the country into account, and one where there is ample consultation with local teachers and schools.

Overall, the papers in this section not only provide informative observations and useful research results, they also provide a great deal of food for thought. With globalisation, the curricula of different countries are becoming more similar. But the findings in these papers remind us of the importance of cultural values in each country, and the consequences of disregarding cultural roots when we endeavour to bring about curriculum reform.

Chapter 2-1

SOME COMPARATIVE STUDIES BETWEEN FRENCH AND VIETNAMESE CURRICULA

Annie BESSOT[1] and Claude COMITI[2]
[1]Leibniz Laboratory and University of Grenoble; [2]Leibniz Laboratory and IUFM of Grenoble

1. MAIN FEATURES OF THE GENESIS OF THE TEACHING OF MATHEMATICS IN FRENCH SCHOOLS

1.1 From the end of the 15th century to the end of the 18th century: the emergence of mathematics teaching[1]

During this period, mathematics was divided into two parts: pure mathematics, (made up of arithmetic, geometry and algebra) and mixed mathematics (Mechanics, Optics, Astronomy, Geography, Chronology, Military Architecture, Hydrostatics, Hydraulics, Hydrographics or Navigation etc., according to the list given by d'Alembert).

This distinction between pure mathematics and mixed mathematics extended the field of mathematics. Moreover, the role played by mathematics in society increased. Although at first limited to the area of economics, mathematics progressively became part of other institutions, as is demonstrated by two indicators. Firstly, certain contributions were made by mathematicians, such as Descartes' publication on the art of weighing,

[1] Ce paragraphe s'appuie sur Artaud (1998)

Huygens' work on clocks, or Pascal's invention of an arithmetic machine "for calculation without pain and without knowledge". Secondly, the social need for mathematics was evident through the number of works published and the quality of those involved in this publishing activity.

Hence, throughout this period, the social visibility of mathematics increased, founded on a solid cultural and social foundation, arising from the structure of the world of science. This world was made up of several concentric spheres. In the centre were the great scientists such as Galileo, Descartes, Pascal, Huygens, Newton, Leibniz – and later, Bernoulli or Euler: they defended their discoveries jealously and sometimes aggressively, and only communicated among themselves indirectly or in the privileged context of highly select meetings. A crowd of people revolved around them: the organizers of scientific life, who acted as their secretaries (like for instance Father Mersenne in the first half of the 17th century), and also individuals who busied themselves in the corridors of science: people who used Mathematics, who published books about mathematics written by other people, as well as mathematics tutors.

1.1.1 How did the teaching of mathematics emerge in this context?

The teaching system was, at the time, made up of two essential components, universities and religious colleges, as well as the *Collège Royal*, the forerunner of the *Collège de France.* Alongside these two components, there existed "small schools", which were mainly directly dependent on town mayors and clergy, colleges dependent on religious congregations, and specialised schools such as military schools. The university was made up of four faculties: the faculties of law, medecine and theology to which one had access through the faculty of liberal arts, the equivalent of today's secondary system. During the 16th century, Jesuit colleges emerged. They closely resembled the arts faculties, which declined as the Jesuit colleges gained in influence. These colleges provided eight or nine years of study : grammar, rhetoric and dialectic took up the first six years, the last two or three years being given over to the study of philosophy. It was in these colleges that the first chairs of mathematics were created in the early 17th century, and here that the first real teaching of mathematics was introduced. Most of the great mathematicians were educated there (Descartes, for example, attended the College de La Flèche). At the same time the need for the expansion of the State, through the army and the navy, increased this movement through the creation of royal chairs of mathematics and hydrography from the end of the 17th century: it was necessary to train military and naval officers by teaching them the theory of arithmetic, the sphere and rules of navigation. These chairs were held by Jesuits until their expulsion from France in 1762. In the last decades of the 17th century, this process of extension, which sprang from the energy of the Jesuits, encountered the development and consolidation of the State and its hold over land and sea.

However, of the forty Paris universities, only one dispensed a "real" mathematics course: Mazarin college (where Legendre studied and presented a thesis on infinite calculus in 1172).

Hence it can be said that the teaching of mathematics began to emerge at the end of the 18th century.

1.2 From the 1789 Revolution to the mid-20th century

The main ideas concerning children's education emerged shortly before the Revolution, and were formulated in different registers of grievances. These included:

- The organisation of an educational system designed to educate the citizens of the country,
- The extension of teaching to both sexes and to all social classes,
- The creation of a teaching body,
- A central agency to oversee the setting up of a plan throughout the kingdom.

Professional schools, training for knowledge specific to certain professions, were no longer enough to ensure the educational needs of a growing number of citizens.

The idea of education for all went along with the dislocation of religious congregations which had control of schools. The promotion of a scientific culture became a means to oust the elites, especially religious elites whose teaching was directed towards "humanities" based essentially on the learning of classical languages, mainly Latin.

Communal *lycées* and *collèges* were created in the early 19th century under the Consulate.

Under the 3rd Republic, Jules Ferry, the Minister for Public Instruction (1879-1883), set up a secular, free and compulsory primary school system and opened up secondary schools to girls[2]. At that time primary and secondary education were two parallel and opposed teaching orders: the former was free, taking in children of both sexes from all social classes, with utilitarian objectives, the latter was fee-paying, accepting only boys from the

[2] loi de 1882, Art.4

elites which made up the traditional ruling class, and with humanist objectives.

It was not until the beginning of the 20th century[3] that mathematics was accepted as an educational discipline, on the same level as classical languages, and "scientific humanities" were placed alongside classical humanities in secondary teaching.

1.3 From the 1950s to the end of the 20th century

Only a minority of pupils had access to scientific courses until the mid 20th century when higher primary schools (middle schools) were integrated into the secondary system and the secondary system became free. This period was characterised by the emergence of a new educational model that was based on the notion of cause and effect between scientific development and a country's economic development, making the training for the scientific and technical careers needed by France a priority.

Higher education was thoroughly reformed, the curricula of mathematics and science were modernised, and the curricula of human sciences (economy, sociology...) were modelled on those of the exact sciences. This general change improved the standing of mathematics and science in *lycées* and *collèges*. The country was subsequently confronted with the problem of democratic access to secondary education: how were the knowledge and skills which the secondary system were designed to teach to be defined?

1.3.1 The role of the development of research in education in the French educational system

After the failure of the modern mathematics reform in the 1970s, the IREM (*Instituts de Recherches sur l'Enseignement des Mathematiques*) were created, and a community of researchers emerged whose focus was didactics of mathematics – a science of the conditions of the spread and teaching of mathematical knowledge useful for human institutions. This line of work investigated the ways in which knowledge was structured in the thinking of scientists, of students, and within particular situations. Theories of didactics sought to discern and articulate patterns in the ways these structures overlapped and influenced one another. Some studies focused on "didactical transposition" (Chevallard, 1985), others on pupil's scientific knowledge within particular conceptual fields (Vergnaud, 1990), and others on "didactical situations" (Brousseau, 1997).

[3] loi de 1902

One of the effects of the development of this field of research was new views of mathematics teaching and of the teacher's role, and the emergence of crucial questions for the development of new curricula and teaching methods.

The results of research in mathematics education, and the creation of a body of knowledge about the learning of mathematics at primary and secondary school, influenced the sphere of people reflecting on and influencing mathematics, the orientations of the national programme of studies and the organisation of teaching in secondary school (Laborde, 2002):

- Practice of a scientific approach (mathematics as a place for experimenting, modelling and formulating)
- Interplay between points of view and ways of expression and representation
- Use of Information and Communication Technologies
- Interdisciplinary activities and projects
- Mastering mathematical techniques and importance of proof

2. MAIN FEATURES OF THE GENESIS OF THE TEACHING OF MATHEMATICS IN VIETNAMESE SCHOOLS[4]

2.1 The feudal era and the Mandarin competitive examinations

Vietnam gained its independence in the Xth century when it threw off Chinese feudal domination. The Ly dynasty was the first royal dynasty which achieved a centralised monarchy ruling over all the Vietnamese territories, and had to face three major problems: the building and maintenance of a widespread network of dams intended to control the floods of the Red River Delta, the safeguard of National Independence and the tackling of rural struggles.

Therefore, the Ly dynasty set up a centralised administration and, in the XIth century, created the first Mandarin competitive examinations. Their purpose was to recruit high-ranking civil servants; competitions were open

[4] Thanks very much to Professor Ngo Thuc Lanh for authorizing us to use his paper not again published intituled *Vai net ve lich su giao duc toan hoc o Vietnam* (Some features of the teaching of Mathematics in Vietnam).

to everyone except actors and women; the first examination took place in 1075 [5]. Regional juries awarded titles to the candidates who were subsequently allowed to sit for imperial examinations to obtain a "doctorate". The rate of success in regional examinations was very low: barely a hundred out of several thousand candidates. As regards the doctorate, the title was awarded to slightly more than 2000 men in ten centuries. Regional and imperial laureates could be appointed "civil servants" and could become "Mandarins", thus belonging to a body controlling the entire administration of the kingdom.

The monarchy dominated a cluster of very lively rural districts whose collective life was led at a cultural and ideological level by a Confucian *élite* composed of the unfortunate candidates who had failed the State examinations and became preceptors, masters of ceremonies, public writers or schoolmasters.

2.1.1 School and education

One school, One master; the principle applied everywhere; it was paid for by well-off families and left to private initiative; classes were devoted to reading, writing, reciting classical texts by heart, commenting on the traditional themes of the Confucian philosophy, and composing poems in view of the State examinations.

The tests intended to recruit the Mandarins included literary, moral, political essays, writing a poem and official texts, but generally ignored Mathematics, even if the book *Daiu Viet Su Ki Toan Thu*[6] mentions some questions on the subject (years 1077, 1261, 1373, 1404, 1437, 1475, 1477, 1483, 1507, 1722 and 1762). Those tests, probably devised to recruit scribes in charge of taxes, salaries, buildings, mostly dealt with dividing quantities into equal or proportional parts, as well as area or volume calculations. The papers[7] included relevant data but also unnecessary information connected with the real world. The candidates were supposed to answer the questions but equally to comment on the superfluous data unlinked to Mathematics. Basically they had to achieve a literary work.

Furthermore, in this largely rural country whose inhabitants were mostly farmers, the *élite* remained unfamiliar with production problems. In a word,

[5] The last one took place in 1919.

[6] Compilation of historic annals, from Le Van Huu (1912) completed by Ngo Si Lien (1479) and, in the XVIIIth century by Pham Cong Tru and Le Hi.

[7] As we can find them in Pham Huu Chung book published at the beginning of the 1700's.

it can be said that before the colonial era, there was neither teaching nor training of Mathematics at school.

2.2 The colonial era

In the second half of the 19th century Vietnam had to face up to French Colonial Aggression; whereas the Monarchy and the Mandarins compromised with the Enemy, the Confucian *lettrés* (the learned elite) roused the farmers and organised a form of resistance. However bloody defeats inflicted by a foe using modern weapons spelt the end of Confucianism. Meanwhile Chinese translations of Rousseau's and Montesquieu's works brought in notions of science and democracy unknown to Confucian philosophy.

Around the beginning of the 20th century teaching in Chinese was abolished. All the schools of Indochina must use French. New generations of students were moulded in the French schools; they learnt Physics, Algebra, Biology, Geography as well as the Republican Constitution and electoral laws. But they could not play the part of the Confucian *lettrés*, because they were not closely connected enough with the common people. In those days, Education was limited and obscurantist; more than 90% of Vietnamese were illiterate. Elementary schools were few: from 5 to 6 per district or province. The number of *collèges*[8] found in some large towns and provinces amounted to 20. The whole country did not number more than 5 *lycées*[9]: two in the North, one in the Centre and two in the South, out of which two were reserved to students from French origin, or from Mandarin and wealthy families.

The syllabus for Mathematics was the French one, more or less adapted for primary and secondary schools. The textbooks came from France.

The Indochinese Communist Party was created in 1930; Marxism came to Vietnam as a tool for independence, after the failure of the anti-feudal and anti-colonial struggles led by the Confucian *lettrés* and an elite belonging to the middle-class. Thus Marxism would succeed Confucianism while acknowledging it and the work of the past elite as a positive national legacy to be integrated into the new society.

Higher Education was only opened in 1940, since French students could no longer go to France, because of the war. In those days only a few bright grant holders, reached a University level in Mathematics. The number of Vietnamese students who had graduated in Maths hardly reached 10, and only 2 of them obtained a Masters. At the end of their *cursus*, they became teachers in secondary Education. Among them, Pr. Hoàng Xuân Han,

[8] Lower secondary schools

[9] Upper secondary school

student at the Paris *Ecole Polytechnique*, regarded as the founder of the Vietnamese scientific culture, suggested a translation of scientific terms into Vietnamese. In 1942, he published a glossary *Danh Tu Khoa Hoc* which would allow the textbooks in Mathematics, physics and chemistry at all levels to be written in Vietnamese in the future.

2.3 The time of independence (after 1945)

2.3.1 The French influence on the evolution of Vietnamese teaching of mathematics

From April to June 1945, Pr Hoàng Xuân Han, Minister of Education and Arts supervised the curriculum of secondary schools in Vietnamese. This curriculum bears his name. The Math syllabus was common to both sections A (Classical literature) and B (Modern literature). In fact, it took up the syllabus of the *lycée* of colonial times, though modified on some points.

After the revolution of August 1945, the new curricula were modified to face the needs and difficulties of the time and eliminate what was regarded as "against democratic institutions". In wartime in 1950, school attendance was reduced from 10 to 9 years. Concerning Mathematics, it led to the suppression of several subjects of the French syllabus: Probabilities, Analytic Geometry, Analysis, Arithmetic, Descriptive Geometry, Trigonometry, Mechanics and Astronomy. This light syllabus would remain in force until 1956.

2.3.2 The influence of the Soviet system of education

From the fifties, the Soviet Union and the communist countries supported the training of students and scientists in various subjects, among others in Mathematics. They offered Vietnamese students numerous grants for higher Education (BAs, Masters, PhDs).

In 1956, Education in North Vietnam was reformed for the second time to make it closer to the Soviet curriculum prevailing at the time[10]. In Vietnamese high schools the teaching of Mathematics was inspired by the Soviet system and the textbooks were influenced by the Russian experiences. Since schooling in the Soviet Union only lasted 10 years the syllabus in Mathematics was lighter than the French one. For instance, Arithmetic was studied in forms 5 and 6, but was absent in the top form (which is not the

[10] This reform did not reach South Vietnam over American influence at that time.

case in France). Furthermore, the secondary school curriculum did not include Analysis, Mechanics or Astronomy. However the Vietnamese maintained the elements of Analysis such as functions, limits, derivatives and primitives in the top form (10). In the North, this 10-year curriculum close to the Soviet one would remain unchanged from 1956 to 1980.

Similarly, the teaching of Mathematics at university level, introduced in 1951 at the time of the Vietnamese resistance, whose aims had been widened after Geneva's agreement in 1954, was now largely influenced by Soviet directions and experiences.

2.3.3 Years of reunification (after 1975)

In 1975, the war was over and Vietnam was reunited. Mathematicians from the South trained in western countries such as France, USA, Germany, along with those trained in the Soviet Union and in Eastern countries now built together a Vietnamese culture in Mathematics.

3. BRIEF COMPARATIVE STUDY OF ORGANIZATION OF SCHOOL AND OF MATHEMATICAL PROGRAMS

3.1 Two centralized systems

In France, as in Vietnam, the teaching is organized at a national level, in particular for each subject matter the number of hours as well as the programme of studies is decided at a national level and is followed in all schools across the country.

In the two countries, the school system is now divided into three parts

- Elementary school: 6 year old to 11 year old students, grade 1 to 5
- Lower secondary school (called *Collège* in France): 11 year old to 15 year old students, grade 6 to 9
- Upper secondary school (called *Lycée* in France): 15 year old to 18 year old students, grade 10 to 12.

In France, there is only one unified type of class for all students from the age of 6 to the age of 16, called *tronc commun*, so mathematics teaching is the same for all students until they have completed grade 10. The number of hours varied over time and is currently smaller than it used to be. Teachers complain about the lack of time for allowing a deep learning even if the content to be taught has been rather decreasing in size.

In Vietnam, the government promotes the students "gifted" in sciences and particularly in Mathematics from the beginning of lower secondary schools (11 year old), where selection is carried out from marks obtained during primary *cursus*: when they attend upper secondary schools, they can be recruited, on a competitive selection basis, in gifted classes opened in the best upper secondary schools, in which the teaching staff is carefully selected and the number of students reduced compared to normal classes.

It is only when attending grade 11 that French students start following different tracks: they may choose then between Literature and Language orientation, Economic and Social orientation, and Scientific orientation, in the general *lycées*, or between four tracks, Tertiary Science and Technology, Industrial Science and Technology, Laboratory Science and Technology, Medical and social Sciences in the technological *lycées* or between several tracks in vocational *lycées*.

In France as in Vietnam, secondary school ends with a national examination called in France *Baccalauréat* (success rate was 78.3% in 1999 in France and 90% in 2002 in VN). But if, in France, only the general or technological *Baccalauréat* is needed to enter university, in Vietnam, students need to pass competitive exams to enter universities and the success rate is very low: about 18.6%.

3.2 The two national programs

3.2.1 France [11]

The programme of studies in France has been changing at a regular pace (about every ten years) since the modern mathematics reform at the beginning of the seventies. This programme is designed by a group of experts: mathematics teachers, mathematics teacher-educators, inspectors, as well as mathematics university lecturers.

As Laborde (2002) writes: "A specific epistemological approach and learning hypothesis underlies the national curriculum (...): mathematics by essence is abstract, and to acquire mathematical concepts, students must themselves achieve these abstraction processes by manipulating, experimenting and observing. Mathematical objects will be constructed by students

[11] d'après C. Laborde (2002)

as emerging from these experiments in a process of abstraction, eliminating all irrelevant aspects linked to the context of emergence. (…)”

In order to help teachers to improve students learning, the programme of studies is not simply the presentation of the main objectives and orientations of mathematics teaching, and the list of notions to be taught at each grade level. It also provides documents with comments and examples of classroom activities giving indications and trends about ways to organize the teaching in order to favour learning with developed examples.

3.2.2 Vietnam

In Vietnam, the system of Education recently underwent the two following reforms: in 1990, Scientific, Techno-scientific and Literary final forms were created at the beginning of upper secondary school; vectorial method in geometry and introduction to computational science were introduced in form 10 and analytical geometry, basics of combinatorics, of integral calculus, and mathematical statistics and probability in form 12.

In 1998, the distinctions between these three sections were abolished and a return was made to a final unique form before University. However this latest reform is partly formal, since competitive examinations giving access to the University are based upon the former distinctions; thus upper secondary schools distribute the students among pseudo-sections, with a common curriculum but applied at different levels.

A new curriculum is currently being developed: it plans a return to the division of pupils into two streams: science and humanities from 10th grade.

The aim of this third reform is to lighten theoretical input by removing contents and techniques considered to be too complex and replace them with activities and problem solving.

In class, the teacher uses a single textbook, supplemented by a teacher’s guide and a workbook of related exercises with corrections.

This reform seems to focus more on teaching content and organisation than on the transformation of teaching practices.

4. TWO COMPARATIVE STUDIES AT THE SECONDARY LEVEL

4.1 Introduction and methodology

In spite of specific genesis, the present Vietnamese mathematics syllabus seems similar to those of Western Countries, but the arrangement of the content and the approach to particular topics could be very different because of this genesis.

If we consider curriculum as a process which goes from official text (Official Curriculum) to the implementation in the classrooms and effective learning by students (Real Curriculum), the main factors which influence this process are those attached to the educational system actors: teachers, pedagogical advisors, who act directly on student learning. So we are interested in analysing the means provided for this process development and the "products" of process.

Therefore we will try to propose this comparative study not only in terms of educational strategy, but also and mostly in terms of produced and observed teaching and learning in the classroom. We can make the hypothesis that the distortions which appear between real curriculum and official curriculum are the key to certain problems and questions which must be identified and understood.

However, it is not enough to gather data in the field to understand the origin of the problems encountered. Our methodology is based on the conducting and linking of three studies connected to the process of didactic transposition (Chevallard 1985):

- **An epistemological study** of the mathematical concept at stake. This study allows the characterisation of the problems which give meaning to the concept, the relative position of elements of this knowledge in the wider conceptual field in which it is contained, and also the variability of this data in relation to periods and institutions. Its aim is to determine the conditions which allowed the passage from one stage to another in the historical evolution of the knowledge or on the other hand, the obstacles which slowed its development.

- **A comparative study of the historical evolution of curricula and textbooks.** This study draws out and discusses the stages of the processes which culminated in the state of the knowledge to be taught in the two educational systems (France and Vietnam) in the period studied. Successive, often local, modifications and reorganisations, were essentially due to constraints within the teaching institution.

- **An experimental study** of students' and teachers' personal relationship to the knowledge at stake, designed to test the hypotheses which were formulated through the two preceding studies, about the institutional relationship to knowledge in each country, as well as foreseeable difficultties. These personal relationships are due both to constraints inherent to the institution studied, and the interactive process of the negotiation of the knowledge between the teacher and the students, a process which is conditioned by the culture and traditions of the countries studied.

The research below has been conducted[12] in this context:

- Didactical and Epistemological Study about Teaching Vector at Upper Secondary School, in France and in Vietnam (Lê Thi Hoai Chau[13], 1997)
- A didactic study of links between functions and equations in French and Vietnamese Upper Secondary Education (Lê Van Tien[14], 2001)

We now develop some of this Studies' results and articulate them with history and underlying values of Vietnamese society.

4.2 Introduction of vectorial geometry in form 10, and the subsequent results on student's performances

4.2.1 Context of the study

Very briefly, we shall say that the French approach to Mathematics is to divide the subject between "pure" and "applied"; emphasis is laid more on its formative value than on its social usefulness.

Mathematics in secondary schools should basically train the students "to learn a scientific approach through experimental abilities, power of reasoning, imagination and acute critical analysis (...) abilities, organisations and

[12] These studies have been conducted in the context of a co-operative activity between Laboratory Leibniz (University of Grenoble) and the mathematics department of the University of Ho Chi Minh.

[13] PhD (Grenoble and Vietnam Universities) under the co-direction of Tran V. T. and C. Comiti .

[14] PhD (Grenoble and Vietnam Universities) under the direction of Nguyen B. K. and A. Bessot

communications (...)"[15]. Regarding the usefulness of Mathematics, it is simply mentioned casually as if it were taken for granted.

In Vietnam, absolute abstraction unconnected with reality goes against the principles of Vietnamese Education. According to uncle HO's maxim "training and practice are always linked", the Vietnamese schools always attempt to link theory and reality.

The choices for introducing the study of vectors in form 10 seem to have been due to this pragmatic approach. It took place in 1990 when the curricula were greatly modified and vectors appeared for the first time in the first year of the upper secondary school as a new Mathematical object.

We shall use this example of vectors being studied in form 10 to compare the weights of different cultural and Mathematical traditions in both countries. The real question will be to account for the difficulties encountered by French and Vietnamese students tackling vectors for the first time: are they due to differences in culturally different teaching methods or "more simply" induced by the intrinsic difficulty of the subject?

4.2.2 Comparing the French and Vietnamese approaches to introducing the teaching of vectors in the 1990's

In France, the starting point is translations (which are not introduced as a transformation of the whole plane, but only through its effects on specific geometrical figures). The object 'vector' is inextricably linked with information made up of three inseparable components (direction, orientation and length) through which this translation is entirely defined. In this case, the equality vectorAB = vectorCD means that the first point of each pair (A,B) and (C,D) has the second for its image through the same translation. Therefore, right from the beginning, the vector does not appear through only one of its representations. This choice allows the algebraic nature of the geometrical vector to be taken into account while it diminishes the risk of confusing a vector and one of its representations, an essential step towards the idea of class of equivalence.

The authors of the Vietnamese syllabus and textbooks dismiss the French approach to the teaching of vectors on the following ground: "The French definition of vectors is too far from the notions of Force and Speed in Physics. Therefore, the students will be confronted with difficulties when having to apply vectors in Physics"[16]. The vector is basically regarded as an oriented segment, both ends of which are defined. According to the authors:

[15] Extract from the official comments of the french curriculum.

[16] Van Nhu Cong and al., p.10

"Even if this definition is imperfect, it does not harm the following knowledge of geometry whereas it favours the understanding of such notions of Physics as force and speed", the main aim of vectors being thus: "to build up a tool to study geometry without trying to deepen this knowledge as a Mathematical object"[17].

Yet, in order for skills on vectors to be an effective tool in studying geometry, it is necessary that one be able to calculate with vectors which have different starting points. With the given definition, there is the problem of the determination of the result of vectorial operations. Therefore, it is clear that if one wants to introduce vectorial methods in geometry, it is indispensable to introduce free vectors (or classes of equivalence of equipollent bi-points or directed line segments).

In spite of these different approaches, each country maintains the three basic features of a vector: direction, orientation, length.

In French textbooks and syllabus, a vector direction is clearly connected with the direction of two parallel lines or one straight line (slope). Furthermore, the vector is linked with a translation, and the vectorial equality is expressed by a parallelogram; that helps the students to distinguish between the "orientation"[18] of a vector, and the meaning of "orientation" (*sens*) in common language.

On the contrary, Vietnamese students are not taught what a vector direction is. When meaning that two vectors AB end CD have the same "orientation" they are reduced to saying that: "Lines AB and CD are parallel or merged, and the orientation from A to B is similar to the one from C to D". To justify this choice, the Vietnamese authors put forward the following reasons: "there is no need to mention the 'orientation of a vector' because 'orientation', 'same orientation', 'opposite orientation' are familiar to students. These notions belong to their daily experience: two friends go together from home to school and come back from school to home. They have met those situations in common life as well as in lower secondary school problems"[19].

4.2.3 Are the students' performances influenced by these different approaches?

We shall compare the results of the same test dealing with vectors equality, between Vietnamese students (form 10) and French students (form 10).

[17] ibid. p.11

[18] Called *sens* in French

[19] Van Nhu Cong and al., p.12

Both groups seem reluctant to leave the metrical approach when confronted with the first question; nearly 25% of the students tested in any Group[20] uses length equality as the only vector criterion. However, it is more obvious with the Vietnamese students than with the French ones (comparing vector lengths): numerous Vietnamese students who supply a correct answer to this first question come back to a logic entirely based on lengths when the following questions of the test become more complex.

The historical study of the concept of vector and of the growth of vectorial calculus shows how metrical geometry hindered the genesis of vectorial calculus. It can be guessed that this type of error has epistemological roots: whatever the different types of teaching, it is not easy to move from scalar magnitudes to the notion of orientation of vectorial entities, and the difficulty cannot be avoided. Nevertheless the Vietnamese choice (introducing the notion of vector from the oriented segment) seems to worsen the mental process out of the metrical model.

Results show that 65% of the Vietnamese students and 32% of the French ones have not properly mastered the criteria of direction and orientation; for instance, they compare the orientation of two vectors whose directions are different, deciding on the equality of 2 vectors because: "they have the same orientation", this orientation being as well 'from left to right' as 'from top to down' or 'that of the hands of a watch, or confuse the two criteria in their answers – for example, proving the non-equality of two vectors by: "because they have neither the same orientation nor the same direction".

The epistemological study[21] carried out in the thesis shows how it is possible to move from the metrical model to the vectorial model in two stages. The first one will lead the student from the metrical model to a uni-directional oriented model including two orientations on the same direction. This process is similar to moving from the natural numbers model to the Integers one. The second stage will be shifting from an oriented uni-directional model to the vectorial model, it requires giving up the image of the straight line and accepting the concept of several directions oriented and their links in a plane or in space.

Thus, the concept of "orientation" should come first, before the concept of "direction". To give up the metrical model, it is necessary to "comprehend" the oriented straight line first. Then the oriented uni-directional model

[20] 28% of Vietnamese students n the standard forms and 11% in the "Maths specialised forms" versus 36% of French students which will access to non scientific form 11, but 0% we shall find next year in Scientific form 11.

[21] Lê T.H.C. 1997, chapitre B1, pp.79-108.

must be overcome to reach the vectorial model including all the oriented directions. Integrating both direction and orientation in a single concept raises very complex epistemological problems.

Again, it can be said that epistemology accounts for the difficult transfer from the scalar magnitudes to the features of orientation of the vectorial magnitudes. However, the study shows the importance of the didactic choices, linked to cultural and mathematical tradition in each country as we see above, which help to overcome this difficulty, since in the selected samples, the number of Vietnamese students in trouble was (at the time of our study) twice as high as the one found among French students.

4.3 Graph in analysis in secondary schools in France and Vietnam

Teaching in both countries can be analysed through two reforms: 1980 in France, 1990 in Vietnam.

4.3.1 The French "counter-reform"

The French "counter-reform" shows a sharp break with the 1970 reform inspired by Bourbaki. Mathematics is no longer regarded as a world of structures, but as a field where problems are solved. In terms of teaching it means promoting the experimental feature of Maths first, and stressing the need for the linking of an experimental and a theoretical stage, regarded as distinct. The experimental stage highlights the student's activity regarded as essential, in agreement with the constructivist theories of the moment. To the minds of the people who initiated the counter-reform, experimental activity must be grounded in problem solving and observation of numbers and/or graphs; it must emerge into terms expressing conjectures which will therefore have to be demonstrated later on. Such an approach changes the part played by drawings[22] in the course of teaching: in the 70's they were scarce and restricted to an ancillary role of illustration, interpretation, or synthesis of theoretical results. With the counter-reform, they are massively brought in at the level of the part they are expected to play in experimental activity.

In this context, solving equations appears to be a privileged ground for approximation methods which assign a major role to function graphs and

[22] As graphical representations of functions in Calculus or drawings in Geometry.

numbers. Then Calculus is understood by the leaders of the counter-reform as the realm of approximation and techniques of *majorations – minorations*[23].

However, coming into this Calculus supposes a break from the Algebraic approach and serious difficulties are to be foreseen before carrying out such a teaching. In this connection, Legrand (1993) mentions an epistemological obstacle. French curricula seem to take these difficulties into account:

- some themes involving *majorations – minorations* have been suppressed and approximation complexity has been reduced in upper secondary schools,
- the notion of experimentation is altered by the growing importance of numerical and graph records, having lost their conjectural nature.

We shall now try to see how these observations affect the practice of both students and teachers when they tackle equation solving and examine the part assigned to the graph in this operation.

4.3.2 What are the Vietnamese orientations?

The basic epistemological choices of the reform of the years 1990[24] are neither those of the French reform of modern Mathematics, nor those of the counter-reform. Its aim is to promote the methodical feature of the Mathematical universe founded on rational rules: "To form and develop among the students the abilities of logical, algorithmic, and dialectical thought through providing them with basic knowledge of methods and rules of reasoning, demonstration, (algebraic) calculus, resolution of equations, transformations of algebraic expressions, graphic representations of functions, as well as some laws of dialectic…"[25]

With such a context Calculus is greatly influenced by Algebra. The institutional relation to the concept of equation is built on the algebraic resolution of fundamental equations. In such a Calculus the graph plays a minor part in the study of elementary functions; it provides a synthesis of results obtained theoretically and helps to visualise the properties of the function studied.

[23] Setting an upper bound - setting a lower bound.

[24] Lê V. T. analysed official documents as syllabus, pedagogic guide of handbook, the book of Nguyen B.K. et al. (1994) for 4th form of upper training schools. He also interviewed M. Tran Van Hao, mathematician and leader of a team of handbooks' authors during the 1990 reform.

[25] Extract from the comments of 1990 curriculum

However, problems including graphic resolutions of equations are now found in the competitive exams for upper secondary school entrance (though the subject does not appear in 9th year textbooks); therefore we may predict its unofficial teaching at the college level.

The study of both school systems leads us to wonder about the part played by graphs in student's and teacher's real practice of Calculus. We will simply restrict the question to the study of the role of the graph in demonstrating the existence of the solutions of an equation[26].

4.3.3 Results of a survey conducted among teachers and students of forms 11 and 12

The survey shows clear differences about the influence of graphs on the student theoretical justifications between French and Vietnamese schools.

France

Most French teachers reject the graph as a sole means to validate a demonstration; however the majority of students (74%) regard the graph when it is present in the terms of the paper as an opportunity to use it to demonstrate or justify. It is obvious that some students' answers demonstrate that the "graph" method and the "theoretical" one ("bijection theorem") are equally valid in their eyes.

One of the possible explanations for this observation could be the alteration of the didactic notion of experimentation. In accordance with institutional attempts, the French teacher regards the graph as a tool for illustration or conjecture, but never uses it for a demonstration; he trains the students to be wary of the graph as a proof when he has the opportunity of doing so. But the present curriculum offers few opportunities to grant results obtained by graphs the nature of mere conjectures. Experimental activity is reduced to the benefit of graphic records and theoretical activity is reduced to the benefit of generalisation of terms; therefore, students are more formally and less strongly taught to beware of graph results as a means of demonstration.

[26] In France, the study took place after teaching the notion of derivative and the "bijection theorem": "If a function f is derivable and strictly monotonically decreasing or monotonically increasing in a closed interval [a, b] and if f(a).f(b) <0, then it exists one and only one number c in]a,b[at which f(c)=0". The notion of continuity is missing in Syllabus. In Vietnam, the study took place after teaching the notion of continuity and the "intermediate value theorem": "If a function f is defined and continuous in a closed interval [a, b] and if f(a).f(b) <0, then there is at least one number c in [a,b] at which f(c)=0."

Vietnam

Graphs are much less used to prove the existence of equation solutions than in France. Many students make a link between graph records and demonstrations. However, the influence of graphs on justifications is increasing (though it is much less than in France: 52% against 97%), when the graph is available, and the equation to solve is of the form $f(x) = k$ ($k \neq 0$) instead of the usual form $f(x) = 0$.

When the graph is not available and when the "intermediate value theorem" is unable to provide a complete solution, 31% of the students use the graph: it means that they regard the graph as an available substitute for missing or latent theoretical data.

The use of graphs as a validating tool shows a contradiction between Vietnamese teachers and students, which is not the same as in France.

Whereas the majority of Vietnamese teachers are willing to accept the graph as a total or partial proof of the existence of solutions of an equation in a standard situation, Vietnamese students will be reluctant to use the graph as a demonstration.

This paradox can be accounted for by the minor part played by graphs in Vietnamese curricula, especially in the study of the link between functions and equations. In fact, graph resolution of equations is taught in lower secondary schools; in upper secondary schools the students are familiar with them, and they have made the acquaintance of the graphs. But in Vietnam, lower secondary school and upper secondary school teachers are trained in different training schools. Thus there is a possibility for upper secondary school teachers to know little about the use of graphs in lower secondary teaching (applied to the resolution of equations). They are silent about a break they are not aware of. The students may translate that silence about the graph as an official prohibition banning it as a validating tool.

REFERENCES

Artaud, M., 1998, Conditions et contraintes de l'existence des mathématiques dans l'enseignement général, *Petit x*, n°50, IREM de Grenoble, pp.23-38.

Brousseau, G., 1997, *Theory of Didactical Situations in Mathematics*, Kluwer.

Belhoste, B. and al., 1996, *Les sciences au lycée. Un siècle de réforme des mathématiques et de la physique, en France et à l'étranger.* Vuibert, Paris.

Chevallard, Y., 1985, *La transposition didactique*, éd. La Pensée Sauvage, Grenoble.

Dorier, J-L., 1997, *L'algèbre en question.* Coll. Travaux et thèses en Didactique des Mathématiques, éd. La Pensée Sauvage, Grenoble..

Hoàng, Xuân, H., 1942, *Danh Tu Khoa Hoc.*

Laborde C., 2002, The Mathematics Curriculum in France at the Beginning of the New Century, Chongqing Conference, China.

Lê Thi, H.C., 1997, *Didactical and Epistemological Study about Teaching Vector at Upper Secondary School, in France and in Vietnam.* PhD, University Joseph Fourier and University of Viêt-nam.

Lê, T.H.C., 2001, Difficultés d'apprentissage de la notion de vecteur pour des élèves de première année de lycée en France et au Viêt-nam. *Recherche en Didactique des Mathématiques*, 21(1.2.), éd. La Pensée Sauvage, Grenoble, pp.157–188.

Lê, V.T., 2001, *A didactic study of links between functions and equations in French and Vietnamese Upper Secondary Education.* PhD, University Joseph Fourier and University of Viêt-nam.

Ngô Thuc, L., in press, *Vai net ve lich su giao duc toan hoc o Vietnam.*

Nguyen, K.V., 1962, Confucianisme et marxisme au Viêt-nam. *Expériences vietnamiennes.* Paris, pp.201-232.

Pham, H.C., beginning of the 1700's, *Cuu chuong lap thanh tinh phap.*

Nguyen, B.K., & al., 1994, *Méthodologie de l'enseignement des mathématiques – Enseignement des contenus mathématiques fondamentaux.* Maison de l'Education, Hanoi, Viêt-nam.

Legrand, M., 1993, Débat scientifique en cours de mathématiques et spécificité de l'analyse. *Repères* – IREM, **10**:9–12.

Van Nhu, C., et al., 1990, *Géométrie de 10e*. Maison de l'éducation, Hanoi, Viêt-nam.

Vergnaud, G., 1990, La théorie des champs conceptuels, *Recherches en Didactique des Mathématiques,* 10 (2.3), éd. La Pensée Sauvage, Grenoble, pp.133-170.

Chapter 2-2

AN OVERVIEW OF THE MATHEMATICS CURRICULA IN THE WEST AND EAST

Discussions on the Findings of the Chongqing Paper

Margaret WU[1] and ZHANG Dianzhou[2]
[1]*University of Melbourne;* [2]*East China Normal University*

1. INTRODUCTION

An international conference on mathematics curricula was held during 16-20 August 2002 in Chongqing City, China. Ten countries participated in the conference: United States, France, Netherlands, Russia, Australia, Brazil, Japan, Korea, Singapore, and China (mainland). Four were from the East, and six were from the West. The speakers from ten countries spoke about recent reforms in mathematics curricula. A summary of findings from this conference, known as the Chongqing paper, was presented at the ICMI study conference. The curriculum and textbooks working group discussed the findings of this paper. Below we present a summary of the discussions on the Chongqing paper.

2. DETERMINANTS OF CURRICULUM

In an international comparison of national curricula, the first difficulty arises in the establishment of a basis for comparison. For example, in many countries there is not a single national curriculum, but there are many different curricula at the state level. Whether there is a single national body for determining a country's curriculum, or multiple authorities to design curricula for different states, we can gain a glimpse of the strengths of control, the processes and the goals of education authorities by studying the determinants of curriculum in each country. We can ask a number of questions:

- Is the curriculum centralised or decentralised?
- Who are those involved in the design of the curriculum: policy makers or educators?
- Is the design of the curriculum widely consulted? Does the consultation go beyond the national context?
- What are the mechanisms of curriculum development? Are there trials in schools before a curriculum is adopted? Is teacher feedback valued?

In the following sections, we attempt to examine some of these points.

3. CENTRALISED AND DECENTRALISED NATIONAL CURRICULUM

The Chongqing paper gave the following classification in relation to whether a country has a centralised or decentralised curriculum.

Table 2-2-1. Centralised and Decentralised Curriculum Systems

Centralised	Decentralised
Japan	United States
Korea	Australia
Singapore	Netherlands
China	Russia
France	Brazil

Table 1 shows that France and the four Eastern countries mentioned above have a centralised national mathematics curriculum. However, such a dichotomy of centralised/non-centralised classification does not sufficiently convey the distinction between the two systems. A continuous scale, showing the degree of centralisation may be more useful. For example, in some countries, while there is a single curriculum document, there is a great deal of flexibility and freedom for schools to implement the curriculum. On the other hand, while there may be numerous curricula at the regional levels, these could be very similar so that essentially there is only one curriculum. Russia, for example, was classified as having a decentralised curriculum. However, the government has a strong control over the materials that must be taught in schools. In Japan and Korea, while there are many different secondary school textbooks published by private publishers, the Ministry of Education stipulates the size, the number of pages, and the number of colours to be used in the textbooks so that all the textbooks look more or less identical.

While it is true that the CHC (Confucian Heritage Culture) countries (Hong Kong, Taiwan, Japan, Korea, Singapore) all have centralised curricula, countries like France and England also have centralised curricula. It was suggested that some governments want to have a tight control over the national curriculum to ensure equity of education where all students receive the 'same' instructions in classrooms and all schools follow the same curriculum and textbooks. In addition, there is a 'distrust' of commercial/private enterprises whose goals are generally profit oriented. While cultural traditions are likely to play a part in determining whether a country has a centralised or non-centralised curriculum, other factors such as the homogeneity of the population, sense of nationalism and political orientations may also impact on the control of the specificity of curriculum documents.

A centralised national curriculum carries a good deal of authority. It is usually revised after a period of several years. If this curriculum is excellent, it can benefit all pupils in this country. However, if it has quite a few flaws, the consequences can be serious. For example, before 1999, the Chinese mathematics curriculum did not include probability and statistics (data analysis) so that many Chinese are unable to understand the random events in daily life. This situation has not changed a great deal (Zhang, 2002).

4. CURRICULUM DEVELOPMENT IN VARIOUS COUNTRIES

In the United States, where there is a decentralised curriculum, the NCTM (National Council of Teachers of Mathematics) standards set the directions for all state-wide curricula. In particular, it should be noted that:

- 49 states have state standards or frameworks.
- 43 states have "adopted or substantially incorporated the recommendations from the national standards into their own state standards and curriculum frameworks".
- there are continuing efforts to study the impact and influence of standards and standards-based materials in the NCTM Standards Impact Research Group.

In Russia, after the former USSR collapsed, there have been many political parties in Russia resulting in a proliferation of different curricula and textbooks (Sharyjin, 2002). Some of these textbooks are not always of the best quality. "One needs to separate the weeds from the seeds", Sharyjin said. The so-called education reform, or modernisation of mathematical education, does not necessarily lead to improvements in education. There are both supporters and skeptics of this modernisation movement in Russia (Sharyjin, 2002).

In Brazil, there is a Brazilian PCN (National Curricular Parameters) for the Middle School Mathematics (1998) which, in addition to being the first nation wide curriculum proposal, was based on a shift from previous curricula that focused mainly on content, to a curriculum design to emphasise the ideas of competences, abilities, interdisciplinary studies and transversal themes (Maria, 2002). Maria found a discrepancy between the intended curriculum (PCN) and the implemented curriculum in the classrooms. More specifically, Maria found that there is too much emphasis on the exploration of mathematical ideas and a lack of organisation and systematisation of mathematics, when the PCN is implemented in the classrooms.

In the East, governments recognise that a centralised system is unable to meet the different needs of students. Consequently, more optional courses are designed in the new curricula of the 21st century.

In Korea, the Ministry of Education released the seventh amendment of the national curriculum in 1997. The core of the seventh curriculum is the implementation of a "differentiated curriculum", emphasising the need to consider individual students' interest, ability, aptitude and other attributes (Choe, 2002). The new curriculum is divided into "Level Based Differentiated Curriculum" (LBDC), "Enrichment and Supplement Differentiated Curriculum" (ESDC), and "Subject Selection Differentiated Curriculum" (SSDC). The seventh mathematics curriculum offers the following subjects at the grades 11 and 12 levels: Practical Mathematics, Mathematics I, Mathematics II, Calculus, Probability and Statistics, and Discrete Mathematics. The latter three of these subjects were added to meet a variety of needs of students.

In China, the new mathematics curriculum, modeled on the French system, sets five tracks for upper high school students for the first time in 2002: natural sciences, social sciences, art and physical, vocational, and mathematical specialisations. This curriculum also reflects the view of a "differentiated education". The designers of the new mathematics curriculum reported a process of information gathering, from both Chinese and foreign sources, as well as through a series of conferences in which mathematicians and educators participated, for the preparation of the document "Standards of Senior High School Mathematics' (Wang et al., 2002). The new mathematics curriculum includes new topics such as mathematical modeling, mathematical researching and culture of mathe-matics, calculus, statistics and probability.

5. CHALLENGES FACED IN CURRICULUM DEVELOPMENT

Curriculum and examination culture

In many countries, examinations drive education. In particular, major examinations play a key role as a determinant of (at least) the implemented curriculum in the East. An examination culture is part of the Confucian culture that is deep-rooted in Japan, Korea, Singapore, Vietnam and China. Examinations are also regarded as the "fairest" method of assessment as every student takes the same test and tests are "objectively" scored. This view has strong support in the East, where the fear of corruption and distrust of officials are prevalent in people's minds.

The problem of an examination-driven curricula is accentuated in mathematics education. Hirabayashi (2002) states that:

> Mathematics is the key subject of entrance examinations to higher education and a higher education qualification is believed to lead to a higher status in the future society. (Page 134)

"Teach to the test!" "Teach what will be examined!" These are often the ways in which teachers organise their lessons. While there are many impressive and grand goals in the Chinese national curriculum such as "Use technology for mathematics learning!", "To foster creative abilities!", or "Take on the problem solving approach", most mathematics teachers have to teach routine problems so students can pass examinations. As is well known in China, test questions in the examinations are just basic routine problems. The use of calculators or computers is not allowed in the examinations, so there is little point in teaching these in the classrooms when they do not improve students' examination scores.

The Chinese mathematics curriculum provides two options of mathematics strands for upper high school students: A. No calculus. B. Containing calculus. However, in the past 20 years, nearly all schools and all students chose Strand A, because calculus is not part of the entrance examination program. The Chinese tradition, which emphasises learning by rote and stresses the importance of examinations, encourages Chinese students to study for examinations. "They forget what they have learned once the test is over." (Zhang, 2002)

Examinations do not play such an important role in other countries. For example, in the United States, the SAT does not seem to have any significant influence on curriculum development.

It is a difficult balancing act for the teachers to help students pass examinations, and at the same time, not make students slaves of examinations.

6. TENSIONS BETWEEN MATHEMATICS EDUCATORS AND POLICY MAKERS

The much-publicised "Math War" in the United States (Becker & Jacob, 1998) highlights the tension between the education authorities and the mathematics communities. While both sides of this debate agree about the importance of mathematics education, the differences in opinion lie in the way mathematics is taught and the emphasis placed on specific mathematics topics.

Another interesting story concerns the French minister of science, research and education in 2000, Claude Allègre. It was reported (Browder, 2002) that Allègre believed that, with the advance of technology and computers, people no longer required mathematics skills to do science, and that mathematics was over-emphasised in French education. Naturally, Allègre's view was met with strong opposition from teachers' unions. In a socialist regime like France, teachers' unions are powerful and Allègre eventually had to step down from his ministerial position.

In the East, however, government authorities appear to have a stronger hold on education policies. Nevertheless, there are numerous reports of discontent regarding mathematics education policies. Sawada (2002) reported the anxiety expressed by mathematics educators about the declining mathematics ability of Japanese children, citing results of international studies and a survey of computation abilities of primary school children. They attribute this decline to the reduction in the number of mathematics lessons in schools. However, the Monbusho (Ministry of Education) maintains that students' mathematics ability has not fallen. Sawada made the following comment:

> We Japanese have some bad habits. We have a tendency to jump at new things as soon as they are introduced, and dump the old. When "modernisation in mathematics" was proposed, we all seized on that idea. But once the novelty wears off, they are utterly forgotten. Whether something is "good or bad" instead of "old or new" should be used as a yardstick to judge what to get rid of and what to save. I remember the proverb that we learn a lesson from the past.

In Korea, Choe (2002) reported that 77% of primary school teachers and 85% of high school teachers are unhappy with the new curriculum. They prefer to postpone, if not abandon entirely, the implementation of the new curriculum. Choe stated that:

> Even though Korea has changed the mathematics contents of school curriculum more than seven times, the teachers in the primary, middle and high schools never fully understand the necessity or reason.

Zhang (2002) reported that there is also a 'math war' in China. Many mathematicians do not like the new curriculum. The leaders of the Chinese mathematics communities are concerned that the reform will weaken the students' mathematics ability. Some mathematicians believe that there is no need for any change, because the old curriculum is an excellent one. Proponents of the old system believe that this can be seen from the outstanding performances of Chinese students in international mathematics assessments.

There is no doubt that any reforms will create tensions between the various stakeholders. In mathematics curriculum reforms, the process needs to involve different sectors of the community, to bridge the gap between the views of officials and academics. Lee and Fan (2002) point out:

> In reform, we need the participation of university professors and we also expect the university professors who advise us to have a global view beyond their narrow academic interest.

7. TRENDS IN CURRICULUM DEVELOPMENT

7.1 Diminishing differences between countries

From the papers presented in the Chongqing conference, and from the discussions at the ICMI study conference, it appears that there are a number of similarities between the curriculum reforms in various countries. This is not surprising, as there is an increasing international collaboration between countries, both at government level and at research institutions. The common trends in curriculum reform can be summarised as follows:

- There is an emphasis on differentiated education. More optional courses are offered, particularly for senior high school students.
- There is a reduction of mathematics lessons taught in schools, particularly in Eastern countries, to reduce the burden on students' learning.
- Probability and statistics are introduced, but there is a reduction of formal geometry proofs.

- There is an emphasis on technology in schools. Calculators and computers are incorporated into the curriculum.
- There is an increase in problem solving and mathematical modeling activities, and in general, an emphasis on improving higher-order thinking skills and independent learning.

The first point on the differentiated curriculum has been covered in the first section of this paper. We will discuss the other four points in more detail.

7.2 Reducing the burden of learning

The Chongqing conference reported that governments of most countries in East Asia propose to alleviate students' burden in mathematics learning. The following shows estimated proportions of reduction in school hours:

Japan	30%
Korea	30%
Singapore	1/3
China (Shanghai)	1/3

Sawada (2002) reported that, in Japan, the new Course of Study published by Monbusho will be adopted in schools from the 2002 academic year. Saturday classes will be abolished and students will attend a five-day school week. In mathematics, the total number of class periods will be reduced by 30% as compared to the "New Mathematics Curriculum" in the 1970s.

In Korea, the high standard of mathematics expected of students is blamed for the growing trend of private lessons across the nation. People are of the opinion that lowering standards will alleviate pressure on students, and thus reduce the stress level caused by private tutoring (Choe, 2002). The new curriculum proposes a reduction of 30% in mathematics content. Despite concerns expressed by mathematics educators, the administrators are firm on their decision to reduce the mathematics content.

In China, the head of Shanghai Education Committee called for a 1/3 reduction in every school subject from the 1999 curriculum.

In contrast, no mathematics academic credits have been reduced in the United States in recent years. According to a report from Japan (Sawada, 2002), the number of school hours of arithmetic and language in the United States has increased sharply, and the NCTM Standards expect all students to study mathematics during each of the four years that they are enrolled in high school (p. 288).

7.3 Changes in mathematics content: more statistics, but less deductive geometry

Probability and statistics have received special attention in many curriculum reforms. Korea, France, China and many other countries have all noted the importance of the role of statistics in the 21st century. Those who advocate the inclusion of statistics often argue that statistical skills are important not only in our daily lives but also in other fields such as economics and social studies. It was pointed out that in some countries, probability is not included in the syllabus because traditionally, probability is linked to games and gambling which are forbidden in cultures such as Islam. Therefore, cultural values and religious beliefs certainly play a part in the determinants of the contents of a curriculum. One difficulty with the introduction of statistics is that most teachers have not received adequate training in this area. There is an urgent call for providing support for teachers to teach topics that are relatively new to them.

It was reported that by 2000, there were only very few countries that still retained a deductive geometry topic in the mathematics curriculum of grades 6-9. It is likely that deductive geometry may disappear altogether in mathematics courses in the 21st century. Nevertheless, mathematics educators in China and France still stress the importance of formal proofs, and they believe that there is still value in teaching students such skills.

7.4 Technology in the curriculum

While most countries reported the inclusion of technology in the mathematics curriculum, the degree to which technology was actually utilised in the classrooms varied a great deal between countries.

In the United States, the NCTM Standards (1989) emphasised the importance of technology:

> Technology, including calculators, computers, and videos, should be used when appropriate. These devices and formats free students from tedious computations and allow them to concentrate on problem solving and other important content. They also give them new means to explore content. As paper-and-pencil computation becomes less important, the skills and understanding required to make proficient use of calculators and computers become more important.

Laborde (2002) reported that in France, ICT (information and communication technology) is integrated into many disciplines of study. In particular, the French Programme of Studies documents[1] stress the necessity of this integration as expressed in the following excerpt:

> All students now have access to calculators, and mathematics teaching must take it into account. [...] The new curricula show the necessity of working with calculators by ensuring that everyone acquires the ability of mental and written computation" *(Accompanying documents of Middle school programme of studies)*

Scientific and programming calculators, mathematics software, spreadsheets, graphing tools are used at collège and lycée levels. The following mathematics content areas are linked to the use of specific technology:

Statistics	Calculator, Spreadsheet
Geometry – Transformations	Dynamic Geometry Software
Sequences – Approximation	Calculator, Computer Algebra System
Functions	Spreadsheet, Graphing tool

The use of ICT in Singapore is largely under the influence of the government's IT master plan (Lee & Fan, 2002). IT has been increasingly used in mathematics teaching. Scientific calculators have been a standard tool for secondary schools since the early 1990s. Graphing calculators are increasingly used as an optional tool in secondary schools and junior colleges.

In Korea, the problem is about finding ways to implement the use of computers and calculators in mathematics curriculum and in each classroom. The sixth curriculum already mentioned and encouraged the use of calculators for some specific mathematics content. However, in reality, schools do not allow active use of such technologies. Thus, one of the main points of the seventh curriculum will be the combination of the use of calculators and computers with the mathematics curriculum (Choe, 2002).

In China in the 1990's, the mathematics curriculum encouraged students to use calculators. But classroom teachers rarely put the calculator to use. Why? The key problem is that the examination policy forbids students to bring any calculator into testing places. An exceptional case was noted in Shanghai. After 2000, students were permitted to use scientific calculators (not graphic calculators) in key examinations. In addition, about 100 schools are trying to acquire graphic calculators of type TI-92. More recently, the

[1] These documents are available as printed texts or CDROMs, as well as electronic (sometimes animated) documents on the Internet, for example, http://www.educnet.education.fr/math/ and smf.emath.fr/Enseignements/

new curriculum emphasised the importance of technology and recommended students in senior high school to use scientific calculators in mathematics learning (Zhang, 2002).

7.5 Problem solving and mathematical modeling

Life-long Learning and Learning To Learn are the new slogans of the 21st century. In recent years, educators have called for education changes that improve students' higher-order thinking skills and independent learning skills.

In Singapore, problem solving is at the centre of mathematics education. The pentagon, as a framework of the mathematics curriculum, is known to all teachers. It has mathematical problem solving in the centre, surrounded by five inter-related components – concepts, skills, processes, attitudes and metacognition (Lee & Fan, 2002).

In France and Singapore, inter-disciplinary project work is introduced to encourage the development of students' sense of autonomy and thinking skills (Lee & Fan, 2002; Laborde, 2002).

Sawada (2002) reported that the new Course of Study in Japan stresses "the zest for living" in all subjects. Spontaneous problem solving is emphasised in mathematics. In all levels of mathematics, 'mathematical activities' are introduced for the first time. 'Enjoyment' of mathematics is described in the curriculum at elementary and lower secondary levels, and fostering 'creativity' is described at the senior secondary level.

In China, the new "Standards" document includes topics like "mathematical exploration," "mathematics modeling", "mathematics reading", "mathematics activity", and so on. It is useful for students to develop active and pro-active approaches to the study of mathematics. The purpose of the "Standards" document is to develop students' interests to learn, to encourage students to participate, explore, cooperate and communicate, and to help students to develop the habit of thinking independently (Wang, 2002).

In France, mathematics is viewed as a place for experimenting, modeling and formulating. One of the objectives of mathematical teaching as expressed in the current national texts is to introduce students to a scientific approach which includes the following steps: experimenting, reasoning, imagining, analysing, critiquing. This is based on a conception of a "genuine mathematical activity" as described by the Programme of Middle School: identifying a problem, conjecturing a result, experimenting on examples, building an argument, formulating a solution, controlling the results, and evaluating their relevance in relation to the problem under investigation. The idea of experimenting and using mathematics as a modeling tool appears several times in the Programme of Studies and in the accompanying documents of comments. Mathematics is now presented in the same light as other sciences that seek to understand the real world in which we live. Mathematics is thus considered as contributing to the education of the citizen.

In a presentation at the Chongqing conference, Gravemeijer (2002) spoke about the reform in mathematics education in the Netherlands. He focused on a mathematical approach known as "realistic mathematics education", or RME, that originated in The Netherlands.

Proponents of "realistic mathematics education" have gathered a great deal of support in the West. But many mathematicians in the East are cautious about adopting this approach. They still think that rigorous logical proofs are the core of mathematics thinking. They fear that "realistic mathematics education" may result in the decline of students' abilities to carry out formal reasoning. They say, "A conjecture is a conjecture only" (Zhang, 2002). A deductive approach to mathematics is preferred to an inductive approach in Eastern countries.

8. CONCLUSIONS

It is clear from both the Chongqing conference and the ICMI Study Conference that a great deal is taking place in curriculum reform. Furthermore, there are many common approaches and goals across countries in their effort to develop the best possible curriculum. Some trends are emerging in curriculum development. In general, curricula are broadened to cater for individual students' needs. There is an increasing interest in mathematical processes and not just in the outcome of learning. The affective dimension is formally acknowledged in the curricula. The use of technology is clearly made a priority.

In the discussions on curriculum reform, there is also a note of caution. There is a danger of adopting a curriculum by "shopping around". A common approach in curriculum reform is to send a group of people abroad to learn about the latest developments and changes in other countries. The adoption of foreign ideas without critical evaluation can be detrimental to an education system. Unlike teaching and learning, which have deep cultural roots and hence are difficult to change overnight, education curriculum and textbooks are particularly susceptible to significant changes following the recommendations of a few influential individuals.

One question that is of foremost importance is "what is a good curriculum?" Professor Hirabayashi, in his ICMI Study Conference address, reminded us that having a high achievement in international mathematics studies is not the only criterion. Students should enjoy doing mathematics and regard learning mathematics as an art to improve one's well-being. A "good" curriculum is one that fits in with the cultural values of the country, and one that can be achieved realistically, given the educational resources and conditions.

We conclude that, in sharing our experiences in curriculum reforms, we need to be open-minded but also cautious, be collaborative but also critical, and above all, develop a realistic curriculum that is in accordance with our cultural values.

REFERENCES

Becker, J. and Jacob, B., 1998, Math War Developments in the United States (California). *ICMI Bulletin*, No.44, June 1998.

Browder, F., 2002, Reflections on the Future of Mathematics. *Notices of AMS*, 49(6), June/July.

Choe, Y.H., 2002, *The Seventh Curriculum of Mathematics in Korea*. Pre-Proceedings of Chongqing conference.

David, M., 2002, *Recent Curriculum Reform in Brazil and some Implications in Textbook Materials and in the Systematization of Mathematics in the Classroom*. Pre-Proceedings of Chongqing conference.

Gravemeijer, K., 2002, *How to Support Reform Mathematics Education*. Pre-Proceedings of Chongqing conference.

Hirabayashi, I., 2002, *A Traditional Aspect of Mathematics Education in Japan: Mathematics as GEI (art), its Jutsu (technique) and Do (way)*. Paper presented at the ICMI comparative study conference, the University of Hong Kong, 20th – 25th October, 2002.

Laborde, C., 2002, *The Mathematics Curriculum in France at the Beginning of the New Century*. Pre-Proceedings of Chongqing Conference. 2002.

Lee, P.Y. and Fan, L., 2002. *The Development of Singapore Mathematics Curriculum*. Pre-Proceedings of Chongqing Conference.

NCTM, 1989, *Curriculum and Evaluation Standards for School Mathematics*. National Council of Teachers of Mathematics. U.S.A.

Sawada, T., 2002, *The Reform of Mathematics Curriculum and its Issues in Japan*. Pre-Proceedings of Chongqing conference.

Sharygin, I., 2002, *On One Project of a System of Mathematics Textbooks for High School*. Pre-Proceedings of Chongqing conference.

Wang, S., Wang, L, Li, J., and Zhang, D., 2002, *The Conception of the Framework of Chinese Standards of Senior High School Mathematics*. Pre-Proceedings of Chongqing conference.

Zhang, D., 2002, *An Overview of the Mathematics Curricula in the West and East: Summary of Findings at the Chongqing Conference in 2002*. Paper presented at the ICMI comparative study conference, the University of Hong Kong, 20th – 25th October.

Chapter 2-3

CLASSIFICATION AND FRAMING OF MATHEMATICAL KNOWLEDGE IN HONG KONG, MAINLAND CHINA, SINGAPORE, AND THE UNITED STATES

An Analysis of Textbooks in Socio-Cultural Contexts

LI Yeping[1] and Mark B. GINSBURG[2]
[1]University of New Hampshire; [2]University of Pittsburgh

Curriculum – that is, the selection, organization, and transmission of educational knowledge – can contribute to students' cognitive development, and at the same time, transmit specific social-cultural values and regulatory norms. With a focus on identifying curricular influence on students' academic achievement,[1] researchers have come to understand that cross-system variations in curriculum can provide a partial explanation of cross-system variations in students' academic performance, especially in mathematics (Fuson, Stigler and Bartsch, 1988; Schmidt, McKnight and Raizen,1997; Schmidt, McKnight Valverde, Houang and Wiley, 1997;

[1] It is important to consider whether performance on standardized examinations, emphasizing content knowledge, should be the only or main measure of the quality of education. For example, it is argued that mathematics curriculum and teaching in China tends to focus on students' acquisition of traditional content knowledge, which may be emphasized in international studies of achievement, while in the United States more stress may be devoted to developing students' skills of solving practical problems in everyday life [see James W. Stigler, & M. Perry, "Cross cultural studies of mathematics teaching and learning: Recent findings and new directions," in D.A. Grouws, T.J. Cooney, & D. Jones (eds.), *Effective mathematics teaching*, (Reston, VA: NCTM, 1988), pp.104-223]. Moreover, as J. Wang ["TIMMS primary and middle school data: some technical concerns," *Educational researcher* 30(6) (August/September 2001):17-21] observes concerning TIMMS middle school level achievement data, there are "technical problems that can alter the comparative results, undercutting the reliability of TIMMS benchmarking" (p.17).

Westbury, 1992). In particular, when compared to the curriculum materials from some high-achieving education systems in East Asia, research has revealed that the U.S. curriculum materials failed to provide challenging mathematics content (Li, 1999; Schmidt et al., 1997a; Mayer, Sims and Tajika, 1995). "Both the Second International Mathematics Study (McKnight et al., 1989) ... and the Third International Mathematics and Sciences Study (Schmidt, McKnight, Valverde, Houang and Wiley, 1997)[5] ... found American textbooks to be more fragmented and superficial than texts in most other countries." (Kennedy, 1997:8) Findings such as these have provided a basis for considering curricular changes in the United States. (Mathematical Sciences Education Board, 1990; Silver, 1998)

Meanwhile, curriculum also embodies the values and norms, including those concerned with authority relations, of a specific society. Thus, curriculum can serve as a window through which to examine the cultural features that support or resist instructional and social changes in a given educational system and societal context. However, when comparative studies focus primarily on the impact of mathematics curriculum on students' cognitive outcomes, the social-cultural substance reflected in curricula are overlooked or trivialized as merely surface phenomena (Reid, 2000). Fortunately, this stance has been undermined as recent research in mathematics education has examined how students' learning in classrooms is also a process of social construction (Bussi, 1996; Lerman, 1998).

This study is not aimed to explore the potential relationships between curriculum and students' achievement. Rather, the purpose of this study is to examine mathematics curriculum as a system and societal artifact that reflects both culturally valued knowledge and principles of social control. Specifically, because textbooks are often the curricular materials that are the most influential to what happens in classrooms (Eisner, 1987; McKnight et al., 1989), this study examines mathematics textbooks from the United States and three education systems in East Asia (Hong Kong, mainland China, and Singapore) to reveal cross-system variations in knowledge selection and organization as well as in envisioned pedagogical relationships between teacher and students. The results, in turn, may also provide a basis to predict the feasibility of possible proposals for curricular change.

1. CONCEPTUALIZATION

Our conceptualization is based on the work of Basil Bernstein's (1975a)[2] sociological analysis of the curriculum, which centers on two key concepts, classification and frame. Bernstein defines *classification* in terms of the degree of "differentiation between contents," whether contents are traditional subject areas (mathematics, economics, physics, etc.) or are topics or sections within one subject area (Bernstein, 1975a:88). "Where classification is strong, contents are well insulated from each other by boundaries." (Bernstein, 1975a) According to Bernstein, *frame* "refers to the strength of the boundary between what may be transmitted and what may not be transmitted in the pedagogical relationship ... to the degree of control teacher and[/or] pupil possess over ... the knowledge transmitted and received in a pedagogical relationship." (Bernstein, 1975a) The stronger the frame, the less flexibility teacher and students have to control what is taught and learned in the context of the pedagogical relationship. Moreover, the concept of frame has another aspect that addresses "the strength of boundary between educational knowledge and everyday community knowledge." (Bernstein, 1975a:89).

Importantly, as Bernstein clarifies, the tension between strongly or weakly classified and framed curricula "is not simply a question of what is to be taught but a tension arising out of quite different patterns of authority, quite different concepts of order and control." (Bernstein, 1975b) As Apple observes, "the logic and modes of control ... are entering the school through the form the curriculum takes, not only its content." (Apple, 1981:30) Thus, cross-system differences in content classification and framing may reflect differences in authority patterns characterizing the societies' political cultures. Likewise, intra-system differences in content classification and framing may correspond to differences in the nature of authority relations of groups (for example, social classes) which predominate in particular educational tracks.

[2] For discussions of Bernstein's framework, see Ann and Harold Berlak, *Dilemmas of Schooling: Teaching and Social Change* (New York: Methuen, 1981); John Eggleston, *The Sociology of the Curriculum* (London: Routledge and Kegan Paul, 1977); Alan Sadovnik (ed.), *Knowledge and Pedagogy: The Sociology of Basil Bernstein* (Norwood, NJ: Ablex, 1995); Geoff Whitty, *Sociology and School Knowledge: Curriculum Theory, Research and Politics* (London: Methuen, 1985); Michael F.D. Young, "An Approach to the Study of Curricula as Socially organized Knowledge," in M. Young (ed.) *Knowledge and Control* (London: Collier-Macmillan, 1971).

2. METHODS

This study focuses on the United States and three high-achieving educational systems in East Asia: Hong Kong,[3] mainland China, and Singapore. Nine mathematics textbooks developed for eighth graders in these four education systems were analyzed in this study, the same books that were used in the "Third International Mathematics and Sciences Study" (TIMSS) (Schmidt et al., 1997b). Among these nine textbooks, five textbooks were from the United States, one from Hong Kong, two (actually a two-volume set) from mainland China, and one from Singapore.[4] Among the five U.S. textbooks, one was an algebra-specific textbook (hereafter, US-Algebra or US-A) and four others were popular non-algebra-specific mathematics textbooks de-signed for use in the eighth grade (hereafter, US-Non-Algebra or US-NA).[5]

To assess the degree of classification of mathematical knowledge in each system the student versions of these texts were content analyzed. Two text levels were used in the analysis: (1) chapters within a book and (2) sections within a chapter. For each of these levels, the degree of separation of content was assessed with respect to two content contrasts: (a) between mathematics and non-mathematics content topics and (b) between algebra and non-algebra content.[6]

To determine the strength of framing, two types of information were collected. The first type of information focused on the organization of these

[3] Note that the textbooks from Hong Kong used in this study were published in 1992 – before, but during the period anticipating, Hong Kong's change from being a British colony to becoming again a part of China.

[4] *Mathematics for Hong Kong Book 2* (Hong Kong: Canotta Publishing Co., Ltd., 1992); *Daishu/Algebra*, Vol. 3 (Beijing: People's Education Press, 1993); *Jihe/Geometry*, Vol. 1 (Beijing: People's Education Press, 1992); *New Syllabus: Mathematics 2* (Singapore: Shing Lee Publishers Ltd., 1987); *Addison-Wesley Mathematics*, Grade 8 (USA: Addison-Wesley, 1993); *Exploring Mathematics*, Grade 8 (USA: Scott Foresman, 1991); *Mathematics in Action*, Grade 8 (USA: MacMillan/McGraw Hill, 1992); *Mathematics - Exploring Your World*, Grade 8 (USA: Silver Burdett and Ginn, 1991); *Algebra: Structure and Method Book 1* (USA: Houghton Mifflin Co., 1990).

[5] For other details about the textbooks studied, see Yeping Li, *An analysis of algebra content, content organization and presentation, and to-be-solved problems in eighth-grade mathematics textbooks from Hong Kong, Mainland China, Singapore, and the United States*.

[6] Content classifications at the levels of both chapter and section were determined by their titles and content included (at least 75% page space were devoted to the content category classified). The TIMSS curriculum framework [see Robitaille et al., *Curriculum frameworks for mathematics and science* (Vancouver: Pacific Educational Press, 1993)] was adopted for differentiating mathematics content presented in the textbooks.

four educational systems, which would help to reveal the degree of control that teacher and students may have over the selection and transmission of educational knowledge. In particular, the existence of system-wide syllabi and the degree of individual teachers' autonomy in selecting and/or using mathematics textbooks were examined. Moreover, the teacher's versions of these texts were content analyzed, with attention being focused on guidelines that encouraged or discouraged teachers and/or students to bring in knowledge from outside the official curriculum as examples, etc.

Finally, secondary sources were consulted to characterize the political cultures, especially norms concerning authority, in each society.[7]

3. RESULTS

The findings are presented below, first focusing on the classification of mathematical knowledge and second portraying the framing of such knowledge.

3.1 Classification

The following table summarizes the classification of contents exhibited in eighth grade textbooks across four educational systems: Hong Kong, mainland China, Singapore, and the United States.

Level	*Content Contrast*	*Hong Kong*	*PR China*	*Singapore*	*United States*
Chapters within a Book	*Math vs. Non-Math content topics*	100% math	100% math	100% math	US-A: 100% math US-NA: 100% math
	Algebra vs. Non-Algebra	Both (29% algebra chapters)	Both (33% or 75% algebra chapters)*	Both (50% algebra chapters)	US-A: Both (83% algebra chapters) US-NA: Both (13%, 13%, 14%, & 21% algebra chapters)
Sections within a Chapter	*Math vs. Non-Math content topics*	100% math	100% math	100% math	US-A: 100% math US-NA: Both (72%-96% math topic section contained in a chapter)
	Algebra vs. Non-Algebra	100% algebra or non-algebra	100% algebra or non-algebra	100% algebra or non-algebra	US-A: Both (95% algebra sections in algebra chapters)

[7] For example, see M.A. Brimer, "Hong Kong"; C.C. Dong, "China, People's Republic of"; R. Murray Thomas, "Singapore".

Level	*Content Contrast*	*Hong Kong*	*PR China*	*Singapore*	*United States*
					US-NA: Both (76%, 80%, 86%, & 96% algebra sections in algebra chapters)

* Note: In Mainland China, two texts (actually a two-volume set) were used together for eight graders. One is an Algebra text (containing 4 chapters, three of which are identified as chapters on algebra content), and the other is a Geometry text (containing 5 chapters). Thus, if we counted in terms of the whole eighth grade as we did for other educational systems, there were three out of nine chapters (33%) on algebra. If counted in terms of the book that contains algebra content, three out of four (75%) chapters in Algebra text are on algebra content.

The results show that the selected texts from these four education systems are similar in that all of the texts have 100% of their chapters focused generally on math content topics. That is, in terms of the contrast between math and non-math topics, our analysis examining the content focus at the level of chapters indicates that all the textbooks examined are strongly classified. A weaker classification would be evidenced if a textbook had at least some chapters devoted primarily to content other than mathematics.

However, as we continue to examine the results of the chapter-level analysis, we observe important cross-national (as well as some intra-U.S.) differences in the classification of mathematical knowledge. In particular, the textbooks from Hong Kong, Mainland China, and Singapore contain approximately one-third to one-half (respectively, 29%, 33%, and 50%) of their chapters that focus primarily on Algebra topics, while the US-Non-Algebra textbooks contain approximately one-seventh to one-fifth (13%-21%) of their chapters that present mainly algebra topics. In contrast, more than four-fifth (83%) of the chapters in the US-Algebra textbooks have a primary focus on Algebra topics.

When we review the findings for the analyses at the level of sections within a chapter, the cross-system differences in degree of classification of mathematical knowledge become even more apparent. Specifically, whether we consider contrasts in content with respect to math vs. non-math topics or algebra vs. non-algebra, the textbooks from the three Asian education systems are very strongly classified. That is, the sections in a given chapter of a textbook all contain a homogeneous set of content topics. The Asian textbooks differ from those used in the U.S., especially the non-algebra texts, which tend to be much more weakly classified. All the U.S. textbooks differ from the Asian textbooks, in that they contain heterogeneous contents within a chapter with respect to the algebra vs. non-algebra content contrast. Note however, that 95% of the sections in algebra chapters in the US algebra text focus on algebra topics, while the percentages for the U.S. non-algebra texts tend to be lower (76%, 80%, 86%, and 96%).

In examining the math vs. non-math topic contrast, we note that the U.S. non-algebra textbooks also differ from the Asian system textbooks, in that the former texts have a weaker classification of mathematical knowledge than the latter texts. That is, unlike the eighth grade texts used in the Asian systems, the chapters in the U.S. non-algebra textbooks contain a mixture of section topics, including math as well as non-math topics such as "Curriculum Connection: Art" and "Enrichment" (Stevenson and Stigler, 1992).[8] In this content contrast, however, the U.S. algebra textbook resembles the Asian system textbooks, evincing a strong classification of mathematical knowledge (100% of its content sections are focused on math content topics).

3.2 Framing

Hong Kong, mainland China, and Singapore have a centralized education system, and all three differ from the United States that has a decentralized education system.[9] In particular, curriculum guides and textbooks used in these three Asian education systems are required to bear an approval from a national or system-level authority. Teachers and students system-wide are required to cover the same content that is specified in syllabi and textbooks. In contrast, the United States leaves such responsibilities of developing curriculum guides and selecting textbooks to states, local school districts, schools, or even individual teachers.[10] In contrast to

[8] Stevenson and Stigler elaborate based on a similar finding; they note that in contrast to textbooks used in elementary schools in mainland China, Taiwan, and Japan, the U.S. texts for mathematics and other subjects contain many "colorful illustrations, photographs, drawings, or figures [, which] ... along with digressions into historical and biographical material, ... introduced to engage children's interest, ... may instead distract attention from the central purpose of the lesson" (p.139).

[9] A. Beaton, I. Mullis, M. Martin, E. Gonzalez, D. Kelly, & T. Smith, *Mathematics achievement in the middle school year: IEA's Third International Mathematics and Science Study (TIMSS)* (Chestnut Hill, MA: TIMSS International Study Center, Boston College, 1996).

[10] It should be noted, however, that despite the opportunities for decentralized curricular decisions, there is a standardizing of texts in the United States. This occurs because of the major role played by a small and decreasing number of textbook publishers (generally part of multinational corporations) and their profit-motivated efforts to design texts that will be adopted in large states (e.g., California, New York, and Texas), which have state-level adoption procedures. See, for example, Michael Apple and Linda Christian-Smith (eds.), *The Politics of the Textbook* (New York: Routledge, 1991). At the same time the combination of (multi)national publishers and local/state text adoption means that (at least) eighth-grade mathematics textbooks tend to "include the content specified by the

what is often said in China that the "textbook is the base for classroom instruction," it is not uncommon in the United States for teachers not to follow or use any specific textbooks as they are published for many different school districts. Therefore, in these three Asian education systems, teachers and students have almost no flexibility in determining what is taught and learned (except some modifications on content topic sequencing and pacing). But teachers (and perhaps students) in the United States have had, until recently, somewhat more say in shaping the curriculum followed and the textbooks used, indicating a weaker framing of mathematics knowledge.

The texts from Hong Kong and Singapore do not have accompanying teacher's instructional guides or teacher's version of texts. Teachers were assumed to use the same textbooks as their students. The mainland China texts have accompanying teacher's guides, which provide instructional suggestions for teaching each chapter. In general, the guide contained (1) general instructional requirements, (2) textbook analysis and instructional suggestions, (3) answer keys or hints for the exercise problems, (4) appendices with additional information related to the content contained in a given chapter in the student's version text. The guide is very condensed with most of its page space devoted to textbook analysis, instructional suggest-ions and answer keys. Even for the textbook analysis and instructional suggestions, the guide mainly highlights the key and/or difficult concepts in each chapter and offers suggestions for teaching these concepts. There is no specific encouragement for teachers or students to bring in knowledge from outside the official curriculum. Although mainland China differs from Hong Kong and Singapore in providing instructional guides for teachers, all three are centralized educational systems. In such contexts, teachers are required to teach students the knowledge presented in textbooks.

In contrast, all U.S. texts have teacher's versions, organized similarly as an expanded version of the student's text. The format presents instructional suggestions alongside the content material that appears in the student's text. Furthermore, they all include detailed instructional suggestions and extra examples for almost every lesson to aid the teacher's use of the textbook. The student's versions of the five U.S. texts contain higher percentages of problems situated within real world contexts (e.g., about 11-18% in algebra

---

guidelines from a number of different states" [Lois Peak, *Pursuing excellence: A study of U.S. eighth-grade mathematics and science teaching, learning, curriculum, and achievement in international context* (Washington, DC: U.S. Department of Education, National Center for Education Statistics, 1996), p.36]. This may partly explain the finding (reported later) that eighth-grade mathematics textbooks in the US, especially those not designed for Algebra classes, exhibit a lower degree of classification and framing than is the case for eighth-grade texts in Hong Kong, mainland China, and Singapore.

chapters), compared to the three Asian texts (1.5-5.7% in algebra chapters). The frame around mathematical knowledge is further weakened in the teacher's version of the U.S. texts, in which many more examples related to real world or other non-math content areas are provided. Moreover, the texts present clear and strong suggestions to teachers (and, through teachers, to students) to bring in knowledge from outside the curricular knowledge specified in the text. For example, in the section on "Writing Algebraic Sentence" in "Mathematics in Action" (student edition) published by MacMillan/McGraw-Hill, the content was introduced in the context of comparing two persons' money sums in a foreign currency. In its teacher edition, it was suggested for teachers to start the lesson by asking students whether they have lived or traveled in a foreign country. If so, ask students to identify the currency used. Then the following problem was suggested to the teacher to continue the discussion:

> If a person from the United States went to a bank to change $1,000 for foreign currency, would the person be getting 1,000 units of the foreign currency? Why or why not? (p.92)

Likewise, in the section on "Greatest Common Factor" in "Exploring Mathematics" (student edition) published by Scott, Foresman and Company, the content was introduced in a pure mathematics context in the student's text. In the teacher edition, it was suggested to teachers to motivate students through proposing the following situation to students:

> You are in charge of designing a banner for your school, and it is to have a border of squares in the school colors. (p.168)

This would be followed by questions:

> If the banner is 45x40 inches, what is the largest square that can be used for the border? [5 sq in.] How many squares will fit along the 40 in. side? [8 squares] (p. 168, the italic and parentheses were original.)

We should note, however, that compared to the teacher's editions of the U.S. non-algebra texts, those for the U.S. algebra text indicate a stronger frame around mathematical curricular knowledge. In the teacher's edition of the latter text, far fewer instructional suggestions were given in the margin of the text. Except for some suggestions/additions of pure mathematical problems that are similar to what are given in the student's text, no explicit examples or suggestions were given to teachers (and their students) to bring in everyday, out-of-school knowledge in their lessons.

4. DISCUSSION

In interpreting the findings we will focus on three sets of contrasts: a) the three East Asian systems versus the U.S., b) similarities and differences among East Asian systems, and c) algebra versus non-algebra texts used in the United States (most likely with students from, respectively, upper middle and middle class backgrounds. versus lower middle and working class backgrounds).

4.1 Differences among East Asia and the U.S.

According to the results, the three Asian systems' mathematics textbooks exhibited a higher degree of classification and framing than either the US algebra or the US non-algebra books.[11] In attempting to understand the higher degree of classification and framing of mathematics knowledge in Hong Kong, mainland China, and Singapore (compared to the United States), we can point to the differences in authority relations in the dominant political cultures of these societies.[12] Employing Wilson's concept of "compliance ideologies,"[13] we can say that East Asian political cultures tend to be dominated by a *positional* compliance ideology, which stresses "forms of control that emanate ... from the community" and relationships in which

[11] Similarly, Bernstein ["On the classification and framing," p. 92] concluded that, compared to England and continental Europe, the "course-based, non-specialized USA [curriculum pattern had] ... the weakest classification and framing."

[12] Here we focus on what have been described as the dominant political culture extant in the contexts of the respective educational systems. This is not to deny the existence of one or more subordinate political cultures or subcultures in each setting. Similarly, while Hong Kong, Mainland China, and Singapore share a cultural root and the majority of residents in each setting are Chinese, there are a variety of minority groups living in the three settings. And although the population in the United States is dominated (numerically and politically) by people who emigrated from Europe, an increasing proportion of the population have their cultural origins in Africa, Asia, and Latin America. [See M.A. Brimer, "Hong Kong," in T.N. Postlethwaite (ed.) *The encyclopedia of comparative education and national systems of education* (Elmsford, NY: Pergamon Press, 1988), pp.332-38; C.C. Dong, "China, People's Republic of," in Postlethwaite (ed.) *The Encyclopedia*, pp.197-201; R. Murray Thomas, "Singapore," in Posthlethwaite, *The encyclopedia*, pp. 594-597.]

[13] Note that Wilson's conception of positional versus contractual "compliance ideologies" or forms of authority associated with different political cultures is paralleled by Bernstein's conception of positional versus personal forms of authority associated with family and classroom cultures of different social classes (working versus middle, respectively), which we will discuss below. See Basil Bernstein, "Social class, language and socialization," in *Class, Codes and Control, Volume I: Theoretical Studies towards a Sociology of Language* (London: Routledge and Kegan Paul, 1971), pp.170-89.

"duties are matched against rights in terms of one's place in society." (Wilson, 1992:89) In contrast, the United States tends to be dominated by a *contractual* compliance ideology, which emphasizes "defined limits of authority, the intrinsic value of the individual, and legal guarantees regarding negotiating processes." (Wilson, 1992:89)

According to Yee (1999), a fundamental element of the political culture of China is an emphasis on "paternalistic-dependency" relations between leaders and their followers:

The hallmark of Chinese political culture ... is the 'displays of deference by subordinates and grace in asserting command by superiors.' (Pye, 1998:32) Confucianism was the mainstream ideology of imperial China because 'it fitted the ideals and needs of both the rulers in their political realm and the common people in their family and clan settings.' (Pye, 1998:34) The hierarchical order of a Confucian society thus dominated the relationship between rulers and the followers in traditional China." (Pye, 1985:205)

Similarly, Hong Kong exhibits one important aspect of traditional Chinese political culture, a strong paternalistic orientation "reminiscent of [Confucianism-influenced] traditional Chinese expectations of the government: 73 percent of respondents agreed, and 8.2 percent agreed very much, with the statement that 'the government should treat the people like a father treats his children.'" (Pye, 1985:205) And the culture of authority manifest in Singapore's political system also seems to be more in line with what Wilson labels as a positional (rather than a contractual) form of compliance ideology:

[C]ore values of Singapore are defined in Confucian terms. ... Confucian rulers are expected to exercise power hierarchically, yet with decorum and respect for their followers. [And followers are expected to demonstrate] ... [r]espect for superiors[, which] often manifests itself as unquestioning, even obsequious, behavior toward those in authority (Neher, 1999:47-8).

In contrast to mainland China, Hong Kong, and Singapore, which have been strongly influenced by Confucianism, the dominant political culture of the United States seems to be more in line with what Wilson terms a contractual form of compliance ideology. As Verba and Nye explain in their often-cited book on *The Civic Culture*:

In the United States ... independent government began with republican institutions, and a mood that rejected the majesty and sacredness of traditional institutions, and without a privileged aristocratic class [in contrast to Great Britain]. ... In an even broader sense ... the general pattern of authority in American social systems, including the family, tended to stress political competence and participation rather than obedience to legitimate authority (Almond and Verba, 1963/1989:35-36).

The contrast in authority relations in the political cultures of mainland China, Hong Kong, and Singapore, on the one side, and the United States, on the other, also obtains with respect to the organization of their educational systems that have been mentioned before. In particular, the three East Asian educational systems are highly centralized, while the United States has a relatively decentralized education system (Beaton, Mullis, Martin, Gonzalez, Kelly and Smith, 1996). Moreover, teachers of eighth grade mathematics are more likely to have specialized content knowledge training in the Asian systems than in the U.S., which is in line with expectations that teachers' specialization would be greater in systems with strongly classified and framed curricula (what Bernstein terms a "collection code") (Bernstein, 1975:116-56) than in systems with weakly classified and framed curricula (what Bernstein terms an "integrated code") (Bernstein, 1975:116-56). That is, middle school teachers in Hong Kong, mainland China, and Singapore are required to be content specialists, while middle school teachers in the United States, until recently, were not required to have specific certification to teach mathematics.[14]

Different strength patterns of knowledge framing are also consistent with authority relationships that can be observed in classrooms between the U.S. and the three Asian education systems. In the United States, teachers are encouraged to make use of a variety of curriculum materials and pedagogical approaches in their classrooms. In particular, student cooperative learning in small groups is a popular approach in U.S. mathematics classrooms. This allows teacher and students themselves to bring everyday knowledge into the teaching and learning process. In contrast, mathematics classroom instruction in Hong Kong, mainland China, and Singapore is often didactic with the teacher assuming the role of lecturer. As teachers themselves have very limited autonomy in determining what is taught, students are assumed only to follow the teacher's instruction and requirements. The consistency between knowledge framing in texts and authority relationships in classrooms shows steps of social reproduction through schooling and classroom instruction.

4.2 Similarities and differences among East Asian systems

Our study showed that there are more similarities than differences among these three East Asian systems. All three Asian texts are strongly classified and, although the evidence with regard to framing is less clear in Hong Kong

[14] As part of efforts to improve U.S. students' performance in mathematics, there has been a trend in the U.S. for middle school mathematics teachers to become content specialists.

and Singapore texts, it appears that the framing of mathematical knowledge is similarly strong in all three Asian systems. This finding is consistent with the general social-cultural context that these three educational systems share.

However, we should also note that the math textbooks from Hong Kong and Singapore were written in English (versus Chinese), a residue of the fact that Hong Kong and Singapore (but not mainland China) were British colonial territories (Brimer, 1988; Thomas, 1988). Moreover, Hong Kong and Singapore contain a mixture of cultural elements from the East and West. In particular, "Singaporeans practice Western competitive individualism in the economic pursuits." (Haas, 1999:1, 3) Hong Kong also has been described as having "a mixed political culture with characteristics resembling traditional Chinese political culture and those 'imported' from the West."[15] Because of the role played by the English language and because of the influence of non-Confucian (i.e., western, individualist) culture, one might expect the textbooks from both Hong Kong and Singapore to be somewhat more like the ones used in the United States, in terms of classification and framing, than is the case for mainland China's textbooks. This suggests that further studies are needed to examine possible subtle differences embedded in curriculum that reflect social-cultural influences.

4.3 Differences between algebra and non-algebra texts in the U.S.

Our findings indicate that the US-Algebra textbook exhibits a higher degree of classification and framing than the US-Non-Algebra textbooks. Because of the social class stratified curricular tracks in many schools in the United States (Oaks, 1985; Persell, 1977; Spring, 2000), this probably means that textbooks with stronger classifi-cation and framing (what Bernstein terms a *collection code*) are more likely to be used with eighth graders from middle and upper middle class families, while the textbooks exhibiting weaker classification and framing (what Bernstein terms an *integrated code*) would be used in classrooms populated predominantly by students from lower middle and working class families.

This is interesting in light of previous theoretical and empirical work based in England. For example, Bernstein (Bernstein, 1975:116-56) theorized that a strongly classified and framed curriculum is likely to be preferred by families that have explicit, pre-defined, formal (*position-oriented*) authority relations, as tends to be the norm among members of the

[15] Yee, p.12; drawing upon and quoting from Lau Siukai and Kuan Hsinchi, *The Ethos of the Hong Kong Chinese* (Hong Kong: The Chinese University Press, 1988), pp.71-74.

old (or lower) middle class and working class. In contrast, Bernstein postulated that new (or upper) middle class families, which tend to manifest implicit, continually negotiated, informal (*person-oriented*) authority relations, are likely to prefer weakly classified and framed curricula.[16] However, MacDonald (MacDonald, 1977:33) observed that in schools in England in the 1960s and 1970s a weakly classified curriculum was more likely to be introduced in secondary modern schools, which were disproportionately populated by students of working class and lower middle class families, which might be characterized as having position-oriented authority relations.

Thus, our findings here appear to be in line with McDonald's observations and seem to contradict Bernstein's theorizing. This conclusion, though, must be considered in light of the differences in the status of knowledge as organized in mathematics or other subjects within the "competitive academic curriculum." (Connell, 1985) Perhaps family preferences for the type of knowledge code vary depending on the status of the subject area or perhaps the traditional ideas regarding the control and organization of knowledge by academic experts outweighs the preferences of families.

5. CONCLUSION

The finding that the differences in the societal norms of authority match with differences in the degree of classification and framing evidenced in textbooks suggests that it may be very difficult to change the curriculum and pedagogy organized through textbooks without also pursuing the more challenging task of altering societal norms regarding the nature of authority relations in a given society. At the same time, the findings that there are variations in classification and framing among Asian nations and (more so) among different texts used in the United States suggest that there may be some room for creative intervention, even in the context of countervailing norms regarding authority. The study illustrates the complexity of curriculum reform and suggests the importance of examining the socio-cultural characteristics before seeking to adopt curriculum practices from another system. Otherwise, without full consideration of the "conditions under which

[16] Note that Bernstein's conception of position-oriented versus person-oriented authority relations in families is parallel to Wilson's conception of positional and contractual compliance ideologies relations associated with societal level political cultures (see earlier discussion).

certain foreign practices deliver desirable results,"[17] [17] (Noah, 1986:162; LeTendre, Baker, Akiba, Goesling and Wiseman, 2001) simple adoption of foreign practices may constitute an "abuse" of comparative education. Meanwhile, this study shows the value of examining knowledge codes signaled by the organization and control of curriculum content knowledge. It opens another window through which we can examine possible opportunities and resistances along the process of knowledge selection, transmission, and evaluation in a society.

REFERENCES

Almond, Gabriel and Verba, Sidney, 1963/1989, *The Civic Culture: Political Attitudes and Democracies in Five Nations*, Sage Publications, Newbury Park, CA.

Apple, Michael, 1981, Reproduction, Contestation and Curriculum: An Essay in Self Criticism, *Interchange*, (Fall):30.

Beaton, A., Mullis, I., Martin, M., Gonzalez, E., Kelly, D., and Smith, T., 1996, *Mathematics Achievement in the Middle School Year: IEA's Third International Mathematics and Science Study (TIMSS)*, TIMSS International Study Center, Boston College, Chestnut Hill, MA.

Bernstein, Basil, 1975a, On the Classification and Framing of Educational Knowledge, in: *Class, Codes and Control, Volume 3: Towards a Theory of Educational Transmission*, Routledge and Kegan Paul, London.

Bernstein, Basil, 1975b, On the Curriculum, in: *Class, Codes and Control, Volume 3: Towards a Theory of Educational Transmission*, Routledge and Kegan Paul, London.

Bernstein, Basil, 1975c, Class and Pedagogies: Visible and Invisible, in: *Class, Codes and Control, Volume 3: Towards a Theory of Educational Transmission*, Routledge and Kegan Paul, London, pp.116-156.

Brimer, M.A., 1988, Hong Kong, *The Encyclopedia of Comparative Education and National Systems of Education*, in: T.N. Postlethwaite, ed., Pergamon Press, Elmsford, NY, pp.332-38;

Bussi, M.G. Bartolini, 1996, Mathematical Discussion and Perspective Drawing in Primary School, *Educational Studies in Mathematics*, **31**:11-41.

Connell, Robert , 1985, *Teachers Work*, George Allen and Unwin, Sydney, Australia.

Eisner, E., 1987, Why the Textbook Influences the Curriculum, *Curriculum Review*, **26**(3): 11-13.

Fuson, Karen, Stigler, James and Bartsch, Karen, 1988, Brief Report: Grade Placement of Addition and Subtraction Topics in Japan, Mainland China, the Soviet Union, Taiwan, and the United States, *Journal for Research in Mathematics Education*, **19**(5):449-456.

Haas, Michael, 1999, The Singapore Puzzle, in: *The Singapore Puzzle*, Michael Haas, ed., Praeger, Westport, CN, 1999.

[17] Note with caution, however, "that the unspecified assumptions inherent in an image of national cultures leads to too much idealization and celebration of national differences" (p. 13), when in fact global dynamics yield degrees of similarity across nations with respect to educational and societal structures and cultures.

Kennedy, Mary, 1997, The Connection between Research and Practice, *Educational Researcher*, **26**(7):8.

Lerman, S., 1998, Research on Sociocultural Perspectives of Mathematics Teaching and Learning, in: *Mathematics Education as a Research Domain: A Search for Identity: An ICMI study*, Book 2, A. Sierpinska and J. Kilpatrick, eds., Kluwer, (Dordrecht, The Netherlands, pp.333-350.

LeTendre, G., Baker, D., Akiba, M., Goesling, B., and Wiseman, A., 2001, Teachers' Work: Institutional Isomorphism and Cultural Variation in the U.S. Germany, and Japan, *Educational Researcher*, **30**(6) (August/September):3-15.

Li, Yeping, 1999, *An Analysis of Algebra Content, Content Organization and Presentation, and To-be-solved Problems in Eighth-grade Mathematics Textbooks from Hong Kong, Mainland China, Singapore, and the United States*, Unpublished University of Pittsburgh doctoral dissertation, Pittsburgh, PA.

MacDonald, Madeline, 1977, Culture, Class and the Curriculum, in: *Culture, Class and the Curriculum and the Politics of Curricular Reform* [Schooling and Society, Block III: Knowledge, Ideology and the Curriculum], Roger Dale and Madeline MacDonald, eds., Open University Press, Milton Keynes, England.

Mathematical Sciences Education Board, 1998, *Reshaping School Mathematics: A Philosophy and Framework of Curriculum*, National Academy Press, Washington, DC.

Mayer, Richard, Sims, V., and Tajika, H., 1995, A Comparison of How Textbooks Teach Mathematical Problem Solving in Japan and the United States, *American Educational Research Journal*, **32**:443-460.

McKnight, C.C., Crosswhite, F.J., Dossey, J.A., Kifer, E., Swafford, J.O., Travers, K.J., and Cooney, T.J., 1989, *The Underachieving Curriculum: Assessing U.S. School Mathematics from an International Perspective*, Champaign, Stipes, IL.

Neher, Clark, 1999, The Case for Singapore, in: *The Singapore Puzzle*, Michael Haas, ed., Praeger, Westport, CN.

Noah, Harold J., 1986, The Uses and Abuses of Comparative Education, in: *New Approaches to Comparative Education*, Philip Altbach and Gail Kelly, eds., University of Chicago Press, Chicago, IL.

Oaks, Jeannie, 1985, *Keeping Track: How Schools Structure Inequality*, Yale University Press, New Haven, CN.

Persell, Caroline Hodges, 1977, *Education and Inequality: The Roots and Results of Stratification in America's Schools*, The Free Press, New York.

Pye, Lucian, 1998, *The Mandarin and the Cadre: China's Political Cultures*, Center for Chinese Studies, University of Michigan, Ann Arbor, MI.

Pye, Lucian, 1985, *Asian Power and Politics: The Cultural Dimensions of Authority*, Harvard University Press, Cambridge, MA.

Reid, W.A., 2000, Curriculum as an Expression of National Identity, *Journal of Curriculum and Supervision*, **15**:113-122.

Schmidt, W.H., McKnight, C.C., and Raizen, S.A., 1997a, *A Splintered Vision: An Investigation of U.S. Science and Mathematics Education*, Kluwer Academic Press, Dordrecht, the Netherlands.

Schmidt, W.H., McKnight, C.C., Valverde, G.A., Houang, R.T., and Wiley, D.E., 1997b, *Many Visions, Many Aims (Volume 1): A Cross-national Investigation of Curricular Intentions in School Mathematics*, Kluwer Academic Press, Dordrecht, the Netherlands.

Schmidt, W.H., McKnight, C.C., Valverde, G.A., Houang, R.T., and Wiley, D.E., 1997c, *Characterizing Pedagogical Flow: An Investigation of Mathematics and Science Teaching in Six Countries*, Kluwer Academic Press, Dordrecht, the Netherlands.

Silver, Edward, 1998, *Improving Mathematics in Middle School: Lessons from TIMSS and Related Research*, U.S. Department of Education, Washington, D.C.

Spring, Joel, 2000, The Social Structure and American Education, in: *American Education*, 9th edition, McGraw Hill, New York.

Stevenson, H., and Stigler, J., 1992, *The Learning Gap: Why our Schools are Failing and What We can Learn from Japanese and Chinese Education*, Summit Books, New York.

Thomas, R. Murray, 1988, Singapore, in: *The Encyclopedia of Comparative Education and National Systems of Education*, in: T.N. Postlethwaite, ed., Pergamon Press, Elmsford, NY, pp.594-597.

Westbury, Ian, 1992, Comparing American and Japanese Achievement: Is the United States really a Low Achiever?, *Educational Researcher*, (June-July):18-24.

Wilson, Richard, 1992, *Rethinking Political Culture*, Cambridge University Press, Cambridge, UK

Yee, Hebert, 1999, *The Political Culture of China's University Students: A Comparative Study of University Students in Mainland China, Hong Kong, Macao, and Taiwan*, Nova Science Publishers, Commack, NY.

Chapter 2-4

COMPARATIVE STUDY OF ARITHMETIC PROBLEMS IN SINGAPOREAN AND AMERICAN MATHEMATICS TEXTBOOKS

YEAP Ban-Har[1]; Beverly J. FERRUCCI[2] and Jack A. CARTER[3]
[1]Nanyang Technological University; [2]Keene State College; [3]California State University

1. INTRODUCTION

Mathematical problem solving is a focus of school mathematics internationally. In the United States of America (USA) problem solving received renewed attention in the 1980s after the publication of *An Agenda for Action* (National Council of Teachers of Mathematics, 1980). Nearly a decade later, NCTM (National Council of Teachers of Mathematics, 1989) continued to emphasize that "problem solving should be the central focus of the mathematics curriculum" (p.23). The recent *Principles and Standards for School Mathematics* (National Council of Teachers of Mathematics, 2000) reinforced a vision of school mathematics where students are "flexible and resourceful problem solvers" (p.3). Halfway around the world in Singapore, the primary aim of the mathematics curriculum for the last two decades has also been to enable students to develop their ability in mathematical problem solving (Ministry of Education of Singapore, 1990, 2000).

This paper reports part of a larger study that compared elementary school mathematics textbooks in Singapore and the USA. The purpose of the research reported in this paper was to compare arithmetic problems included in elementary school textbooks used in the two countries as well as to investigate how the textbooks portrayed the teaching of problem solving. Previous research has indicated that textbooks affect both what and how teachers teach (Robitaille & Travers, 1992). The *Third International*

Mathematics and Science Study (Beaton, Mullis, Martin, Gonzales, Kelly, & Smith, 1996) found that textbooks were a main written source that teachers used to plan and conduct their lessons.

2. METHODOLOGY

2.1 The textbooks

The present study analyzed textbooks used in Primary 5 in Singapore and in Grade 5 in the USA. The American textbook was recently published for use in the state of California. The Singapore textbook was a recent edition that is currently used in most of Singapore's primary schools. Also, both textbooks are for children in the same age group, contain many similar topics, and are from publishers whose texts are widely used in the respective countries.

The American textbook comprised 629 pages, while the Singapore textbook was subdivided into a series of four booklets, two of which were workbooks. The Singapore textbook had a total of 400 pages. Chapters in both books that contained the same arithmetic topics were identified for analysis. In all, 286 pages out of 400 (72%) of the Singapore textbook and 355 pages out of 629 (56%) of the American textbook were selected for analysis. All the worked examples and practice tasks on these pages were analyzed.

2.2 Analysis framework

Researchers have defined a problem as a situation that needs to be resolved and a way to achieve that resolution is not immediately obvious (Kilpatrick, 1985; Krulick & Rudnick, 1987). Moreover, problems are distinct from routine exercises. In routine exercises, a procedure or way to resolve the situation that is presented in the exercise is immediately obvious (National Council of Teachers of Mathematics, 1969). Each worked example and practice task in the selected parts of the textbooks was first classified either as an exercise or a problem. Problems among the worked examples and practice tasks were identified for subsequent analysis. Computation exercises were excluded. The subsequent analysis framework for the worked examples comprised three aspects, while that for the practice tasks consisted of four aspects.

Analysis of Worked Examples: Worked examples that satisfied the criteria of being mathematical problems were further analyzed in three aspects. Firstly, the worked examples were examined with respect to how they modeled problem-solving stages. The solutions to these worked examples were analyzed to determine if the problem-solving stages of understanding, planning, executing and looking back (Polya, 1957) were modeled. Table 1 gives the operational definition of the occurrence of each of these four stages in a worked example.

Secondly, the problem-solving heuristics used in the worked examples were identified. It was also noted if the heuristics used were explicitly named. Heuristics that occurred more than once were coded. These included using number sentences, drawing a diagram, drawing a 'model', using reasoning, and observing a pattern.

Thirdly, the worked examples were also analyzed to assess if they effectively modeled (a) teaching of problem solving by emphasizing the problem-solving processes, (b) teaching for problem solving by focusing on the application of mathematical content, and (c) teaching through problem solving by emphasizing the use of problems to teach the content.

Table 2-4-1. Operational Definitions of Problem-Solving Stages

Stage	Operational Definition
Understanding	This stage is explicitly mentioned. Or comprehension questions are asked. Or there is an explicit encouragement for students to engage in this stage.
Planning	This stage is explicitly mentioned. Or possible plans are suggested. Or there is an explicit encouragement for students to engage in this stage.
Executing	This stage is explicitly mentioned. Or a solution is given. Or there is an explicit encouragement for students to engage in this stage.
Looking Back	This stage is explicitly mentioned. Or there is some reconsideration of the solution such as solving the problem in a different way or checking the answers. Or there is an explicit encouragement for students to engage in this stage.

Analysis of Practice Tasks: The tasks were initially classified as word problems, non-word problems, computation exercises, communication tasks or practical activities. The problems were analyzed in four aspects. Firstly, each problem text was analyzed to determine if there was sufficient, insufficient or excess information for a solution.

Secondly, each set of tasks was analyzed to determine if there were isomorphic problems within the same set of tasks. Problems were isomorphic if they contained exactly the same semantic relations (Marshall, 1995) and the same combinations of givens and unknowns. Table 2 provides a brief explanation of the types of semantic relations in arithmetic word problems.

Table 2-4-2. Semantic Relation in Arithmetic Word Problems

Relation	Example
Change	Chris had 623 stamps. He gave his friend 572 stamps. How many stamps does Chris have now?
Group	Ginny baked 315 chocolate cookies. She also baked 59 vanilla cookies. How many cookies did she bake?
Restate	Roslan has 316 marbles. Ray has 49 marbles more than Roslan. How many marbles does Ray have?
Vary	There are 25 cookies. Vani puts the cookies equally into 5 boxes. How many cookies are there in each box?
Compare	Ming has 120 stamps. Mei has 115 stamps. Who has more stamps?

The two word problems in Figure 1 were coded as isomorphic. Both contain the same semantic relation. The unknowns in both problems are of the same nature.

In an examination, 40 out of 44 pupils passed. What fraction of the pupils passed the examination?

Minah earns $350 a month. She saves $70 each month. What fraction of her earnings does she save?

(p. 44, CPDD, 1999a)

Figure 1. Isomorphic problems

Thirdly, a semantic analysis revealed single or multiple relations as well as algebraic or arithmetic structures. A multi-relation problem contained two or more semantic relations. Each problem was also coded as either having algebraic or arithmetic structures. In a word problem with arithmetic structure, there is only one unknown at any step of the problem solving process. In a word problem with algebraic structure, there is more than one unknown at some stage of the problem solving process. Table 3 shows examples of arithmetic and algebraic problems, as well as single-relation and multi-relation problems. Previous studies (Yeap & Kaur, 2001) have found that multi-relation problems were more difficult than single-relation ones, and problems with algebraic structure were more difficult than those with arithmetic structure.

Table 2-4-3. Word Problems with Different Semantic Structures

Word Problem	Semantic Structure	
There are 1400 pupils in a school. $^1/_4$ of the pupils wear spectacles. $^2/_7$ of those who wear spectacles are boys. How many boys in the school wear spectacles? (p. 66, CPDD, 1999c)	Arithmetic	Single-relation
Mr Chen had \$600. He gave $^3/_5$ of it to his wife and spent $^3/_8$ of the remainder. How much did he spend? (p. 68, CPDD, 1999c)	Arithmetic	Multi-relation
Mrs Chen made some tarts. She sold $^3/_5$ of them in the morning and $^1/_4$ of the remainder in the afternoon. If she sold 200 more tarts in the morning than in the afternoon, how many tarts did she make? (p. 70, CPDD, 1999c)	Algebraic	Multi-relation

Finally, the problems were analyzed to determine if they have a propensity to model reality within the solution process. A problem was said to have modeled reality if the data used in the problems were from an authentic source or if the solution required the problem solver to make sense of a computation. For example, the problem "A bag of cat food contains 35 cups of food. Each day, a pet store uses 3 cups of food. How many days will one bag of cat food last?" (p. 381, Houghton Mifflin, 2001) was considered to have modeled reality because some qualitative reasoning was required for the solution. Computation alone would not yield a sensible solution. A typical computation performed to solve this word problem is $35 \div 3 = 11$ remainder 2 or $11\frac{2}{3}$. Pupils need to consider the remainder in the computation and the exact way the 3 cups of food are given to the pets each day to offer a realistic answer.

3. FINDINGS

3.1 Case study on worked examples

Worked examples that were selected for further analysis included those that were considered to be problems and excluded routine computations. Using this criterion to select worked examples for analysis, 55 worked examples were analyzed in the Singapore textbook while 62 were analyzed in the American textbook.

Did the worked examples model the problem-solving process? Table 4 shows how the worked examples modeled the problem-solving process. In the Singapore textbook, 45% of all worked examples demonstrated only the Executing stage. None of the worked examples in the American textbook showed only this stage.

The findings suggested that the problem-solving process was more explicit in the American textbook. The Singapore textbook tended to place much greater emphasis on the execution part of the problem-solving process. Nearly a quarter of the worked examples in the American textbook modeled all the four stages of Polya's model.

Table 2-4-4. The Occurrence of Polya's Problem-Solving Stages in the Worked Examples

Polya's Stages	Singapore Textbook	American Textbook
Number of examples that modeled all the four stages.	0 (0%)	14 (23%)
Number of examples that modeled Understanding stage.	13 (24%)	15 (24%)
Number of examples that modeled Planning stage.	12 (22%)	59 (95%)
Number of examples that modeled Executing stage.	55 (100%)	62 (100%)
Number of examples that modeled Looking Back stage.	8 (15%)	25 (40%)

What heuristics were used in the worked examples? Table 5 shows the frequency of occurrence of problem-solving heuristics in the worked examples.

Table 2-4-5. Problem-Solving Heuristics Used in Worked Examples

Heuristics	Singapore Textbook	American Textbook
Writing a number sentence	30 (43%)	41 (64%)
Drawing a picture / diagram	11 (16%)	4 (6%)
Drawing a 'model'	22 (31%)	1 (2%)
Using reasoning	7 (10%)	2 (3%)
Observing a pattern	0 (0%)	2 (3%)
Others	0 (0%)	14 (22%)

The 'model' heuristic was a characteristic feature of the Singapore textbook (Ferrucci, Yeap & Carter, 2001, 2002). It was observed only once among the worked examples in the American textbook. One feature of the 'model' heuristic is the algebraic thinking involved. Figure 2 shows how the 'model' heuristic was used in an example in the Singapore textbook. In the American textbook, heuristics were explicitly named in about a third of the worked examples while in the Singapore textbook heuristics were not named.

Encik Hassan gave $^{2}/_{5}$ of his money to his wife and spent $^{1}/_{2}$ of the remainder. If he had \$300 left, how much money did he have at first?

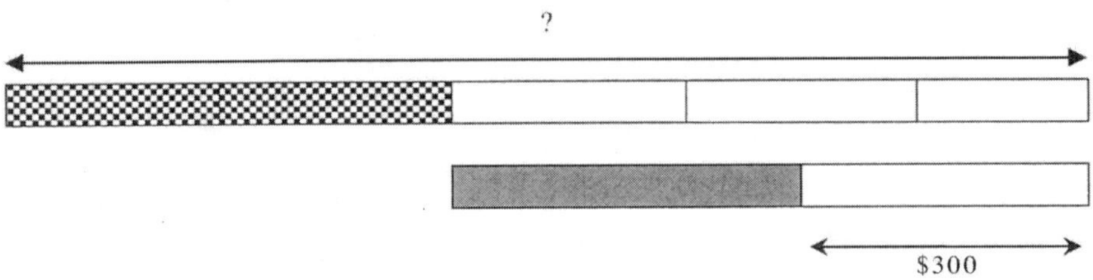

Figure 2. Using the 'model' heuristic

The Singapore textbook included four heuristics and each of them was used repeatedly. The American textbook included about twenty heuristics and more than ten of these heuristics were used only once. Both textbooks placed significant emphasis on the use of number sentences.

One notable difference between the two textbooks was the emphasis of the Singapore textbook on heuristics that are diagrammatic in nature. The American textbook did not place an emphasis on diagrammatic heuristics. Another difference was that while the American textbook provided experiences with a large number of heuristics, the Singapore textbook tended to focus on a few heuristics.

What were the purposes of the worked examples? The purpose of the worked examples were broadly classified as (a) teaching of problem solving by emphasizing processes, (b) teaching for problem solving by focusing on the use of mathematics content to solve problems, and (c) teaching through problem solving by emphasizing the use of problems to teach mathematics. Table 6 shows the number and proportion of worked examples that exemplified each of these three purposes. In the American textbook, two in three worked examples served the purpose of contextualizing new mathematical concepts. In the Singapore textbook, new concepts were more often presented in the bare essential form. The worked examples that followed frequently showed how the concepts were used in word problems.

Table 2-4-6. The Purpose of the Worked Examples

Purpose	Singapore Textbook	American Textbook
Teaching of problem solving	12 (22%)	21 (34%)
Teaching for problem solving	28 (51%)	0 (0%)
Teaching through problem solving	15 (27%)	41 (66%)

3.2 Case study on fraction problems

In this paper, only the analysis on fraction problems is presented. About 20% of all tasks in the Singapore textbook were word problems (72 out of 343 tasks). About 28% of all tasks in the American textbook were word problems. Problems other than word problems were almost totally absent in the two textbooks.

Were the word problems semantically complex? Table 7 shows the number and proportion of problems with different semantic structures.

Table 2-4-7. Semantic Structures of Word Problems

Complexity of Problems	Singapore Textbook	American Textbook
Single-Relation Arithmetic	38 (56%)	39 (74%)
Multi-Relation Arithmetic	20 (29%)	13 (25%)
Multi-Relation Algebraic	10 (15%)	1 (1%)

Although algebra is not formally taught to Primary 5 students in Singapore, there were many algebraic problems that required the use of the 'model' method. The word problems in the Singapore textbook tended to be semantically more complex than those in the American textbook.

Did the word problems require sense making in their solutions? Nearly all the word problems in both textbooks did not require students to interpret answers. There were a few problems in the American textbook that did so, but none of the problems in the Singapore textbook required the consideration of contextual factors in the solution. Similarly, the use of data from authentic sources in word problems was absent.

Were the problems isomorphic to others in the same set of problems? In half of the 16 sets of problems in the Singapore textbook there were isomorphic problems. The American textbook rarely had isomorphic problems within a set of problems. This analysis revealed another major difference between the textbooks. The problems in the American textbook tended to be related by context. For example, problems in the same set in the American textbook were based on a theme such as sports. Sets of problems in the Singapore textbook tended to contain problems with the same structure with some kind of variation.

Table 8 shows some word problems taken from the same problem set. Among the ten word problems, there were only two types of semantic structure. One of these structures was found in five of the ten problems. Although all the five word problems were based on the same combination of semantic relations, there was always some kind of variation as pupils progress through the set of problems. For example, the problems used both discrete quantities such as stickers as well as continuous quantities such as kilograms and dollars. For the Group semantic relation, the total quantity was given in some problems (Questions 1, 3 and 6) and was the unknown in

Table 2-4-8. Variation in Word Problems with the Same Semantic Structure

Semantic Structure	Word Problem
Practice 1D, Question 1	John is 15 kg heavier than Peter (*Restate*). Their total weight is 127 kg (*Group*). Find John's weight.
Practice 1D, Question 2	There are 3 times as many boys as girls (*Restate*). If there are 24 more boys than girls (*Restate*), how many children are there altogether (*Group*)?
Practice 1D, Question 3	The total weight of Peter, David and Henry is 123 kg (*Group*). Peter is 15 kg heavier than David (*Restate*). David is 3 kg lighter than Henry (*Restate*). Find Henry's weight.
Practice 1D, Question 6	Peter has twice as many stickers as Ali (*Restate*). Ali has 40 more stickers than Lihua (*Restate*). They have 300 stickers altogether (*Group*). How many stickers does Peter have?
Practice 1D, Question 10	John and Paul spent \$45 altogether (*Group*). John and Henry spent \$65 altogether (*Group*). If Henry spent 3 times as much as Paul (*Restate*), how much did John spend?

others (Questions 2 and 10). For the Restate semantic relation, the relationship between quantities was additive (15 kg heavier) in some problems (Questions 1 and 3) and multiplicative (3 times as much) in others (Question 10). In yet other problems, both additive and multiplicative relationships were included (Questions 2 and 6). Some variations were slight. For example, the only difference between Questions 1 and 3 was the number of quantities involved. In Question 1, there were two quantities in the relation, while in Question 3 there were three.

Were there word problems with insufficient information in the text of the problems? In the Singapore textbook, there was always sufficient information in the text for a solution. In the American textbook, there were instances (about 10% of the problems) when students were required to locate information from a table or a graph. The American textbook also explicitly showed examples where there was insufficient or extraneous information. This was absent in the Singapore textbook.

4. CONCLUSION & DISCUSSION

This paper is based on a larger study that investigated a selection of elementary school textbooks used in the United States of America and Singapore. The case study on worked examples indicated that (a) the teaching of problem solving was more explicit in the American textbooks, (b) the Singapore textbook used a larger proportion of diagram-based heuristics, (c) the American textbooks provided opportunities for students to know more heuristics, and (d) the American textbook often employed word problems to teach mathematics, while the Singapore textbook frequently used word problems as a means to apply mathematics.

The case study on fraction problems indicated that (a) the Singapore textbook included a greater proportion of semantically complex problems, including algebraic problems, (b) word problems in the American textbook tended to be related by story context, while those in the Singapore textbook tended to be similar in semantic structure, (c) word problems in the Singapore textbook did not require students to handle superfluous or insufficient information, while those in the American textbook did.

The differences in the way textbooks portrayed problem solving in the two cultures cannot be attributed to fundamental differences in the way problem solving is represented in the curriculum. The Singapore mathematics curriculum, which primarily aims to develop ability in problem solving, was developed by mathematics educators who were more familiar with the idea of problem solving as reflected in NCTM's documents. The notion of problem solving in the Singapore curriculum has its roots in the

American, British and Australian mathematical problem-solving literature. There is no distinctive difference between the representation of problem solving in the Singapore curriculum and in NCTM's documents. This cannot be said for some Asian countries. The Japanese, for example, developed and implemented their own notion of problem solving in the *open-ended approach* (Becker & Shimada, 1997).

It is, therefore, interesting that textbook authors in Singapore and the USA translated similar notions of problem solving in distinct ways. The authors of the Singapore textbook seemed to have retained several beliefs about mathematics learning that are characteristically East Asian. These included the emphasis on practice, mastery of mathematics content and examination preparation. The authors of the American textbook seemed to be guided by the prevalent rhetoric of emphasizing problem-solving processes and links to everyday experiences.

The emphasis on practice resulted in sets of arithmetic problems with gradual variation in semantic structure in the Singapore textbook. The idea of practice amongst Singaporean, and perhaps many other East Asian, teachers is to provide a collection of similar tasks with some kind of variation. In contrast, problem sets in the American textbook tended to be grouped thematically, rather than semantically. Such a thematic approach, as well as the use of problems to introduce new content, reflected the importance placed on relating mathematics to everyday experiences in the American textbook.

The high expectation typically placed on East Asian pupils to master mathematics content probably explains the inclusion of semantically complex problems in the Singapore textbook. The American textbook provided opportunities for pupils to review content done in earlier grades. This resulted in a relatively higher proportion of single-relation word problems in the Grade 5 American textbook. Singaporean pupils were not provided with similar opportunities as the assumption was that they have mastered the content taught in the earlier grades. Thus, most of the word problems in the Singapore textbook were of the complexity level for that grade level.

High-stake examinations have been an important feature of schooling in East Asian countries. Such examinations are invariably of the paper-and-pencil type, consisting of tasks with one correct answer. Such tasks are, therefore, carefully written in an unambiguous manner. The word problems in the Singapore textbook were written in this manner. Thus, unlike some problems in the American textbook, those in the Singapore textbook did not require pupils to handle superfluous or insufficient information.

While the 1980s was the problem-solving decade in the USA, the Singapore curriculum only included problem solving in the 1990s. The rhetoric of problem solving is still not pervasive among Singapore mathematics teachers. This time lag probably explains the absence of explicit modeling of the problem-solving process and the inclusion of only a few heuristics in the Singapore textbook. It is expected that the Singapore textbook will explicitly model the problem-solving process and include a wider range of heuristics.

It will be interesting to investigate, in a few years, if textbooks in the two cultures specifically, and in different cultures generally, tend to be more similar than different. If they do, perhaps it is because of more extensive exchanges among mathematics educators from both the East and the West. If they do not, perhaps it is because there exist textbook features that are anchored in long-held beliefs that are resistant to change.

REFERENCES

Beaton, A.E., Mullis, I.V.S., Martin, M.O., Gonzales., E.J., Kelly, D.L., and Smith, T.A., 1996, *Mathematics Achievement in Middle School Years: IEA's Third International Mathematics and Science Study (TIMMS)*. Center for the Study of Testing, Evaluation and Educational Policy, Boston College, Boston, MA.

Becker, J.P., and Shimada, S., eds., 1997, *The Open-ended Approach: A New Proposal for Teaching Mathematics*. National Council of Teachers of Mathematics, Reston, VA.

Curriculum Planning and Development Division (CPDD), 1999a, *Primary Mathematics 5A*. Federal Publications, Singapore.

Curriculum Planning and Development Division (CPDD), 1999b, *Primary Mathematics 5B*, Federal Publications, Singapore.

Curriculum Planning and Development Division (CPDD), 1999c, *Primary Mathematics 5A Workbook*. Federal Publications, Singapore.

Curriculum Planning and Development Division (CPDD), 1999d, *Primary Mathematics 5B Workbook*. Federal Publications, Singapore.

Ferrucci, B.J., Yeap, B.H., and Carter, J.A., 2001, Achieving Understanding in Solving Word Problems. *Pythagoras*, **54**(April):26-29.

Ferrucci, B.J., Yeap, B.H., & Carter, J.A., 2002, Developing Algebraic Thinking, *Mathematics Teaching*, *178*(March).

Houghton Mifflin, 2001, *Houghton Mifflin Mathematics (California Edition)*, Houghton Mifflin, Morris Plain, NJ.

Kilpatrick, J., 1985, A Retrospective Account of the past 25 years of Research on Teaching Mathematical Problem Solving, in: *Teaching and Learning Mathematical Problem Solving: Multiple Research Perspectives*, E.A. Silver, ed., Lawrence Erlbaum, Hillsdale, NJ.

Krulik, S., and Rudnick, J.A., 1987, *Problem Solving: A Handbook for Teachers (Second Edition)*, Allyn and Bacon, Boston.

Marshall, S.P., 1995, *Schemas in Problem Solving*. Cambridge University Press, New York.

Ministry of Education of Singapore, 1990, *Mathematics Syllabus: Primary*. Ministry of Education, Singapore.

Ministry of Education of Singapore, 2000. *Mathematics syllabus: Primary*. Curriculum Planning and Development Division, Singapore.

National Council of Teachers of Mathematics, 1969, *Hints for Problem Solving*. National Council of Teachers of Mathematics, Reston, VA.

National Council of Teachers of Mathematics, 1980, *Agenda for Action: Recommendations for School Mathematics for the 1980s*. National Council of Teachers of Mathematics, Reston, VA.

National Council of Teachers of Mathematics, 1989, *Curriculum and Evaluation Standards for School Mathematics*. National Council of Teachers of Mathematics, Reston, VA.

National Council of Teachers of Mathematics, 2000, *Principles and Standards for School Mathematics*. National Council of Teachers of Mathematics, Reston, VA.

Polya, G., 1957, *How to Solve it*. Second Edition. Anchor Books, New York.

Robitaille, D.F., and Travers, K.J., 1992, International Studies of Achievement in Mathematics, in: *Handbook of Research on Mathematics Teaching and Learning*, D.A. Grouws, ed., Macmillan, New York, pp.687-709.

Yeap, B.H., and Kaur, B., 2001, Semantic Characteristics that make Arithmetic Word Problems Difficult, in: *Numeracy and Beyond*, J. Bobbis, B. Perry, and M. Mitchelmore, eds., MERGA Inc., Sydney, pp.555-562.

Chapter 2-5

A COMPARATIVE STUDY OF THE MATHEMATICS TEXTBOOKS OF CHINA, ENGLAND, JAPAN, KOREA, AND THE UNITED STATES

PARK Kyungmee[1] and LEUNG Koon Shing Frederick[2]
[1]Hongik University; [2]The University of Hong Kong

1. INTRODUCTION

The purpose of this study is to compare the eighth-grade mathematics textbooks of China, Japan and Korea on the one hand, and those of England and the United States on the other, and to explore the implications for mathematics education in East Asia and the West.

There have been a number of international studies in mathematics education such as TIMSS and IAEP where the mathematics curricula (and student achievements) in various countries were compared (Beaton *et al,* 1996; Fuson *et al*, 1988; Mullis *et al,* 1997, 2000, Lapointe *et al*, 1992). Such "macro" comparison usually focus on the curriculum or curricula that are followed by the majority of the student population, and hence by their very nature tend to be crude. On the other hand, classroom studies (e.g. Leung, 1995) typically focus on activities led by a single or a handful of teachers out of tens of thousands in the population (the TIMSS Video Survey (Stigler and Hiebert, 1999) may be an exception), and is by their nature a "micro" approach. But for textbook studies, since each country rarely has a very large number of mathematics textbooks, if we select a popular series of textbooks in each country to study, as is the practice of the present study, it will cover a fair amount of the student population already. So textbook studies are somewhere between a macro and a micro approach in terms of scale.

The present study is based on an analysis of a series of eighth-grade mathematics textbooks in each of China (People's Education Press, 1999), Japan (Fukumori, 1997), Korea (Kim and Kim, 1997), England (Briggs, 1996; McGuire, P. and Smith, K. 1996), and the United States (Billstein and Williamson, 1999). Each of the East Asian series has the largest market share in its respective country. The English and American series are also popular ones, but since there is such a large number of textbooks with different market shares, no single textbook can be said to be dominant.

2. MAJOR FINDINGS

2.1 Textbook development and publication policies: u*niformity vs. diversity*

Textbook development and publication policies differ from country to country. In each of China, Japan and Korea, there is a uniform curriculum determined at the national level. In China, before 1992, there was only one series of textbooks published by the People's Education Press, the official publisher appointed by the Government, for the entire country. From 1992, selected publishers in different provinces have been commissioned to publish textbooks that serve different needs.

The textbooks for elementary schools in Korea are developed and published by the Ministry of Education, and there is only one series of mathematics textbooks in the country. The textbooks for the secondary schools in Korea and all the textbooks in Japan are published by private companies upon obtaining prior approval from the Ministry of Education. Not only does the content reflect closely the national curriculum, but the terms and symbols used mostly adhere to the curriculum as well. The Ministry of Education even stipulates the size, the number of pages, and the number of colors to be used in those privately published textbooks. Thus, textbooks in Korea and Japan look more or less identical.

On the other hand, the approval processes by the national educational authorities of England and the United States for commercially published textbooks are less strict than those in Korea and Japan. Publishing companies exercise their own discretion and publish a broad spectrum of textbooks, ranging from "conservative" (e.g. Saxon Publishers of the United States publishes traditional textbooks under the principle of "methods of incremental development and continual practice and review (http://www2. Philosophy.html saxonpub.com/corp)) to "radical" (e.g. the *Contemporary*

Mathematics in Context, consequential on the Core-plus Mathematics Project, with content structured according to themes rather than mathematical domains) (cf. Stevenson and Stigler, 1992, pp.138-9).

2.2 Choice of content: essential vs. discretionary

An obvious difference between the textbooks in the two Western countries and the East Asian countries is that there are more elective elements in the Western textbooks. This is of course just a reflection of the differences in the curricula of the respective countries mentioned above. In the East Asian countries, students, irrespective of their abilities and inclinations, are expected to follow the same curriculum and to learn the same content. In Western countries, there are often a number of different curricula to cater for the different needs and interests of the students. Even within the same curriculum, there is more flexibility built in so that teachers and students are presented with more choices in the teaching and learning. These different emphases are reflected in the Western textbooks where there are either different contents catering for students of different abilities or ample elective elements for teachers and students to choose from. For example, the textbooks of England's Oxford University Press are published in three different strands (higher, intermediate and foundation) to accommodate various levels of individual ability. And in the American textbook, repetitive and exploratory activities are given to help the lower ability students. At the same time, there are higher-level problems to challenge and motivate the more able ones.

As a result of this difference, the American and English textbooks are larger and thicker than the East Asian ones (see Table 1). This is also a consequence of the fact that the contents in the East Asian textbooks are introduced in a compressed way. In Asian countries, textbooks are regarded as a body of the minimum and essential knowledge that everyone must learn and understand. However, in Western countries, textbooks are rather like "little encyclopaedia" that contain numerous and various contents, from which appropriate topics are selected as desired.

This difference is related to the typical class size and teaching practice in the two cultures. In Asian countries, teaching is conducted in classes with large class size, and teachers speak to the entire class most of the time instead of focusing on individual students. In contrast, the class size in Western countries is much smaller, and individualized teaching or small group activities are more commonly adopted. These different practices may be a reflection of the different values in the educational systems of the two societies in handling individual differences. This will be elaborated in the Discussion Section later.

2.3 The role of textbooks in teaching and learning: *absolute vs. relative*

In many East Asian countries, teachers and students regard the textbook as a "bible" which contains all the essential knowledge. This is related to the last point on the availability of elective elements in the textbooks, and also a consequence of the adherence of the textbooks to the national curriculum mentioned in the first Section above. Since the public examinations in these countries adhere closely to the national curriculum, students rely on the textbooks heavily in order to pass the public examinations to gain access to the next stage of schooling. The common mind-set in these counties is that the textbooks must be learned from cover to cover (Stevenson and Stigler, 1992, p.141). In fact, textbooks are virtually the sole teaching tool around which class activities are organized. In China, for example, textbooks are referred to as "jiaocai", which literally means "teaching materials". So the textbook is equated with *the* teaching materials used in class.

It is true that textbooks in the United States and England are an important source of knowledge for teaching as well, but they are not accorded the same importance as in the East Asian countries. Class activities may or may not follow those suggested in the textbooks, and even when they do, they may not be in the same order as appear in the textbooks. Teachers sometimes adapt and modify the content in the textbooks, taking into consideration students' level of understanding or their interest, and very often teachers use materials outside textbooks in their teaching.

2.4 Physical appearance of textbooks: *plain vs. colourful*

Different physical appearances of textbooks suggest that they are meant to be used in different ways, and thus the physical appearance of textbooks has implications for pedagogical strategies and approaches (Schmidt *et al*, 1997, pp.37-38).

There are marked differences in the physical appearances of textbooks in different countries. Chinese, Japanese and Korean textbooks are small, thin volumes with few pictures that are mostly in black and white. This contrasts with the visually attractive textbooks in England and the United States, which feature various page layouts and many pictures in full colour. The various physical features of typical textbooks in different countries are summarized in Table 1 below:

The visually unimpressive Asian textbooks may fail to attract student interest. On the other hand, while Western textbooks may be able to arouse student interest through their interesting pictures and stories, these features may also cause some distraction to students in their learning of mathematics

(Stevenson and Stigler, 1992, p.139). This point will be elaborated further in the Discussion Section later.

Table 2-5-1. Physical Features of Textbooks in Different Countries

	China	England	Japan	Korea	The United States
Size (mm)	130×185	210×275	148×210	*148×210*	*223×265*
Number of pages	268	*240*	*214*	*323*	*620*
Pictures	few	many	few	*few*	*many*
Colour	black and white	full colour	mainly black and white	mainly black and white	*full colour*

2.5 Characteristics of the content

2.5.1 Strand and linear vs. thematic and spiral

Firstly, for the East Asian textbooks, each chapter is titled according to a mathematical curricular strand (e.g. Equations first, followed by Functions, and then Probability), and each chapter is composed of homogeneous content within the strand. In contrast, the American textbook in this study consists of modules centred around themes such as "Making choices" or "Search and rescue" and each module has several sections in which heterogeneous mathematical concepts are introduced (e.g. Algebra, Function and Statistics all in one section) (There are other American textbooks which follow the "strand" approach too.). This is also the case for the English textbook. For example, chapter titles with themes such as "High flyers", "Beadcraft", 'Cycling" and "Sea life" are used to introduce various mathematical contents of that chapter.

Secondly, East Asian textbooks have a linear structure within each grade, and one strand is usually covered by one chapter only (across grades, the structure may still be considered as a spiral one, for example, Linear Equations in the 7th grade, System of Linear equations in the 8th grade, and Quadratic Equations in the 9th grade). On the other hand, Western textbooks typically have several strands within one module, and any one specific strand is usually repeated over several modules. In contrast to the within grade linear structure seen in East Asian textbooks, there are many small spirals even within a grade.

Lastly, textbooks in East Asian countries consist of a limited number of components or features, and concentrate mostly on explanations, examples, and exercises. American and English textbooks, on the other hand, have various features (such as projects, practice and application exercises, technology, student self-assessments, and career connections, etc) that are rarely found in East Asian countries.

2.5.2 deductive vs. inductive

East Asian textbooks in general take a deductive approach in which general concepts are introduced before specific examples are given. For example, the 8th grade textbook in China presents the following rigorous proof in Euclidean Geometry (People's Education Press, 1999, p. 66):

> **Theorem on the property of isosceles triangles: The two base angles of an isosceles triangle are equal.**
>
> Given: In ΔABC, AB=AC.
>
> To prove: $\angle B = \angle C$
>
> **Proof**: Construct the angle bisector AD of the apex angle (mark the two angles $\angle 1$ and $\angle 2$ in the diagram).
>
> In ΔBAD and ΔCAD,
>
> AB = AC (given)
>
> $\angle 1 = \angle 2$ (by construction)
>
> AD = AD (common side)
>
> $\therefore \Delta BAD \cong \Delta CAD$ (SAS)
>
> $\therefore \angle B = \angle C$ (corresponding angles of congruent triangles equal)

The approach above is in marked contrast to the inductive approach that the American and English textbooks take. Typically, Western textbooks first present various activities and explorations in a realistic context, which serve as the "scaffolding" based on which students grasp the meanings of abstract mathematical concepts step by step. For example, the American textbook introduces the concept of trapezoid gradually by asking students to identify the common features of the geometric figures in the Food Guide Pyramid (except the triangle at the top). This is followed by the definition of the trapezoid after students' explorations (Billstein and Williamson, 1999, pp. 512-513). Such an approach takes into consideration students' psychological state and different levels of understanding, and allows students to familiarize themselves with new concepts step by step in a rather easy and natural manner. On the other hand, the Eastern approach demands more effort on the part of the students because the content is presented in a compact and efficient way.

These deductive and inductive approaches might have their roots in the Eastern and Western ways of thinking. Eastern students are taught not to doubt the teachings of their ancestors and great men of past generations, let alone argue against them. Thus, it is more likely for Eastern students to accept mathematical contents presented in a deductive way. However, this approach does not fit well with students of the West, who are used to accepting the facts only after ample evidence is given. In this regard, an inductive approach is more persuasive than a deductive approach for Western students.

2.5.3 content vs. context

East Asian textbooks usually contain minimal real world contexts, and give great weight to the mathematical content itself. On the other hand, mathematical contents in Western textbooks are usually presented in the context of various real world situations. For instance, the Korean textbook presents the concepts of Proposition, Hypothesis, and Conclusion immediately after only one simple question (Kim and Kim, 1997, pp. 202-203):

> Chapter 4 Properties of Geometric Figures
> 1. Proposition and Proof
> §1. Proposition
> Question: Which one is a true sentence?
> (1) The Earth is moving (2) 2-2>1
> (3) A triangle with two equal internal angles is a regular triangle
> The sentence in which we can decide whether it is true or false is a *proposition*.
> Question: Decide whether the following sentence is a proposition.
> When a, b, c are numbers, if a=b, then ac=bc.
> Usually, the part before "then" is a *hypothesis* and the part after "then" is a *conclusion*. For example, in the above proposition, a=b is a hypothesis and ac=bc is a conclusion.

Students are not led to think about why a proposition is needed, but are "forced" to internalize the new concepts of proposition, hypothesis, and conclusion. In contrast, Western textbooks typically present various introductory activities and explorations that are quite lengthy. Although they are supposed to set the mathematics content "in context", the "context" is not always effectively related to the mathematical concepts concerned, and the activities do not always guarantee that the mathematics is meaningful to the students. Sometimes they serve only as the "scaffolding" and students may fail to go beyond the scaffolding to grasp the mathematical ideas.

For instance, "Search and Rescue" is the theme of the second module of the American textbook. It begins with a paragraph quoted from *Hatchet* written by Gray Paulsen: A thirteen-year old boy was in a double-seater plane, when suddenly the pilot had a heart attack and the plane crashed. This section deals with several terms related to the concept of Angle such as angle, ray, vertex, degree, right angle, straight angle, acute angle, obtuse angle etc. The core mathematical concept in the section is only vaguely related to the example of the plane crash, i.e., through "reading a compass". Given the fact that a compass reads clockwise while angle measures in Geometry are usually done counter-clockwise, this example may even confuse students on how to read an angle and may lead to confusion that cannot be easily rectifiable.

2.5.4 Contrived vs. realistic

In China, Japan and Korea, calculators are rarely used in class. According to Mullis *et al* (2000), the percentages of Korean and Japanese students having access to calculators in class are 28% and 34% respectively, which were the lowest among the TIMSS-R countries. This situation affects the nature of the problems presented in the textbooks. Eastern textbooks include relatively few application problems, and even when application problems are presented in a real life context, the calculations involved are usually simple. For example, the following problem dealing with a system of linear equations is one of the application problems included in the Japanese textbook (Fukumori, 1997, p.36):

The prices of a rose and a lily are 200 yen and 300 yen respectively. How many flowers did you buy if you paid 2400 yen for 10 flowers combining roses and lilies?

Numerical values in the problem are artificially simplified because students have to do pencil-and-paper calculations. Thus the problems are contrived, even though they appear to be given in a real life context. This is why students in East Asian countries tend to regard mathematics as "petrified" knowledge only limited to the textbook.

In contrast, calculators are widely used in the Western classroom. Textbooks thus have more leeway to include various real life problems, unrestricted by the complexity of calculations. For instance, one of the tasks included in the English textbook presents the real exchange rates (up to the third decimal point) for the ten European countries (this was before the Euro replaced the respective European currencies), and asks about the amount of money exchanged from one currency to another (Briggs *et al*, 1996, p.126-127). An authentic data set can be used because the complexity of the calculations is not a concern. Through solving problems with realistic data,

Western students are relatively more apt to realize the usefulness of mathematics in their real lives.

3. DISCUSSION

3.1 Different views on the nature of mathematics: *"naked" mathematics* vs. *"dressed-up" mathematics*

East Asian textbooks tend to "force" students to learn the noble logical system of mathematics by presenting a combination of concepts, symbols, and algorithms in a decontextualised way. On the other hand, Western textbooks first lead students through various activities and explorative examples, and based on these, mathematical concepts are introduced within profound contexts. This difference may be a reflection of the different views on the nature of mathematics. If a Platonic view of mathematics is taken, and mathematics is considered as absolute truth in the realm of ideas, the teacher's role is naturally one of presenting mathematics concepts clearly and helping the "ignorant" students to acquire the mathematics. Real life examples may help arouse students' interest in the learning, but in the final analysis, the truth of mathematics does not depend on these real life examples. They may even distract students from the mathematical truth. But the contrasting fallibilist position sees mathematics as a human endeavour, with all the limitations and errors shared by all other human activities. Mathematics is born out of human activities, and so it is fitting that mathematics be learned "in context". The abstract mathematical concepts need to be "dressed up" in realistic situations again, not only to aid their learning, but also to return them to their true nature.

However, even if a fallibilist view is taken, such dressing-up of mathematics is a two-edged sword. The dressing-up is supposed to take into account student's different levels of understanding, which is of course desirable. On the other hand, such an approach may distract or mislead students when the scaffolding process, with various introductory activities, is only vaguely connected to the main content of mathematics. Sooner or later, students will have to face the hard mathematics that is concealed under the comfortable fancy outlook. The rich context contained in the Western textbooks may be an effective tool to draw students' interest momentarily. But they have their limitations since a continuing interest in mathematics is mostly acquired through earnest efforts to understand essentially difficult mathematical concepts.

In short, the difference between East Asian and Western textbooks derives from whether difficult, abstract mathematical concepts are presented in a "naked" or a "dressed-up" way. East Asian textbooks encourage students to encounter abstract ideas directly without side tracking while Western textbooks "embellish" the ideas in order to reach the students.

3.2 Different didactic phenomena in teaching and learning: *formal abidance vs. meta-cognitive shift*

As hinted at in the last Section, different views on the nature of mathematics have implications for different views on teaching and learning, which may in turn lead to different didactic phenomena. The didactic phenomenon which is likely to occur in the teaching and leaning with East Asian textbooks may be classified as "formal abidance", while the possible phenomenon which may occur in the teaching and learning with Western textbooks can be categorized as "meta-cognitive shift" (Brousseau, 1984, 1997; Kang, 1990).

East Asian textbooks tend to minimize the metaphorical use of knowledge and are full of logical presentation of formulated knowledge. Thus, the extreme didactic phenomena associated with East Asian textbooks is formal abidance, which is a consequence of de-emphasizing or ignoring students' personalization and contextualization while they are exposed to mathematical knowledge.

In contrast, meta-cognitive shift is a phenomenon which takes place when the process of personalization and contextualization of mathematical knowledge is overemphasized. In some cases, the didactical efforts of Westerns textbooks are shifted from the mathematical knowledge to a didactical device. For instance, in the example cited previously in Section 5C, there is a meta-cognitive shift from the content (angle) to the context (rescue situation). The problem with a meta-cognitive shift is that while it is desirable to draw students' interest with an appealing setting (context), the interest may not extend to the mathematical concept (content). Instead, students' interest may just linger on the situation itself and they may never get to learning the concept.

3.3 Different views of man: social vs. individual

The difference in the availability of elective elements mentioned in Section 2 is perhaps a reflection of a deeper philosophical difference between the two cultures. The East Asian culture believes in orthodoxy, and students are expected to adhere to the orthodoxy despite their individual differences. In the Western culture however, the individual is of paramount importance. Hence the curriculum has to adjust to the needs of the individual rather than the individual adjusting to an orthodox curriculum.

This philosophical difference hinges on different views on the nature of man. In the typical East Asian "social orientation" philosophy, in contrast to the "individual orientation" in Western cultures (Yang 1981), integration and social harmony is of prime importance. Man is defined in relation to other human beings in the community, and it is the obligation of the individual to adjust himself or herself to the social hierarchy. On the other hand, the Western "individual orientation" philosophy treasures the value of the individual. Each individual has his or her own rights, and the social system itself is expected to cater for the needs of the individual.

This difference has resulted in different approaches; East Asian countries introduce content with less consideration for individual differences while Western countries take pains to cater for individual differences.

4. CONCLUSION

As the textbook is the medium that articulates what should be taught in the curriculum, a study of the mathematics textbooks will reveal critical characteristics of both the intended and the implemented curricula in different countries, which may in turn reflect important differences between the cultural values in these countries.

As shown above, differences in various aspects of the textbooks in East Asia and the West reflect dissimilar social and cultural values. As the textbook is a powerful means through which students acquire both knowledge and values, these textbook differences may in turn reinforce the underlying cultural differences. The discussions above also show that each of the Eastern and Western approaches has its own strengths and weaknesses. For instance, textbooks in the West may help students realize how useful mathematics can be in their lives, but if the link between a mathematical concept and the corresponding real life situation is not made clear, sometimes students may not be able to completely grasp the mathematical concept. By contrast, the East Asian textbooks may succeed in conveying ideas in an economical way, but they often fail to motivate students to learn. Therefore it is important to take a critical view of each approach. It is through a critical understanding of the differences between different cultures that we are able to learn from each other and to put the results of such comparative studies to better use in the future.

REFERENCES

Beaton, A.E., Mullis, I.V.S., Martin, M.O., Gonzalez, E.J., Kelly, D.L. and Smith, T.A., 1996, *Mathematics Achievement in the Middle School Years*, Centre for the Study of Testing, Evaluation and Educational Policy, Boston College, Boston.

Briggs, S., 1996, *Oxford Mathematics (Foundation Link, Higher Link)*. Oxford University Press, Oxford.

Billstein, R., and Williamson, J., 1999, *Middle Grades MATH Thematics*. McDougal Littell Inc, Evanston, IL.

Brousseau, G., 1984, The Crucial Role of the Didactical Contract in the Analysis and Construction of Situations in Teaching and Learning Mathematics, in: *Theory of Mathematics Education (TME)*, H.G. Steiner, ed., (ICME 5-Topic Area and Miniconference: Adelaide, Austrailia, pp.110-119). Institute für Didacktik der Mathematik der Universität Bielefeld, Bielefeld, F. R.

Brousseau, G., 1997, *Theory of Didactical Situations in Mathematics*. Kluwer Academic Publishers, Norwell, MA.

Fuson, K., Stigler, J., and Bartsch, K., 1988, Grade Placement of Addition and Subtraction Topics in Japan, Mainland China, the Soviet Union, Taiwan, and the United States, *Journal for Research in Mathematics Education*, **19**:449-456.

Fukumori, N., 1997, *Lower Secondary School Mathematics Textbook*. Keirinkan, Osaka. (in Japanese)

Kang, W., 1990, *Didactic Transposition of Mathematical Knowledge in Textbooks*. Unpublished Doctoral Dissertation, University of Georgia.

Kim, Y.S., and Kim, H.K., 1997, *Mathematics, the Grade 8*. Dusan Donga Publishing Company, Seoul. (in Korean)

Lapointe, A.E., Mead, N.A. and Askew, J.W., eds., 1992, *The International Assessment of Educational Progress Report No.22-CAEP-01: Learning Mathematics*. The Centre for the Assessment of Educational Progress, Educational Testing Service, New Jersey.

Leung F.K.S., 1995, The Mathematics Classroom in Beijing, Hong Kong and London, *Educational Studies in Mathematics*, **29**:297-325.

McGuire, P. and Smith, K., 1996, *Oxford Mathematics: Link Books - Intermediate Link Year 9*. Oxford University Press, Oxford.

Mullis, I.V.S., Martin, M.O., Beaton, A.E., Gonzalez, E.J., Kelly, D.L. and Smith, T.A., 1997, *Mathematics Achievement in the Primary School Years*, Center for the Study of Testing, Evaluation and Educational Policy, Boston College, Boston.

Mullis, I.V.S., Martin, M.O, Gonzalez, E.J., Gregory, K.D., Garden, R.A., O'Connor, K.M., Chrostowski, S.J. and Smith, T.A., 2000, *TIMSS 1999 International Mathematics Report*, International Study Center, Lynch School of Education, Boston College, Boston.

People's Education Press, 1999; *Geometry, Book 2*. People's Education Press, Wuhan. (in Chinese)

Schmidt W.H., McKnight, C.C., Valverde, G.A., Houang, R.T. and Wiley, D.E., 1997, *Many Visions, Many Aims (Volume 1)*. Kluwer Academic Publishers, Dordrecht.

Stevenson, H.W. and Stigler, J.W., 1992, *The Learning Gap*. Simon & Schuster, New York.

Stigler, J.W. and Hiebert, J., 1999, *The Teaching Gap*, Free Press, New York.

Yang, K.S., 1981, Social Orientation and Individual Modernity among Chinese Students in Taiwan, *Journal of Social Psychology*, **113**:159-170.

Chapter 2-6

A COMPARISON OF MATHEMATICS PERFORMANCE BETWEEN EAST AND WEST: WHAT PISA AND TIMSS CAN TELL US

Margaret WU
University of Melbourne

1. INTRODUCTION

International studies such as the Trends in Mathematics and Science Study (TIMSS) and the Programme for International Student Assessment (PISA) provide results of relative mathematics achievement outcomes of countries. While the ranking of countries in terms of mathematics performance is interesting for both policy makers and the scientific community, what is more important for the mathematics education community is to identify factors that underlie the differences between countries. More specifically, while we have an interest in overall assessments of how students perform, we are perhaps more interested in understanding the nature of the differences between countries.

In this paper, we report a more focused comparison of Eastern and Western countries in students' performance in the PISA 2000 mathematics assessment. We examine the cognitive demands of each item and identify item characteristics that influence, in different ways, students' performance across the Eastern and Western cultures. In particular, we look for patterns and similarities within a group of Eastern countries and a group of Western countries. However, before delving into the item analysis of the PISA assessment, we first take a look at the differences between TIMSS and PISA results. We observe a pattern of differences that leads us to form a hypothesis about the differential item functioning (DIF) in Eastern and Western countries.

2. BACKGROUND OF PISA AND TIMSS

PISA is an international comparative study conducted by the OECD. The main aim of the project is to assess 15 year-old students' knowledge and skills in a number of subject domains, with an emphasis on these students' preparedness for life (OECD, 1999). PISA is intended to be an on-going study, with data collection conducted every three years. The first cycle of PISA (PISA 2000) spanned four years, from 1998 to 2001, with the main study data collection conducted in the year 2000. Thirty-two countries participated in this survey. PISA 2000 assessed reading, mathematics and science, with reading as the major focus. For mathematics, there were 60 minutes of testing material in the assessment, but only five-ninths of the students were administered mathematics items, and each of these students received 30 minutes of mathematics items in a rotated-forms test design (Adams and Wu, 2002).

TIMSS (Third International Mathematics and Science Study) was an IEA (International Association for the Evaluation of Educational Achievement) study first conducted in 1994-1995, with 41 countries participating at 5 grade levels. TIMSS 1999, also known as TIMSS Repeat or TIMSS-R, is a replication of TIMSS at the lower secondary school level – the eighth grade level in most countries, with an average student age of 14.4 years.

A key difference between PISA and TIMSS is that PISA has a "literacy" based orientation with a goal of "assessing the extent to which young people have acquired the wider knowledge and skills that they will need in adult life" (OECD, 1999). The PISA mathematics framework states that

> The term *literacy* has been chosen to emphasise that mathematical knowledge and skills as defined within the traditional school mathematics curriculum do not constitute the primary focus of OECD/PISA. Instead, the emphasis is on mathematical knowledge put to functional use in a multitude of different contexts and a variety of ways that call for reflection and insight.

TIMSS, on the other hand, starts the development of the assessment framework by surveying the mathematics curricula of all participating countries (Mullis *et al.*, 2001), although the TIMSS assessment framework is not solely based on the overlap of the curricula of participating countries. While the starting points of PISA and TIMSS are different, there is no doubt that the PISA assessment has considerable overlap with TIMSS, as the designers of curricula generally also aim for preparing students for skills needed in their future life. The differences between PISA and TIMSS are mainly in the emphasis of various skills and the manner in which questions are posed, rather than any fundamental differences in mathematics content.

As mathematics was a minor assessment domain in PISA 2000, only two areas of mathematics applications, referred to as *big ideas* in PISA, were chosen for the assessment: *change and growth*, and *space and shape*. The PISA mathematics framework gives the following rationale for the selection of these two big ideas:

> First, these two domains cover a wide range of subjects from the content strands. Second, these domains offer an adequate coverage of existing curricula. Quantitative reasoning was omitted from the first survey cycle because of the concern that it would lead to an over-representation of typical number skills.

The fact that PISA avoided the inclusion of purely computational items reflected the general thinking of the expert group that advised on the PISA mathematics assessment. Few PISA items required only the recall of knowledge. Most PISA items focused on "analysing, reasoning and communicating ideas". So the PISA items were largely concerned with the application of mathematical ideas and making sense of mathematics, not only about knowing algorithms or computational procedures. In analysing PISA data and in assessing the differences between the results of PISA and TIMSS, it is important that we bear in mind the differences in the conceptualisation of these two projects.

3. OVERALL MATHEMATICS PERFORMANCE IN PISA

As is generally the case in international studies of mathematics, PISA showed that Asian countries outperformed western countries when mean mathematics proficiencies at the country level are compared. Figure 1 shows the country mean scores and 95% confidence intervals (OECD, 2001). In PISA, only two countries are from East Asia: Japan and Korea. It can be seen from Figure 1 that Japan and Korea outperformed all other countries in mathematics in PISA. In general terms, we summarised the PISA results as follows: Students in Asian countries had the highest average scores, followed by students in English-speaking countries, northern European countries, eastern European countries, southern European countries, and then central and south American countries.

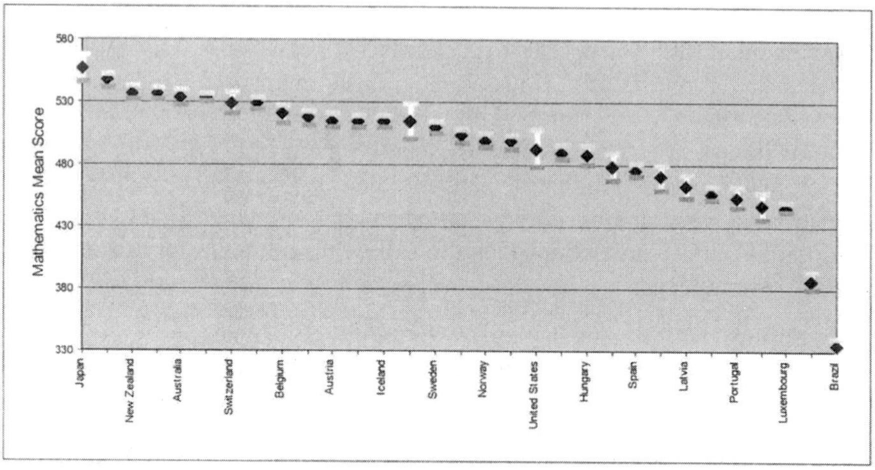

Figure 2-6-1. PISA Mathematics Mean Scores

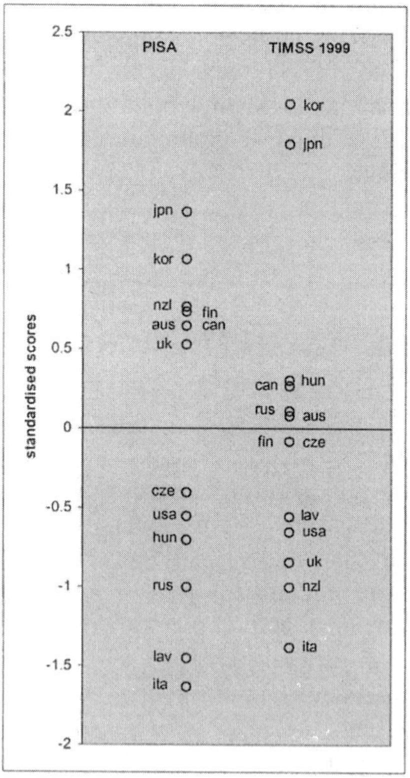

Figure 2-6-2. Standardised PISA and TIMSS 1999 Scores

When PISA 2000 results are compared to TIMSS 1999 (Mullis *et al.*, 2000) results, a number of striking differences can be seen. Figure 2 shows a comparison of PISA 2000 and TIMSS 1999 results. In this figure, we include the 13 countries that took part in both PISA 2000 and TIMSS 1999: Australia, Canada, Czech Republic, United Kingdom, Finland, Hungary, Italy, Japan, Korea, Latvia, New Zealand, Russia and the United States.

The PISA 2000 and TIMSS 1999 results are not scored on the same scale. They need to be standardised to make them comparable. The country mean scores were standardised in the following way. We computed the means and standard deviations of the 13 country means for each study, and computed the standardised score for each country by subtracting the mean and dividing by the standard deviation of the 13 mean scores in each study. For example, a standardised PISA score of 1.37 for Japan means that Japanese PISA mean score is 1.37 standard deviations away from the overall mean of the 13 PISA country mean scores. Similarly, we see that Korea is 2.05 standard deviations away from the mean of the 13 TIMSS 1999 country means. In this way Figure 2 shows the relative standing of each of the 13 countries in PISA 2000 and in TIMSS 1999, in terms of the number of standard deviations from the mean of the group of countries under comparison.

We can make three observations about Figure 2. Firstly, in TIMSS 1999, there is a large gap between the Asian countries (Japan, Korean) and the rest of the countries. In PISA, the gap is narrowed. Secondly, English-speaking countries performed well in PISA 2000 and were not too far behind Japan and Korea. However, in TIMSS 1999, English-speaking countries performed relatively poorly. Thirdly, eastern European countries performed poorly in PISA 2000 as compared to their performance in TIMSS 1999.

Figure 3 displays a scatter plot of standardised PISA 2000 and TIMSS 1999 scores for the 13 countries. This plot shows even more clearly the difference between PISA 2000 and TIMSS 1999 scores between two groups of countries.

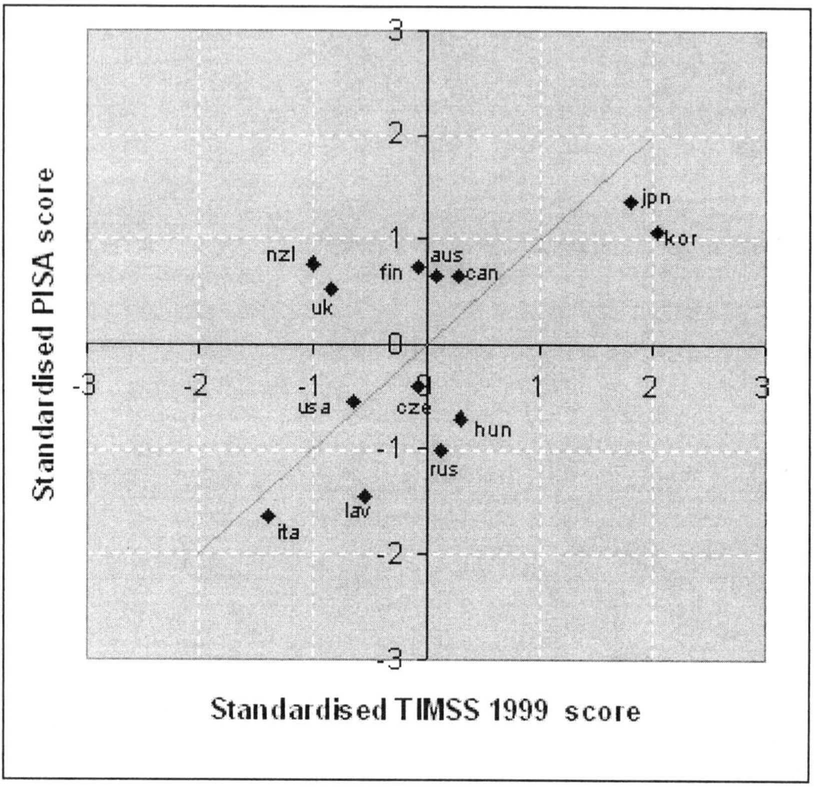

Figure 2-6-3. Standardised PISA 2000 score versus standardised TIMSS 1999 score

A 45° line is drawn to indicate the line of equality between standardised PISA 2000 score and TIMSS 1999 score. If we group Japan and Korea together with eastern European countries and define this as Eastern countries, then it is clear that Western countries are above the line of equality, and Eastern countries are below the line, with the exception of Italy. This means that Western countries performed relatively better in PISA 2000 than they performed in TIMSS 1999, and, generally speaking, Eastern countries performed relatively better in TIMSS 1999 than in PISA 2000. From the point of view of mathematics curriculum design, eastern European countries have more similarities with Japan and Korea than with Western countries, in terms of the content of traditional and formal mathematics taught in schools. Therefore, this finding of the two groups is not surprising.

The rank orders of countries are also quite different between PISA and TIMSS-R. Table 1 shows the rank orders of the 13 countries for PISA 2000 and for TIMSS 1999:

Table 2-6-1. Rank orders of countries in PISA 2000 and TIMSS 1999

	PISA 2000 rank	TIMSS 1999 rank
Japan	1	2
Korea	2	1
New Zealand	3	12
Finland	4	8
Canada	5	4
Australia	6	6
England	7	11
Czech Republic	8	7
United States	9	10
Hungary	10	3
Russia	11	5
Latvia	12	9
Italy	13	13

New Zealand has moved from third place (out of 13 places) in PISA 2000, down to 12th place in TIMSS 1999, while Hungary has moved from 10th place (out of 13 places) in PISA 2000, to third place in TIMSS 1999. Such discrepancies will no doubt raise questions like "which set of results is more *valid*?" and "why are there such differences?" These questions are not easy to answer. We are unlikely to find a simple answer to the question of assessing the validity of the results. But we can uncover some of the reasons for the differences. Three variables might be causes of the differences: population definition, time of the survey, and test content. The target populations in the two surveys are different in that PISA is age-based and TIMSS is grade-based. The average age of the samples of students for PISA is about one year older than the average age of the samples for TIMSS. The PISA survey occurred about one year after the TIMSS 1999 survey. In some sense, PISA 2000 and TIMSS 1999 essentially tested the same cohort of students in the countries, although one may argue that an age-based sample captures a slightly different cohort from a grade-based sample due to factors like retention. It is also possible, but unlikely, that Eastern countries had a program that accelerated students' learning from 14 to 15 years-old. We do not think that the age definition and the time of survey are likely to have caused the differences we observed. In this paper, we will examine the third variable, test content, in more detail. To try and better understand how the test content can affect performance, we need to examine item characteristics of each assessment. Our first hypothesis is based on the conceptual difference between PISA and TIMSS, as described earlier. The main difference is that PISA is not curriculum-based. The mathematics curricula in the 13 countries differ in varying degrees to the PISA mathematics framework. PISA's approach of assessing applications of mathematics may present more challenges to students who are used to learning mathematics in a more formal way. In the following sections, we examine students' performance on different types of items. In doing so, we hope to identify performance patterns that can be related to item characteristics.

4. PERFORMANCE AT THE ITEM LEVEL

To keep the interpretation of results manageable, we start with the analysis of four countries only. As this paper is primarily concerned with a comparison of the East and the West, where East is defined as East Asian countries, we include Japan, Korea, Australia and the U.S.A. in our first analysis. We use PISA 2000 assessment data to carry out this analysis, as we are familiar with the items, having been involved in the test development process. Owing to the embargo on a number of items that are used for linking purposes for future PISA cycles, we are not able to describe in detail all the items used in this assessment. We will illustrate our findings using some released items, which are included in the Appendix.

There were 31 mathematics items in the PISA 2000 database. One item was deleted in Japan owing to translation errors, so we will include only 30 items that were common to all countries in the following analyses. Table 2 and Figure 4 show the estimates of item facilities for these 30 items by country.

Of the four countries, Japan scored the highest country mean score, Korea is the next highest, followed by Australia, then the United States. If all items behave in a similar way in the four countries, we expect to see the percentages correct for each item also following the same order: Japan, Korea, Australia, the United States. For example, item 2 (M034Q01T) shows the four facilities (56.5, 51.3, 43.9, 29) in the order we expect, according to the order of country mean scores. Similarly, items 3, 4, 5, 7, all show the same ordering in terms of percentages correct. The same pattern can be seen in Figure 4, where the percentages correct are displayed visually. The fact that Figure 4 shows percentages correct moving up and down across items reflects the item difficulties of the items. In general, when an item is relatively difficult, the percentages correct are low for all four countries. Similarly, when an item is easy for one country, it is usually easy for all other countries. We see the four percentages for each item moving in relative unison across most, but not all, items.

For some items, we observe that the ordering of percentages correct is not quite as expected. For example, item 27 (M179Q01T) shows that Australia and the United States performed better than Japan and Korea. Figure 4 also shows that for some items, the four percentages correct are far apart from each other, and for other items, the percentages correct are close together. When items exhibit varying relative difficulties across countries, we say that there is Differential Item Functioning (DIF) on these items.

Table 2-6-2. Percentages correct of PISA 2000 mathematics items by country

	M033Q01	M034Q01T	M037Q01T	M037Q02T	M124Q01
Item No.	1	2	3	4	5
jpn	81.5	56.5	81.6	85.6	46.1
kor	74.0	51.3	70.9	80.0	41.2
aus	77.4	43.9	66.8	63.4	30.8
usa	72.5	29.0	46.4	59.8	25.5

	M124Q03T	M136Q01T	M136Q02T	M136Q03T	M144Q01T
Item No.	6	7	8	9	10
jpn	37.2	81.5	50.8	21.1	84.9
kor	11.7	73.4	60.6	30.4	78.6
aus	19.9	61.7	25.4	19.3	72.9
usa	17.6	53.4	23.9	14.8	52.7

	M144Q02T	M144Q03	M144Q04T	M145Q01T	M148Q02T
Item No.	11	12	13	14	15
jpn	41.7	85.8	49.9	72.6	23.3
kor	35.5	78.8	49.8	63.7	15.2
aus	29.6	86.1	43.0	64.6	26.7
usa	12.4	74.3	34.8	52.4	21.1

	M150Q01	M150Q02T	M150Q03T	M155Q02T	M155Q03T
Item No.	16	17	18	19	20
jpn	76.6	77.5	45	63.5	22.5
kor	77.5	86.7	48.3	68.0	22.9
aus	64.3	73.1	63.7	73.5	19.2
usa	50.8	61.2	57.2	64.2	18.1

	M155Q04T	M159Q01	M159Q02	M159Q03	M159Q05
Item No.	21	22	23	24	25
jpn	62.6	82.2	90.2	87.9	53.9
kor	60.2	75.5	90.9	86.9	32.8
aus	59.8	75.4	90.7	88.9	36.0
usa	54.3	62.3	83.2	81.6	22.6

	M161Q01	M179Q01T	M192Q01T	M266Q01T	M273Q01T
Item No.	26	27	28	29	30
jpn	72.4	24.9	59.0	42.6	68.3
kor	63.1	28.0	52.0	35.9	57.3
aus	60.3	37.5	46.0	24.4	56.4
usa	45.5	30.0	28.0	12.8	47.6

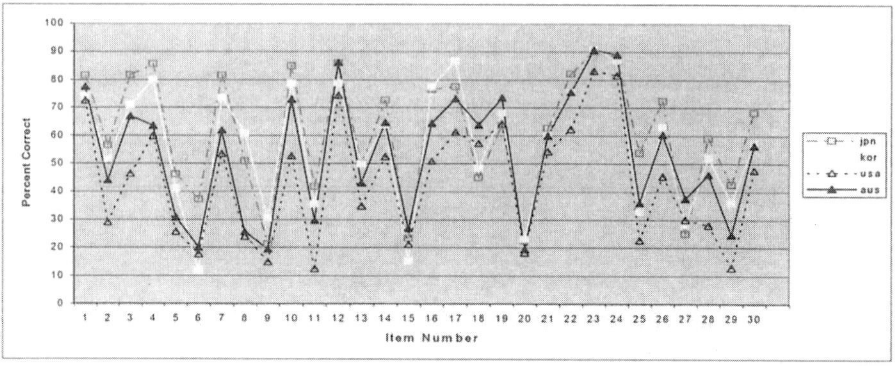

Figure 2-6-4. Plot of percentages correct by item and by country

5. DIFFERENTIAL ITEM FUNCTIONING

In this section, we examine the items that exhibit DIF. Further, assuming that there are differences between East and West traditions in mathematics, we test the hypothesis that when there is DIF, Japan and Korea are likely to be performing similarly as a group, and Australia and United States are performing similarly in another group. That is, the DIF is between these two groups of countries, rather than between the four individual countries.

To show DIF, we need to compare item difficulties across countries after making adjustments for the ability of students in those countries. That is, since we know that Japanese students are generally more proficient in mathematics than students in Australia, we expect that the percentages correct for Japan will be higher than those for Australia. Such differences in percentages correct are not an indication of DIF. However, after adjusting for ability, that is, comparing item difficulties given a student's ability, any observed differences in percentages correct would indicate the presence of DIF. We say that there is DIF when a student with the same ability in Japan would find an item more difficult, or easier, than a student of the same ability in Australia.

In addition, percentages correct, as a metric for measuring item difficulty, are known to have problems, mainly because of the bounded nature (between 0 and 1) of these measures. Therefore, for our analysis, we use item difficulty estimates obtained from calibrations of PISA data through item response modelling (IRM)[1] to examine DIF.

IRT analysis was carried out for each country separately. That is, the items were calibrated country by country. Table 3 gives the IRT item difficulty estimates by country. These item difficulty estimates are obtained after making adjustments for the ability level of students in a country. A high (more positive) value of item difficulty estimate indicates that the item is

Table 2-6-3. IRT item difficulty estimates (logits) by country

Item Code	Item No.	Japan	Korea	Australia	United States	Mean
M033Q01	1	-1.197	-0.925	-1.456	-1.598	-1.294
M034Q01T	2	0.315	0.281	0.595	0.852	0.511
M037Q01T	3	-1.399	-0.839	-0.867	-0.446	-0.888
M037Q02T	4	-1.700	-1.514	-0.543	-1.173	-1.233
M124Q01	5	0.602	0.804	1.013	0.802	0.805
M124Q03T	6	0.953	2.334	1.475	1.482	1.561
M136Q01T	7	-1.208	-0.909	-0.450	-0.669	-0.809
M136Q02T	8	0.536	-0.088	1.558	1.086	0.773
M136Q03T	9	2.015	1.201	1.668	1.355	1.560
M144Q01T	10	-1.707	-1.332	-1.137	-0.854	-1.258
M144Q02T	11	0.832	1.021	1.024	1.914	1.198
M144Q03	12	-1.865	-1.389	-2.267	-2.046	-1.892
M144Q04T	13	0.417	0.272	0.332	0.27	0.323
M145Q01T	14	-0.301	-0.244	-0.365	-0.271	-0.295
M148Q02T	15	1.915	2.256	1.441	1.297	1.727
M150Q01	16	-0.694	-0.936	-0.449	-0.321	-0.600
M150Q02T	17	-0.384	-1.016	-0.878	-0.896	-0.794
M150Q03T	18	0.980	0.617	-0.349	-0.620	0.157
M155Q02T	19	0.114	-0.423	-0.857	-1.015	-0.545
M155Q03T	20	1.829	1.616	1.938	1.528	1.728
M155Q04T	21	0.121	-0.172	-0.106	-0.424	-0.145
M159Q01	22	-0.919	-0.874	-1.017	-0.774	-0.896
M159Q02	23	-1.786	-2.12	-2.247	-2.205	-2.090
M159Q03	24	-1.479	-1.641	-2.098	-2.068	-1.822
M159Q05	25	0.617	1.455	1.189	1.456	1.179
M161Q01	26	-0.189	-0.100	-0.011	-0.002	-0.076
M179Q01T	27	2.167	1.520	1.086	1.020	1.448
M192Q01T	28	0.257	0.284	0.509	1.016	0.517
M266Q01T	29	1.237	1.279	1.869	2.160	1.636
M273Q01T	30	-0.081	0.179	0.220	-0.015	0.076

[1] Item response theory (IRT) was used in calibrating item difficulties in PISA. In particular, the generalised Rasch model (Wu, Adams & Wilson, 1997) was applied to item response data to estimate item difficulty parameters.

difficult, whilst a low (more negative) value indicates that the item is easy, for a person with some fixed ability. The item difficulty estimates are essentially a non-linear, monotonic transformation of the percentage correct, but adjusted for the ability level. The unit of the item difficulty estimate is 'logit', short for 'log of the odds'. If there is no DIF on an item, then we would expect the item difficulty estimates to be the same, within measurement errors, across the four countries. When the differences between the item difficulty estimates across the four countries are so great that they could not be explained simply by measurement errors, we would then conclude that DIF exists for this item. For example, for item 1, a student in Japan would find the item easier than a student with the same ability in Korea, but a student in Australia or the United States would be more likely than a Japanese student of the same ability to be successful.

Without formally carrying out statistical significance tests for DIF, we examine graphically the deviation of item difficulty estimates for each country from the mean item difficulty of the four countries for each item. Figure 5 shows the results. For example, the four points (-0.47, -0.28, 0.06, 0.69) plotted for item 4 shows that for Japan, the item difficulty is 0.47 logits lower than the average item difficulty for this item, after adjusting for ability level. For Korea, the item difficulty is 0.28 lower than the average. For the United States, the item difficulty is very close to the average item difficulty. But for Australia, students find this item relatively difficult as compared to students of similar ability in other countries, and one needs to add 0.69 logits to the average item difficulty to obtain the item difficulty estimate for Australia.

A wide range of points plotted for an item in Figure 5 shows that the item is not functioning in the same way in all four countries, while a clustering of the four points for an item indicates the item is functioning in the same way in all four countries.

When an item shows DIF in Figure 5, it is interesting to observe that the Eastern countries (Japan and Korea) and Western Countries (the United States and Australia) tend to group together. We use square symbols to indicate Eastern countries, and triangle symbols to indicate Western countries. For item 1, we see that the square symbols are on one side, and the triangle symbols are on the other side. We observe this pattern for many of the 30 items. For example, Eastern countries find items 2, 4, 7, 8, 10, 16, 28, 29 easier. Western countries find items 1, 12, 15, 18, 19, 23, 24, 27 easier. This is an indication that when there are deviations from the mean item difficulty, Japan and Korea tend to have the same kind of deviation, while the United States and Australia tend to have similar deviation as well.

Two questions come to mind regarding these observations: (1) Is the clustering of Eastern countries and Western countries happening by chance?

(2) When the clustering happens on opposite sides, that is, when Eastern countries find an item easier than Western countries, and when Western countries find an item easier than Eastern countries, can we identify item characteristics relating to these two directions of deviation?

To answer (1), we give an approximate assessment of the chance of observing 16 items out of 30 items showing clustering of Eastern countries and Western countries. Assuming that the clustering of four countries can happen in any order by chance, that is, it is equally likely to observe [(1,2), (3,4)], [(1,3), (2,4)], [(1,4), (2,3)], the chance of observing the clustering of Eastern countries and Western countries is one in three. From a binomial distribution with p=0.333 and n=30 items, the probability of observing 16 items or more with Eastern and Western clusters is 0.019. That is, there is only a 2% chance of observing 16 or more items with the East and West clustering. This is a small probability, so we conclude that the observed clustering of East and West countries is not likely to happen by chance.

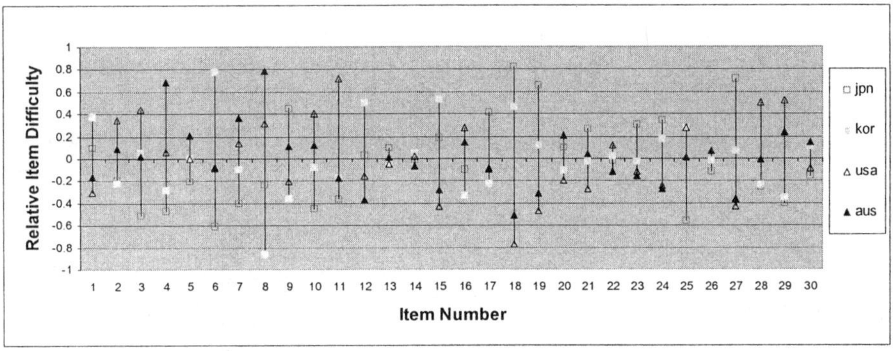

Figure 2-6-5. Deviations of item difficulty estimates from mean item difficulty

6. CLUSTER ANALYSIS

Another method to evaluate the "distances" between the four countries in terms of the item difficulty estimates is to carry out a cluster analysis. The data set for the cluster analysis consists of a matrix of 4 cases (corresponding to the 4 countries) and 30 variables (30 item difficulty estimates). A hierarchical cluster analysis is carried out to cluster cases (countries). A Dendrogram is produced as shown in Figure 6.

```
* * * H I E R A R C H I C A L   C L U S T E R   A N A L Y S I S * * *


 Dendrogram using Average Linkage (Between Groups)

                         Rescaled Distance Cluster Combine

   C A S E      0         5        10        15        20        25
  Label  Num    +---------+---------+---------+---------+---------+

  aus       1
⇩×⇩⇩⇩⇩⇩⇩⇩⇩⇩⇩⇩⇩⇩⇩⇩⇩⇩⇩⇩⇩⇩⇩⇩⇩⇩⇩⇩⇩⇩⇩⇩⇩⇩⇩⇩⇩⇩⇩⇩⇩⇩⇩⇩⇩⇩⇩⬂
  usa       4   ⇩⬀                                                 ⇔
  jpn       2
⇩⇩⇩⇩⇩⇩⇩⇩⇩⇩⇩⇩⇩⇩⇩⇩⇩⇩⇩⇩×⇩⇩⇩⇩⇩⇩⇩⇩⇩⇩⇩⇩⇩⇩⇩⇩⇩⇩⇩⇩⇩⇩⇩⇩⇩⇩⇩⬀
  kor       3   ⇩⇩⇩⇩⇩⇩⇩⇩⇩⇩⇩⇩⇩⇩⇩⇩⇩⇩⇩⇩⇩⬀
```

Figure 2-6-6. Dendrogram from cluster analysis with 4 countries

Figure 6 shows that Australia and the United States are very close in their patterns of item difficulty estimates, with a distance of about 1 unit between the two countries, so these two countries form cluster 1. Next closest is Japan and Korea, at a distance of about 11 units, and these two countries form cluster 2. At a distance of around 25 units, the four countries form one large cluster. In summary, the cluster analysis identifies two clusters, with West countries in one cluster, and East countries in another cluster. Furthermore, the distance between the two West countries is much closer than the distance between the two East countries. But the two clusters are at least as far apart as the distances between countries within each cluster.

This result provides encouraging support for the hypothesis that Eastern and Western countries have consistent differences in their patterns of item facilities. Furthermore, these differences are evident in the PISA assessment and the differences are in some sense measurable. In view of the similarities between eastern European countries and Japan and Korea, as shown earlier in the comparison between PISA and TIMSS, we carried out a further cluster analysis with eight countries: Australia, Germany, Great Britain, Hungary, Japan, Korea, Russia, and the United States. The results of the cluster analysis are shown in Figure 7.

```
* * * H I E R A R C H I C A L   C L U S T E R   A N A L Y S I S * * *

 Dendrogram using Average Linkage (Between Groups)

                         Rescaled Distance Cluster Combine

   C A S E      0         5        10        15        20        25
  Label   Num   +---------+---------+---------+---------+---------+

  aus       1    ⇩×⇩⇩⇩⬊
  gbr       7    ⇩⬃    ▫⇩⇩⇩⇩⇩⇩⇩⇩⇩⇩⇩⇩⬊
  usa       8    ⇩⇩⇩⇩⇩⬃              ▫⇩⇩⇩⇩⇩⇩⇩⇩⇩⇩⇩⇩⬊
  ger       2    ⇩⇩⇩⇩⇩⇩⇩⇩⇩⇩⇩⇩⇩⇩⇩⇩⇩⇩⬃
▫⇩⇩⇩⇩⇩⇩⇩⇩⇩⇩⇩⇩⇩⇩⇩⇩⇩⇩⇩⬊
  hun       3    ⇩⇩⇩⇩⇩⇩⇩⇩⇩⇩⇩⇩⇩⇩⇩⇩⇩⇩⇩⇩⇩⇩⇩⇩⇩⇩⇩⇩⇩⬃
⇔
  jpn       4
⇩⇩⇩⇩⇩⇩⇩⇩⇩⇩⇩⇩⇩⇩⇩⇩⇩⇩⇩⇩⇩⇩⇩×⇩⇩⇩⇩⇩⇩⇩⇩⇩⇩⇩⇩⇩⇩⇩⇩⇩⇩⇩⇩⇩⇩⬊   ⇔
  kor       5    ⇩⇩⇩⇩⇩⇩⇩⇩⇩⇩⇩⇩⇩⇩⇩⇩⇩⇩⇩⇩⇩⇩⇩⬃
▫⇩⇩⇩⬃
  rus       6
⇩⇩⇩⇩⇩⇩⇩⇩⇩⇩⇩⇩⇩⇩⇩⇩⇩⇩⇩⇩⇩⇩⇩⇩⇩⇩⇩⇩⇩⇩⇩⇩⇩⇩⇩⇩⇩⇩⇩⇩⇩⇩⇩⇩⇩⬃
```

Figure 2-6-7. Dendrogram from cluster analysis with 8 countries

Figure 7 shows the "distances" between the eight countries. The closest distance is between Australia and Great Britain. The United States joins this group next, followed by Germany, followed by Hungary. The other group starts with Japan and Korea in the cluster. They are joined by Russia before joining the other group to form one big cluster. This cluster analysis again shows that the grouping of Eastern and Western traditions is evident.

7. LINKING ITEM CHARACTERISTICS TO CULTURAL DIFFERENCES

Having established that there are indeed differences between Eastern and Western countries, we return to the question of whether we can identify item characteristics that can be linked to these differences. Table 4 is a table of items which Eastern countries find easier, and Table 5 shows a table of items which Western countries find easier.

The classifications of the items are mostly self-explanatory. We give only a few remarks. In the column headed "Big Idea (class)", the number in the brackets is the competency class number. There are three competency classes in PISA (OECD, 1999). These are:

Class 1: reproduction, definitions, and computations.
Class 2: connections and integration for problem solving.
Class 3: mathematical thinking, generalisation and insight.

Table 2-6-4. Items Eastern countries find easier

Item No.	Item Code	Item Name	Item format	Response type (process)	Big Idea (class)	Mathematics Strand	Formal Mathematics
2	M034Q01T	Not released	Closed Constructed Response	Numeric answer (counting)	Space and Shape (2)	Geometry	No
4	M037Q02T	Farms Q2	Closed Constructed Response	Numeric answer (geometric property)	Space and Shape (2)	Measurement	Yes
7	M136Q01T	Apples Q1	Closed Constructed Response	Numeric answer (pattern)	Growth and Change (2)	Algebra	Some
8	M136Q02T	Apples Q2	Closed Constructed Response	Numeric answer (equation)	Growth and Change (2)	Algebra	Yes
10	M144Q01T	Not released	Closed Constructed Response	Numeric answer (counting)	Space and Shape (1)	Geometry	No
16	M150Q01	Not released	Closed Constructed Response	Numeric answer (read graph)	Growth and Change (1)	Number	Some
28	M192Q01T	Not released	Multiple Choice	Match function	Growth and Change (2)	Measurement	Yes
29	M266Q01T	Not released	Multiple Choice	Assess property of shapes	Space and Shape (2)	Measurement	Yes

In the column headed "Formal Mathematics", we asked four experts to make judgments on whether an item contains mostly formal, curriculum-based content, or non-curriculum mathematics that nevertheless calls for sense making of real-world problems using mathematics. The judgments of the experts are averaged and summarised as *Yes*, *No* or *Some*. This exercise was not carried out with a stringent experimental design and control. It was done merely to seek some indications of item content. A wider consultation of this kind is necessary to have a fuller, and more accurate, evaluation of the processes involved in the items.

Table 2-6-5. Items Western countries find easier

Item No.	Item Code	Item Name	Item format	Response type (process)	Big Idea (class)	Mathematics Strand	Formal Mathematics
1	M033Q01	Not released	Multiple Choice	Spatial orientation	Space and Shape (1)	Geometry	No
12	M144Q03	Not released	Multiple Choice	Numeric Answer (counting)	Space and Shape (2)	Geometry	No
15	M148Q02T	Continent Area	Closed Constructed Response	Numeric Answer (estimation)	Space and Shape (2)	Measurement	Some
18	M150Q03T	Not released	Open Constructed Response	Verbal explanation of graph	Growth and Change (2)	Statistics	Some
19	M155Q02T	Not released	Closed Constructed Response	Numeric Answer (read unconventional graph)	Growth and Change (2)	Statistics	Some
23	M159Q02	Racing Car	Multiple Choice	Interpret graph	Growth and Change (1)	Functions	Some
24	M159Q03	Racing Car	Multiple Choice	Interpret graph	Growth and Change (1)	Functions	Some
27	M179Q01T	Not released	Open Constructed Response	Verbal explanation of graph	Growth and Change (2)	Functions	Some

What conclusions can we draw from Table 4 and Table 5? First, we note that Western countries are likely to perform better when the item content involves less formal mathematics. Second, Eastern countries perform well when an item involves numeric computation related to curriculum-based content, but they do not perform as well when an item calls for verbal explanations or interpretations of graphs. So the response type appears to have an impact on the performance between Eastern and Western countries. The item format (multiple choice or constructed) does not appear to make any difference in the relative performance between Eastern and Western countries; neither do Big Ideas nor Competency Classes. There may be a suggestion that Eastern countries do not perform as well in Statistics.

We give two examples to illustrate the key distinctions between the performance of Eastern and Western countries. Two items show large differences between Eastern and Western countries (see Figure 5): item 8 and item 15. These two items are given in the Appendix.

Item 8 (Apples Q2) requires students to form an equation and solve it, although students can use trial-and-error method as well. It is clear that both Japan and Korea performed extremely well on this item as compared to other items. This item calls for the use of formal mathematics learned in schools, including the use of symbolic representations of quantities.

Item 15 (Continent Area Q2) asks students to make an estimation of an irregular area. Many methods can be used. There is no single correct answer. Students are open to innovative ideas. They can use estimation methods learned in the classroom, or draw on their own experience in real-life to solve this problem. While they do need to understand the concept of area and scale units, the estimation method is completely open to their own creative resourcefulness.

These two examples highlight the item characteristics that make a difference to the performance of Eastern and Western countries. There are, however, many factors that have an impact on students' performance. Unfortunately, with the embargo on some of the items, it is difficult to illustrate these factors fully.

8. CONCLUSIONS

This paper demonstrates that there is indeed an Eastern tradition and a Western tradition in mathematics education. Further, these traditions are reflected in international comparative studies, and some characteristics of these traditions can be identified. What are the implications of these findings? We return to a question raised earlier about the validity of international studies. Clearly, having found the distinguishing features between Eastern and Western countries in their performance in mathematics, one can manipulate the content of an assessment to change the rankings of countries, particularly in relation to Eastern and Western cultures. PISA 2000 and TIMSS 1999 results suggest that differences in the balance of test material may result in a re-ordering of country performances. There are two implications of this observation. Firstly, the interpretation of international study results must be made in the light of the construct that is being tested. The term "Mathematics" has many different meanings to individuals, education specialists and policy makers. It is only meaningful when we report a country's relative standing in mathematics achievement when we clearly articulate what is being assessed. Secondly, in constructing any assessment, one must be careful about the nature of the items included and about the balance of the different kinds of items, as these can have a profound impact on the results. Mathematics educators must take an active part in deciding, reasoning and debating the kinds of mathematics competencies valued by the society in the 21st century. The revolution in information and communication technology has changed the world, and no doubt will continue to change the demands of skills and competencies in the workplace and in the home. Mathematics educators need to continue to adjust their goals to meet the demands of the changing world. What is the relative importance of being able to carry out formal mathematics procedures, or being able to communicate results to others, or being able to make sense of mathematical problems? These are questions we need to find answers to, before we can improve mathematics teaching and learning in schools.

Finally, this paper demonstrates that comparative studies can help us identify each country's strengths and weaknesses. Without international collaboration, we will not be able to make significant progress in making changes to educational practices. We also hope that the methodological approaches described in this paper, together with our findings, will stimulate further research in the area of international comparison.

REFERENCES

Adams, R.J. and Wu, M.L., 2002, *PISA 2000 Technical Report*. OECD. Paris.

Mullis, I.V.S., Martin, O.M., Gonzalez, E.J., Gregory, K.D., Garden, R.A., O'Connor, K.M., et al, 2000, *TIMSS 1999. International Mathematics Report*. International Study Centre, Boston College, Chestnut Hill.

Mullis, I.V.S., Martin, O.M., Smith, T.A., Garden, R.A., Gregory, K.D., Gonzalez, E.J., et al, 2001, *TIMSS Assessment Frameworks and Specifications 2003*. Boston College. Chestnut Hill.

OECD, 1999, *Measuring Student Knowledge and Skills: A New Framework for Assessment*. OECD Publications, Paris.

OECD, 2001. Knowledge and Skills for Life – First Results from PISA 2000. OECD Publications, Paris.

Wu, M.L., Adams, R.J., and Wilson, M.R., 1997, *ConQuest: Multi-Aspect Test Software.*, [computer program] Australian Council for Educational Research, Camberwell.

APPENDIX: SELECTED RELEASED PISA 2000 ITEMS

A farmer plants apple trees in a square pattern. In order to protect the apple trees against the wind he plants conifer trees all around the orchard.

Here you see a diagram of this situation where you can see the pattern of apple trees and conifer trees for any number (n) of rows of apple trees:

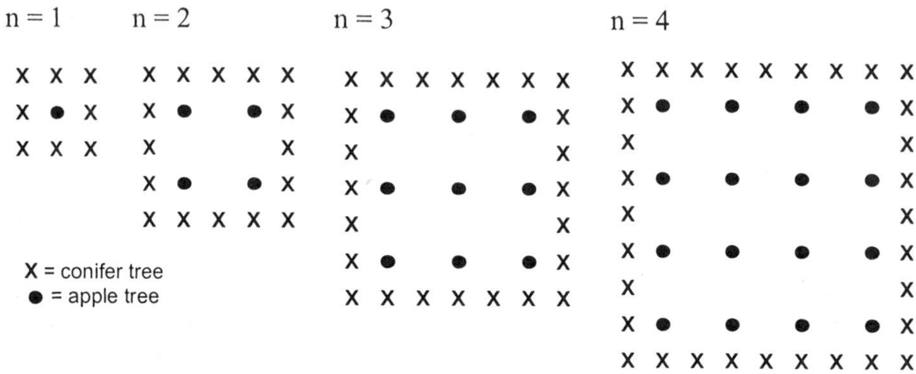

APPLES QUESTION 1 (ITEM 7, M136Q01)

Complete the table:

n	*Number of apple trees*	*Number of conifer trees*
1	1	8
2	4	
3		
4		
5		

APPLES QUESTION 2 (ITEM 8, M136Q02)

There are two formulae you can use to calculate the number of apple trees and the number of conifer trees for the pattern described above:

Number of apple trees = n^2

Number of conifer trees = $8n$

where n is the number of rows of apple trees.

There is a value of n for which the number of apple trees equals the number of conifer trees. Find the value of n and show your method of calculating this.

CONTINENT AREA (ITEM 15, M148Q02)

Below is a map of Antarctica

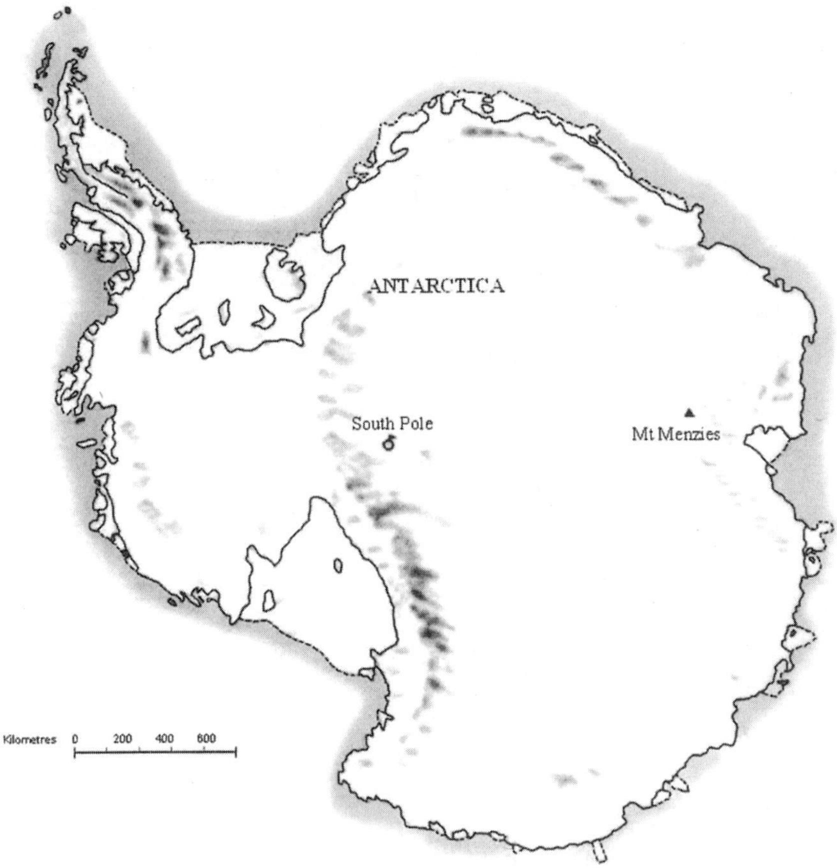

Estimate the area of Antarctica using the map scale.
Show your working out and explain how you made your estimate. (You can draw over the map if it helps you with your estimation)

Chapter 2-7

CASE STUDIES ON MATHEMATICS ASSESSMENT PRACTICES IN AUSTRALIAN AND CHINESE PRIMARY SCHOOLS

ZHAO Da-Cheng, Joanne MULLIGAN and Michael MITCHELMORE
Macquarie University, Australia

International comparisons of mathematics achievement have informed much recent research in mathematics education. A number of cross-cultural studies, especially large-scale assessments (Lapointe, Mead, & Askew, 1992; Mullis, Mattin, Beaton, Gonzalez, Kelly, & Smith, 1997; Mullis et al., 2000; Robitaille & Garden, 1989), have documented that overall East Asian students outperform their Western counterparts in mathematics. Researchers have described various factors that contribute towards this learning gap, such as differences in cultural contexts (Leung, 1995; Stevenson & Stigler, 1992); differences in amount of time spent on learning mathematics (Robitaille & Garden, 1989); differences in content and standards in mathematics curricula (Lindquist, 2001); differences in teachers' mathematics knowledge (Ma, 1999); and differences in mathematics classroom teaching (Stigler & Hiebert, 1999).

However, recent research (Cai, 2000) has suggested that the superior performance of East Asian students is limited to certain aspects of mathematics achievement, including basic skills of computation and solving routine problems. Western students perform better than their Asian counterparts in aspects such as using visual and graphical representations and solving open-ended problems.

Given this paradox, there is a need for more in-depth qualitative studies to identify those factors contributing to these differences (Leung, 2001). There is especially a need for comparative studies giving descriptive accounts of assessment practices to examine similarities and differences in authentic settings. Except for a few large-scale surveys at international level, such as the Third International Mathematics and Science Study (TIMSS), few comparative studies of assessment practices between East Asian and Western countries have been conducted at school level. Ethnographic research using case studies may be valuable in uncovering explanations for differences in assessment practices.

The major purpose of the study reported in this paper was to collect data on primary school mathematics assessment practices in China and Australia, to examine similarities and differences, and to identify underlying factors that may account for these.

The selection of case studies in Australia and China was fortuitous: The first author is from China and was studying under the supervision of the second and third authors in Australia. Although it is widely acknowledged that considerable cultural differences exist between China (representing East Asian culture) and Australia (representing Western culture), the two education systems have been showing an increasing interest in each other. On the one hand, Australia has endeavoured to raise standards in mathematics for the 21st century by considering strategies and practices from East Asian countries (Australian Council for Educational Research, 1999). On the other hand, Chinese educators have been advocating the reform of Chinese education through the adoption of practices from Western countries, such as America and Australia, in order to foster greater creativity and problem solving skills among students (Zhao, 1999).

1. CHINESE AND AUSTRALIAN APPROACHES TO MATHEMATICS ASSESSMENT

Assessment practices have assumed an increasingly important role in mathematics education both in China and Australia (Zhao, 2000). Driven by an outcomes-based approach, the purposes of assessment in Australia range from providing information to assist teachers to improve student learning, to plotting a national strategy to improve mathematics education for the nation (Department of Education, Training and Youth Affairs, 2000). A number of researchers (Black & Wiliam, 1998; Leder, 1992; Lokan & Doig, 1997) have reported that developing pedagogical knowledge through assessment of students' mathematical learning has a substantial impact on teachers' instruction and leads to improvements in students' learning. According to Little (1996), the emphasis on assessment is now evident worldwide, both in developing and developed countries. It has been commonly accepted that reform in assessment is a necessary condition for any reform of the mathematics curriculum. It is also widely acknowledged that the major purpose of assessment should be the improvement of students' learning

(DETYA, 2000; Chinese Ministry of Education, 2000a; National Council of Teaching Mathematics, 2000) by integrating assessment with teaching and learning; supporting curriculum rather than driving curriculum.

The Australian *National Statement* (Australian Education Council, 1991) stipulated two general principles of assessment: Assessment should reflect all of the goals of the school mathematics curriculum, and assessment should be demonstrable fair, valid and reliable. In comparison, in China, the *Syllabus* (Chinese Ministry of Education, 2000b) asserts that mathematics assessment should be based on the teaching objectives and basic requirements of the syllabus, and that mathematics assessment should not only assess students' understanding and mastery of basic mathematical knowledge but also students' mathematics ability. Analysing these principles, it is clear that both Chinese and Australian assessment policy-makers put the alignment of assessment practice with syllabus outcomes as an underlying principle for mathematics assessment.

Mathematics curricula in both China and Australia also suggest strategies for assessment. For example, considering the fairness, validity and reliability of assessment, the *National Statement* recommends that more developmental work is needed on useful, practical and fair assessment strategies. The New South Wales *Mathematics K-6* (NSW Department of Education, 1989) encourages a variety of assessment strategies including pen-and-paper tests, observation, listing, structured interviews, student-teacher discussions, student explanation and demonstration, samples of student work, and practical investigations. Similarly, the Chinese curriculum documents pay great attention to assessment strategies. The *Syllabus* suggests that, although the primary method of mathematics assessment is pen-and-paper test, oral examinations and practical tasks are also to be encouraged. Besides the end-of-term examination, the *Syllabus* suggests that more emphasis should be put on daily classroom teaching and homework in order to achieve a better understanding of students' performance. Teachers are accordingly encouraged to use these sources of information to improve teaching and students' learning. The *Standards* (Chinese Ministry of Education, 2000a) recommends a variety of approaches to assessment (such as pen-and-paper examinations, thematic activities, written essays, group activities, self-assessment and daily observations) so that the teacher can synthesise these data to make valid assessment judgements. It is obvious that both systems recommend a variety of strategies for mathematics assessment.

We shall refer to the principles and practices recommended by official curriculum documents as the *intended assessment.* A second aim of the present study was to compare the intended mathematics assessment in China and Australia with what actually happens – the *implemented assessment.*

2. METHODOLOGY

Case studies can offer a means of investigating complex social factors that are multi-faceted and of potential importance in understanding observed differences. The case study model adopted in this study is described by Yin (1994) as "multiple cases containing embedded cases". The multiple cases selected were three government primary schools (one from China and two from Australia), and the embedded cases involved classroom examples of assessment practices. Two Australian cases were selected from the Sydney metropolitan region (one urban and one semi-rural), and a Chinese case was selected from the Guangzhou metropolitan region. Data were collected between August 1999 and August 2002.

Classroom observations, interviews with teachers and parents, and document analyses were employed as instruments for collecting and analysing data. Six teachers from each school participated, one from each year level at each school, as well as one parent from each class, the school principals, and other administrators. Classroom observations focussed on specific features of assessment such as students' responses, adjusting lessons according to students' responses, students' strategies, and students' explanations about the process of problem solving. Interviews explored participants' beliefs about and attitudes to assessment. Documents analysed included the school website, newsletters, school policies and mathematics programs, professional development programs, students' portfolios, school reports, mathematics work samples, mathematics texts, examination papers, and homework.

After an initial review of documents, analysis of data began with the first teacher interview and the first classroom observation. Emerging insights, themes, and tentative hypotheses were formulated and constructed to inform proceeding interviews and observations. Tape-recorded interviews were transcribed verbatim, and classroom observations, school programs, and work samples were collected so that a holistic approach to data analysis could be conducted. The findings from the two case studies are outlined as follows.

Because of space limitations, only data on the case studies of the Australian urban school (School A) and the Chinese school (School C) are reported here.

3. MAIN FINDINGS FROM CASE STUDIES

3.1 Background information on schools

School A was a large government primary school with a high Asian population (40% of its students were from a Chinese background) located in the northern part of the Sydney metropolitan area in New South Wales, Australia. School A showed a high average level of academic achievement; it not only outperformed other schools in Basic Skills Tests, but also did extremely well in selective high school examinations and mathematics competitions.

School C was a large government primary school situated in the eastern part of the Guangzhou metropolitan area in Guangdong province, China. School C also showed a high average level of academic achievement; it not only outperformed other schools in the region in external examinations, but also did extremely well in mathematics competitions.

Both schools shared considerable similarities, including the number of classes (in 2001, they both had 29 classes), high academic achievement in mathematics, and high socio-economic levels of their community. In addition, a considerable number of Chinese background students at School A were from the Guangzhou metropolitan area.

One obvious difference between the two schools was the physical setting and the number of students in the classrooms. At School A, the classroom furniture arrangement was flexible, and students' desks and chairs were not fixed in rows. From Kindergarten to Year Two, students sat on the floor or in small groups. From Year Three to Year Six, students sat individually, in pairs or in small groups. The average class size at School A was 30 students. In contrast, except for six out of forty lessons where students sat in small groups, students at School C sat in rows facing the teacher. The average class size for School C was 50 students, including Years One and Two. Consistent with the classroom setting, teaching at School A was more flexible than at School C. Lessons at School C were found to be much more structured and teacher-directed than lessons at School A. School C also employed special teachers for mathematics in all classes, whereas in School A the classroom teacher taught all the main key learning areas.

3.2 Findings from classroom observation of assessment practices

A total of 80 lessons were observed at the two schools: 60 mathematics lessons (usually five lessons per teacher) and 20 other lessons. An analysis of these classroom observations revealed major differences related to mathematics assessment practices.

At School A, teachers were expected to follow an outcomes-based approach to teaching, learning, and assessing. The interview data suggested that outcomes-based education had strongly influenced teaching, learning, and assessment practices at the school. Since textbooks were not used in mathematics classes, the mathematics teachers made decisions as to the particular sequence of teaching content (but linked to syllabus outcomes). During lessons, the teacher usually gave students copies of worksheets, most of them copied from a range of different textbooks. Teachers usually checked students' classroom assignments individually during classroom time. Since portfolios were used as the major approach to assessment and reporting of mathematics achievement at School A, teachers paid close attention to collecting students' work samples regularly. In 3 of the 30 mathematics lessons observed, after students had finished their seatwork teachers required them to check their own answers and then put their work samples into their folders.

However, there was a considerable mismatch between the observed assessment practices at School A and the intended assessment advocated by the syllabus. First, there was little attention paid to the Working Mathematically strand of the curriculum, where students should be asked to solve open-ended questions and explain their answers. Second, even when students were sitting in groups, they were rarely assessed for their participation in the group-work process. Third, few teachers used authentic assessment in which mathematics was presented to students through real-life problems, including open-ended investigations or projects or integrated activities to explore within or after school.

In contrast, teachers at School C were expected to follow a content-based approach to teaching, learning, and assessing. It was found that teaching content and students' seatwork were both centred on textbooks. All the teachers indicated that they were required to follow the specifications of the mathematics *Syllabus* (Chinese Ministry of Education, 2000b) and therefore based their teaching and assessment on the content of textbooks. They also indicated that they followed a centralised mathematics assessment practice using written examinations with papers supplied about once every month by the local education office. In 3 out of the 30 mathematics lessons observed, students took a formal written examination in class to test their grasp of the content of the unit they had just finished. According to the teachers, the major purpose of these examinations was for ranking the students and reporting their achievement.

As in School A, it was clear that there was a considerable mismatch between the intended assessment advocated by the official curriculum documents and the actual assessment practices at School C. First, the classroom teaching was completely dominated by the content tested by external examinations. Teachers seldom changed their pre-arranged teaching sequence to respond to the needs of their students. Second, classroom teaching in most lessons was dominated by the teacher; there was no active role played by students in the lesson. Third, few teachers used assessment in which mathematics was presented to students through real-life problems, including open-ended investigative questions or giving students projects or integrated activities to explore their school mathematics or homework.

3.3 Findings from the analysis of students' work samples and examination papers

At School A, 36 mathematics work samples were selected from student portfolios. At School C, 60 mathematics examination papers were selected from teachers' collections of student examination scripts. In both cases, the work generally came from one student per year level. From analyses of these materials, two major differences were found.

The main difference was the form of the tasks presented. Most items included in the work samples of School A were presented as pictures or drawings, while most test items included in the examination papers of School C were presented in words or mathematical symbols, even in Years 1 and 2.

In terms of assessment content, the most important difference was that assessment tasks at School A focused on content related to students' daily lives, while assessment tasks at School C focused on the four operations (including fractions and decimals) and word problems. In School A there were more examples on Space and Measurement than in School C, but examples on Number dominated in both cases.

3.4 Findings from interviews with teachers

At each school, semi-structured individual interviews were conducted with six teachers after at least one lesson had been observed. Considerable differences existed in their views on assessment practices:

First, in terms of the purpose of assessment, all the teachers interviewed at School A indicated that the major purpose of assessment was to gather information about students' learning and use it for improving their mathematics teaching. Although most teachers interviewed at School C recognised the importance of assessment for their teaching, they all believed the major purpose of assessment was to inspect students' mathematics learning in order to stimulate students' motivation to improve their achievement level. Two teachers explained that in the Chinese tradition teachers attributed achievement (or lack thereof) to students' motivation and parents deeply respected and supported the teacher, and that the teacher should not be "blamed" for poor achievement. But all teachers at School C indicated that local education officers used students' achievement as an indicator of teaching accountability. In contrast, two teachers at School A saw assessment as focussed on teachers rather than students and indicated that parents believed teachers were responsible for poor achievement.

Second, in terms of mathematics assessment strategies and reporting, all the teachers interviewed at School A reported that they used portfolios for reporting students' achievement and they thought it was advantageous for teachers to communicate to parents about students' learning. They recognised that to make portfolios representative of all aspects of students' mathematics learning, there was a need to use a variety of assessment strategies. However, they indicated that lack of time, resources, and training were the major factors hindering them from effective teaching and assessing in mathematics. In contrast, all the teachers interviewed at School C reported that they used examination papers and a grading system to record and report students' achievement. Although they recognised that mathematics achievement cannot only be assessed by written examinations, they thought pen-and-paper examinations were highly valued by parents. They indicated that they gleaned informal information on student learning from classroom observation, checking students' seat work and homework, but that large class sizes, constraints of the current examination system, traditional cultural views, and lack of resources and professional training hindered them from trying out the new methods of mathematics assessment recommended by the *Syllabus* and the *Standards* (Chinese Ministry of Education, 2000a).

3.5 Findings from interviews with parents

At School A, three of the six parents interviewed were from Mainland China, one was from Hong Kong, and the other two were native English speakers. At School C, all parents interviewed were Chinese.

A major finding was that all Chinese parents, both at School A and School C, shared similar views related to the importance of mathematics learning: mathematics was a priority for logical and scientific thinking. All the Chinese parents had high expectations of their children's academic lives (they all encouraged their children to work hard to enable entry to selective schools) and valued their children's mathematics achievement. They also paid great attention to their children's mathematics learning (they always communicated with teachers about their children's mathematics learning about twice a month) and valued especially after-school learning. They all adopted a similar approach for their children's after-school learning, either sending them to coaching schools or coaching them themselves. They reported that the average after-school time their children spent on mathematics learning was about four hours per week. A Year Six parent at School A reported:

> Before we came to Australia, we brought a series of Chinese mathematics textbooks and coached our son by ourselves. ... We guided him every day, first let him do the normal homework, then we checked what errors he had made so that we taught him accordingly.

In contrast, the two native English-speaking Australian parents indicated they did not push their children to study hard. The Year One parent explained: "I feel my girl probably wouldn't be able to go into a selective school, because it accepts a limited number of children. So I told her, don't worry." The two parents also had similar views on after-school tutoring. They reported that their children only spent half an hour per week on after-school mathematics learning – strikingly different from the Chinese parents' views. The Year One parent at School A remarked:

> I don't think they need go to coaching school. I think they do enough in school. I know different nationalities may have different views of children's learning. Lots of people like me think the children are young and their school and after-school time should be play time. Therefore, we must do everything to make our children happy.

4. DISCUSSION

Based on these case study data, two major 'gaps' have emerged. The first gap exists between the intended and the implemented assessment in both Australia and China. The second gap is between the implemented assessment at School A and School C.

4.1 The gap between intended and implemented assessment

In regard to the first gap, both Chinese and Australian assessment policy makers recommend integrating assessment with teaching and learning and using a variety of strategies to assess students' mathematics learning (Australia Education Council, 1991; Chinese Ministry of Education, 2000a). Teachers at both schools saw formal examinations as limited, but traditional assessment practices still dominated classroom implementation at both School A and School C. In School A, even though portfolios were used, pen-and-paper examinations still played the major role in assessment. Limitations in time and lack of professional training in assessment were cited as the major reasons for the gap at School A. For School C, large class sizes, restrictions imposed by traditional culture, and lack of professional training in assessment were mentioned as the principle reasons causing the gap.

4.2 The gap between assessment practices at School A and School C

This study found there was a gap between assessment practices across School A and School C. The gap can be attributed to cultural differences in beliefs about mathematics education and assessment and about intrinsic versus extrinsic motivation.

4.2.1 Beliefs about mathematics education and assessment

In this study, it was found that Chinese background parents had very high expectations of their children's mathematics learning and established high academic standards for their children. In terms of the rationale for mathematics learning, a Chinese background parent with a child in Year 5 at School A explained:

> I come from China with a high degree of education in science so naturally I look at mathematics. There is a general saying in China that "once you have grasped the mathematics, physics and chemistry, you will be bold to do everything else". Grasping the essence of mathematics is good for the logical thinking and can be beneficial to other subjects, widening the knowledge coverage.

Chinese mathematics curriculum developers (Zhang et al., 1994) define mathematics "as a branch of science concerning the relationship of space and number in which reasoning is based on these relationships" (p.4). As a science, mathematics consists of concepts, rules and laws characterised by abstraction and rigor. It is clear that Chinese curriculum developers adopt a purist view of mathematics and mathematics learning which addresses the importance of concepts, rules, and laws of mathematics and understanding the structure of mathematics itself. The intended mathematics curriculum in China also emphasises the importance of mathematics in terms of its relationship with further studies. According to the *Syllabus* (Chinese Ministry of Education, 2000b), mathematics is significant for further learning and is also the foundation for studying science and technology. Recognising the abstract and rigorous characteristics of mathematics, Chinese mathematics educators also assert that mathematics can be used as a useful tool in training students' logical ability (Cao, 1996).

In contrast, according to Australian documents mathematics is a science of patterns and relationships (Australian Education Council, 1991, p.4). It is a way of thinking characterised by processes such as exploring, manipulating, discovering, and provides a powerful, precise, and concise means of communication. At a fundamental level, it is concerned with practical applications in many branches of human activity (NSW Depart-ment of Education, 1989, p.2). These views suggest that an underlying rationale for Australian mathematics curricula is based on the philosophy of pragmatism and a constructivist approach to learning.

Most teachers at School C suggested that Chinese assessment traditions explain current practices. The Year Six teacher at School C stated:

> In China, the examination system is called 'the baton' (*zhi hui bang*), directing the teaching and learning at which it points. The National Unified Entrance Examination for Institutions of Higher Education is the most powerful assessment in students' lives, it also affected those students studying at the primary schools and their parents. The form and content of this sort of examination has powerful influence on what the mathematics is and how it should be assessed.

This viewpoint was confirmed by most of the other teachers and parents at school C. For example, the Year Five parent reported:

> I think mathematics plays an important role in developing human being's thinking. It is also one of the three core subjects assessed by exams applied in high schools and universities. It is one of my son's favorite courses, so we have focused more on cultivating and developing his ability in this subject, expecting him to be a specialist in the future. We have paid much attention on encouraging his interest since he was in Year 1. It is the tradition in China that parents encourage their children to study hard only courses that are examined.

In contrast, Australian teachers at School A had different views on mathematics education especially different beliefs about mathematics assessment. Generally they regarded mathematics as one of the six key learning areas and assessment as the process of gathering information about students' learning. As the Year One teacher indicated:

> By assessing, you see whether the students have learned the content you tried to teach them or not. You will also know if they have achieved the outcomes or not. If they have grasped your teaching, they can learn more, if not, we should go back to teach them again.

The two native English-speaking Australian parents also reported that their views on mathematics education were not influenced by assessment practices because they did not expect their children to go to a selective high school. As the Year One parent indicated:

> I care most about English. I think that in primary school the most important thing is reading and writing. I just generally hope they can learn enough to prepare them for the future. They can get their job and work in the community.

These data suggest that Australian teachers and parents' beliefs about mathematics assessment differ dramatically from those of Chinese background teachers and parents. These differences could well be a factor in explaining the observed difference in assessment practices at the two schools.

4.2.2 Intrinsic versus extrinsic motivation

According to Leung (2001), assessment practices may be different in East Asian and Western societies because the two cultures have different views on extrinsic versus intrinsic motivation. It is suggested that educators in the West place a high value on intrinsic motivation in learning mathematics, whereas educators in the East Asian countries recognise extrinsic motivation (such as pressure from assessment) as an acceptable source of motivation for student learning. Preference for extrinsic motivation is well established in Chinese culture, where the fairness and objectivity of pen-and-paper examinations are highly trusted and valued – therefore, the achievement of a high score on an examination is a powerful motivation for teaching and learning. To ensure a high score, teachers and parents in China pay great attention to the content and form of assessment and spend considerable time preparing students for assessment. In this way, Chinese parents assume responsibility for their children's high achievement in mathematics.

The case studies found that Chinese background parents at both School A and School C all expected their children to work hard to achieve success in their mathematics learning. As the Year 6 parent at School C argued:

> Of course, I wish that she could study in a good high school and renowned university in the future. My perception is that the importance of individual diligence in study is above all in comparison with school teaching; her future mainly depends on her own efforts.

In contrast, the native English-speaking Australian teachers and parents did not push their students or children to study hard for high mathematics achievement and entry to selective schools. They emphasised the students' intrinsic motivation and innate ability more than other factors such as personal effort. It was accepted that students are either very able or not, and they did not expect less able children to do well academically. As the Year Four parent at School A indicated:

> I suppose the selective school is good. If your child is gifted and you know they really like English and Mathematics, you can encourage them to attend a selective school rather than stay in a normal school, held back by others in the majority of the class. … Like my children, if they got to a selective school, there would be too much to learn. It is very hard for them.

This view was shared by Australian teachers and parents, which suggests that Australian teachers and parents value intrinsic motivation and ability to learn mathematics more highly than Chinese background teachers and parents do. This difference could be a second factor explaining the difference between mathematics assessment practices at the two schools.

5. CONCLUSION

To sum up, the case studies have demonstrated a gap between the intended and the implemented mathematics assessment – both in the Australian School A and the Chinese School C. The data suggest that limitations in time, resources, professional training, and parental support were the main reasons for this gap in School A. For School C, large class sizes, restrictions imposed by traditional culture, and lack of professional training in assessment were the principle reasons.

Considerable differences in mathematics assessment practices between School A and School C were also found. This gap is at least partially explained by differences in beliefs about mathematics education and assessment, and in different values attributed to intrinsic and extrinsic motivation in Australian and Chinese cultures.

These findings suggest that the relatively high mathematics achievement of Chinese children cannot only be attributed to the teaching they receive. Other factors such as motivation to achieve, parental help, and after-school mathematics study certainly play a significant role.

REFERENCES

Australian Education Council, 1991, *A National Statement on Mathematics for Australian Schools.* Author, Melbourne.

Australian Council for Educational Research, 1999, *Raising Australian Standards in Mathematics and Science: Insights from TIMSS.* Author, Melbourne.

Black, P., & Wiliam, D., 1998, Inside the Black Box: Raising Standards through Classroom Assessment, *Phi Delta Kappan*, **80**:139-148.

Cai, J.F., 2000, Mathematical Thinking involved in US and Chinese students' Solving Process-constrained and Process-open Problems, *Mathematics Thinking and Learning*, **2**:209-240.

Cao, F.Y., 1996, *Collected Works on Reform in Primary School Mathematics Education.* The People's Education Press, Beijing.

Chinese Ministry of Education, 2000a. *National Mathematics Curriculum Standards for Nine-year Compulsory Education.* Beijing Normal University Press, Beijing.

Chinese Ministry of Education, 2000b, *The Mathematics Syllabus for Primary Schools for Nine-year Compulsory Education* (revised edition). The People's Education Press, Beijing.

Department of Education, Training and Youth Affairs, 2000, *Numeracy, a Priority for all: Challenges for Australian schools*, Author, Canberra.

Lapointe, A.E., Mead, N.A., and Askew, J.M., 1992, *Learning Mathematics.* Educational Testing Service, Princeton, NJ.

Leder, G.C., eds., 1992, *Assessment and Learning of Mathematics.* Australian Council for Educational Research, Melbourne.

Leung, F.K.S., 1995, The Mathematics Classroom in Beijing, Hong Kong and London, *Educational Studies in Mathematics*, **29**:297-325.

Leung, F.K.S., 2001, In Search of An East Asian Identity in Mathematics Education, *Educational Studies in Mathematics*, **47:**35-51.

Lindquist, M., 2001, NAEP, TIMSS, and PSSM: Entangled Influences, *School Science and Mathematics*, **101**:289-291.

Little, A., and Wolf, A., eds., 1996, *Assessment in Transition: Learning, Monitoring and Selection in International Perspective*, Pergamon, Oxford.

Lokan, J., and Doig, B., eds., 1997, *Learning from Children: Mathematics from a Classroom Perspective.* Australian Council for Educational Research, Melbourne.

Lokan, J., Ford, P., and Greenwood, L., 1997, *Maths and Science on the Line: Australian Middle Primary Students' Performance in the Third International Mathematics and Science Study.* Australian Council for Educational Research, Melbourne.

Ma, L.P., 1999, *Knowing and Teaching Elementary Mathematics: Teachers' Understanding of Fundamental Mathematics in China and the United States*, Lawrence Erlbaum, Mahwah, N.J.

Mullis, I., Mattin, M., Beaton, A., Gonzalez, E., Kelly, D., and Smith, T., 1997, *Mathematics Achievement in the Primary School Years: IEA's Third International Mathematics and Science Study*. Center for the Study of Testing, Evaluation, and educational Policy, Boston College, Boston, MA.

Mullis, I., Martin, M., Gonzalez, E., Garden, R., O'Connor, K., Chrostowski, S., and Smith, T., 2000, *TIMSS 1999: International Mathematics Reports. Findings from IEA's repeat of the Third International Mathematics and Science Study at the Eighth Grade*. Center for the Study of Testing, Evaluation, and Educational Policy, Boston College, Boston, MA.

National Council of Teachers of Mathematics, 2000, *Principles and Standards for School Mathematics*, Author, Reston, VA.

New South Wales Department of Education, 1989, *Mathematics K-6*. Author, Sydney.

Robitaille, D.F., and Garden, R.A., 1989, *The IEA study of Mathematics II: Contexts and Outcomes of School Mathematics*, Pergamon, Oxford.

Stevenson, H.W., & Stigler, J.W., 1992, *The Learning Gap: Why our Schools are Failing and What We can Learn from Japanese and Chinese education*, Summit Books, New York.

Stigler, J.W., and Hiebert, J., 1999, *The Teaching Gap: Best Ideas from the World's Teachers for Improving Education in the Classroom*. Free Press, New York.

Yin, R.K., 1994, *Case Study Research: Design and Methods*. Sage, London.

Zhang, W.G., Wang, Y.L., Gu, A.Z., Mai, Y.C., and Li, J.H., 1994, *Teaching Materials and Methods for Primary School Mathematics*, The People's Education Press, Beijing.

Zhao, D.-C., 1999, For the Students' Future: A Review of the Reform on High School Assessment in New South Wales, *Studies in Foreign Education*, **26**(4):18-22.

Zhao, D.-C., 2000, Establishing an Evaluation System Suitable for Quality Education: A Review of the Evaluation System of Primary School Pupils in Australia, *Studies in Foreign Education*, **27**(1):39-43.

Chapter 2-8

PHILIPPINE PERSPECTIVE ON THE ICMI COMPARATIVE STUDY

Bienvenido F. NEBRES, S.J.
Ateneo de Manila University

1. SOCIAL CONTEXT

The first impression of a visiting mathematics educator from countries with a stronger mathematics education tradition in discussions with counterparts from the Philippines might be that of similarities in situations. As solutions begin to be discussed, however, he might begin to realize that beneath these similarities are greater differences. The dominant reality in a country like the Philippines is the scarcity of resources, both human and material. Five or six students have to share a textbook. Many schools lack classrooms, so classrooms meant for 40 students are crammed with 80 students. Or schools have double sessions, in some cases triple sessions a day. Teachers are poorly trained and have to teach in a very difficult environment.[1]

2. DEPENDENCE ON WESTERN COUNTRIES

The paradoxical aspect of these differences, in particular the scarcity of human and material resources, is that instead of isolating us from developments in advanced countries, they make us more vulnerable to them. This is because we have to depend on Western mathematics educators and Western textbooks. We do not have the necessary number of experts nor the funds to develop our own textbooks. In the 1960s, for example, our Department of

Education invited Peace Corps Volunteers from the U.S. to bring in the 'New Mathematics' into Philippine schools. In the late 1980s and early 1990s, the Secondary Education Development Project, which developed new mathematics textbooks and teacher training, was funded by the World Bank with foreign consultants and advisers.

3. NO SIGNIFICANT IMPROVEMENT

However, after several decades of curricular reform in mathematics education, we have not seen significant improvement in the achievement of our students. The challenge then is to reflect on our methods of mathematics education reform and ask if we can find better ways.

4. DOMINANT APPROACH

The dominant approach has been to:

1. Bring in a new approach, usually theory-derived and usually from the U.S. This was the method in bringing in the New Mathematics in the 1960s and in subsequent reforms – newer trends such as back-to-basics, problem-solving, constructivism.
2. Develop materials based on these approaches.
3. Do pilot studies on small, selected scales, which usually say that the new approach is better.
4. Then, given that necessary funding is available, implement on a national scale. In this implementation, teacher training is done following what is called the cascade model:

The school system is organized into regions, which are divided into divisions, then into districts and finally into individual schools. The training program cascades as follows:[2]

- National level training for regional trainers
- Regional level training for division trainers
- Division level training for district trainers
- District level training for school trainers
- School level training for teachers in the schools

5. SHORT TIME FRAME FOR REFORMS

One of the major constraints in World Bank or OECF funded projects is their short timeframe, namely about 5 years. The teacher training part usually takes place in the last year or so. Because of the number of students and teachers in the Philippine school system and financial and time constraints, the training periods tended to become shorter and shorter as the training cascaded down until, at the level of school teachers, the training was just too short. While the training of the regional trainers might be for six months, by the time the training gets to the schoolteachers it might be just two weeks. Worse, because the training had to be compressed into such tight schedules, harassed administrators would send teachers for the training just to comply with quotas, even if they were not going to be teaching mathematics.

In talks I have given, I have compared the impact to that of a flash flood, too much in too short a time. The new curriculum and textbooks wipe out the past, but they are not absorbed.

As one reflects on this mode of mathematics education reform, one notes the following: the focus is on the intended curriculum. The greatest amount of time is given to the development of the textbooks and materials and the higher-level trainers. It is also from the top, from mathematics education experts from universities and from abroad. The time frame is too short.

6. TYPICAL OF WORLD BANK FUNDED OR FOREIGN-ASSISTED REFORM INITIATIVES

This is typical of World Bank and other Overseas Development Assistance Education Reforms in Developing Countries. The 5-year time frame of the ODA funding might work for building school-buildings, but in a large country like the Philippines or Indonesia it is too short for academic reform to be absorbed down to the individual classroom.

There seems to be an underlying assumption that there is an absolute best way of teaching and learning mathematics (usually the one espoused by the experts hired by the project). The method is to incorporate it into the new textbooks and materials and cascade it through the rapid teacher training.

Subsequent studies, of course, show that there is not much measurable improvement in the teaching and learning. The reason always given is the inadequacy of the teachers. Since this is the recurrent refrain, one wonders why the money is not simply used to address the inadequacy of the teachers rather than embarking on another curricular reform.

7. SEARCHING FOR MODELS IN EAST AND SOUTHEAST ASIA

Because of my work since the early 1970s in mathematics and mathematics education in East and Southeast Asia, I began to ask if there might not be a different way. I had noticed already in the 1970s that Singapore and Hong Kong did not simply drop their old curriculum and take in the New Mathematics as a whole (as we did), but only took certain parts and preserved much of the traditional mathematics.[3]

7.1 Role of ICMI Comparative Study

Thus, what is the role of this ICMI Comparative Study on mathematics education reform in a country like the Philippines? I offer the following reflections:

1. It relativizes dominant country influences (the United States for us) and helps us see alternative ways. In particular, that there are no off-the-shelf solutions and no absolute best way. What is good or best has to be seen in a particular situation and culture.
2. Weaknesses in mathematics education are not just due to lack of money or other resources. Resources are needed (e.g. textbooks), but if the deeper underlying factors are not understood, the resources will not be well used. For example, while new textbooks may be well and good, if the adequacy of teachers and sufficiently long teacher-training are not taken care of, not much improvement will come from the investment in new textbooks.
3. For us in the Philippines, a deeper appreciation of the importance of culture and values in mathematics achievement may help us look more closely at the different cultures in our own country. For example, we all know that the students coming from Chinese-Filipino schools are outstanding in mathematics performance. Our mathematics educators might consider studying these schools and benchmarking with them.
4. We could consider effective cooperation and benchmarking with schools in other countries. We have started to do this with the Grade School and High School of Ateneo de Manila. For example, our high school has been visiting and learning best practices from Anglo-Chinese High School and Chinese High School in Singapore. Before going there, we asked our visiting administrators and teachers to first read and discuss Stevenson and Stigler's “Learning Gap” and Liping Ma's "Knowing and Teaching Elementary Mathematics". These helped our visiting team look into areas they never looked at before.

The conclusion is not to copy practices blindly, but to reflect on the goals and values and to ask what practices (they may be the same) in our culture might achieve them. As one of the papers in this ICMI study says, we seek not blueprints, but mirrors.

8. A DIFFERENT MODE OF SCHOOL MATHEMATICS REFORM: FOCUSING ON THE IMPLEMENTED CURRICULUM

In talks to various groups in the Philippines, I have been discussing a different mode of school mathematics reform. The usual way (as described above) has been to focus on the intended curriculum, following major trends in the West. Then, to develop new textbooks and learning materials and do a pilot project which shows that the new approach is more effective (pilot projects always give this result) and then to seek to implement in the larger school system.

I have compared this with the longer 10 to 12 year cycle of school mathematics reform in Japan, where

a. Immediately on implementing a new reform, a process begins of feedback on the textbooks, materials, etc. from teachers and classrooms
b. This feedback is then processed and sifted through reports, conferences, discussions at different levels of the school system
c. Then policies and decisions are made on the main lines of the next cycle of reform
d. These are carried out in guidelines for new textbooks, books are written, pilot-tested
e. And implementation begins for the new cycle.

The main feature I have pointed out is that reform begins from the classroom, the implemented curriculum, and ideally the key players are the classroom teachers and school leaders. It is also a more evolutionary, rather than a revolutionary approach. We begin with what we have and improve on it, rather than wipe it out and totally replace it.

9. RECENT EFFORTS FOLLOWING THIS DIFFERENT MODE

We have been following these reform approaches at two levels:

1. Ateneo de Manila Grade School and High School

   Here we have asked our teachers to learn best practices from other schools in the Philippines (notably the Chinese-Filipino schools) and from partner schools in Singapore. We have also introduced them to comparative studies such as those of Stevenson and Stigler, Stigler and Hiebert, and Liping Ma. We use these "mirrors" to help us in continually improving our materials and our teaching.

2. On a larger scale, I led a group that was asked to help in improving mathematics teaching in the larger public school system (12.3 million elementary school students, 5 million high school students, 36,579 elementary schools, 4,629 high schools) in school-years 2001-2002 and 2002-2003. It is a huge and complex system.[4]

I will describe mainly the work we did for high schools as the work with elementary schools is just starting.[5] The first goals were to address the lack of textbooks (several students had to share one textbook) and the inadequacy of teachers and to do this on a large scale.

10. TEXTBOOKS AND LESSON GUIDES

We decided to go back to the more traditional discipline-based approach, rather than the spiral approach, to high school mathematics. The four-year high school series would thus be: Elementary Algebra, Intermediate Algebra, Geometry, Algebra and Trigonometry (with some Statistics). It is not that the spiral approach is not good in itself, but it demands more ideal conditions for its success (better trained teachers, ability to cover most of the book, etc.)

It was also emphasized that it was important not to introduce too many innovations to the teachers, to stay with what they were familiar with and improve on them patiently, and to develop teacher-training modules that help them with their actual textbook and day-to-day teaching.

In practice, this meant two things: We stayed with their actual textbooks, but reorganized the material with some transition sections. And we worked with the master teachers of the Department of Education together with some private high school teachers to develop detailed lesson guides for the teachers. This was done for the first three years of high school in 2001-2002 for implementation in 2002-2003 and for fourth year high school in 2002-2003 for implementation in 2003-2004.

11. TEACHER TRAINING

Teacher training for 1500 mathematics teachers for first to third year high school was done in April 2003 using the textbooks and the lesson guides. The training was given by high school teachers, master teachers from the public schools and selected private high school teachers. We found that it was more effective to have high school teachers with actual classroom experience deliver the training, rather than have college teachers. These 1500 teachers in turn provided training in their divisions and districts to the larger group of teachers using the same lesson guides and textbooks.

The same process was followed for fourth year high school with teacher training for 500 teachers in April 2003 and these teachers providing training for the others in May-June 2003.

12. INITIAL FEEDBACK AND AREAS FOR IMPROVEMENT

In terms of levels of mathematics, it is a small start. But in terms of scale, it is a large initiative. Students in high schools expressed appreciation that for the first time they each had a textbook. Teachers expressed appreciation that they had detailed lesson guides corresponding to their syllabus and their textbooks.

Areas needing improvement also surfaced quickly. In the first national consultation in school-year 2002-2003, third year high school teachers pointed out that the third year Geometry book was quite weak and needed a lot of work. This is true and we noted it when we looked at the Geometry content. But then this was the content that had always been there in the past 10 years or so. Their inadequacy was just not noted, because the material was distributed over several years in the spiral approach. This will have to be an area of follow-up this school-year.

13. FOCUS ON SCHOOLTEACHERS AND IMPLEMENTED CURRICULUM

If there is any point to be emphasized in these initiatives, it is the focus on the classroom and schoolteachers and on the implemented curriculum. Throughout the process, the leadership and work was carried out by high school teachers. The feedback on reform of the Geometry book has come from the classroom teachers.

It is a beginning effort, following a more evolutionary process, centering on the classroom and teachers. But given the scale of our school system, it is a very challenging effort.

The grade school work is even more challenging, given the size of the system and dispersal of schools in remote parts of the country.

NOTES

1. The reflections in this section and the next were already made in Nebres (1980) after the Berkeley International Congress on Mathematics Education.
2. More data on various interventions and reform initiatives may be found in Talisayon (1998). The description of the cascading model of teacher-training is on pp. 125 ff.
3. This was noted in Nebres (1988), the plenary address I gave at the Budapest ICME which compared East and Southeast Asian values, beliefs and practices in mathematics education with the Philippine experience from the United States.
4. Nebres (2003) is a report given to the National Academy of Science and Technology on the scale and complex problems of our public elementary and secondary school system.
5. A more detailed report is given in Oracion (2003).

REFERENCES

Ma, Liping, 1999, *Knowing and Teaching Elementary Mathematics*. Lawrence Erlbaum Associates, New Jersey.

Nebres, B.F., 1980, Major Trends from ICME IV: A Southeast Asian Perspective, in: *Teaching Teachers, Teaching Students*, Steen and Albers, eds., Birkhauser, Boston, pp.56-67.

Nebres, B.F., 1988, School Mathematics in the 1990s: Recent Trends and the Challenge to Developing Countries, in: *Proceedings of the Sixth International Congress on Mathematical Education*, Ann & Keith Hirst, eds., Budapest, Hungary.

Nebres, B.F., 2003, Demographics and Basic Education, paper presented at Annual Scientific Meeting of the National Academy of Science and Technology, July 10, 2003.

Oracion, C., 2003, A Report on Initiatives towards Improving Achievement in Mathematics at the Basic Education Level, report prepared for the Department of Education.

Stevenson, H.W. and Stigler, J.W., 1992, *The Learning Gap*, Simon and Schuster, New York.

Stigler, J.W. and Hiebert, J, 1999, *The Teaching Gap*, Simon and Schuster, New York.

Talisayon, V.M. et al, 1998, Materials and Methods in Basic Education and In-Service Teacher Training in Science and Mathematics, in: *Science Education in the Philippines: Challenges for Development, Technical Papers (Vol. I)*, E.B. Ogena and F.G. Brawner, eds., pp.107-147. Manila: NAST, SEI & CIDS.

Section 3

TEACHING AND LEARNING
Introduction

Colette LABORDE
University of Grenoble

Despite its brevity, the laconic title of this section refers to the two main kinds of processes involved in mathematics education that are related to all the components of the mathematics education system. Teaching and learning can be seen as the two extremes of a continuum of the education system. At one of the endpoints of this continuum, society pursues educational objectives and goals by organizing teaching content and teaching structures; at the other extreme, learning can be considered as resulting from the teaching activity. Therefore several issues and questions discussed in this section are linked to the other sections of the book.

In this book, teaching and learning refer to what happens in the mathematics classrooms, in relation to the teacher and the students, the mathematical content at stake, the organisation, the nature and content of the teacher activity as well as of student activity, and the exchanges among all participants in the classroom: teacher-students or student-student interaction. The title "Teaching and Learning" also expresses the intention to consider these two core processes of education as interrelated. As the TIMSS Video Studies concentrated on teaching, as Neubrand notices in her chapter, several chapters of this section focus on students' activities, inter-individual communications in relation to the teaching structure and content.

The range of studies addressed by the chapters of this section reflects the diversity expressed by the title from several perspectives:

- the type of contribution: research studies, innovative teaching or learning experiments, paper expressing a standpoint
- the focus of the study: teacher practice, lesson organisation, learners' activity and solving strategies

- the notion of classroom itself: some classrooms are “virtual” classrooms organized through distance communication using Internet possibilities or videoconferencing.

It is worth noticing that the starting intention of the ICMI study to focus on the cultural diversity in different countries from East Asia and the West led to a study of a much larger diversity of classroom activities and practices than in usual studies in mathematics education. We are faced with a seemingly paradoxical situation that the decision to study and explain the diversity of mathematics education by focusing on culture results in a greater variety of contexts for the studies and finally in an increasing diversity. But this is not very surprising, as even in a single country with a national programme of studies, the everyday classroom may be a place of large variation with respect to the curriculum. The topic of the section is probably emphasizing the variability of mathematics education.

From the multifaceted view on the influence of culture in mathematics education given by the section, some general issues and related questions emerge:

- To what extent are the differences of classroom practices in different cultural traditions more important than the variations that can be observed within the same culture, even within a single classroom?
- With respect to the gap between everyday classroom situations that are investigated in comparative studies and the distance teaching and learning experiments done between two or more countries: what conclusions can be drawn from those episodic but challenging experiments? what could be the role of technology in promoting awareness among teachers and students of the cultural diversity of mathematics classrooms?
- The contribution of these types of studies on a better understanding of not only mathematics classroom practices but also the mathematical content itself, which is taught and learned.

These three issues are discussed below.

1. A TWO FACETED PERSPECTIVE ON THE TEACHING LEARNING DIVERSITY

It is now well known and established by research that even a subject matter such as mathematics may give rise to a large range of possible classroom practices, teacher and student behaviours and attitudes. The role of research on the cultural dimension is two faceted in such situations: on the one hand, it may aim at seeking regularities and trends in this diversity, on the other hand it may focus on the diversity itself. In both cases, the ultimate goal is to develop some rationale for both observed regularities or variations. At first glance, both perspectives are represented in this section but if one tries to go deeper into the developed argumentation of papers, it appears that each paper contains in itself the two facets mentioned above.

All the papers share indeed the common perspective of the existence of a strong cultural determination of mathematics lessons. Two trends in the chapters of this section can be distinguished. In the first one, a central hypothesis is that the differences within a country are less than the differences between the countries (Neubrand in this volume). Studies related to that trend aim at comparing lessons in different countries or tasks given by the teacher (Hino et al, Neubrand). In the other trend, the cultural diversity is analyzed in the lesson itself, as for example the heterogeneity of the student community (Clarke et al. in this volume). But even within the comparative paradigm, ten years after the first TIMSS Video study of 1995, mathematics lessons in a country are no longer considered as homogeneous. Various poles in mathematics teaching are distinguished and the comparison between lessons or textbooks are carried out on each of those poles (Hino et al.). The type of knowledge involved in problems posed in the TIMSS Video study 1995 is analyzed by Neubrand and a comparative analysis of these problems is carried out between lessons in US, Germany and Japan coming from the Videostudy.

The extent to which the cultural tradition changes, and the way it does, is an object of investigation in Zheng Yuxin's paper: it describes the Chinese mathematics teaching and its changes when it was affected by a Western influence. The paper shows how the process of change depends itself on the Chinese tradition of conceiving teaching.

It appears that after more than ten years of comparative studies aiming at showing convergences within a cultural community and differences between communities, a finer approach is developed: a larger number of parameters are taken into account, processes are also the object of studies, variability is also investigated in places that were in a first approach considered as homogenous.

Therefore research methods changed, since they must allow fine observations of the classes. Hino et al. as well as Clarke et al. developed participant observations based on careful observations of classes and ethnographical research approaches including discussion with teachers and participation to various events related to teaching.

2. INNOVATIVE TEACHING EXPERIMENTS DEVELOPING EXCHANGES BETWEEN CLASSES FROM DIFFERENT COUNTRIES AND CULTURES

The section contains three papers (Isoda et al., Xu Fei, and Graf & Moriya) describing distance learning activities between students belonging to countries respectively influenced by West and East traditions. The common issue of these papers deals with the role of distance technology (Internet) allowing communication between teachers and/or students who usually have no opportunities of collaborative work. In two papers (Isoda et al., Xu Fei), the distance technology was used to organize exchanges among students of two countries with different traditions: about Painting and Geometry between Japan and China, and about various mathematics problems to be solved (in arithmetic and geometry) by Australian and Japanese students in a written communication. Both papers show how this experience allowed students to develop awareness of other existing conceptions of mathematical notions or other solving strategies than their own ones. The third paper, reporting on several teaching experiments between Germany and Japan, shows also the growth of awareness of each students' community about a different tradition in the other country. It also reveals the great curiosity of all students of a country to know more about the partners of the other country. Simultaneously some wrong ideas about stereotypical behaviours in the other country could be destabilized by such experience. The papers are enthusiastic about these experiments. The unusual nature of these distance exchanges (except for the Internet exchanges between Australia and Japan) constitutes both a strength and a weakness of the experiments. It shows how technology can act as a catalyst. Technology could enhance communication and understanding about mathematics between students and teachers. It could stimulate the awareness of not only the diversity of mathematics teaching but also of the existence of fundamental common invariants in mathematics: German and Japanese students could discover that geometry may bring everywhere solutions to perspective drawing, even if those solutions differ according to countries. Students succeeded in cooperating on a mathematical project by using the same mathematical words. The weakness of such experiments lies in the cost in terms of technological tools and organization. For the time being, such distance teaching cannot be the everyday teaching. However, as mentioned in the paper of Graf & Moriya, technological progress might reduce the cost in the future.

In all the three experiments, organizing communication between classrooms contributed to increase the autonomy of students, mainly for secondary education. Students decided to become the teachers of the partners from the other country. They were faced with a situation in which they had to develop linguistic means and reflections on their actions and solving processes, in order to be able not only to communicate but also to explain what they did. An additional effect of such experiments is a deepening of mathematical knowledge. Generally speaking all the papers of this section about Teaching and Learning mathematics contribute to develop a better understanding of mathematics.

3. A BETTER UNDERSTANDING OF MATHEMATICS

Mathematical notions or processes themselves are taken into account as variables subject to cultural variation: comparison of teaching is based on the assumption that the cultural dimension of teaching takes different forms according to the content of teaching. The epistemic orientation is, for example, the object of comparison in the paper by Hino et al. In their study of the cultural determination of teaching and learning, the chapters of this section investigate the ways and forms of presenting the mathematical content (Hino et al.), the kind of examples used to illustrate the notions (Graf & Moriya), the type of problems and of solving processes promoted by teaching (Neubrand, Clarke et al.), and the actual solving processes of the students (Clarke et al., Isoda et al.). All these aspects contribute to shape an image of mathematics, more complex and more versatile than the stereotypical image widespread in the common opinion. We would like to claim that, from this study, we learned not only about teaching and learning across the world but also about mathematics.

Chapter 3-1

THE TIMSS 1995 AND 1999 VIDEO STUDIES

Johanna NEUBRAND
University of Vechta

1. ORIGINS AND AIMS OF THE TIMSS VIDEO STUDIES

From the late 60's on, the IEA (International Association for the Evaluation of Educational Achievement) conducted international comparative studies of achievement in the fields of mathematics and science, the so-called First and Second International Mathematics Study (FIMS and SIMS), and the Six Subject Studies (Husén, 1967; Travers & Westbury, 1989). TIMSS, the Third International Mathematics and Science Study in 1994, was the biggest one in that series (Beaton & al., 1997). Experiences from all those studies have shown that different structural patterns in achievement exist in different countries. Consequently, the question arose, where did such different patterns came from. One of the origins of the differences in achievement could be different ways of teaching mathematics. But if this is so, then the question is to what extent can the observation of classroom processes be a source of information. That is, what are good databases to draw conclusions from?

A classical way is, for example, to develop questionnaires about teaching practices, as was already done as a part of SIMS. But words and concepts used in questionnaires may be understood in different ways, even between teachers in one nation, the more so across countries. Even if there are organizing elements in lessons that seem to be formally the same, they may have different modes of use in different teaching cultures. In the "Survey of Mathematics and Science Opportunity" over six countries (SMSO, Schmidt & al., 1996), a predecessor of TIMSS, it was documented that, for example, "seatwork" refers to different situations and functions in the countries. One of the reasons is that teaching tradition is often communicated within the community of teachers of one country and is passed along from one to the next generation of teachers (Hiebert, Stigler & Manaster, 1999).

How then is it possible to study teaching across cultures? The rising interest in classroom research in recent years and the availability of video tools, including the technical facilities of inspecting and examining videos in an effective way, are some of the reasons to set up video studies in international comparisons of teaching mathematics. Since videotapes could form good bases for analyses, and also for repeated analyses, under objective and various theoretical perspectives, the aim was to collect big enough and representative samples of everyday lessons in selected countries in a video-survey. The size of the TIMSS achievement study 1995 gave a unique opportunity to realize that idea. TIMSS-Video functioned therefore as one of the additional studies in the TIMSS project (Neubrand & Neubrand, 1999). Meanwhile, there were two videotape projects within or in cooperation with TIMSS: the TIMSS 1995 Video Study and video study conducted in connection to TIMSS-R[1] 1999.[2]

The two TIMSS Video Studies investigated teaching practices in eighth-grade in mathematics (and science) classes. One perspective is common to both studies. Classroom events were observed as events of *teaching*. All documents collected in the studies concentrate on teaching, and the ways that a certain lesson was constructed and organized by the teacher. Thus, the videos capture the classroom from a total camera angle, the collected materials contain the preparation of the lesson by the teacher; the lesson tables drawn later from the video source were written as on overview of what was going on at the classroom level etc. Individual processes of learning, student activities as far as they are not directly bound to the teaching intentions, inter-individual communication in the classroom etc., are not in the central focus.

2. THE TIMSS 1995 VIDEO STUDY

The TIMSS 1995 Video Study was administered in 1994 and 1995. Of the 41 TIMSS countries, Germany, Japan and the US were chosen, since

[1] TIMSS-R = "TIMSS-Repeated", a successor study to TIMSS.

[2] There are some more video studies of mathematics teaching around the world administered by other research groups, as is also mentioned in this book, e.g. the Learners' Perspective Study.

Germany and Japan were "viewed as important economic competitors of the US" (Kawanaka, Stigler & Hiebert, 1999, p 89). There was a special interest towards both non-American countries. Japan has repeatedly scored at the top of international comparisons of mathematics achievement, and Germany was considered from the American perspective as a country with a highly valued tradition in mathematics education. However, it became clear later that for the benefit of improving the construction of mathematics lessons there was much more to observe in the Japanese lessons than in the German ones, since Germany turned out not to be a high achieving country after the TIMSS achievement test (Baumert & al., 1997).

The video sampling was realized as a sub-sample of the TIMSS main study: 100 classrooms were videotaped in Germany, 81 classrooms participated in the United States, and 50 video lessons were produced in Japan (Kawanaka, Stigler & Hiebert, 1999). Besides videotaping, supplementary materials were collected, such as copies of textbook pages, worksheets and preparation sheets of the teachers. In addition, teachers were asked to identify the goal of the lesson in a questionnaire, and to judge how typical was the lesson she or he had just delivered.

As mentioned before, analyzing activities depends on an easy access to the video and the other data. Videotaped lessons therefore were converted into digital formats, linking by suitable time codes the video sequences and the transcribed lesson text, with an English translation if necessary. Also the lesson tables, serving as an overview of the lesson, can be linked to that data by time codes. Finally, a multimedia database has been set up and is now available.[3] Besides the videotapes, the study provides several types of data. There are qualitative documents (lesson materials, teacher's preparation etc.) and the quantitative data (questionnaires and codes to analyze the videotapes).

A central question was, what should be coded from the lessons already at the first cycle of analysis of the lesson. Since the goal of the study was to reconstruct the instructional quality, and to yield informative descriptions of how mathematics lessons were taught in the three cultures, three dimensions of the lessons were seen as the most important, according to Kawanaka, Stigler & Hiebert (1999):

- the work environment (number of students in class, groups or individual learning, access and use of books and materials, interruptions etc.)

[3] Some of the lessons are free accessible for demonstration purposes, the rest is available for re-analyses upon request; see the TIMSS home pages for details.

- the involvement of students in class (skills, problem solving, level of mathematics, inner coherence etc.)
- the methods teachers use (structuring the lessons, classwork vs. seatwork, teachers' roles in the class at various occasions, discourse in the class, performance expectations etc.)

The salient question however is to translate the impressions one gets from the inspection of the lessons into reliable codes. This has a specific meaning when comparing lessons from different teaching cultures. The codes must be as objective as possible with respect to capturing the "right" meaning in any of the three (or more) teaching cultures. Therefore, in a rather complex process before collecting the data, code developers from all the participating countries collaborated, watched preliminary sample lessons, discussed and constructed suitable codes. In the further development of the study, these codes were gradually extended and refined, and served as the basis for the official report of the TIMSS 1995 Video Study (Stigler & al., 1999).

This short overview has already pointed to the main characteristics of TIMSS 1995 Video Study. The most general conclusion was that there are apparent differences between teaching, captured by the concept of "lesson-scripts", and that the differences within a country are less than the differences between the countries. Teaching therefore appeared "as a cultural activity" (Stigler & Hiebert, 1998, 1999).

3. A RE-ANALYSIS OF THE TIMSS 1995 VIDEO STUDY ON THE BASIS OF FEATURES OF THE MATHEMATICAL TASKS

One of the main advantages of a video based data collection is that these data can be re-analyzed, and even re-coded in order to vary the aspects under which the lessons should be considered. Such re-analyzes were indeed realized in the TIMSS 1995 Video Study. These two re-analyzes will be reported in the following sections. They both use the tasks given in the videotaped lessons as the units of analysis.

Using tasks as units of analysis makes sense, since features of mathematical tasks can be a rather objective means of analysis, provided the features taken into consideration are related to cognitive processes of mathematics learning. Furthermore, mathematical tasks unify two different strands in the teaching and learning process: they serve the teacher to construct his or her lesson, and they serve as well the students to construct their mathematical knowledge (Stein & Henningsen 1996; Christiansen & Walther 1986; Schoenfeld 1987). Tasks are central instruments for the teacher, since in the form of tasks most often the mathematical content is transformed into teaching structures (Bromme, Seeger & Steinbring, 1990). The consideration of tasks used in the classroom is therefore close to the content and its instructional realization, being the central issue when comparing lessons from different cultures.

Margaret Smith's system of analyzing the mathematical tasks, and its use to exhibit the differences between the countries at a content-based level (Smith 2000), became part of the report of the TIMSS 1999 Video Study. This approach is therefore described in detail in the following section.

This section explicates how the classification system of the author was used to explain that even at the level of the tasks posed and worked on in the classroom, the differences between the three countries (the US, Japan and Germany) became perceptible in the TIMSS 1995 Video Study (Neubrand, 2002). One of the main differences from the official report (Stigler & al., 1999) consists in a thorough differentiation between the two subject areas algebra and geometry, since the analysis showed that the teaching behavior, at least in Japan, is highly dependent on the choice of the topic. The second guiding idea is to differentiate the cognitive activities that are the basis of the problems posed and worked on in the classroom. This means that distinctions are made which are recognized as central and distinctive for mathematics.

For this re-analysis of the TIMSS 1995 Video Study, 22 lessons in each country were selected. This sub-sample contains lessons equally from algebra and geometry, and focused especially on lessons which more or less extended seatwork phases. These 22 lessons are therefore suitable to reveal the content structure of the lesson, and to make visible on which basis the students are expected to construct their knowledge on their own. The observation of the change between seatwork and classwork in the lesson can additionally show to which tasks students are exposed to work independently, an essential source to describe the pedagogical intention of the lesson.

The bases of the re-analysis are the lesson tables; schematic descriptions of the lessons, which show the sequences of problems, social settings, activities etc. in the classrooms (see Stigler & al., 1999, for examples of that instrument). However, the sequences of the problems in the lessons, and the relations between the problems, were completely newly defined and coded, so that a sample of 1153 problems forms the basis for the analysis. The instrument used is an extended classification system for mathematical tasks and problems, developed for this purpose (Neubrand 2002). Only some features of this classification system are used in this report. The classification system allows quantitative and qualitative analyses of the data. In

Neubrand (2002) methods of multivariate analysis of variance were essentially used to analyze the quantitative data.

3.1 Number of problems and their implementation in the lessons

Assuming that tasks play similar roles for the teaching in the countries, one should expect, that in any mathematics lesson, the number of problems would be rather constant. However, the numbers of problems posed in a lesson differ significantly in the three nations, Japan showing the least number, the US the highest. Fig. 1 shows this on the basis of the medians (boxplot). The same picture appears in algebra and geometry lessons.

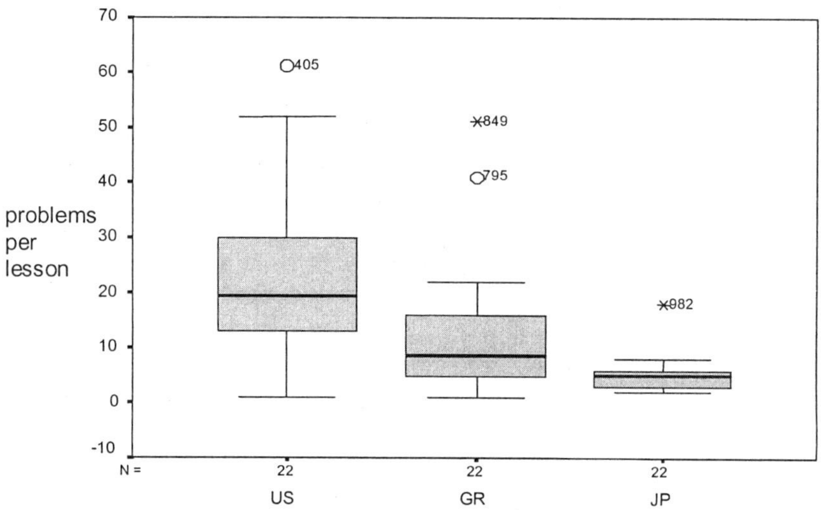

Figure 3-1-1. Number of tasks per lesson (medians)

The differences in the number of problems can be explained when looking at different situations in the lessons. Five situations can be distinguished to implement a problem in a lesson:

- the problem is worked on in Seatwork (SW), but not shared in Classwork (CW);
- the problem is posed, or only checked in CW;
- the problem is worked on and solved in SW, and shared in CW;

- the problem is worked on and solved in both SW and CW;
- the problem is worked on, solved and shared completely in CW.

Fig. 2 shows that the differences in the number of problems in a lesson result from the amount of problems given for seatwork without sharing them in classwork, as well as from those posed or only checked in classwork. The picture of Fig. 2 is not affected by the subject areas, algebra or geometry.

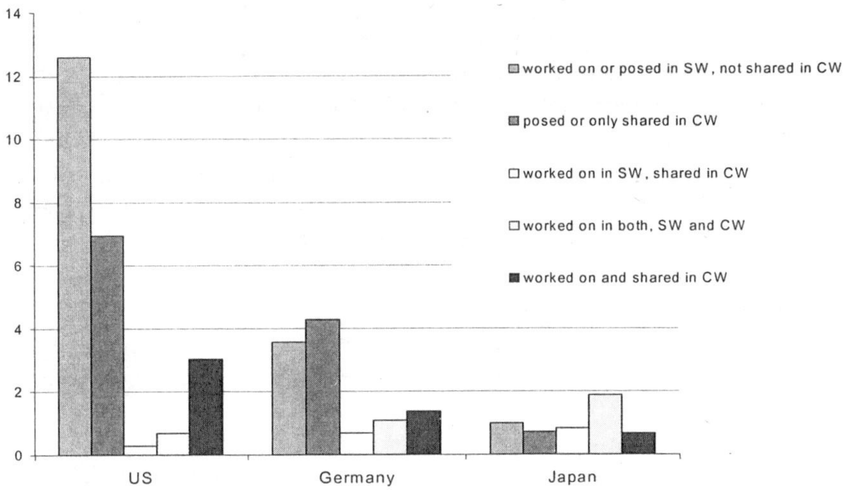

Figure 3-1-2. Number of tasks per lesson (means) in different implementations into the lesson

The last three situations, shown as the three bars at the right in Fig. 2, describe the problems which are completely solved during the lesson. These are the problems on which the teaching in the class is based. These problems determine the classroom dynamics since they are totally worked out and made public in the classroom and influence the learning processes of the students explicitly. Therefore they will be collected later as the problems "worked on *or* shared in Classwork". These groups of tasks establish a "cultural script" (in the sense of Stigler & al., 1999) like this (Figure 2): There are nearly equal numbers of such problems in the three nations, approx. 3 in GR and JP, 4 in the US. However, problems of this kind mostly include seatwork phases in Japan, but seldom do so in the US, whereas in Germany both cases occur.

3.2 Types of problem, according to the cognitive activities they require

Cognitive activities can have their origin in both the demands of the problem posed to the class, in whatever implementation form, and in the ways the posed task is worked on, either in class or in seatwork. The distinction between the problems as posed, and the problem as worked on and solved, is critical to all analyses of mathematical tasks. Several authors point to these differences (Stein, Grover & Henningsen, 1996; Smith, 2000; Neubrand 1999, 2002). In the TIMSS 1999 Video Study this distinction comes also into the analysis (see later in this chapter). The analysis in this section, however, concentrates on the problems posed to the class.

From the point of view of mathematics education, the salient question then is, of which nature are the problems posed? In the classification system developed to analyze the mathematical tasks (Neubrand, 2002), the problems are characterized by the following most central aspects:

- *"character of knowledge":* Is the knowledge needed to solve the problem *procedural* (here, this includes applying algorithms as well as stating or recalling concepts) or *conceptual* (in the sense of Hiebert, 1986), or a *combination* of the two?
- *"complexity of knowledge":* Is a modeling process or a problem-solving process necessary to solve the problem, and how complex is this process; i.e. are there *one or several* units of knowledge *explicitly* given in the task, or are they *implicit* in the task and have to be extracted by the solver from her or his knowledge base?
- *"application":* Is there an *extra*-mathematical or an *intra*-mathematical application in the task, or is *no* application at all present?

These three aspects are considered central because they allow, using the tasks as indicators, to characterize which general mathematical guidelines a specific lesson is based on. They refer to both content and processes of mathematics, and therefore can describe what the "Standards and Principles" call "a comprehensive set of instructional goals and activities" (NCTM 2000).

Combining these three central features specifically, several "Types of problem" can be defined. An analysis of the frequency of the configurations (according to Krauth 1993) produced 12 types of task. The following examples illustrate the nature of these types.

3.2.1 Type 1: "Procedural" and "explicit" tasks

These tasks rest on procedural knowledge, no modeling or problem-solving process is required, and there is only one unit of knowledge to be activated. An application is not present.

Examples

- $x^2 + x - 6 = 0$. – Solve the equation.
- Given AB = 7 cm; $\alpha = 60°$; $\beta = 70°$. – Draw the triangle ABC using ruler and compass.
- $y = 3x + 2$. – Draw the graph of this function.

3.2.2 Types 2 to 12: "Advanced" tasks

These tasks rely on conceptual knowledge or a combination of both procedural and conceptual knowledge. In any case, a modeling or a problem-solving process is needed to transfer the given situation into a solvable piece of mathematics. There may be one or several units of knowledge to be activated. This may happen in any application, extra- or intra-mathematical, or even without any application.

Example (in this case the characteristic features are: conceptual thinking; modeling process needed, since there is no explicit formula to work on; one unit of knowledge, namely the properties of parallel lines; intra-mathematical context) – taken from the lesson JP10 – TIMSS 1995 Video Study:

▪

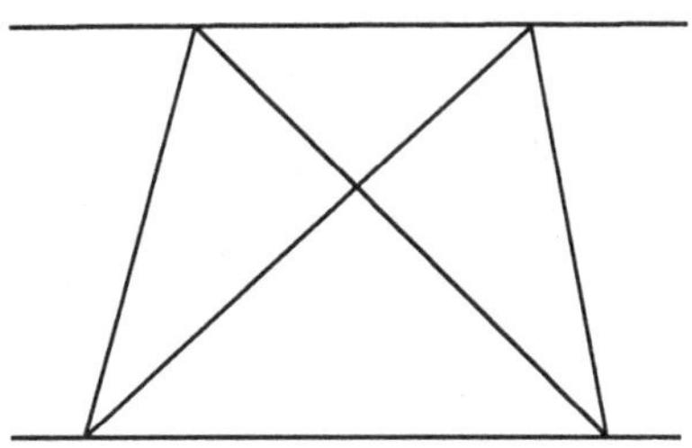

(drawing on the blackboard)

When ℓ and m are parallel lines, find the relationship between triangle ABC and triangle A'BC. – "You may talk with your neighbors".

Table 1 shows how these types are distributed over countries and subject areas. Some of the combinations of the three centrals features of the tasks do not occur at random, but characteristically in greater frequency in some of the countries. The occurrence of the types in the countries points to the goals and aims of the teaching in that country.

Table 3-1-1. Types of task - characterization, subject areas and nation (absolute numbers)

				US		GR		JP	
Type	**character of knowledge**	**complexity of knowledge**	**application**	**Alg**	**Geo**	**Alg**	**Geo**	**Alg**	**Geo**
1	procedural	one-explicitly	without	147	77	108	76	25	2
%				83%	75%	90%	85%	71%	4%
2	conceptual	one- implicitly	intra-math					1	7
3	conceptual	sev- expl./impl.	intra-math	2					1
4	conceptual	sev- implicitly	intra-math					1	13
5	combination	one- implicitly	extra-math	4	6	6	3	5	
6	combination	one- implicitly	intra-math		2	3			8
7	combination	one- implicitly	without	14	11	2			6
8	combination	sev- explicitly	extra-math				5		
9	combination	sev- expl./impl.	extra-math	1			3		
10	combination	sev- expl./impl.	intra-math		5	1	1		
11	combination	sev- implicitly	extra-math				2	3	
12	combination	sev-implicitly	intra-math		1				8
2 - 12				31	25	12	13	10	43
%				17%	25%	10%	15%	29%	96%
total of 569				**178**	**102**	**120**	**89**	**35**	**45**

The type of the problem, the subject areas, and the nation, interact with each other, (hierarchical loglinear analysis). Therefore, the main results are:

- In *Japan* one observes a distinction between algebra and geometry with respect to the selection of the tasks. In algebra those problems predominate (at a rate of 71%) which require procedural knowledge, but no modeling process. In geometry however the conceptual problem types with intra-mathematical application play the main role. Only less than 4% of the problems in geometry are of Type 1.
- In *Germany* the highest percentages of tasks of Type 1 are observed in both areas, 90% in algebra and 85% in geometry.
- In the *US* 83% of algebra problems and 75% of the geometry problems belong to Type 1, and as in Germany this type is by far the most frequent.

3.3 Types of problem worked on or shared in classwork

From the point of view of the lesson dynamics, it is interesting to investigate how the different types of problem, as defined in the previous section, distribute over the teaching situations. To get a picture of what happens in the classrooms, the problems worked on or shared in classwork are the most influential. However, the process of working on is still not described, but the nature of the problems stated in the different teaching situations. It can be distinguished whether self-activity of the students, meaning phases of seatwork, are included in the solving process or not, either on the Type-1-problems or on the Advanced Problems (Types 2-12 in Table 1). Figures 3 and 4 show how the countries behave, in algebra and geometry lessons respectively.

The analysis can be summarized as follows:

- In the *US*, most of the problems totally worked on or shared in classwork are Type 1 problems (procedural, no modeling or problem-solving necessary), in both algebra and geometry. The few more advanced problems are worked on in seatwork in algebra, worked on in classwork in geometry. So, classwork in the US is devoted mostly to problems requiring procedural thinking when no modeling activity is expected. (An additional analysis showed that often advanced problems are only posed in seatwork and not shared in classwork.)

- In *Germany* a predominance of Type-1 problems in algebra is also observed. In geometry there is an equal distribution of the two types. However, the advanced problems (conceptual knowledge and modeling activities) are mostly implemented in both situations, partial working on in seatwork and sharing and explanation in classwork.

- Lessons in *Japan* show an emphasis on independent and demanding work of the students that, in contrast to the US, influences the structure of the whole lesson. The selection of problems is strictly different in the two subject areas. There are still many, though less than in GR and the US, procedural problems in algebra; however, the more complex types of algebra problems include seatwork. In the geometry lessons the variation in the types is the biggest. Students are confronted in seatwork with a great variety of conceptual problems requiring intra-mathematical application; classwork is up to 80% devoted to tasks that come from seatwork into classwork, all of them being the Advanced Type.

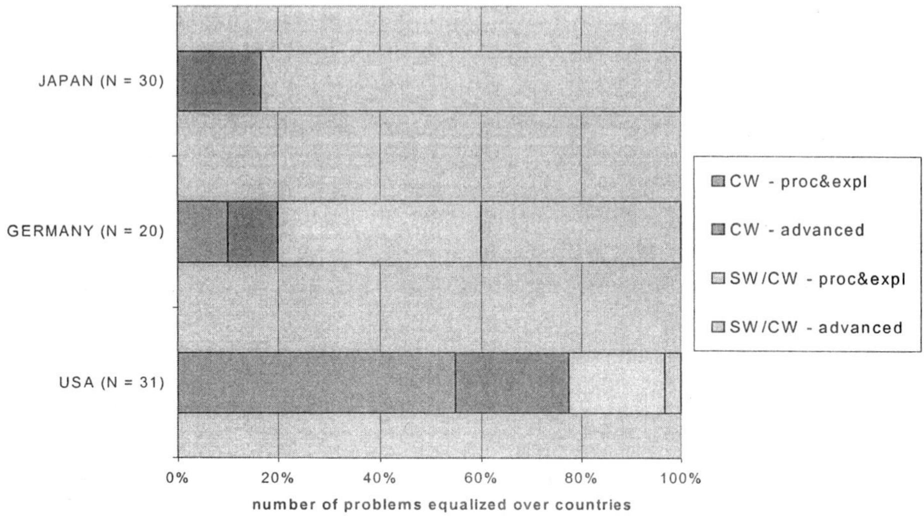

Figure 3-1-3. Problems worked on or shared in Classwork: Geometry (implementation & cognitive demand)

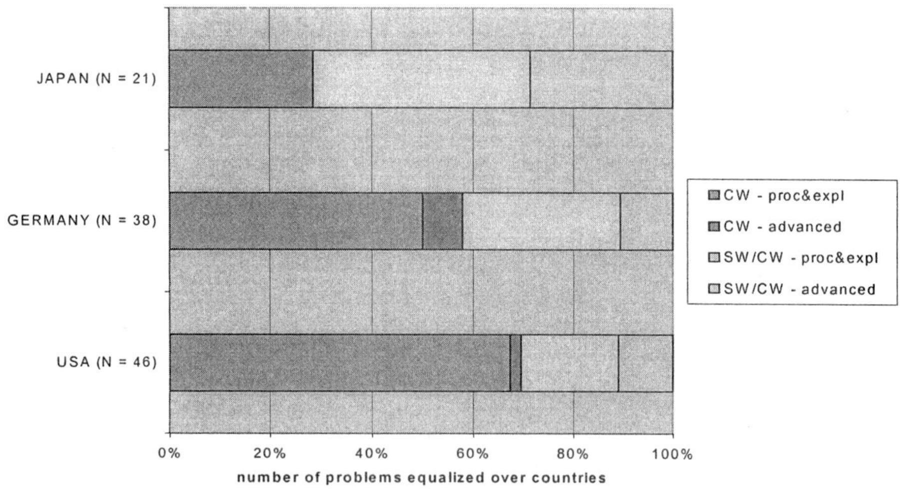

Figure 3-1-4. Problems worked on or shared in Classwork: Algebra (implementation & cognitive demand)

3.4 Conclusion: Ways of organizing lessons, by the means of tasks, point to different mathematical demands

As the means to organize the construction of mathematical knowledge in a lesson, mathematical problems play in fact different roles in the three nations' classrooms. The Japanese, German and US American videotaped lessons are distinct with respect to

- the inner structure of the lessons,
- the selection of the problems posed in seatwork for independent work, and
- the characteristic discrimination of the subject areas.

Thus, the cultural script idea says more than just other kinds of pedagogical organization of a lesson. The differences between the US, Germany and Japan are deeply rooted in different mathematical demands on the posed problems. The most surprising result is that geometry problems in the three countries are so different: There were few problems with procedural knowledge in Japanese geometry classrooms in grade 8, and consequently the "thinking methods" Stigler & al. (1999) observed in Japanese lessons, have their origin to a great extent in the way geometry was taught.

4. FINDINGS OF THE TIMSS 1999 VIDEO STUDY ON POSED AND SOLVED MATHEMATICAL PROBLEMS

Since the results of the TIMSS 1995 Video Study, in the way they were presented in the previous section, bring out that the countries behave so differently with respect to the mathematical content and the cognitive demands, two questions can serve as an orientation for the analysis in this chapter:

- Could it be that there is a general cultural difference in teaching mathematics between "the East and the West"?
- Or alternatively, is it a matter of the high achieving countries having a lesson and task structure similar to the one in Japan?

The second question was one of the goals of the TIMSS 1999 Video Study. Four basic ideas motivated the set-up of this study: There is more than one country with high achievement in TIMSS, science lessons were included, analytic tools should be developed, and later on, there should be a more extensive public use library at hand e.g. for teacher education purposes. Additionally, the videotapes were recorded across the school year to capture a broader range of topics. Seven countries participated: Australia, Czech Republic, Hong Kong SAR, Japan[4], the Netherlands, Switzerland and the US (Hiebert & al., 2004). As the final data source, the TIMSS 1999 Video Study has collected 638 eighth-grade lessons in mathematics from all seven participating countries.

The analysis of the data focused first, being not so different from the analysis of the 1995 data, on the lesson structure in a rather general pedagogical sense, discussing issues like time spent on content and task, clarity and flow of information in the classroom, interruptions, role of homework, etc. In a similar way, the nature of the mathematical content treated in the lessons was described, e.g. what distribution of contents occurred in the countries, what learning opportunities were given, etc. (for details see Hiebert & al., 2003). However, in addition to the first analysis of the TIMSS 1995 Video Study, an extensive part was devoted to instructional practices, relying directly on the tasks students had to work on. To this purpose, the mathematical problems themselves were considered as the units that create the instructional process.

4.1 Problems posed and stated

A basic distinction, already mentioned in the previous section, has to be made in advance: Tasks are selected by the teacher, posed or stated, and then given to the class for working on. How the nature of the posed problems can be described was exhibited in detail in the previous section. Also, to describe the quality of the problems worked on and solved, Neubrand (1999, 2002) made up two basic categories. On the one hand, working on can be of procedural nature. The routines of the algorithm, the performance of single steps, even learning by heart, can then be the observable features. On the other hand, a task can also be worked on with understanding. Then features one has to look at are pointing to the meaning of a solution idea, the drawing of connections, reflective and metacognitive activities. In Neubrand (1999, 2002) some detailed codes to distinguish procedure-oriented working on a task from an understanding-oriented access are constructed. However, the character of the problem statement does not need to remain the same when the problems are going to be solved in class.

[4] The Japanese lessons were collected in 1995 and reanalyzed under the views of the 1999 Study.

Stein, Grover & Henningsen (1996), Smith (2000), Neubrand (1999, 2002), Hiebert & Handa, (2004), and others noted that mathematical tasks can change their nature after being worked on in class. This can happen in different directions, e.g. a problem originally posed as a "doing mathematics" problem can be broken up into single steps when given to the class, as Stein, Grover & Henningsen (1996) showed on several instances. Also the reverse occurs. Procedural tasks can be interrupted by conceptual sub-problems, related to the original task, so that the character of the task as a whole changes. Neubrand (2002) observed this in Japanese algebra lessons[5]. More than half of the *sub*-problems given to seatwork in these lessons activate conceptual knowledge (Table 2), in contrast to the fact that procedural knowledge predominates to 71% on the basis of the core tasks (cf. Table 1).

Table 3-1-2. Procedural and conceptual knowledge in sub-tasks of algebra lessons, worked on in seatwork (Neubrand, 2002, Fig. 17.7(excerpt), transformed into the categories of Smith (2000) - see 4.2.)

	USA	Germany	Japan
procedural knowledge & stating concepts	9	7	17
making connections	4	4	23

A similar observation is reported by Hiebert & Handa (2004). They describe a Hong Kong lesson, in which just by inserting "connecting" remarks and references, the character of the procedural topic changed into a topic rich with connections and meanings. Tasks can be worked on, so that conceptual knowledge even within a procedural task can be made visible to the students.

Consequently, it is an interesting and necessary feature to make a difference between a *problem stated*, and a *problem solved*. This is also one of the basic ideas in Smith's work (Smith, 2000). In the following analysis the categories of Smith (2000) used in the TIMSS 1999 Video Study (Hiebert & al., 2003) are the basis of the analysis[6].

[5] As mentioned earlier, the Japanese lessons in the TIMSS 1999 Video Study are those collected in 1995.

[6] The next paragraph follow closely Chapter 5 in Hiebert & al. (2004).

4.2 Mathematical processes suggested by stated problems

In the TIMSS 1999 Video Study, the statements of mathematical problems were classified along the kind of mathematical processes they imply. The following processes were distinguished (quoted from Hiebert & al., 2003):

- "*Using procedures*: Problem statements that suggested that the problem was typically solved by applying a procedure or set of procedures. These include arithmetic with whole numbers, fractions, and decimals, manipulating algebraic symbols to simplify expressions and solve equations, finding areas and perimeters of simple plane figures, and so on. Problem statements such as "Solve for x in the equation 2x + 5 = 6 - x" were classified as using procedures.

- *Stating concepts*: Problem statements that called for a mathematical convention or an example of a mathematical concept. Problem statements such as "Plot the point (3, 2) on a coordinate plane" or "Draw an isosceles right triangle" were classified as stating concepts.

- *Making connections*: Problem statements that implied that the problem would focus on constructing relationships among mathematical ideas, facts, or procedures. Often, the problem statement suggested that students would engage in special forms of mathematical reasoning such as conjecturing, generalizing, and verifying. Problem statements such as "Graph the equations y = 2x + 3, 2y = x - 2, and y = -4x, and examine the role played by the numbers in determining the position and slope of the associated lines" were classified as making connections."

(Hiebert & al., 2003, p 98)

On the basis of these categories for the stated problems, Fig. 5 (Hiebert & al., 2003, p 99, Fig. 5.8) shows the predominance of the "Using procedures" problems in all countries, except Japan. Japan on the other side presents the largest percentage of posed and stated problems on "Making connections", followed by the Netherlands.

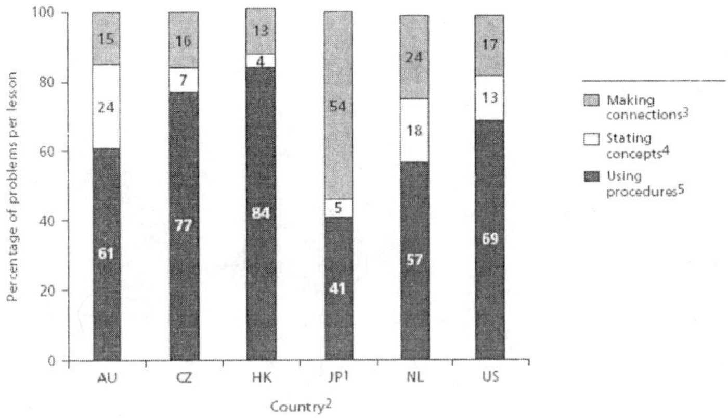

[1]Japanese mathematics data were collected in 1995.
[2]AU=Australia; CZ=Czech Republic; HK=Hong Kong SAR; JP=Japan; NL=Netherlands; and US=United States.
[3]Making connections: JP>AU, CZ, HK, US
[4]Stating concepts: AU>CZ, HK, JP; NL, US>HK, JP.
[5]Using procedures: CZ>JP, NL; HK>AU, JP, NL, US; US>JP

NOTE: Analyses do not include answered-only problems (i.e., problems that were completed prior to the videotaped lesson and only their answers were shared). For each country, average percentage was calculated as the sum of the percentage within each lesson, divided by the number of lessons. English transcriptions of Swiss lessons were not available for mathematical processes analyses. Percentages may not sum to 100 because of rounding. The tests for significance take into account the standard error for the reported differences. Thus, a difference between averages of two countries may be significant while the same difference between two other countries may not be significant.
SOURCE: U.S. Department of Education, National Center for Education Statistics, Third International Mathematics and Science Study (TIMSS), Video Study, 1999

Figure 3-1-5. Average percentage of problems per eighth-grade mathematics lesson: Problems stated, by country (copy of Hiebert & al., 2003, Fig. 5.8)

As far as Japan is concerned, this result is coherent with the analysis of the TIMSS 1995 Video Study, reported in Chapter 3-1. The first two categories in Fig. 5 are the "procedural tasks" in the analysis of Neubrand (2002) since "procedural" in her classification system contains also the knowing, recalling or stating of facts (without connecting them to meaning or other concepts). The "making connections" problems however are the "conceptual tasks" in Neubrand's analysis in the previous chapter, or the tasks which contain both characters of knowledge, procedural *and* conceptual.

However, one of the main results in the previous section was that, within the countries, there exist considerable differences in the teaching of algebra and geometry. In fact, these differences can be made even more visible by the following figure (Neubrand, 2002, Fig.17.4[7]) The Japanese lessons especially show a strictly distinct selection of problems in algebra and geometry (Fig.6).

[7] The basis of Fig. 6 are all (core-) problems (N_{total}=869), not only those belonging to the ones aggregated in the "Types" in Tab.1.

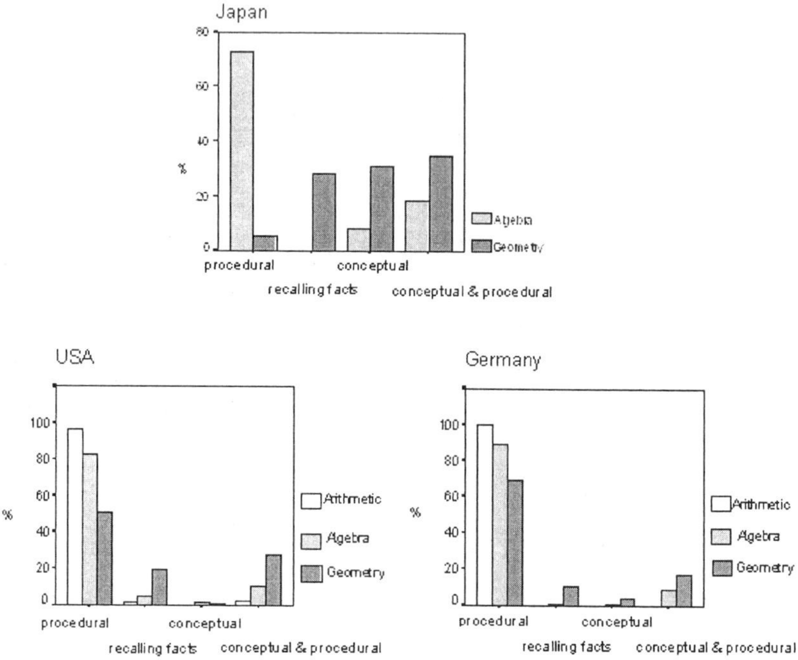

Figure 3-1-6. Distribution of all problems, by country and subject area (Neubrand, 2002, Fig. 17.4). $N_{Japan} = 94$; $N_{Germany} = 247$; $N_{US} = 528$.

As a matter of fact, in the TIMSS 1999 Video Study, tasks are not equally distributed among the subject areas (Table 3). To compare the results of the TIMSS 1999 Video Study with the comparison of the three countries in the TIMSS 1995 Video Study in the last section, the distribution over the subject areas has to be regarded (Table 3).

Table 3-1-3. Average percentage of problems per eighth-grade mathematics lesson within each subject category, by country (excerpt from Hiebert & al. (2003), Tab.4.1).

	AU	CZ	HK	JP	NL	SW	US
Arithmetic	36	27	18	-	16	42	30
Geometry	29	26	24	84	32	33	22
Algebra	22	43	40	12	41	22	41
other, including Statistics and Trigonometry	9	4	16 (14 Trig)		10 (10 Stat)	3	7

Since there are many more geometry than algebra problems in the Japanese sample, the procedural problems in Fig.5 could be associated with the algebra problems. On the other hand, there are less geometry problems in the other countries, so that the percentage of conceptual tasks could be small just on the basis of a smaller percentage of geometry tasks sampled. Conversely, the Netherlands seem to tend from the sheer distribution of the subject areas towards "making connection" problems, since geometry as well as statistics calls for more conceptual thinking. The question in general is if most of the "using procedures" tasks come indeed from algebra. These questions must be open as long the data do not report the distinction of the subjects. However taking that away, the overall picture remains well-matched with the results of the TIMSS 1995 Video Study: Japan shows a considerable high percentage of "making connections" tasks.

Based on this comparison, one can only plausibly guess that being in favor of "making connections" tasks seems to be more a characteristic of the high achieving countries than a characteristic East-West difference. However, the sample would deserve an intense re-analysis containing a differentiation of the fields of subjects. The more likely interpretation then could be that among the Eastern countries, as well among the high achieving countries, there are different, but characteristic ways of teaching mathematics, at least with a higher percentage of "making connections" problems in some fields of subjects.

4.3 Mathematical processes used when solving problems

To what extent will this picture be the same when it is considered how the problems were really worked on in the class? The categories of the problems are only recognized when the solution is made or stated publicly in the class. Therefore, an extra category has to be formed, that only the answer to the problem, not the solution path, is made public. Hiebert & al. (2003) distinguished the nature of the *solved problems* in four categories:

- "*Giving results only*: The public work consisted solely of stating an answer to the problem without any discussion of how or why it was attained.
- *Using procedures*: The problem was completed algorithmically, with the discussion focusing on steps and rules rather than underlying mathematical concepts.
- *Stating concepts:* Mathematical properties or definitions were identified while solving the problem, with no discussion about mathematical relationships or reasoning. This included, for example, stating the name of a property as the justification for a response, but not stating why this property would be appropriate for the current situation.
- *Making connections*: Explicit references were made to the mathematical relationships and/or mathematical reasoning involved while solving the problem.

Each problem was classified into exactly one of the four categories based on the mathematical processes that were made explicit during the problem solving phase. This phase began after the problem was stated and lasted until the discussion about the problem ended."

(Hiebert & al., 2003, p 99-100)

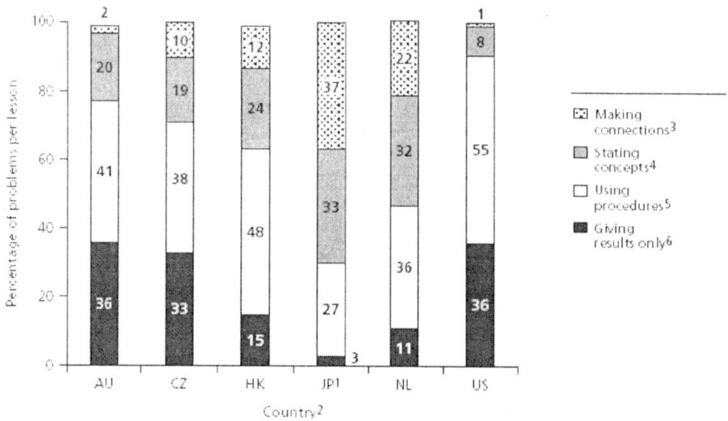

[1]Japanese mathematics data were collected in 1995.
[2]AU=Australia. CZ=Czech Republic; HK=Hong Kong SAR; JP=Japan; NL=Netherlands; and US=United States.
[3]Making connections. CZ, HK, NL>AU, US JP>AU, CZ, HK, US.
[4]Stating concepts. AU, CZ, HK, JP>US; NL>CZ, US.
[5]Using procedures: HK>JP; US>CZ, JP, NL.
[6]Giving results only: AU, CZ, US>HK, JP, NL; HK, NL>JP.
NOTE: Analyses only include problems with a publicly presented solution. Analyses do not include answered-only problems (i.e., problems that were completed prior to the videotaped lesson and only their answers were shared). For each country, average percentage was calculated as the sum of the percentage within each lesson, divided by the number of lessons. English transcriptions of Swiss lessons were not available for mathematical processes analyses. Percentages may not sum to 100 because of rounding.
SOURCE: U.S. Department of Education, National Center for Education Statistics, Third International Mathematics and Science Study (TIMSS), Video Study, 1999.

Figure 3-1-7. Average percentage of problems per eighth-grade mathematics lesson: Problems solved, by country (copy of Hiebert & al., 2003, Fig. 5.9)

The results in Fig. 7 (Hiebert & al., 2003, p.101, Fig. 5.9) point to the fact that giving results only has large percentages in Australia, Czech Republic and the US, to a lesser degree in Hong Kong and the Netherlands, and to only three percent in Japan and it is *vice-versa* with respect to making connections while solving the problems.

Australia and the US changed their 15% or 17% of making connections problems when the problems were stated to 1% or 3% when the problems were solved. Thus one characteristic of these two countries is that there were no connections made while speaking about the solution of the problems. A further analysis shows what kind of problem statements were changing.

4.4 From stated to solved problems

To what extent do the problems keep their characteristics when they are going to be worked on in the class? Hiebert & al. (2003, Fig. 5.10 and 5.11) investigated the three categories of stated problems. For two of them, the using procedures problems and the making connections problems, Fig. 8 shows the character of the solving processes in the countries.

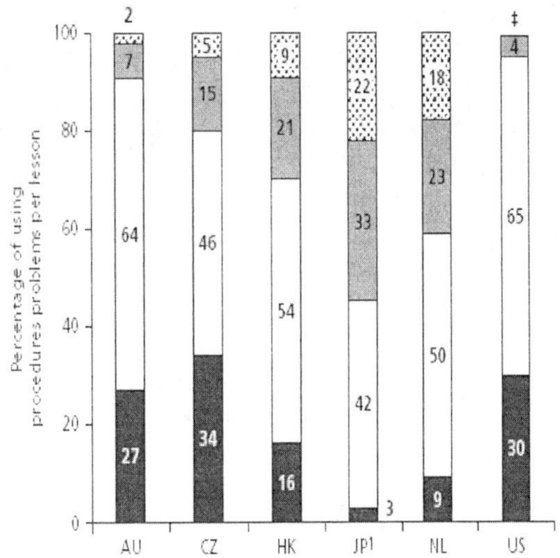

"using procedures" problems

‡Reporting standards not met. Too few cases to be reported.
1Japanese mathematics data were collected in 1995.
2AU=Australia; CZ=Czech Republic; HK=Hong Kong SAR; JP=Japan; NL=Netherlands; and US=United States.
3Making connections: HK, JP, NL>AU; JP, NL>CZ.
4Stating concepts: HK, JP, NL>AU, US; CZ>US.
5Using procedures: US>CZ, JP.
6Giving results only: AU>JP; CZ>HK, JP, NL; HK>JP; US>HK, JP, NL.
NOTE: Analyses only include problems with a publicly presented solution. Analyses do not include answered-only problems (i.e., problems that were completed prior to the videotaped lesson and only their answers were shared). For each country, average percentage was calculated as the sum of the percentage within each lesson, divided by the number of lessons. English transcriptions of Swiss lessons were not available for mathematical processes analyses. Percentages may not sum to 100 because of rounding and data not reported. Lessons with no using procedures problem statements were excluded from these analyses.

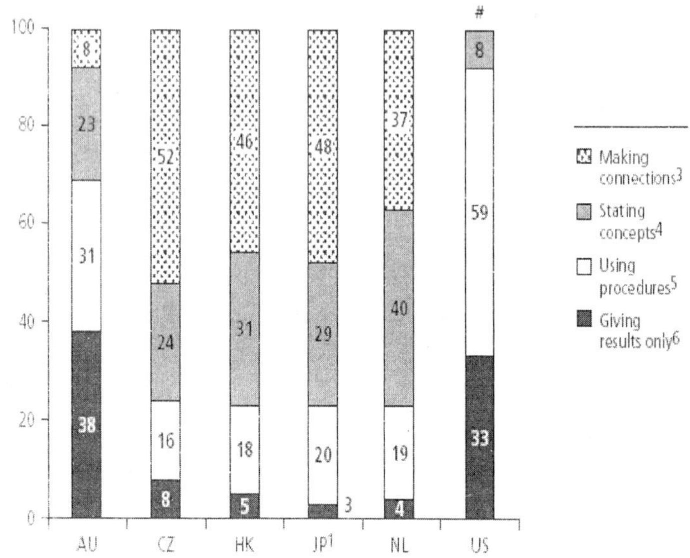

"making connections" problems

#Rounds to zero.

1Japanese mathematics data were collected in 1995.

2AU=Australia; CZ=Czech Republic; HK=Hong Kong SAR; JP=Japan; NL=Netherlands; and US=United States.

3Making connections: CZ, HK, JP, NL>AU, US.

4Stating concepts: JP, NL>US.

5Using procedures: US>CZ, HK, JP, NL.

6Giving results only: AU, US>CZ, HK, JP, NL.

NOTE: Analyses only include problems with a publicly presented solution. Analyses do not include answered-only problems (i.e., problems that were completed prior to the videotaped lesson and only their answers were shared). Lessons with no making connections problem statements were excluded from these analyses. For each country, average percentage was calculated as the sum of the percentage within each lesson, divided by the number of lessons. English transcriptions of Swiss lessons were not available for mathematical processes analyses. Percentages may not sum to 100 because of rounding. The tests for significance take into account the standard error for the reported differences. Thus, a difference between averages of two countries may be significant while the same difference between two other countries may not be significant.

Figure 3-1-8. Average percentage of stated problems, solved by explicitly using processes of each type (excerpts from Hiebert & al., 2003, Fig. 5.10 and Fig 5.12).

4.4.1 Mathematical processes used when solving "using procedures" problems:

22% of the Japanese problems stated as "using procedures" turn into "making connections", which underlines the observations given in Table 2. This kind of transition however is also present in the Netherlands, and in parts also in Hong Kong, as the example given by Hiebert & Handa (2004) illustrates (see 4.1), but not in Australia and the US. It may also be seen as an instance of teaching in a high achieving country. In all countries the percentage of problems "giving results only" remains nearly the same as for "all problems solved" (cf. Fig. 7).

4.4.2 Mathematical processes used when solving "making connections" problems:

The most striking result is that all "making connections" characteristics disappear in the US when it comes to solve a "making connections" problem. In part, this also holds in Australia. Making connections problems remain as such in Hong Kong, Czech Republic, Japan and the Netherlands and don't change in "giving results only" like in Australia and the US. The difference between Australia and the US is that in the US many more problems change to "using procedure problems".

4.5 Mathematical problems within private work time

The re-analysis of the TIMSS 1995 Video Study as presented in section 3 of this article (Neubrand, 2002) already showed how different from the content and cognitive view the problems given to the seatwork of the students were (Fig. 3 and 4). Also the TIMSS 1999 Video Study investigated this question, however only with respect to how far repeating procedures dominate the private work of the students. The results, shown in Fig. 9, support again the findings in the 1995 Study. It is only observed in Japan that students are confronted to such an extent with not repeating, e.g. as Fig.3 and 4. show, advanced, problem-solving oriented, conceptually enriched tasks. In this respect, Japanese mathematics teaching seems to be unique, both as a high achieving and an Eastern country.

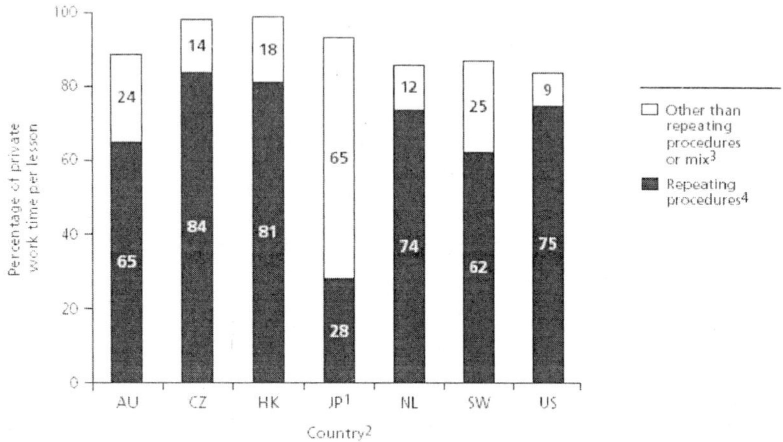

[1]Japanese mathematics data were collected in 1995.
[2]AU=Australia; CZ=Czech Republic; HK=Hong Kong SAR; JP=Japan; NL=Netherlands; SW=Switzerland; and US=United States.
[3]Other than repeating procedures or mix. JP>AU, CZ, HK, NL, SW, US; AU, SW>US.
[4]Repeating procedures: AU, NL, SW, US>JP; CZ, HK>JP, SW.
NOTE: For each country, average percentage was calculated as the sum of the percentage within each lesson, divided by the number of lessons. Percentages may not sum to 100 because some private work segments were marked as "not able to make judgment."
SOURCE: U.S. Department of Education, National Center for Education Statistics, Third International Mathematics and Science Study (TIMSS), Video Study, 1999

Figure 3-1-9. Average percentage of private work time per lesson devoted to repeating procedures (copy from Hiebert & al., 2003, Fig. 5.13)

4.6 Summary

The first question, if there are characteristic differences between the Eastern and the Western countries, cannot be answered by the TIMSS 1999 Video Study. Hong Kong and Japan are the only Eastern countries participating. However, they have a quite different history. It is not clear how far Hong Kong's education system may be influenced by its status as a former Crown Colony of England, since it is known that such traditions last long in the teaching systems. Recently, this could also be observed in Germany. Years after the reunification of East and West Germany, there still existed different patterns to teach mathematics, as Neubrand (2002) analyzed from the data of the TIMSS 1995 Video Study. Also the achievement data collected in the PISA-2000[8] Study exhibited that even 15 years after the reunification the structure of mathematical achievement remains different in the two parts of the country (Neubrand & Neubrand, 2004). The different

[8] PISA = "Programme for International Student Assessment".

educational systems for forty years in Germany seemed to stay stable for long periods.

On the other hand, East-West similarities are not implausible, as other research reported in this book shows. Also, the work of Becker (1992), Liping Ma (1999) and others point to commonalities in the Eastern countries. Liping Ma's investigations of the professional knowledge of US American and Chinese teachers (Ma, 1999) show differences between Chinese and US American teachers. These occurred in the cognitive structure of their knowledge, in the cognitive demands they put on their students, and the explanations they could give to arithmetic and geometric problems. To finally describe if there is an "Eastern style" of teaching, more than the two countries Hong Kong and Japan should be studied by video methods.

Towards the second question, if high achieving countries show common teaching structures, one observation was repeated in several instances. Japan is unique, but the fact has to be considered that the nature of the subject area, e.g. algebra or geometry, strongly influences the teaching in Japan. In two respects however, the high achieving countries look similar. "Making connections" problems remain to a considerable extent as such, and 20% to 55% of the "using procedures" problems change into "making connections" and "stating concepts" problems, when the problems move from the statement into the solving process. "Making connections" therefore could be a central distinctive issue of teaching in high achieving countries.

5. FINDINGS AND CONCLUSIONS

Video studies meanwhile turned out to be an effective method to document and analyze teaching. Video studies are especially suited to detect teaching "as a cultural activity" (Stigler & Hiebert, 1998), provided the analyzing instruments are able to detect cultural differences on a rather objective basis avoiding cultural biases. Mathematical problems, classified according to cognitive activities that are central and distinctive for mathematics, were proved to be suitable analyzing units. On the basis of the task analyses described in this article, observing "others" mirrors back to recognizing "one's own", insofar as it can point to the various possibilities a teacher of mathematics has to construct lessons around the content. Thus, videos are powerful tools to discover oneself by observing, discovering and analyzing the teaching in other cultures. Since teaching is cultural, cross-cultural conversations can be initiated.

The findings in the TIMSS 1995 and 1999 Video Studies presented here supported each other, and one of the main results surely is that the content itself is handled so differently in the different cultures. However, besides the fact that Japanese mathematics teaching seems to be rather unique, some of the issues observed point to characteristics of high achieving countries, like the "not-to-lose the making connections" tendency. A second tendency, still to be investigated further, is that as observed in Japan, effective teaching differentiates the subject areas, choosing different kinds of task for algebra than for geometry. However, to decide the question whether East and West characteristically behave differently, more Eastern countries should be recorded.

The focus on mathematical problems as the units of analysis, and even more precisely on the implementation of the problems into the lessons, has also a practical effect. Problem implementation is one of the most directly accessible points to start off for improving teaching.

Of special interest is to recognize that students work on their own responsibility, in seatwork or private work. In the reality of different teaching cultures, a couple of options were found for tasks to be given into seatwork. These options became visible on the basis of the analysis of characteristic features of mathematical problems.

REFERENCES

Baumert, J., Lehmann, R, Lehrke, M., Schmitz, B., Clausen, M., Hosenfeld, I., Köller, O. and Neubrand, J., 1997, TIMSS – Mathematisch-naturwissenschaftlicher Unterricht im internationalen Vergleich: Deskriptive Befunde. Leske & Budrich, Opladen.

Beaton, A.E., Mullis, I.V.S., Martin, M.O., Gonzales E.J., Kelly D.L. and Smith T.A., 1996, *Mathematics Achievement in the Middle School Years: IEA's Third International Mathematics and Science Study*. Boston College, Chestnut Hill, MA.

Becker, J., 1992, Cross National Mathematicsl Achievement Results and Observations concerning Problem Solving and Creativity of American and Japanese Students, *Journal für Mathematik-Didaktik*, **13**(2/3):99-141.

Bromme, R., Seeger, F and Steinbring, H., 1990, Aufgaben als Anforderungen an Lehrer und Schüler (= IDM-Untersuchungen zum Mathematikunterricht, Bd. 14). Köln: Aulis.

Christiansen, B. and Walther, G., 1986, Task and Activity, in: *Perspectives on Mathematics Education*, B. Christiansen, A.G. Howson and M. Otte, eds., D. Reidel Publishing Company, Dordrecht/Boston/ Lancaster/Tokyo, pp.243-307.

Hiebert, J., ed., 1986, *Conceptual and Procedural Knowledge: The Case of Mathematics*. Lawrence Erlbaum, Hillsdale, NJ.

Hiebert, J., Gallimore, R., Garnier, H., Bogard Givvin, K., Hollingsworth, H., Jacobs, J., Miu-Ying Chui, A., Wearne, D., Smith, M., Kersting, N., Manaster, A., Tseng, E., Etterbeek, W., Manaster, C., Gonzales, P., and Stigler, J., 2003, *Teaching Mathematics in Seven Countries: Results from the TIMSS 1999 Video Study* (NCES, 2003-013), U.S. Department of Education, National Center for Education Statistics, Washington, DC.

Hiebert, J., Stigler, J., and Manaster, A., 1999, Mathematical Features of Lessons in the TIMSS Video Study. *Zentralblatt für Didaktik der Mathematik*-Articles Section- **31**(6): 196-201.

Hiebert, J. and Handa, Y., 2004, A Modest Proposal for Reconceptualizing the Activity of Larning Mthematical Pocedures. Paper presented at the Annual Meeting of the American Educational Research Association, San Diego.

Husén, T., 1967, *International Study of Achievement in Mathematics*, Vol.1, Vol.2, Stockholm: Almqvist & Wiksell und New York, London, Sidney: John Wiley & Sons.

Kawanaka, T, Stigler, J., and Hiebert, J., 1999, Studying Mathematics classrooms in Germany, Japan and the United States: Lessons from TIMSS Videotape Study, in: *International Comparisons in Mathematics Education*, G. Kaiser, E. Luna, and I. Huntley, eds., Falmer Press, London, Philadelphia, pp.86–203.

Krauth, J., 1993, Einführung in die Konfigurationsfrequenzanalyse (KFA): Ein multivariates Verfahren zum Nachweis und zur Interpretation von Typen und Syndromen. Weinheim, Basel: Beltz.

Ma, L., 1999, *Knowing and Teaching Elementary Mathematics: Teachers' Understanding of Fundamental Mathematics in China and the United States*, Lawrence Erlbaum, Hillsdale, NJ.

NCTM, ed., 2000, *Principles and Standards for School Mathematics*, NCTM, Reston VA.

Neubrand, J. and Neubrand, M., eds., 1999/2000, Special Aspects of TIMSS related to Mathematics Education. Part 1, Part 2. Thematic Issues of *Zentralblatt für Didaktik der Mathematik* **31**(6):166-201 / **32**(1):1-24.

Neubrand, J., 1999, Eine Klassifikation mathematischer Aufgaben zur Analyse von Unterrichtssituationen. Schülerarbeitsphasen und Selbsttätigkeit in den Stunden der TIMSS Video- Studie. Doctoral dissertation, Freie Universitaet Berlin.

Neubrand, J., 2002, Eine Klassifikation mathematischer Aufgaben zur Analyse von Unterrichtssituationen. Selbsttätiges Arbeiten in Schülerarbeitsphasen in den Stunden der TIMSS Video- Studie. Franzbecker, Hildesheim.

Neubrand, J. and Neubrand, M., 2004, Innere Strukturen mathematischer Leistung im PISA-2000-Test, in: M. Neubrand (Hrsg.), *Mathematische Kompetenzen von Schülerinnen und Schülern in Deutschland: Vertiefende Analysen im Rahmen von PISA-2000* (S.87-107). Wiesbaden: VS - Verlag für Sozialwissenschaften.

Schmidt, W.H. and al., 1996, *Characterising Pedagogical Flow: An Investigation of Mathematics and Science Teaching in Six Countries*, Kluwer Academic Publishers, Dordrecht, Boston, London.

Schoenfeld, A.H., ed., 1987, *Cognitive Science and Mathematics Education*. Lawrence Erlbaum, Hillsdale, N.J.

Smith, M., 2000, A Comparison of the Types of Mathematics Tasks and How they were Completed during Eighth-grade Mathematics Instruction in Germany, Japan, and the United States. Doctoral dissertation, University of Delaware.

Stein, M.K., Grover, B.W. & Henningsen, M., 1996, Building Student Capacity for Mathematical Thinking and Reasoning: An Analysis of Mathematical Tasks used in Reform Classes, *American Educational Research Journal*, **33**:455 - 488.

Stigler, J.W. and Hiebert, J., 1997, Understanding and Improving Classroom Mathematics Instruction: An Overview of the TIMSS Video Study. *Phi-Delta-Kappan*, **79**(1):14-21.

Stigler, J.W. and Hiebert, J., 1998, Teaching is a Cultural Activity. *American Educator* (Winter):4-6.

Stigler, J.W., and Hiebert, J., 1999, *The Teaching Gap: Best Ideas From the World's Teachers for Improving Education in the Classroom*. Free Press, New York.

Stigler, J.W., Gallimore, R., and Hiebert, J., 2000. Using Video Surveys to Compare Classrooms and Teaching Across Cultures: Examples and Lessons From the TIMSS Video Studies. *Educational Psychologist*, **35**(2):87–100.

Stigler, J.W., Gonzales, P., Kawanaka, T., Knoll, St. and Serrano, A., 1999, *The TIMSS Video Tape Classroom Study: Methods and Findings from an Explanatory Research Project on Eighth-grade Mathematics Instruction in Germany, Japan and the United States* (Publ. NCES 1999-074). National Center for Education Statistics (NCES), Washington D.C.

Travers, K.J. and Westbury, I., 1990, The IEA Study of Mathematics II: Contexts and Outcomes of School Mathematics. Pergamon Press, Oxford.

Chapter 3-2

PROPOSAL FOR A FRAMEWORK TO ANALYSE MATHEMATICS EDUCATION IN EASTERN AND WESTERN TRADITIONS

Looking at England, France, Germany and Japan

Gabriele KAISER[1], Keiko HINO[2] and Christine KNIPPING[3]
[1]University of Hamburg; [2]Nara University of Education; [3]Carl-von-Ossietzy University

1. INTRODUCTION

Empirical surveys within the framework of large international comparison studies, such as the TIMSS accompanying "Survey of Mathematics and Science Opportunities" (SMSO), are indicating a strong cultural determination of lessons, even in subjects like mathematics and science. Thus the authors of the SMSO state:

> "Mathematics and science, unlike culturally embedded subjects such as history and language, are often thought to be a-cultural. For example, many believe 'numeration is numeration' – the concept is the same across all contexts. ... One can argue that if there is something universal about mathematics and science content, there should be something universal about the way this content is presented to students. Our results, of course, suggest that this second assumption needs re-evaluating ... Countries have developed their own ways of engaging students in the substance of mathematics and science. There appear to be strong cultural components, even national ideologies, in the teaching of these subjects." (Schmidt et al. 1996, p.132)

Following their analyses French teaching of mathematics emphasises formal knowledge, because French mathematics teachers identify themselves strongly with their disciplinary counterparts at universities. US educators appear bent on prolonging childhood as long as possible, at least as evidenced by the tendency to exposure to more basic and early-introduced topics in mathematics and science well into lower secondary school level.

Pepin (1997) reveals in her analyses also a strong dependence on cultural traditions for mathematics teaching in England, Germany and France. She states that of course it is interesting and exciting to read descriptions of mathematics teaching in other countries, but the aim of comparative education must be to find explanations for the observed differences and similarities in order to benefit from comparing teaching and learning mathematics in different countries. Thus she concludes that comparative education in mathematics should answer the following questions:

> "How can we understand teachers' practices in the light of what we see and experience. How can we understand teachers' practices in the light of what we know about the different countries? If we believe that the teaching and learning of mathematics is 'culturally embedded', what are the cultural and intellectual underpinnings that influence the teaching and learning of mathematics? Where do the cultural and educational traditions stem from, and how do they feed into the classroom? These and more questions have to be posed and answered, if we want to benefit from comparing teaching and learning mathematics in different countries." (Pepin 2002, p. 246)

These questions have not been answered yet, neither in older nor in recent comparative studies, but seem to be an unalterable step for the development of an educational theory on comparative education. They show the direction in which theory development in comparative education should move in order to be successful in explaining cultural differences in education and proposing effective measures for change.

In the following we will describe a proposal that permits a description or even a classification of mathematics teaching in different countries, in Eastern and Western countries as well, within a broader framework. By means of this proposal for a classification system questions about the reasons for the origins and philosophical bases of the differences in mathematics teaching in various countries will not yet be answered. However, it helps to recognise differences and similarities in mathematics teaching of different countries. Based on further studies this classification system might help to clarify how far these differences are influenced by educational philosophics or external economical and social aspects. Within the framework of a comparative study on German and English mathematics teaching we have set forth exemplarily such an approach (see Kaiser, 2002). The study describes the educational conceptions and philosophies that have been developed during the last centuries in England and Germany and shows their impact on the national approaches to teaching mathematics. Such classificatory attempts for educational systems have a long tradition in comparative education and seem suitable to be continued as a reference for the above mentioned developments (for an overview of such classifications see Holmes 1981).

The framework proposed in this paper focuses on central orientations of mathematics teaching on which a comparison of mathematics teaching can be based. This means concretely that the suggested classification system characterises mathematics teaching according to its epistemic orientation – i.e. the position of the subject's structure and the role of mathematical theory – and the arising aspects that contribute to the construction of an understanding of mathematics. These aspects are the lessons' structures and the approaches to mathematics, the position of formulae and algorithms or calculations, the relevance of the introduction of new concepts, the role of proof, the importance of mathematical language, the consideration of real world aspects and the central teaching-and-learning styles.

The classification system developed here is based on empirical research of the comparative study of mathematics teaching in England and Germany (see Kaiser 1999). For this comparison, mathematics teaching of both countries has been described as idealised types which were developed from observed empirical phenomena but do not exist in this pure way in reality (for details concerning this 'ideal typus' approach see section 2). Furthermore, mathematics teaching of both countries has been described through contrasting characterisations. In another study dealing with mathematics teaching in France and Germany characteristics of mathematics teaching are developed through focusing on the teaching of proofs (see Knipping 2003). A comparison of the results of these two studies makes clear that the polarisation of German and English mathematics teaching is not suitable anymore if regarded from a new perspective: concerning the understanding of mathematics, English and French mathematics must be regarded much more as poles, while position of German mathematics teaching and its orientations is positioned "between" the two countries.

If the perspective of this empirically grounded theory is extended, we must inevitably ask whether the comparative framework is restricted to these three countries or can also be applied to other countries; whether this framework is euro-centric or whether it is appropriate to be used to analyse mathematics teaching of Eastern traditions. These aspects were investigated in a subsequent study which deals with the question of how far the developed characterisations can be used to describe Japanese mathematics teaching. For logistical reasons no classroom observation could be done for this study. Thus, it is based on curricula and textbook analyses, and additional questioning of teachers and discussions with experts were performed. Because the proposed framework is a kind of 'meta-study', different types of data are acceptable, even if the developed descriptions tend to be less close to lessons taught. Further this lower proximity to teaching reality seems not to produce serious problems, because the questioning of the teachers demonstrates how strongly mathematics teaching in Japan is orientated towards textbooks.

2. METHODOLOGICAL APPROACH

The following description of the analysis framework for mathematics teaching in various countries can be described as empirical grounded theory, consisting of a meta-analysis of three studies, which have been carried out successively building on each other.

The first study comparing English and German mathematics teaching (Kaiser 1999) is an ethnographic study embedded in a qualitatively oriented paradigm of the social sciences. In the following, features of the ethnographical approach are briefly described, before details of the concrete methodical procedure are given. Ethnographical methods have been developed from the method of participant observation, which has a long research tradition in social anthropology and ethnology. Participant observation has been understood as a flexible contextualised strategy, which comprises multiple methods. Ethnographical studies are aimed at a description of a social context, in which people live and work. The main effort is to evaluate how different social realities are constructed, i.e. how the situation-related means are used by the actors, within a social situation, in order to construct social phenomena from a participating perspective.

The following three aspects can be formulated as characteristics of ethnographical research: Central is the long-term presence in the research field in order to assume an inside perspective. The process of diving into the research field can be described as process of partial enculturation (Amann & Hirschauer 1997). The second characteristic is the flexible research strategy, i.e. the researcher has to adapt his or her methodical approach to the situation and has to find a balance between the research interest and the requirements of the situation. In order to study the culture of the participants before producing explanations for their behaviour, participant observation and relatively unstructured interviews are the main ethnographical research methods. Formalisations and standardisations of the research procedures are therefore not adequate; by contrast, the methods consider that researcher and research actions are part of the cultural environment that is examined (Hammersley & Atkinson 1983). The third characteristic is ethnographical writing, which is centrally based on detailed field-notes taken during the observations. Field-notes, which are either made on the spot or written up as soon as possible after leaving the field, have to be seen as interpreted reconstructions of the observations. The question how to evaluate ethnographical data, in order to fulfil necessary scientific standards, has therefore become the focus of interest in the discussion of the last few years.

Due to its focus on descriptions of real life and the construction of social phenomena, the ethnographical research approach seems to be especially adequate for the evaluation of mathematics education in England and Germany, its constituents and its determining patterns. Especially bequeathed educational philosophies, which influence the actions of the participants in the educational field significantly, are well known – sometimes even unconsciously – by all actors and are therefore only seldom made explicit. The view into another culture gives us insight into our own teaching culture and the determining constituents. The method of participant observation with its detailed field-notes allows a diving into the field, which is not possible with technically more ambitious research methods. A central basis of this study is, apart from the field-notes of the classroom observations, discussions with teachers, in the staff room during lunch, or after the classroom visits, the participation in school assemblies and discussions with pupils after lesson. Only for the analyses of the teaching-and-learning process, which needed verbatim statements, were audio-tape records made.

In the following we will describe a few more technical details of the study. The study included 17 different schools in England, of which 14 were state-run comprehensive schools and 3 were private schools with selective character. Two of the state schools were grant-maintained. The 14 state schools were comprehensive schools, except for one Grammar School, and 4 of the 14 schools were single-sex schools. The schools were spread all over England. The study is limited to the English school system, as the school systems and the educational philosophies in Scotland and Ireland are quite different from the English ones. As already mentioned, the study relied heavily on classroom observations, apart from the participation into school life, especially in England. 242 lessons were observed from Year 6 to A-level, mainly restricted to Years 9 to 11. In Germany, schools from the three-tier system were included as well as comprehensive schools of different types (using streaming or setting systems). 6 of the schools were situated in the Federal state of Hessen, the others from various regions spread over Germany. 102 lessons were observed from Year 8 to Year 10. The study focussed in both countries on age groups at the end of lower secondary level. Concerning the achievement level of the pupils included, the study put its emphasis on the two higher tiers of the German school system or on the top sets in the English school system. The reason for this choice points to a major problem, well known in comparative qualitative studies: Many teachers of both countries were hesitant about opening the classroom with lower achievement students for observations by a visitor. The classroom observations were mainly carried out from 1990-95. Further research has shown that mathematics teaching in Germany has not changed in a significant manner since then, despite the TIMSS shock and political claims of the necessity for change. The English mathematics teaching has undergone a relevant change since the beginning of the nineties of the last century, especially concerning teaching-and-learning methods and the relevance of the subject structure through the introduction of the National curriculum and the accompanying key stage tests. These changes became visible already during the study and are covered by the classroom observations. Newer change is mainly focussing on primary education, which this study does not deal with.

In general, the study aims – as already stated – at generating general knowledge, based on which pedagogical phenomena might be interpreted and partly explained. Under a narrower perspective, the study aims to generate qualitative hypotheses on the differences between teaching mathematics under the educational systems in England and in Germany. Due to the use of the ethnographical method, the study cannot make any 'lawlike' statements; in contrast, the study refers to the approach of the 'ideal typus' developed by Max Weber (*Webersche Idealtypen*), and describes idealised types of mathematics teaching reconstructed from the classroom observations in England and Germany. That means that typical aspects of mathematics teaching are reconstructed on the basis of the whole qualitative studies rather than on one existing empirical case. The 'ideal typus' does not really describe empirical phenomena, it is constructed by overemphasising typical issues of single phenomena observed and by a combination of different phenomena (for details see Hempel 1971, Weber 1904).

In the second study, comparing French and German mathematics teaching, proving processes in class are described and contrasted (Knipping 2003). Comparative analyses of the processes observed in class illustrate different proving practices. These analyses reveal different functions and roles of proving in mathematics teaching. In addition, the analyses show on the one hand that mathematical theory, mathematical concepts and language have a different status in teaching and on the other hand that real-world problems are important in German teaching, while they are not important in French lessons.

French curricula apply nationally, so all classes in the *Collège* are intended to study the same material. The decision to carry out investigations in different *Collèges* in the Paris region has not resulted in a special sample, with the exception of 2 bilingual classes. In contrast, substantial differences in the topic emphasis of the German curricula can be found not only on a regional level, but also, and more so, among the different school types. While a special value is given to proofs nationally in the *Gymnasium* curricula, in the curricula of comprehensive schools proofs clearly play an inferior role. In the curricula of comprehensive schools this different valuing of proofs is usually reflected in different targets for courses of the upper and lower sets. Analyses of the curricula suggested that it would be difficult to observe proofs and proving processes outside the *Gymnasium* and perhaps the upper sets in comprehensive schools. Consultations with teachers confirmed this, and so, early in the research, the decision was made to choose German classes selectively. It was decided to examine classes in both the *Gymnasium* and the upper sets in comprehensive schools in case there were differences in their classroom proving processes.

The empirical investigation involved proving processes in 6 French and 6 German classes. The data collection was carried out at 6 *Collèges* in the Paris region and 3 *Gymnasien* and 2 comprehensive schools in Hamburg. Two of the observed classes in France are classes in a bilingual stream and are highly selective. French and German curricula, which have been analysed before the beginning of the data collection, list proofs as an explicit topic in geometry for the first time in grade 8. For this reason instructional observations were done in geometry classes at level 8/9 (13-14 year old students). The instructional units were selected according to curricular criteria and cover topics in geometry, including the Pythagorean Theorem. The observations were documented with audio-tape recordings and photos of figures and writing at the blackboard. In addition, observations were recorded in the form of process notes which were made after each session. The tape recordings were transcribed to make detailed analysis of the classroom discourses during the proving processes possible, which was in particular necessary for the reconstruction of argumentations.

In contrast to the first study, the analyses and the comparison of the data were structured by theoretical considerations based on research in the field of proof and argumentation. Analyses of the classroom processes were carried out based on historical and philosophical work (Lakatos 1976; Jahnke 1978; Rav 1999), and research in the field of argumentation (Krummheuer & Brandt 2001), in particular the functional analysis of arguments exposed in the Toulmin model (Toulmin 1958).

Based on Max Weber's methodological concept of the *ideal type,* ideal-typical characterisations of proving processes were developed by comparing processes both on the level of context analyses and on the level of argumentation analyses, with the aim of developing a typology. This involved comparative analyses of all observed episodes "*from an initial interpretation of those episodes to a later theoretical exploration of those episodes*" (Krummheuer & Brandt 2001, p. 78). Prototypical cases or *prototypes* form the basis for the construction of these ideal-typical characterisations of proving processes. A prototype is a case *"in the sense of a concrete model"* (Zerssen 1973, p. 53), not an ideal type, i.e. not an ideally formed theoretical construct. Rather it is a case that can apply to a group as representative in the sense that through it special characteristics of a group of cases become clear (Kluge 1999). Descriptive typical characteristics can be worked out by the characterisation of the prototype. Singling out prototypes forms an intermediate step in the process of constructing ideal types. The comparison of prototypes with further cases is also crucial here. In the light of other cases, typical features become clearer in contrast to individual characteristics. The ideal-typical characterisations developed in this way have a heuristic function, because *"the pure type contains a hypothesis of a possible occurrence"* (Gerhardt 1991, p. 437). The cases discussed below represent prototypes in this sense.

The comparison of prototypical cases has also been an important element of the meta-analyses presented here. These analyses made it possible to specify ideal-typical characterisations of English, French and German mathematics teaching, which are presented here as a proposal for a framework to analyse mathematics teaching. The bi-polar characterisations developed in the two studies have been re-analysed and used to characterise more precisely different poles in mathematics teaching, with respect to the status of mathematical theory, the introduction of mathematical concepts and methods, the position and function of proofs, the role of justifications and examples, the status of precise language and the role of real-world examples.

After developing descriptions of mathematics teaching in England, France and Germany based on these two studies, the framework of a third study was carried out in order to clarify whether this framework would make sense for an analysis of Japanese mathematics teaching. This third study (by Hino) focussed on mathematics education in public schools at lower secondary level (Year 7–Year 9). First, a tentative description was developed based on formal documents such as the National Course of Study and on results of international comparative studies such as the TIMSS-Video-Studies, in combination with discussions with several mathematics educators and an analysis of 6 widespread, common mathematics textbooks. This description was modified and confirmed by the results of a questionnaire carried out with 51 Japanese mathematics teachers at Nara city. In the questionnaire, two aspects of mathematics teaching were studied: the teachers' dependency on textbooks in their daily teaching practice and the relevance of mathematical theory and related issues covered in the framework.

With this approach we got, as already mentioned, two different kinds of data in our analyses of the four countries: data from classroom observations in England, France and Germany and another kind of data referring to recently published descriptions of Japanese teaching. The reason for not carrying out classroom observations was mainly time and capacity restrictions. We consider the description of the Japanese mathematics teaching as highly reliable and close to classroom reality due to the following reasons: The discussions with professors for mathematics education served as expert discussions, because Japanese university professors visit classrooms and observe lessons occasionally as a process of lesson study (see Stigler & Hiebert 1999). They usually have an adequate image of the current trends of classroom teaching. The other reason is the validity of textbook analyses as a means of comparison. In Japan, the content of a textbook is strictly determined by the National Course of Study and authorised by the Ministry of Education. Moreover, there have been data of international comparison such as TIMSS (see National Institute for Educational Research 1998) and OECD/CERI (Shigematsu 1998) that show repeatedly Japanese mathematics teachers' strong dependency on the textbooks in their teaching. The results of the questionnaire confirm this aspect pointing out that the teachers usually rely on textbooks in every major occasion of their teaching: i.e. entering new textbook chapters, introducing new mathematical concepts and procedures, consolidating and summarising the learned content. Therefore, the study of 6 textbooks seemed to be an appropriate means in order to gain insight into mathematics teaching in Japan, especially concerning the relevance and status of mathematical theory and related issues.

3. ANALYSES WITH THE PROPOSED FRAMEWORK

The idealised description of English and German mathematics teaching developed in the first study consisted of polarised descriptions in order to clarify the distinctions made. In the light of the second study – the comparison of French and German mathematics teaching by Knipping (2003) – these descriptions had to be qualified. In an overall description of European educational approaches in mathematics, France and England might be seen as diametrically opposed to each other with German conceptions having an intermediate position. This holds especially with the aspect of understanding mathematical theory. Until now there exist only first attempts for the development of such a frame. The three-country-study of Pepin (1997) covering France, Germany and England limits itself to the perspective of the teacher and does therefore not provide such a frame, but may be taken as an empirical basis for further research.

In the following we will start each aspect by a short description of the characteristics of the two polar mathematics educations, usually England and France, followed by a description of the place of the German and the Japanese mathematics education.

3.1 Understanding of mathematical theory – scientific knowledge versus pragmatic understanding

Two contrasting characteristics of French and English teaching can be reconstructed as contrasting poles relating to the understanding of mathematical theory. Thus **French mathematics** teaching can be described by the ideal type characteristic of a scientific understanding of theory, this means that theoretical mathematical considerations are of great importance. Generally speaking, mathematics teaching in France is characterised by its focus on the subject structure of school mathematics ("savoir enseigné"). This means that theory is made explicit by means of concepts, theorems and formulae.

From an ideal type perspective, **English** mathematics teaching can be described by its pragmatic understanding of theory, which means a practical and purpose-depending handling of theory. Differences between the comprehensive school, the dominant kind of school, and the selective school system, which for the most part consists of private schools, could be recognised. These fundamentally different orientations of French and English mathematics teaching on a level of understanding of theory can be seen from various aspects, such as the introduction of new concepts, the meaning of proof, importance of rules or precise mathematical language, which will be described in the following.

German mathematics teaching is characterised by its focus on the subject structure of mathematics and on mathematical theory. Theoretical reflections emphasising the mathematical subject structure play a dominant role in the higher type of the tripartite school system, still prevailing in Germany, but are of less importance in the other columns of the school system. Theory is often reduced to rules and algorithms, especially in the two lower types of the tripartite school system, subject-related reflections play a more important role in the higher type of the tripartite school system, but are often restricted to remarks by the teacher or remarks in the textbooks.

In **Japan**, the National Course of Study states objectives and content of school mathematics, which emphasises the subject structure of school mathematics. Mathematical theory is made explicit by means of concepts, formulae and theorems and also by means of rules and algorithms although, in the practice of teaching, teachers treat mathematical theory in the classroom in the context of problem solving activities, i.e. teachers spend substantial time on a small number of selected problems. In the current Course of Study (valid since 2002), "mathematical activity" is considered as an important way of learning mathematics. In describing the new Course of Study, Nemoto (1999) states:

> "... for the purpose of making connections with daily life, fostering students' ability of investigating phenomena mathematically and heightening their ability of solving problems by using mathematical ways of viewing and thinking, we tried that students can engage actively in mathematical activities such as finding relationships and rules in the phenomena by means of observations, manipulation and experimentations and reflecting on and thinking of the results once they have reached." (p. 100. original in Japanese, translation by Hino)

3.2 Organisation by subject structure versus spiral-type curriculum

The characteristic of **French mathematics teaching,** a subject-based understanding of theory, leads to a curriculum whose lessons go along with the subject structure of mathematics guided by didactical considerations. In lessons mathematical concepts and methods are taught in a subject-scheduled order as prescribed in the national curriculum. Lessons start from general concepts and rules, and then proceed with special conclusions and applications. The subject-based understanding of theory is given shape also by the great importance of mathematical theorems. The great importance of theorems becomes more obvious in the topic areas of geometry, where the relevance and structure of mathematical theory often shall be demonstrated exemplarily. The units are complete in themselves, but connect subjects including geometric and algebraic issues.

The characteristic of **English mathematics teaching,** which is a pragmatic understanding of theory, is apparent from the spiral-shaped structure of mathematics lessons and curricula, which means that mathematical concepts and methods are introduced quite early but on a more elementary level. Later, in higher classes, they are picked up again. This spiral-shaped approach implies that smaller and easily comprehensible topic areas are discussed, which are not taught in a subject-oriented structure. A fast switching from one topic to another is typical for English mathematics lessons. Sometimes even several topics are dealt with at the same time. Altogether, English mathematics teaching is rarely based on a subject-based systematic. The subject structure of the National Curriculum, which has been obligatory since the beginning of the nineties of the twentieth century, did not lead to subject-structured lessons. As the curricular goals in the National

Curriculum are strongly individual based, there does not exist any obligatory canon of knowledge, to which teachers could refer to for continuing a topic as scheduled in the spiral-shaped curriculum.

This pragmatic understanding of theory, which does not put the subject structure to the foreground, corresponds with the minor emphasis on mathematical formulae, rules and theorems, because the creation of mathematical tools is regarded as being more important than structural analyses. Therefore, theorems like Pythagoras' theorem, which play a central role in a subject-structured curriculum, are called *patterns* in English teaching or they are not taught at all. Theorems and their meaning are not the focus of interest, but rather the constructive aspect of geometrical contents and the algorithmic function of algebraic contents (formulated as rules and formulae) in connection with problem solving.

The focus on theory when teaching mathematics in **Germany** implies a lesson structure that goes along with the subject structure of mathematics. Mathematical theorems, rules and formulae are therefore of high importance. That varies though, with the different kinds of school of the tripartite school system. Bigger coherent topic areas are taught (lasting sometimes months), e.g. fractions, percentages, Pythagoras' theorem and others. These big thematic fields are taught independently of relations to other topic areas, and are later on not referred to again.

In **Japanese** mathematics teaching the mathematics content is classified into three areas, "number and algebraic expressions" "geometrical figures" and "mathematical relations". These areas are located in each grade and taught alternatively. The Course of Study states objectives at each grade level. Not only content but also thinking and interest are considered important. For the objectives of thinking, four levels are distinguished: knowing, understanding, processing, utilising. New trials such as problem situation learning, election of special topics of mathematics, or integrated learning are carried out. Current controversies are focused on what the power and ability is that students should acquire through school learning so as to be useful for their future lives, and how to foster such power and ability. The teaching units are somehow between the French and the English approach covering about 3-5 weeks.

3.3 Introduction of new mathematical concepts and methods

Concerning this aspect, the classification of the various educational systems is different from above, e.g. that the German and the English approach form the poles of the description.

The subject-based understanding of theory of **German mathematics teaching** leads to the high importance of the introduction of new mathematical concepts and the deduction of new methods. Normally, this is planned carefully by the teachers or they refer to detailed introductions from the text book. Often mathematical concepts and methods are illustrated by real-world examples, although the real-world examples depend on their purpose, they often appear artificial and far from real life. Partly, the introduction of concepts refers to basic understandings as representatives of the mathematical "nucleus" of a concept. The introduction of new mathematical concepts is usually done by class discussion, in which the whole learning group participates under the guidance of the teacher. There exist various kinds of teacher guidance. A characteristic of the course of a lesson is that the newly introduced concepts or methods are formulated in detail by phrases or notes on the blackboard, which then is followed by exercises.

Below we will give an example that demonstrates the high value of the introduction of concepts and the connection with exclusively formally meant basic imaginations and the application of introductory real-world examples. This sequence, about the introduction of the concept function, was observed by Kaiser in a *Realschule* of Year eight:

> It starts with a graph about the development of temperature over 24 hours on a worksheet distributed by the teacher to the pupils. The teacher writes on the blackboard "function". Then he asks what the task on the worksheet means and gives the answer by himself by the fact that he writes:
> The connection between time and temperature.
> time→ temperature
> At first various temperatures for given times are determined together by the pupils in a discussion, then the times are noted in a table. After it has been clarified at what time there is the highest and the lowest temperature, the teacher asks: How many temperatures could we declare for one time? A girl answers: one. The teacher comments on this: This we keep in mind. He writes on the blackboard:
> At each time there is only one declaration of temperature.
> He states that for this there is a specific name in mathematics and writes:
> The assignment time→ temperature is defined exactly.
> He continues that this is to be completed by the name which is already written as a headline on the blackboard, and then he writes:
> Exactly defined relations are called functions.
> Then various contrasting examples are discussed in detail. The teacher introduces the notation form of a function and the word equation of a function; then further examples are discussed with all pupils.

The pragmatic understanding of theory of the **English mathematics teaching** influences the importance given to the introduction of new mathematical concepts and methods: Thus English mathematics teaching is characterised by a low importance of the introduction of concepts and methods. This is generally done pragmatically, and often the concepts or methods are given by the teacher just as information or in the style of a recipe. Content-related information is replaced by referring to calculators or mnemonics. Especially new mathematical methods often are explained and demonstrated through experiments, by drawing and measuring. New mathematical concepts normally are not introduced explicitly but implicitly, in connection with exercise sequences. This corresponds with the fact that the introduction of concepts and methods is done in short class discussion sequences or individually with the textbook.

What follows is a description of a sequence of a lesson from English mathematics teaching observed in Year 9 of the top set at a comprehensive school, which deals with the introduction of sine and cosine in right angled triangles. The lesson demonstrates the usage of calculators instead of a content-based understanding:

> During the first lesson the tangent is introduced as follows: Pupils draw individually various right angled triangles with the same angle X, then the length of the two short sides of a rectangular triangle and their ratio are determined from the drawing. Then, in a discussion guided by the teacher, it is clarified that the ratio of the two short sides of each right triangle with equal angle is always stable. The teacher instructs the pupils to enter the angle X into their calculator, then to press the tan key. The pupils recognise that the already calculated ratio is the same as the X value of tan. In this way the tangent of an angle is defined as the ratio of the opposite and adjacent sides. Subsequently the pupils worked individually on further examples.
>
> The next lesson continued with a methodically similar introduction of sine and cosine done by individual work. At the end of the lesson a wide variety among the pupils becomes obvious: Some of them are still busy with the exercises about the tangent, others have already started with further exercises about sine and cosine.

In **France**, as in Germany, both the introduction of new concepts and the deduction of new mathematical methods are of high importance for the teaching of mathematics. They are usually either well prepared by the teacher or follow detailed introductions given in the textbooks. In contrast to German teaching, new mathematical concepts and methods are not motivated by real-world examples, but sometimes prepared by problem situations, with the aim to revise former knowledge. The introduction of new mathematical concepts and methods usually takes place during short periods of class discussion, followed by sophisticated exercises that are supposed to deepen the understanding. The whole teaching is centred towards exercises and individual work of students, while mathematical concepts and methods, which are formulated in the form of definitions or theorems and written down on the blackboard, are equally important.

Within proving processes mathematical concepts and methods are revised and made more precise. Thus proving in class gives value and importance to mathematical concepts and methods. The following example of a French lesson illustrates this.

> In this lesson an arithmetical proof for the Pythagorean Theorem is developed. In class the students and the teacher give detailed reasons why the inner quadrilateral in the proof figure is a square, and finally they note these reasons at the blackboard. As a first step they justify that the inner figure is a rhombus. Based on this conclusion they look for a right angle in the figure and for reasons why BCD is a right angle. In this way they finally conclude that the inner figure is a square and write down together the following proof.

• As ABCD has four sides of the same length, it is a rhombus	• Comme ABCD a quatre côtés de même longueur, c'est un losange.
• The acute angles of a right triangle are complementary.	• Les angles aigus d'un triangle rectangle sont complémentaires.
• The right triangles DHC and BGC are superposable, their corresponding angles are equal.	• Les triangles rectangles DHC et BGC sont superposables, leurs angles sont deux à deux égaux.
• We can deduce that the angles $\angle$ BCG and $\angle$ DCH are complementary.	• On en déduit les angles $\angle$ BCG et $\angle$ DCH sont complémentaires.
• $\angle$ HCG=180° so angle $\angle$ BCD=180 – 90= 90°.	• $\angle$ HCG=180° d'ou l'angle $\angle$ BCD=180 – 90= 90°.
• ABCD is a square.	• ABCD est un carré.

Writing down proofs and solutions of problems underlines the importance of a precise use of terminology and reasoning in class. Explicit and accurate proofs are highly valued in French teaching. In this way the mathematically correct application of concepts and terminology is fostered.

In **Japan**, introduction of new mathematical concepts and methods are of high importance for the teaching of mathematics. They are usually well prepared by the teacher or follow detailed introductions given in the textbooks. New mathematical concepts and methods are often motivated by real-world examples. In textbooks, each chapter often has opening pages that illustrate real-world examples, which is to initiate students to the basic ideas.

In introducing concepts and methods, real-world examples (which are quite often rather artificial) are used to foster students' interest and motivation toward thinking. In class, introduction of new mathematical concepts and methods takes place in a problem-solving situation. It is also the case that the teacher explains the new concepts and methods. Class discussion, during which newly introduced concepts and methods are formulated, is desired but not easily realised. Introduction is followed by exercises. The number of exercises in the textbook is not large. Teachers often allocate students additional workbooks to do more exercises.

When looking at the 6 textbooks in the chapter of linear function (Year 8), all of them started with at least a few situations that contain one or several linear (non-linear) relationships. Rather than just giving detailed instructions, students are asked to do some work along with key questions that initiate them to the basic idea, that is to look at the phenomenon from the perspective of "changing quantities."

> One of the textbooks (Chuugaku suugaku, 2 [Mathematics in lower secondary school 2]. p. 41, published by Osaka-shoseki, 2002) contained the following situation: Water is poured with the speed of 5 cm height per minute in a box-shaped tank with the height of 50 cm that already contains water up to the height of 10 cm. The description is accompanied by the following table:

TIME (MIN)	0	1	2	3	4	5	...
HEIGHT (CM)	10						

> Further, the task contains an illustration with three students talking with each other. In bubbles is written: "Let's look at the relationship between the time spent for pouring water and the height," "I wonder if the relationship would be somewhat different from proportion" and "When will the tank be filled up with water?" Then there is a goal statement "Let's learn the relationship between two quantities that vary together with a fixed rate."

In the questionnaire, 51% of the teachers confirmed that they sometimes ask students to do investigative work with real-world examples in order to initiate them to the basic ideas of linear function (33% said that they do so occasionally). 86% of the teachers reported that they make connections between new mathematical concepts (e.g. linear function) with the concepts the students have already learned (e.g. direct proportions). This result shows the relevance of establishing connections between older and newer concepts and methods within the teaching process.

3.4 The position and function of proofs

The understanding of theory in **French** mathematics teaching can be seen from the strong emphasis put on proof. Proofs are considered important for introducing students into a theoretical understanding of mathematics and for developing their skills in mathematical argumentation. Especially in the context of geometry, proofs are studied and students have to carry out proof problems by themselves. The public justification of mathematical relations and facts is the main function of mathematical proving processes in class. The knowledge of the class, i.e. the concepts, theorems and methods already studied in class that build public accepted knowledge, is extended by new knowledge that is first justified before it becomes part of the accepted public knowledge of the class. Justifying a new theorem means going back to theorems, definitions and methods that are already accepted as public knowledge in class. Justifications, in the form of discussions in class and written texts at the blackboard, that form a discursive culture in class characterise this type of proving process. In problem solving not the solution itself, but the justification of the solution by tracking it back to publicly accepted knowledge, is of primary importance. Written proofs are also a model for justifying solutions of problems. The following example of a French lesson illustrates this.

> In an exercise students have to calculate two sides of a right angled triangle, given one side of the triangle and the length of the median. In order to solve the problem the Pythagorean Theorem has to be applied as well as the circumcircle theorem. The new theorem is explicitly connected with knowledge that was already studied in class and so is inscribed into the knowledge of the class. Writings at the blackboard foster this inscription.

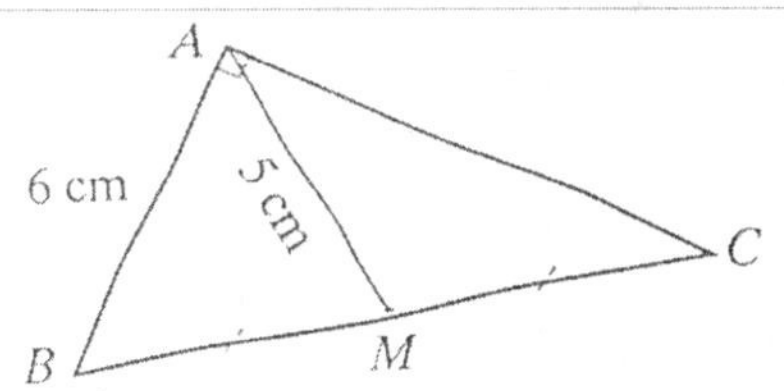

Calculer les longueurs BC et AC.

Hachette, Cinq sur Cinq 1998, p. 161, n° 29

• As ABCD has four sides of the same length, it is a rhombus • The acute angles of a right triangle are complementary. • The right triangles DHC and BGC are superposable, their corresponding angles are equal. • We can deduce that the angles ∠BCG and ∠DCH are complementary. • ∠HCG=180° so angle ∠BCD=180 – 90= 90°. • ABCD is a square. • As ABCD has four sides of the same length, it is a rhombus	• Comme ABCD a quatre côtés de même longueur, c'est un losange. • Les angles aigus d'un triangle rectangle sont complémentaires. • Les triangles rectangles DHC et BGC sont superposables, leurs angles sont deux à deux égaux. • On en déduit les angles ∠BCG et ∠DCH sont complémentaires. • ∠HCG=180° d'ou l'angle ∠BCD=180 – 90= 90°. • ABCD est un carré. • Comme ABCD a quatre côtés de même longueur, c'est un losange.

Formal proof is of low importance **in English mathematics teaching**, both in the selective and in the non-selective school sector. Theorems are often developed experimentally with some examples. Teachers do not make clear that example-related explanations are not sufficient as proof, or that such considerations are not a proof in a formal sense. Very often teachers use the term "proof" for example-related explanations. Consequently, both pupils and teachers do not make a difference between proof and example-related checking. This leads to the fact that many pupils in their own mathematical investigations end their work with example-based checking of formulae or solutions they found, without trying to find out a general explanation. The low importance of proofs corresponds with the fact that mathematical theorems and methods are quite often just announced by the teachers without any attempt to give reasons for them. The following sequence of a lesson observed in Year 10 at the Top Set of a comprehensive school shows exemplarily how proof and example-related checking are not distinguished.

> In the lesson before this one, various theorems on the size of angles of triangles inscribed in circles have been discovered by the pupils themselves through individual work.
>
> At the beginning of the next lesson, after a review, the teacher asks the pupils to start with the practical check of the size of the angle at the circumference of a triangle inscribed in a circle (so-called Umfangswinkelsatz).
>
> "Draw three diagrams with triangles in it, measure the angle and show that it is right, what we said yesterday." He points out that it is important to draw accurately. While the pupils are working individually and the

teacher is walking around and helping some pupils he formulated several times:

"Yesterday we delivered the theory, today we will prove it."

After a couple of minutes most of the pupils have finished three drawings and recognised that the angles in each triangle over a chord are nearly the same. However, as a part of the drawings were done inaccurately, there occurred quite big differences. The teacher asks the pupils what might have been the reason for asking them to do three drawings. Then the following discussion started:

Teacher:	What do you think, why I have asked you to do three drawings?
Pupil:	In order to have three checks.
Teacher:	Do you think I would be satisfied if I would have a theory and just check it with three examples?
Several pupils:	No, no.
Teacher:	So, what do you think, how high is the probability that you all take the same triangle?
Some pupils:	Unlikely.
Teacher:	So we actually have 60 checks. We would be rather sure that if our theory has worked for 60 cases, it will work on the whole.

After this, exercises from the text are done which are to be finished as homework. In this way the efforts for a "proof" ended.

The subject-based understanding of theory in **German mathematics teaching** influences the emphasis put on proof in German mathematics teaching, limited to the *Gymnasium.* A formula-related understanding of proof prevails, while content-related proofs are carried out quite seldom. There are great differences between the various school types, as proof is done less in the *Realschule* than at the *Gymnasium,* and at the *Hauptschule* they almost do not exist at all. Especially in geometry teaching at the *Gymnasium* great meaning is granted to the carrying out of proof. In this connection the importance of proof within the framework of the structures of mathematical theorems is made clear. For this the need of proof for theorems is of high relevance. Thus its meaning is to explain that experimental, practical proofs are not sufficient for the control of the validity of general statements and therefore formal proofs are necessary. These characteristics – the great importance of explaining the need of proof for theorems – is demonstrated by the following example from Year 8 of a *Gymnasium*, which deals with the so-called theorem of Thales.

The lesson starts with the construction of the circle of Thales (drawn with diameter AB) with different triangles, which each pupil does individually. One girl draws this figure on the blackboard.

The teacher asks what is special about it. Some pupils assume that these triangles are always right-angled, others express their doubts about this. The teacher then defines the circle of Thales as a special circle and formulates the theorem of Thales as follows:

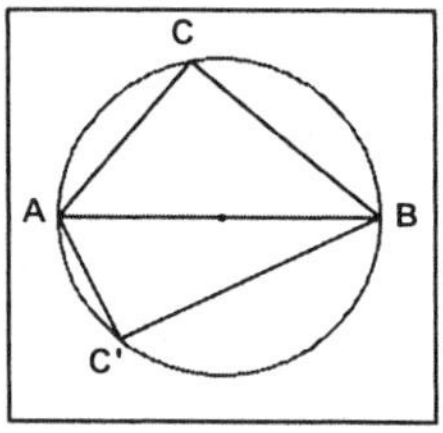

If we connect the points A, B with a point C on the circle of Thales, then we get a triangle with a right angle at point C (theorem of Thales).

The teacher asks whether they may write down this phrase like this. A girl refers to the description of the construction. Then the teacher asks once more what the observation of Thales is and how they checked it. One girl says that she controlled the theorem with 3 or 4 examples. A boy states that it is always true with any triangle he draws. After further contributions to the discussions, the teacher summarises by stating that trying examples does not help, and he asks what to do. A girl suggests that they must argue until everybody believes.

Then, quite suddenly – and without any further inputs from the teacher – the central ideas of the proof are given by two pupils, by drawing the connecting line between the centre of the circle and point C and looking at isosceles triangles.

At the end of the lesson the teacher resumes the discussion as follows: It is important to show that triangles on the circle of Thales have a right angle. He announces the proof for the following lesson.

In **Japan**, proofs of mathematical statements play a high role when teaching mathematics. Proofs are considered important to introduce students into the theoretical frame of mathematics, especially in the context of geometry. Students also carry out proof problems by themselves. In teaching geometry in lower secondary schools, two aspects are considered as important: One is to foster understanding of concepts and properties of geometrical figures and the ability to apply them. The other is to foster a logical way of viewing and thinking, especially knowing and making use of deductive reasoning. Students are also encouraged to make conjectures for general cases about geometrical figures by means of such methods as observation, experimentation or investigation, and verifying the conjecture in a deductive way.

For Japanese students, one major difficulty is to recognise the need for proving. The textbooks often emphasise the importance of proof by repeatedly using such phrases as "explanation of the general case," "giving reason for the phenomenon" and "explanation without using manipulatives" which was the students' familiar way of explanation in elementary school. In addition, as shown below, some textbooks ask students to work with several examples and develop conjectures whether the rule they found always stands as true. Students may also question whether the manipulation in an example is sufficient for justifying the rule. Here, the need for proving is expected to emerge.

> In one of the textbooks (Suugaku, 2 [Mathematics 2], pp.94-95, published by Keirinkan, 2002), "proof" is introduced after working on a problem of folding a tape in the following way:
>
> 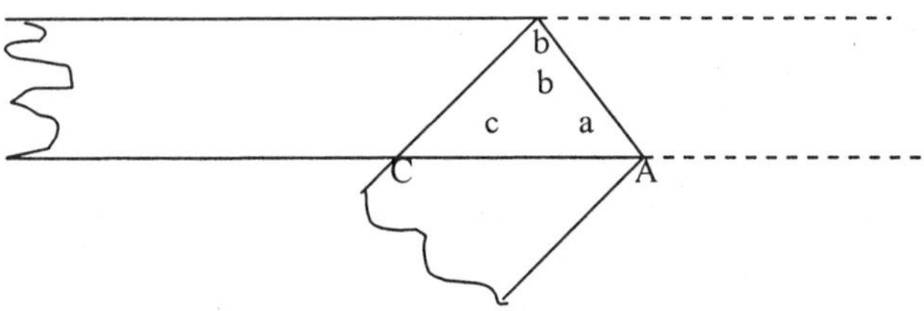
>
>
> The question is "Let's measure the angles a, b and c in the triangle ABC. What do you notice?" After noticing that the measures of ∠a and ∠b are equal, the question continues, "When folding the tape differently, the measurements of the three angles change. But it seems that the equality holds whatever the case." Here students are encouraged to check with different examples. Building on these preparations, there comes a statement, "Let's think how to show that the equality always holds in whatever way the tape is folded." Then the textbook shows the deductive explanation, with a summary statement "In this way of explanation, it shows ∠a = ∠b in general, regardless of the ways of folding the tape. This way is often used when examining the properties of figures."

In the questionnaire, there were only 8% of teachers who replied 'no' when asked if they emphasise the importance of proof in class by saying, for instance, "it explains the general case" or "it gives reason for the phenolmenon."

Students also deal with proof problems by themselves. In the textbooks, aspects of language such as "assumption and conclusion" and "frame of proof" are emphasised. A frame of proof intends to help students to conduct logical proofs in a sort of formal style. In the questionnaire, 71% of the teachers confirmed that they also use "incomplete proofs with boxes" at the beginning. Here students put appropriate sentences in the boxes to complete the proof.

3.5 Focus on justifications or rules versus work with examples

Concerning this aspect there can be no polar description reconstructed:

English mathematics teaching can be characterised by its focus on work with examples, with rules and standard algorithms being of minor importance. Central algebraic theorems, such as the binomial formulae, are labelled as patterns and are not identified as general statements. This minor emphasis on rules and algorithms coincides with the minor importance of generalisation and general solving schemes in contrast to example-related explanations. Thus solutions are often formulated in connection with examples, and there often do not exist special names for general solution formulae as in German mathematics teaching.

German mathematics teaching can be characterised as rule-oriented, in which rules are manifested by algorithms. Exact execution of arithmetic and algebraic algorithms is highly important. However, there are great differences between the different school types, with a strong emphasis on algebraic algorithms in the higher type of the tripartite school system. It is important to many teachers, that pupils are able to execute the central algorithms – e.g. calculation of percentages and interest or term transformations and solving equations – with certainty by heart. Especially in the lower types of the secondary school it is regarded as important that pupils know by heart those algorithms which very often serve as a substitute for a more profound understanding of algorithms. A deeper understanding would in the opinion of many teachers mean overtaxing weak performing pupils. At the Gymnasium a competent usage of the formulae is expected, although in practice even there the transformation of terms and binomial formulae are often reduced to their calculation and practised with plenty of exercises. Besides this orientation towards execution of algorithms by heart some teachers tend to emphasise a content-related understanding of formula and the ability to develop such formulae by themselves.

In **France**, the teaching of mathematics is characterised by its focus on exercises and justification of solutions within the studied theoretical frame of mathematics. The exact and precise processing of arithmetical algorithms is considered as important and functions as a basis for solving more complex problems. Sticking to exactly prescribed procedures and following certain routines when working with algorithms is regarded as important, but for complex problems not always sufficient. This leads to high demands in students' engagement in exercises.

In exercises, extended written solutions are expected. In these solutions justifications are required at a theoretical or general level; an argument based on an example is unlikely to be accepted here. In geometry, proofs are likely to be studied based on a geometrical figure, while mathematical arguments in the proof are expected to abstract from special properties of the figure. Written proofs in class show a certain format, including explanations of applied theorems and concepts. These formats are given as models for further proofs and solutions of problems.

The exact and precise processing of arithmetical and algebraic algorithms is important in **Japanese teaching**. As for essential rules, algorithms and formulae, students have to memorise them. For many students, mathematics is a subject just concerned with memorising rules, algorithms and formulae, which is a big problem in Japanese mathematics education. Rigid and standard solution processes are considered important. In Japan, pressure to pass entrance examinations for upper schools is great. This is especially the case in lower secondary schools. This is one main reason why rigid and standard solutions are emphasised in the lessons.

Rules and formulae are often taught example-bound. For example, in the teaching of computations with square root, 79% of teachers questioned confirmed that they explain what the rule is, and why the rule is approved, by using examples. 69% of the teachers also confirmed that they explain how the rule is effective. It is also the case that many mathematics teachers use students' common mistakes in introducing the rule (84%). Textbooks often encourage students' developing conjectures whether the rule can be made. Here, the calculator is an important tool (especially recently). However, according to the questionnaire, it is still not the case that teachers use calculators in their lessons very often.

3.6 The role of precise language

Precise language and the use of mathematically correct language are considered to be very important when teaching mathematics in **France**. Technical terms are often treated like vocabulary, which have to be learnt by heart. As a consequence, during class discussions, teachers often correct those phrases of the students that are not precise or are slightly incorrect. Concerning correct mathematical arguments neither teachers nor students use terms such as "if ... then" in a strictly logical sense. Nevertheless there is a demand for rigour in thinking and for verbalisation of reasoning in a precise way. Precise language is expected to foster mathematical reasoning, and reasoning in general. Students are asked to verbalize their thinking precisely; in particular, they have to express their reasoning in written solutions of problems in order to make their thinking precise. In school, the role of precise language is made clear to students at an early age and not only in mathematics.

In contrast in **English mathematics teaching** the development of a collectively accepted terminology with reference to the language of mathematics in the context of 'official' communication is only of minor importance. This aspect is strongly connected with the minor relevance of phases of class discussion in contrast to individual teaching-and-learning styles.

German mathematics teaching can be characterised by an intermediate position between the high importance of the usage of correct language in France and the low relevance of this issue in England: On the one hand the usage of precise mathematical language is emphasised in the context of an 'official' classroom discussion, while the grade of usage differs in the tripartite school system. As for the high meaning of results from the class discussion, this part of communication dominates other forms of communication. Therefore, teachers at the *Gymnasium* generally strive for a mathematically precise and formally correct way of speaking, which they correspondingly also demand from their pupils. This often leads to the fact that teachers interrupt the pupils' explanations and that they 'offer' correct and formally exact formulations, in order to enable them to formulate mathematically precisely. At the *Realschule*, in practice it is less uniform how important the usage of a mathematically correct language by the pupils is regarded, while at the *Hauptschule* it is generally less important. Especially at the *Gymnasium*, but at the *Realschule* too, mathematical expressions are taken as vocabulary which must be learned by heart, and sometimes this is explicitly the homework to be done.

In **Japan** precise language and the use of mathematically correct language are considered to be very important when teaching mathematics. Technical terms are often treated like vocabulary, which have to be learned by heart. During the lessons, teachers correct those phrases of the students that are not precise or are incorrect. Still, it is not likely that every teacher just urges students to memorise technical terms and notations. They recognize the importance of meaning of terms including the meaning of Chinese characters and mention conciseness and usefulness of mathematical notations, which is also stated in the textbooks.

3.7 The role of real-world examples

Concerning this aspect England and France can be described as polar approaches, with Germany and Japan in between.

English mathematics teaching can be characterised by the fairly high importance given to real-world examples, which have various educational functions. They serve to introduce and derive new mathematical concepts and methods, but also to impart abilities that enable applying mathematics to solve extra-mathematical problems. These abilities are especially supported through coursework, and by projects within the framework of statistics lessons. These projects, integrated into statistics lessons or coursework, deal with real-world examples, and normally they are realistic examples. However, besides this, 'dressed up' examples are also used for the introduction of new mathematical concepts and methods. A further characteristic of real-world contents in English mathematics teaching is that more recent mathematical topic areas, such as graph theory and network analysis from discrete mathematics with strong application references, are taught. Handling data is taught intensively and embedded into real-world contexts in English mathematics lessons. These real-world examples are often taught through an activity-oriented method, with students doing research tasks they set and then evaluate themselves, often with the aid of computers. The teaching method when dealing with applications strongly depends on the example's function: In connection with the introduction of new mathematical concepts and methods there is found both class discussions and individual work. The prevailing method with application-oriented and more extensive projects is individual work. Generally speaking, many pupils are used to formulate and to solve problems independently – if necessary with their teacher's help.

In contrast to this position real-world examples are of no importance for the teaching of mathematics in **France**. New mathematical concepts and methods are more likely to be introduced by strictly mathematical problems, if teachers decide to motivate them at all. They show students the value of new knowledge for solving problems within mathematics. Geometry, number theory and algebra are traditionally the main topics of mathematics teaching in France. New curricula put more emphasis on statistics, but in class this topic area is still given a subordinate place.

Typical for **German** mathematics teaching is a minor emphasis on real-world and modelling examples, which is nevertheless very different to the situation in France. Real-world examples mainly function as introduction of new mathematical concepts and methods or are used for exercising mathematical methods. More extensive problems, that are meant to promote extra-mathematical goals, e.g. to develop abilities to master everyday life and to solve extra-mathematical problems by means of mathematics, are rather infrequent in everyday school lessons. Normally such problems are only given within the framework of daily or weekly projects. Furthermore, it is typical for German mathematics teaching that real-world examples discussed in lessons are not authentic real-world problems, but made to illustrate mathematical contents. Therefore, these examples give a quite artificial and far from reality impression. Fairly modern mathematical areas, which widely include applications, such as graph theory, until now did not enter German mathematics lessons.

Since teaching is based on textbooks in Japan, the use of real-world examples is influenced by the textbook used. Still, in the actual teaching, the relevance of real-world examples depends on teachers. Real-world examples are often aiming at the introduction of new mathematical concepts and methods, and at exercises of mathematical methods. As described above, in many textbooks, a chapter's opening pages and front pages in textbooks contain illustrations of real-world examples. However, how they are treated and incorporated in the teaching varies according to the teacher.

3.8 Teaching and learning styles

Concerning this aspect German and English mathematics teaching form the polar approaches, with France and Japan in between.

German mathematics lessons are dominated by a teaching and learning style called class discussion - almost all mathematical concepts and methods are introduced during periods of class discussion. Individual work is of fairly low importance, and it can mostly be seen when exercises are worked on. Significant differences are apparent between different types of schools and different years. This means that, in the upper years of a Gymnasium, class discussion is almost exclusively the teaching and learning style used. In Hauptschulen, by contrast, individual work replaces class discussions and, during periods of class discussion, it is essential that students discuss with each other. Therefore, at least temporarily, students refer to each other. Furthermore, significant differences in the extent a teacher guides a class discussion can be seen, ranging from merely guiding to an authoritative directing of the class discussion. The blackboard is the essential medium of a class discussion, and the students sometimes write their solutions on it. All in all, the students shape the class discussion to an appreciable extent.

In **English** mathematics lessons, two main teaching styles are currently recognisable. The first one is more traditional, and it is focused on long periods of individual work. During these, the students work on new mathematical topics by using individual work material, or they practise known terms and methods. These periods of individual work alternate with shorter periods of class discussions, which are rigidly guided by the teacher, during which new topics are introduced or results are compared.

Besides these more traditional teaching and learning methods, another style exists which is more student focused. Its method consists of several problem-solving activities, during which the students carry out investigations and do coursework, often in the form of projects. Generally speaking, in England, class discussions are dominated by the teachers. All of the communication takes place via the teacher, and the students hardly ever refer to each other. Writing down on the blackboard is not important – if something is written down on the blackboard, this is usually done by the teacher. When teaching mathematics in England, individual differentiation often takes place. This is easily possible, since most of the learning material is designed for individual work.

In **France** periods of individual work alternate with shorter periods of class discussions, which are guided by the teacher, during which new topics, concepts, theorems and methods are introduced or results are compared. The blackboard is the essential medium of a class discussion, and the students sometimes write their solutions on it. Generally speaking, in France, class discussions are dominated by the teachers. All of the communication takes place via the teacher, and the students hardly ever refer to each other. In contrast, highly active engagement of students is asked in individual work. Students have to solve a lot of exercises in class and at home. Exercises function as the heart of mathematics teaching in France, they prepare students for new concepts and deepen their understanding of already studied concepts, theorems and formulae in class.

Teaching in **Japan** is based on whole-class teaching. Still, the teacher diligently controls students' activities by shifting types of classroom interaction (Hiebert et al. 2003). Students engage in problem solving while solving a small number of main problems with the teacher in class. Problem solving activities are often carried out firstly on an individual basis. After they get their solutions, students may present their thinking on the blackboard and discuss it. Whole-class discussions are usually guided by the teacher. During individual work, teachers often walk around students' desks and give directions and suggestions. Therefore, compared with the case in Germany, it can be said that individual work plus public interaction rather than discussion among students is emphasised in Japan. Recently, fostering communication skills, including mathematical communication, has been considered important. However, in actual classroom teaching, teachers work hard in order to cover all the content in the textbooks. Time constraints together with pressure of entrance examinations hinder teachers to spend time on class discussion.

4. FURTHER PERSPECTIVES

If we try to understand the differences just described, the influence of cultural traditions on education and educational philosophies have to be considered. Already Michael Sadler, who at the beginning of the 20th century visited Germany with a British expert commission and compared the achievements of the Prussian with the British educational system, described this influence:

> "In studying foreign systems of Education we should not forget that the things outside the schools matter even more than the things inside the schools, and govern and interpret the things inside. ... A national system of Education ... has in it some of the secret workings of national life. It reflects, while it seeks to remedy, the failings of the national character. By instinct, it often lays special emphasis on those parts of training which the national character particularly needs." (Sadler 1900 (1964), p.310).

In the field of comparative education there exists a few comparative studies dealing with educational philosophies and their historical development. One of the first contributions to this was the approach of Lauwerys (1959), who distinguished the attempts of the "Liberal Education", the French "*culture générale*", the German "*Allgemeinbildung*", the American "*General education*" and the Russian "*Polytechnicalization*".

Proceeding from this, McLean (1990) developed various attempts to explain the different traditions of school knowledge, in which he distinguished several European traditions. The encyclopaedic tradition, found predominantly in the French educational system, is historically rooted in the ideals of the French Revolution. McLean characterises this attempt through several principles, from which the principles of universality and rationality are the most convincing ones. Following McLean, the principle of universality means that on the one hand teaching aims to transfer as much knowledge as possible from all important subjects to all learners. On the other hand a certain degree of standardisation and homogeneity of the transferred knowledge shall be reached. Rationality which, following McLean, is the highest objective of the encyclopaedic approach, aims to enable the learners to understand central ideas, structures, logical and ethical systems, for which the understanding of structures and systems created by reasoning, gains great importance. In particular, philosophy and mathematics originally were regarded as the subjects which suited the rational principle most closely. As a second important current of European school systems McLean (1990) defines the “humanist perspective in education” (p. 25), meaning the development of human virtues, which includes not only the development of understanding, sympathy and confession, but courage, intelligence and eloquence. He characterises it by the aim of linking thought and action in ways that would encourage human possibilities in the individual to the fullest. It focused on the individual rather than the social group. It was moral in its emphasis on the development of human virtue but this morality was extended to include aesthetic appreciation and sensibility. This approach, dominant in England and Wales, can be characterised by the principles of morality, individualism and specialism and has a strong relation to pragmatism in philosophy. The description of German educational philosophies was not convincing and we have developed our own description discriminating two different development lines (for details see Kaiser 1999). The educational philosophy dominant for mass education can be characterised as realistic education, i.e. school lessons should be more orientated towards realistic-vocational education and should incorporate concrete knowledge useful in later life. It should especially enable the students to develop social virtues through work. The education for the élite was orientated towards the development of the individual, who should receive a complete formation of humankind. Neither an early specialisation on selected subjects was allowed nor an inner differentiation within the class; all the pupils in the class should progress together at the same speed.

If we now look at the Japanese educational system a high influence of Western, especially European and US-American school traditions, can be recognised. Western schooling traditions were already introduced during the Meiji government in the late 19th century after Japan was forced to open the country to foreign influences. Around the same time, Western mathematics was introduced in Japan. These foreign influences did not come into action as they were, on the contrary they were modified and adopted to the Japanese situation focusing on teaching aspects and the situation in the class. Due to the tradition of the Wasan mathematics, including the aspect of learning elementary arithmetic by means of an abacus, people were able to adjust themselves to Western mathematics. Apart from the Western influences on the Japanese educational system there are special Japanese traditions, which have shaped Japanese teaching. One important influential factor is the tradition of the research lesson and lesson study, already going back to the time of the Meiji government. Lesson study is a collaborative and longitudinal effort (over a couple of months or even a year) of improving classroom teaching by teachers. Their focus is on lessons that they are conducting. They plan, conduct, discuss and improve lessons by studying teaching content, developing teaching materials and discussing the effect by observing each other's lessons. Here, teachers may also learn and discuss theories of teaching as a basis for developing lesson plans. However, they start from the lesson itself, instead of starting from learning theories and then trying to apply it to the classrooms, which is often the case of US American teachers (see Stigler & Hiebert 1999). The development and conduct of lesson plans, together with visits to the classrooms of colleagues, are considered to be important for the reflection of one's own teaching. Such a cultural practice has produced an image of an ideal lesson, including a joint understanding of good mathematical problems to teach certain mathematical ideas and ways the problems should be effectively dealt with in class. This image of an ideal lesson is also reflected in the textbooks, which are written by experienced mathematics educators including experienced school teachers. The structure of the textbooks mirrors the structure of class lessons, which should be like the flow of a river: It begins with cultivating students' interest and proceeds to the solution of problems on their own. This is followed by explanations of mathematical content and ends with exercises (for details see Lewis & Tsuchida 1988).

Furthermore influences of the general style of communication in Japanese culture can be detected in communication processes in teaching. Sekiguchi (2002) examined relationships between argumentation processes and mathematical proofs from a cultural perspective and argued that the teaching of mathematical proofs seems to be conceived in the communication style of a so-called "group" model, common in Japanese communication processes. The model states that cooperation rather than competition is highly valued within a community. According to Sekiguchi, the goal of proof in Japan is to reach a unanimous conclusion, which helps in establishing the harmony in the community. A proof requires one to follow the premises accepted in the community, which helps in keeping the harmony of the community members. Beyond the instruction of mathematical proof, he described that the teaching and learning styles in Japan, especially the importance of exchanging and sharing opinions in a whole-class, follow the group model. The general styles of communication in Japanese culture, in contrast to that in Western culture, seem to reflect the difference described above.

Another important tradition, which is more related to education and general pedagogical aspects, is the aversion towards differentiation of the students in the period of learning basic knowledge and skills. The spirit of giving every child equal opportunities for education since the starting of public education in the late 19th century has been passed on from generation to generation. The extent to which the idea of differentiation is put into practice is one of the controversies in the topical Japanese education reform.

In total, it seems to be a general characteristic of the Japanese debate that it focusses more on the debate concerning teaching and how to improve it, than on reflections concerning educational aspects such as educational philosophies, pedagogical theories and so on.

To summarise, the reflections above show the strong influence of educational and societal philosophies on educational structures as well as on the classroom situation. The framework described in the paper might be seen as a first step to the explanation of differences observed in various studies focussing on classsroom processes. Further, the framework might enable us to see in which parts of the educational processes the different educational systems can learn from each other. Coming back to the introduction and the questions of Pepin, our framework shows the necessity to ask such questions as to the understanding of the teachers' practices, the cultural underpinnings of such practice and the sources of cultural and educational traditions and their influence on the teaching and learning of mathematics. Answers to these questions seem to be necessary in order to come to a real understanding of our own and other educational systems and to allow a reflective "transfer" of effective measures from one system to another.

REFERENCES

Alexander, R., Broadfoot, P., and Phillips, D., eds., 1999, *Learning from Comparing: New Directions in Comparative Educational Research*, Volume 1: Contexts, Classrooms and Outcomes. Symposium Books, Oxford.

Amann, K.; Hirschauer, S., 1997, Die Befremdung der eigenen Kultur. Ein Programm, in: *Die Befremdung der eigenen Kultur.* Zur ethnographischen Herausforderung soziologischer Empirie, S. Hirschauer; K. Amann, eds., Frankfurt, Suhrkamp, pp.7-52.

Gerhardt, U., 1991, *Gesellschaft und Gesundheit. Begründung der Medizinsoziologie.* Frankfurt, Suhrkamp.

Hammersley, M. and Atkinson, P., 1987, *Ethnography: Principles in Practice.* Routledge, London.

Hiebert, J. et al., 2003, *Teaching Mathematics in Seven Countries.* Results from the TIMSS 1999 Video Study. Washington, U.S. Department of Education.

Hino, K., 1997, Cross-cultural Studies of Mathematics Classroom, in: *Rethinking Lesson Organization in School Mathematics*, Japan Society of Mathematical Education, ed., Sangyo-Tosho, Tokyo, pp.3-18.

Holmes, B., 1981, *Comparative Education: Some Considerations of Method*, George Allen & Unwin, London.

Jahnke, H.N., 1978, *Zum Verhältnis von Wissensentwicklung und Begründung in der Mathematik-Beweisen als didaktisches Problem.* Institut für Didaktik der Mathematik, Bielefeld.

Kaiser, G., 1999, Unterrichtswirklichkeit in England und Deutschland. Vergleichende Untersuchungen am Beispiel des Mathematikunterrichts. Weinheim, Beltz - Deutscher Studienverlag.

Kaiser, G., 2002. Educational Philosophies and Their Influence on Mathematics Education – an Ethnographic Study in English and German Mathematics Classrooms, in: Zentralblatt für Didaktik der Mathematik, **34**(6):241-257.

Kluge, S., 1999, Empirisch begründete Typenbildung. Zur Konstruktion von Typen und Typologien in der qualitativen Sozialforschung. Opladen, Leske + Budrich.

Knipping, C., 2001, Towards a Comparative Analysis of Proof Teaching, in: *Proceedings of the 25th Conference of the International Group for the Psychology of Mathematics Education*, van den Heuvel-Panhuizen, M., ed., Utrecht, Utrecht University, **3**:249-256.

Knipping, C., 2003, Beweisprozesse in der Unterrichtspraxis – Vergleichende Analysen von Mathematikunterricht in Deutschland und Frankreich. Hildesheim, Franzbecker Verlag.

Krummheuer, G. and Brandt, B., 2001, Paraphrase und Traduktion. Partizipationstheoretische Elemente einer Interaktionstheorie des Mathematiklernens in der Grundschule. Weinheim, Beltz - Deutscher Studienverlag.

Lakatos, I., 1976, *Proofs and Refutations. The Logic of Mathematical Discovery*. Cambridge University Press, Cambridge.

Lauwerys, J., 1959, The Philosophical Approach to Education, *International Review of Education*, **5**:281-298.

Lewis, C., and Tsuchida, I., 1988, A Lesson is like a Swiftly Flowing River: How Research Lessons improve Japanese Education, *American Educator*, **22**(4): 12-17, 50-52.

McLean, M., 1990, *Britain and a Single Market Europe*, Kogan Page, London.

National Institute for Educational Research, 1998, Chugakko no suugaku kyoiku/rika kyoiku no kokusai hikaku [International comparison of mathematics and science education in lower secondary schools]. Toyokan.

Nemoto, H., 1999, Chugakko sin kyoikukatei no kaisetsu: Sugaku [Explanation of new curriculum in lower secondary school: Mathematics] Daiichi-hoki.

Pepin, B., 1997, Developing an Understanding of Mathematics Teacher in England, France and Germany: An Ethnographic Study. Unpublished doctoral thesis, University of Reading.

Pepin, B., 2002, Different Cultures, Different Meanings, Different Teaching, in: *Teaching mathematics in secondary schools*, L. Haggarty, ed., Routledge, London, pp.245-258.

Rav, Y., 1999, Why do we prove Theorems? *Philosophia Mathematica*, **7**(3): 5-41.

Sadler, M., 1964, How far can We Learn Anything of Practical Value from the Study of Foreign Systems of Education? *Comparative Education Review*, **7**:307-314. Reprinted from: Surrey Advertiser, December 1, 1900.

Schmidt, W. et al., 1996, *Characterizing Pedagogical Flow. An Investigation of Mathematics and Science Teaching in Six Countries*. Kluwer, Dordrecht.

Sekiguchi, Y., 2002, Mathematical Proof, Argumentation, and Classroom Communication: A Cultural Perspective, *Tsukuba Journal of Educational Study in Mathematics*, **21**:11-20.

Shigematsu. K., ed., 1998, *School Case Studies on Implementation of Mathematics Curriculum and Improvement in Method of Education*. Report for Scientific Research Foundation by Ministry of Education. (in Japanese)

Stevenson, H. et al., 1998, *The Educational System in Japan: Case Study Findings*. Washington, U.S. Department of Education.

Stigler, J. et al., 1999, *The TIMSS Videotape Classroom Study: Methods and Findings from an Exploratory Research Project on Eighth-Grade Mathematics Instruction in Germany, Japan, and the United States*. Washington, U.S. Department of Education.

Stigler, J. W. and Hiebert, J., 1999, *The Teaching Gap: Best Ideas from the World's Teachers for Improving Education in the Classroom*. Free Press, New York.

Toulmin, S.E., 1958, *The Uses of Argument*. Cambridge University Press, Cambridge.

Weber, M., 1904, Die "Objektivität" sozialwissenschaftlicher und sozialpolitischer Erkenntnisse, in: J. Winckelmann, Gesammelte Aufsätze zur Wissenschaftslehre. Tübingen. Mohr, pp.146-214.

Zerssen, D., 1973, Methoden der Konstitutions- und Typenforschung, in: Enzyklopädie der geisteswissenschaftlichen Arbeitsmethoden, 9, M. Thiel, ed.,. Lieferung: Methoden der Anthropologie, Anthropologie, Völkerkunde und Religionswissenschaft. München, pp.35-143.

Chapter 3-3

CULTURAL DIVERSITY AND THE LEARNER'S PERSPECTIVE: ATTENDING TO VOICE AND CONTEXT

David CLARKE[1], Yoshinori SHIMIZU[2], Soledad A. ULEP[3], Florenda L. GALLOS[3], Godfrey SETHOLE[4]; Jill ADLER[5] and Renuka VITHAL[6]
[1]University of Melbourne, Australia; [2]Tokyo Gakugei University, Japan; [3]University of the Philippines, The Philippines; [4]Technikon North West, Pretoria, South Africa; [5]University of the Witwatersrand, Johannesburg, South Africa; [6]University of Durban-Westville, Durban, South Africa

1. PROBLEMATISING CULTURAL EXPLANATIONS

In the Discussion Paper for this ICMI study, it is stated, "For this study, culture refers essentially to values and beliefs." The ICMI Study is distinguished from other studies "in that it is specifically concerned with comparing practices in different settings and with trying to interpret these different practices in terms of cultural tradition." The Discussion Paper makes it clear that the ICMI study is "limited to only a selection of cultural traditions" and then argues that "Those based in East Asia and the West seem particularly promising for comparison." In invoking a comparison between East-Asian and Western cultural traditions of mathematics education, the ICMI Study does not "merely refer to geographic areas." Instead, a comparison is made between the "Chinese/Confucian tradition on one side, and the Greek/Latin/Christian tradition on the other." By framing the comparison in this way, the ICMI Study is at risk of oversimplifying the situation by appearing to assume that school systems can be aligned with one 'cultural tradition' or the other.

Of the researchers who contributed to this chapter, the school systems of three draw on more than one cultural tradition and an understanding of mathematics classrooms in these school systems is not advanced by the postulated East Asian-Western (or Confucian/Christian) dichotomy. Australia, in particular, is culturally plural. In the south-eastern suburbs of Melbourne, a class of twenty-five children can include over twenty distinct ethnic backgrounds. This raises the question of the interaction between home culture and school system, and suggests that the identification of a nation with a single culture may be appropriate only rarely in a world that is increasingly internationally mobile. This immediately problematises cultural explanations of international differences in student achievement, since such explanations assume that either the school system or the student body (or both) can be identified with a single culture. The challenge for school systems in countries such as Australia or South Africa is to accommodate and cater to a multiplicity of cultural backgrounds. Perhaps culture itself is not the essential characteristic in distinguishing one school system from another, but rather the differences in how school systems (and classroom practices) have developed in response to either homogeneity or heterogeneity of culture. In this chapter, Japan provides an example of a less culturally-diverse setting. The argument begins with the need to explicitly challenge the identification of nation with culture.

In his re-analysis of data from the Second International Mathematics Study (SIMS), Bracey (1997) suggested that the differences in mathematics performance found at an international level were replicated in a partitioning of the U.S. sample along cultural or ethnic lines. As a simple illustration of this point: Asian-American students, participating in a school system that has been substantially maligned in the U.S. popular press, perform at a level comparable with their high-performing counterparts in schools in Asian countries. This single illustration suggests that differences on particular measures of mathematical performance are at least as attributable to the cultural affiliation of the students as to the particular school system attended. The significance of such internal cultural variation is lost in the aggregation of performance data for countries as culturally plural as the USA, Australia, or Canada. Such analyses also have implications for societies with a small number of substantial ethnically-distinct communities, such as Malaysia and South Africa.

Berliner reiterated this point in an article in the Washington Post (Sunday, January 28, 2001, p.B3). That is, rather than serving an agenda of international competitive comparison, the results of international achievement testing can be analysed to identify members of a nation who are less well served by the school system than others.

> Which America are we talking about? . . . Average scores mislead completely in a country as heterogeneous as ours . . . The TIMSS-R tells us just what is happening. In science, for the items common to both the TIMSS and the TIMSS-R, the scores of white students in the United States were exceeded by only three other nations. But black American school children were beaten by every single nation, and Hispanic kids were beaten by all but two nations. A similar pattern was true of mathematics scores . . . The true message of the TIMSS-R and other international assessments is that the United States will not improve in international standings until our terrible inequalities are fixed.
>
> (Berliner, 2001, B3).

A corollary to this line of reasoning is voiced by Wang (2001) who, in discussing technical concerns with TIMSS, cites Hu (2000, p. 8) as saying, "This study does not break down Americans by race, if they did, Asian Americans would likely score as high as Asians in their home countries, and Whites would rank near top of the European nations." There are several ways to interpret this observation. Berliner's approach seems the most rational and productive. From several perspectives the comparison of national means of student achievement is problematic. Comparisons between sectors of the community within a given country may be more fruitful, within a given state or school system even more so. Such comparisons may at least highlight community groups who are less equal in the benefits they accrue from a school system intended to benefit all students equally. Educational policy can then be framed to address any inequalities. It should be stressed that it is the capacity of the system to benefit learners that is at issue here, and that different learners will experience the school system differently. Some of these differences will be cultural.

In attempting to tease out the patterns of institutional structure and policy evident in international comparative research (particularly in the work of LeTendre, Baker, Akiba, Goesling, and Wiseman, 2001), Anderson-Levitt (2002) noted the "significant national differences in teacher gender, degree of specialization in math, amount of planning time, and duties outside class" (p.19). But these differences co-exist with similarities in school organization, classroom organization, and curriculum content. Anderson-Levitt (2002, p. 20) juxtaposed the statement by LeTendre et al. that "Japanese, German and U.S. teachers all appear to be working from a very similar 'cultural script'" (2001, p.9) with the conclusions of Stigler and Hiebert (1999) that U.S. and Japanese teachers use different cultural scripts for running lessons. The apparent conflict is usefully (if partially) resolved by noting with Anderson, Ryan and Shapiro (1989) that both U.S. and Japanese teachers draw on the same small repertoire of "whole-class, lecture-recitation and seatwork lessons conducted by one teacher with a group of children isolated in a classroom" (Anderson-Levitt, 2002, p.21), but they utilise their options within this repertoire differently.

But what are the implications from the perspective of cultural traditions. The analyses summarized above suggest that the cultural affiliation of the learner (whatever their geographical location) is at least as important as the cultural alignment of the school or school system and certainly should not be simplistically identified with nationality.

This is not intended to challenge the premise that school systems enact cultural values. However, it does challenge the simplistic identification of culture with nationality, and it highlights the possible significance of the cultural affiliation of the learner. Once the confusion of nation with culture has been problematised, then the utility of international comparative research can be considered with greater cultural sensitivity. For example, the identify-cation of international differences and similarities in student mathematical performance has limited utility, except as a form of national report card, unless it is accompanied by data that suggest cultural, societal, or instruct-tional differences that might be used to explain such differences and similarities and then to promote improved mathematical learning and associated performance.

Naively, one might argue that if Asian countries are consistently successful on international measures of mathematics performance, then less-successful non-Asian countries would do well to adapt for their use the instructional practices of Asian classrooms. Such a line of reasoning is grounded in four key assumptions: (i) that the term "Asian" identifies a coherent body of practice; (ii) that the performances valued in international tests constitute an adequate model of mathematics, appropriate to the needs of the less-successful country; (iii) that differences in mathematical performance are attributable to differences in instructional practice (and not to other differences in culture, societal affluence or aspiration, or curriculum); and (iv) that the distinctive instructional practices of more-successful countries (should these exist) can be meaningfully adapted for use by less-successful countries. All such arguments give inadequate attention to issues of cultural heterogeneity, particularly in relation to differences in the students' experience.

Hess and Azuma (1991) assert that formal schooling confronts students with organisational conditions that "are not conducive to learning." They claim that "Teachers deal with these circumstances by encouraging facilitative dispositions in students or by making learning events more appealing" (Hess & Azuma, 1991, p. 2). Most importantly, Hess and Azuma assert that:

> Cultures differ in the emphasis they place on these two strategies. Japanese tend to stress developing adaptive dispositions; Americans try to make the learning context more attractive. National differences in educational achievement may be more completely understood by analysis of cultural differences in student dispositions. The interaction of student characteristics and teacher strategies creates very different classroom climates in the two countries.
>
> (Hess & Azuma, 1991, p. 2).

The curriculum is the embodiment of the aspirations of the school system. It may be that the priority given in many culturally-multiple countries to making the learning context (setting, content, and delivery) more attractive to students is, in fact, the most appropriate curricular response to a student community that does not draw on a single set of values and beliefs for motivation and aspiration. To a significant extent, the teacher is the agent of the system by whose actions the curriculum is put into effect. Teachers, however, interpret the curriculum in idiosyncratic fashion, within the constraints and affordances of both system and culture. Both the curriculum and the teacher have been the focus of recent international comparative study. Among the studies of curriculum and teaching practice, we can lose sight of the student.

> What is absent from nearly all the rhetoric and variables of TIMSS pointing to the future needs of the global economy is indeed this human side: the notion that students themselves are agents. TIMSS makes students from 41 countries into passive objects of 41 bureaucratic gazes, all linked to the seduction of one global economic curriculum.
>
> (Thorsten, 2000, p. 71).

As educational research has increasingly drawn our attention to the importance of the social processes whereby competence is constructed and in which competence is constituted (for both teaching and learning). The agency of the student, the nature of learner practice, and the cultural specificity of that agency and that practice must be accommodated within our research designs. The authors of this chapter are collaborators in research into mathematics classrooms in many countries, with particular emphasis being given in the data collection and analysis to the perspective of the learner.

2. THE LEARNER'S PERSPECTIVE STUDY

The particular results from the Learner's Perspective Study reported here are based on analyses of sequences of ten lessons, documented using three video cameras, and supplemented by the reconstructive accounts of classroom participants obtained in post-lesson video-stimulated interviews. These data are further supplemented by copies of student written materials, by student test performance, and by teacher questionnaires. Details of the study design, the participant countries, and other findings not reported here can be found at www.edfac.unimelb.edu.au/DSME/research/lps. Each participating country in the LPS used the same research design to collect videotaped classroom data for ten consecutive lessons and post-lesson video-stimulated interviews with at least twenty students in each of three participating 8th grade classrooms.

This methodological approach offers an informative complement to the survey-style approach of the TIMSS and TIMSS-R video studies. A research design predicated on a nationally representative sampling of individual lessons, as in TIMSS and TIMSS-R, inevitably reports a statistically-based characterization of the representative lesson. A more fine-grained study of sequences of ten lessons, informed by the reconstructive accounts of the participants, has the potential to address:

- Consistency of lesson structure over a ten lesson sequence
- Degree of variation in lesson structure in the practices of competent teachers
- The extent to which any such variation is linked to the location of the lesson in the instructional sequence and to the teacher's instructional intentions
- Student awareness of the structure of the lesson and how this is related to their perception of significant educational moments in the lesson and to their subsequent learning.

Further details of methodology and research setting are provided in each of the following sections.

The remainder of this chapter addresses three issues:

a) The importance and distinctiveness of the learner's perspective;
b) The need to consider available resources in evaluating practice; and,
c) The relationship between curriculum and societal and political contexts, and how these things are perceived by students.

3. CONTRASTING THE COMPLEMENTARY PERSPECTIVES OF TEACHER AND STUDENT (JAPAN)

The video component of the Third International Mathematics and Science Study (TIMSS) was the first attempt ever made to collect and analyze videotapes from the classrooms of national probability samples of teachers at work (Stigler & Hiebert, 1999; Stigler et al., 1999). Focusing on the actions of teachers, it has provided a rich source of information regarding what goes on inside eighth-grade mathematics classes in Germany, Japan and the United States with certain contrasts among the three countries. The findings of the study include aspects of mathematics lessons as identified with a strong resemblance between Germany and the United States but with Japan seemingly unique. One of the sharp differences between the lessons in Japan and those in the other two countries relates to how lessons were structured and delivered by the teacher. The structure of Japanese lessons was characterized as "structured problem solving."

The Learner's Perspective Study (LPS), on the other hand, is a nine-country study of the practices and associated meanings in "well-taught" eighth-grade mathematics classrooms with a focus on learner practice (Clarke, 2001a, 2001b). In part, the study is motivated by the need to research the postulated cultural specificity of teacher practice and by a strongly felt belief that the characterization of the practices of the mathematics classroom must attend to learner practice with at least the same priority as that accorded to teacher practice.

Among the most interesting analyses afforded by the data collected in the LPS are those related to lesson structure. Analyses of LPS classroom data from Germany, Japan, the USA and Australia have been carried out for the purpose of comparison with the national "lesson patterns" reported as a consequence of the first TIMSS video study (Stigler & Hiebert, 1999). These analyses (papers available at www.edfac.unimelb.edu.au/DSME/research/lps) found little evidence of the reported lesson patterns, and alternative bases for the international comparison of classroom practice have been suggested. An important consideration in any analysis of curriculum implementation or, in this case, lesson structure, is how the intended structure is perceived and experienced by the learners it is intended to benefit. For example, while Japanese teachers may devote considerable effort into the planning and structuring of their lessons, these structures may not be perceived by the students. Discrepancies in perceptions of lesson structures between the teacher and the students will be explored through the analysis of post-lesson interviews with both groups.

The LPS data collection in Japan was conducted at the three public junior high schools in Tokyo. The teachers, one female and two males, roughly represented the population balance of mathematics teachers of the school level. The topic taught in each school corresponded to the three different content areas prescribed in the National Curriculum Guidelines. The first school was located in the old downtown Tokyo. The teacher, Ms. K, who has been teaching mathematics for more than twenty-five years, taught the 8th grade class of thirty-two students. The content taught was "linear function". The teacher intended to achieve the goals of the entire unit. Namely, she tried to have the students learn about the relationship between different representational forms: diagram, graph, and formula, and to think about the domain of change, determine an independent variable against a dependent variable, learn the concept of the rate of change, and so on.

3.1 Different perceptions of lesson structure between the teacher and the students

While Japanese teachers may devote considerable effort into the planning and structuring of their lessons around a "yamaba", a climax of lesson, these structures may be perceived differently, or not at all, by the students. The methodology employed in the LPS offered students the opportunity in post-lesson video-stimulated interviews to "parse" the lesson they had just experienced. That is, the teacher and the students interviewed after lessons, were given control of the video replay and asked to identify and comment upon classroom events of personal importance. It is clearly possible that students identify as significant classroom events quite different from those intended by the teachers.

In the post-lesson interview of the lesson J1-5, the teacher identified nine elements in the lesson to be significant, while each of two students interviewed identified eight and seven elements respectively (Table 1). Although the numbers of elements identified as felt to be significant are similar between the teacher and the two students, their places in the entire lesson were different. Only four elements were identical among three of them. As for the lesson J1-7, the teacher identified twelve elements in the lesson to be significant, while one of the two students identified only eight elements and the other student only three.

Table 3-3-1. Elements in the Lessons Felt to Be Significant

Lesson	Teacher	Student 1	Student 2
J1-5	9	8	7
J1-7	12	8	3

When we look into our data more closely, we can see both conformities and discrepancies in perceptions of classroom events of personal importance between the teacher and the students. The following excerpts from lesson J1-5 relate to the elements in the lesson felt be significant to all three of them.

Ms. K (14:00)

K Can I just pause for a moment?

INT Okay. Here. Um, fourteen minutes, the scene of group discussion.

K Well, this class is comparatively quiet, so I haven't let their desk in groups until today. But I thought it might be useful for students to complete the graph along with the discussion. They can help one another by thinking together, and it allows everyone to see other students' work. Maybe some of them can understand where they couldn't do by themselves.

The two students identified the same event as one of personal significance and commented on the event as follows:

Student 1 (NI) (14:25)

NI Here.
INT Here, right. Fourteen minutes twenty-five seconds. Where you started group work. Ok. Why here?
NI Discussing the graph and the chart in groups and finishing it up.
INT This scene, right? Ok, about this scene.
NI Here, I can share what I thought and what others thought, the answers I thought I had made a mistake on, could actually be right, not a mistake, my mistakes could actually be the answer. Huh?
INT What you thought was a mistake, might actually be correct.
NI Right. What I thought couldn't be incorrect, could actually be a mistake. I can compare my answers with other people and talk it over, so it's nice.

Student 2 (TA) (14:22)

TA Here, in groups.
INT Yes, uh, fourteen minutes and twenty-two seconds, the scene where group learning has started. Ok. What did you think here?

....

INT Where you thought was important.
TA Yes.
INT Important in today's class.

TA In today's class, yes.
INT Um, what were you doing uh, then, together?
TA Oh, together, we first compared the part of the homework I did, and the part my friend did, and we discussed the parts that were different and shouldn't it be this, and such. Then, after that, there was a part we didn't understand and it was the same part, so we were asking each other how we were suppose to solve it, and so we asked the teacher.

Though Student 2 was not so explicit as both the Teacher and Student 1 were in commenting on it, the three of them seemed to share the belief that the learners can help one another by thinking together. The discrepancies in perceptions of classroom events of personal importance between the teacher and the students were most evident in relation to the occasion, later in the lesson, when the teacher did not understand their solution to the problem. The two students, who had worked together in solving the problem, made similar comments as follows.

Student 1 (NI) (37:45)
NI Here, right here.
INT Uh, thirty-seven minutes forty-five seconds. Uh, where she went on to the explanation about the graph, Ni-san, right?
NI The teacher, what the teacher was thinking, and what me and Ta-san were thinking were different, so, we were both trying to explain but the teacher didn't really understand us, and finally, after a while, she understood.
INT The teacher, she wrote down a graph but it was different, right? From what you and Ta-san got, and there was a gap between what you were thinking and what she was thinking, right?

Student 2 (TA) (37:26)
TA Yes.
INT The scene where Ni-san comments on the graph. Thirty-seven minutes twenty-six seconds.
TA Uh, this, the teacher mistook what I said or something and Ni-san pointed out her mistake, but the teacher made mistakes once in a while so it would be nice if someone could say the correct answers in situations like this.
INT Ohh. So, you felt to the teacher, the teacher that your ideas were being interpreted differently, so you were trying somehow to say this to her. Uh, actually she said this for you.
TA Said this for me. Yes.

……

INT That's important. So, then, Ta-san, anything you felt, or rather, thought?

TA Feel ?

INT If you got across what you wanted to say.

TA Oh, what I wanted to get across? I was explaining but Ms. K misunderstood again so what I wanted to say didn't really get across to her but there's a scene after this where I explain something else again.

INT Yes.

TA And, there, uh, I think Ms. K understood the correct answer.

INT Yes, where you went up to the chalkboard. The scene where you're explaining, the teacher understood you correctly.

TA Yes.

Ms. K did not realized until the final part of the entire lesson that she understood differently the explanation made by two students during teaching at their desk as well as in the whole class discussion. All three of them identified the same element around sixteen to seventeen minutes from the start as significant. At this moment, the teacher intended to assess what these two students were doing and she gave a hint to think about. The students commented in the post-lesson interviews on the importance of the teacher's coming and appreciated the teacher's help.

The excerpts from the interview data suggest that the students perceived lesson structure differently from their teacher. As was mentioned earlier, one of the characteristics of Japanese teachers' planning of lessons is the deliberate structuring of the lesson around a climax within a structure. The students in Japanese classrooms can be unaware of the occurrence of these climactic points or their intended significance. The teacher, in turn, can be unaware of what students think of their importance. Essential to any judgment (or understanding) of the effectiveness of an instructional strategy is the meaning that learners attributed to that strategy or instructional act. International comparative research must attend more closely to students' construals of classroom events.

4. AFFORDANCES AND CONSTRAINTS: LARGE CLASS SIZE AND LIMITED RESOURCES (THE PHILIPPINES)

Large class size and lack of resources interplay with other factors to bear upon students' emergent understanding of mathematics, and, most importantly, upon the student's evolving learning style, and the teacher's teaching style, as well as upon the attitudes, values and beliefs of both. Classes in the Philippines are very large ranging from 40 to even 72 pupils per class (Mariñas, 1999). Both class size and the limited availability of instructional resources place significant constraints on teachers and learners. In characterizing classrooms on a national or cultural basis (for the purpose of comparison), constraints such as class size and instructional resources must be taken into account.

4.1 The classroom

Already teaching for 8 years, the LPS teacher discussed here ranked third among those identified as competent by the mathematics supervisor of the public secondary school system in a city that is a business center in the Greater Manila Area. Belonging to the top five sections based on students' average grade in all subjects, out of 44 sections in the second year, her mathematics class consisted of 57 students with 28 boys and 29 girls coming mostly from the average to low socio-economic status. The class met for 40 minutes from 12:10 to 12:50 P.M. from Monday to Friday in a classroom that is 8m x 6m in floor area. A center aisle divided the room into two sets of chairs with 5 rows per set and 5 chairs per row. All the girls sit together on the right side and the boys on the left side of the room, respectively. Touching the back wall is a row of 10 chairs. Daily for ten days, all except for 3 or 4 chairs were filled with students. The only space for convenient movement is in front where the teacher's table is, along the aisle, and in front of the back row of chairs. Except for the teacher's table and the blackboard, there are no other teaching and learning fixtures found in the room.

4.2 The lessons

The lessons were on Geometry covering the basic concepts (point, line, collinear, plane, coplanar, ray, line segment), coordinate of a point, distance, congruent segments, between-ness, midpoint, angles, kinds of angles, points in the interior or exterior of an angle, measuring angles, congruent angles, and Angle Addition Property. The lessons were mainly in English. By mandate of the Department of Education, Mathematics is taught in English although the national language is Filipino.

The lessons were primarily from the unified lesson plan based on a common syllabus being implemented for the first time in the school year 2001-2002 for the entire Greater Manila Area. Pairs of teachers prepared daily lesson plans for their assigned chapters for a given year level. Individual teachers modified the plans as they saw fit as evidenced by handwritten texts and pasted cutouts on the teacher's copy of the daily lesson plans. Changes can be made on the practice exercises and quizzes but not on the long tests and periodical tests because these should be uniform for the entire Greater Manila Area.

4.3 Teacher practice

When the teacher entered the room, she would greet the class "Good afternoon" and in response, the class would stand up and in chorus say "Good afternoon, Ms. Santos. Mabuhay!" At the end of the class, she would say "Good bye." And students would stand up and say together "Goodbye Ms.Santos, Mabuhay!" The introductory activity consisted either of discussion of answers to the assignment, a review of the previous lesson through a quiz or GANAS. In GANAS, the teacher would announce the extra points that may be earned and gives 1 or 2 items that she would read. The introductory activities were also in the form of a visualization problem or guessing game that students enjoyed. During the lesson presentation, the teacher asked questions that were mostly short with specific answers such as factual questions and those answerable by yes or no to which students answered individually or in chorus. These questions did not require much thought for an answer such as those on stating definitions (e.g. reflex angle) and postulates (angle addition property). According to Herrington et al. (1997), since factual questions require short responses, then a large number of students may be called and in a subtle way, the teacher can discipline students without disrupting the flow of the lesson. As such, they are helpful in dealing with large classes.

The teacher also typically used a strategy that involved posing a series of incomplete questions whose answers would eventually lead to a complete idea. She claimed that the students tended to guess the answers to her questions and so, in order to gradually lead them to the correct answer she has to pose several incomplete questions.

4:25 Teacher: Okay. So to get the distance, so say for example we have here [draws on the board] X, Y, Z. Okay. To get the distance of XY, that is get the?

4:42	Class:	Absolute value.
5:01	Teacher:	Okay. Absolute value? Negative 3 minus 0?
5:05	Class:	Negative 3.
5:06	Teacher:	Okay, negative 3. Absolute value of negative 3?
5:08	Class:	3.
5:09	Teacher:	That is how you get the distance. What do we mean by between-ness? So when do we say that a point is in between 2 other points? When do we say that a point is in between 2 other points? Michaela.
5:31	Michaela:	Midpoint.
5:32	Teacher:	If the point is?
5:33	Michaela:	The midpoint.

The teacher sometimes referred to the seat plan to call on particular students as in recitation, if the answer is correct, the student earns points. In a large class, a seat plan is a big help for the teacher to ensure that every student gets the chance of reciting.

The teacher also attempted to explain the lesson well. Based on students' facial expressions, reactions or verbal comments, the teacher would slow down the explanation, repeat the explanation or request a student who understands to make the explanation to the class. During the practice where simpler items were given, the teacher moved around and at times attended to those who needed help. The teacher claimed that it was important to move around because she could get the ideas of students while they are discussing and monitor their work so that she could give hints when needed. However, they would stop talking when she got near them.

After the exercises, a GANAS, which is a little more complicated, may follow. A GANAS is a short test that is sometimes given under time limit. The teacher's reason for such texts is to train students to work under pressure. The test items were taken from the teacher's lesson plan or sometimes from the student's workbook and were similar to those in the practice exercises that required computations or short answers. Quizzes were either orally read by the teacher or written on manila paper and posted on the board. She moved around while the students answered and reminded them of the remaining time. When the time was up, she told them to stop writing, exchange papers (which they did with their seatmates), and to write "corrected by" on the paper that they would mark. She then called students to answer the items or at times she would give the answers. Each correct answer usually earned one point. After the papers were marked, she asked them to return them to the owners. She then surveyed how many students got at least 75% correct answers by asking them to raise their hands. This approach was used by the teacher as a way to make decisions as to whether she should give a remedial lesson or additional seatwork. The teacher collected the papers by having students pass them to the aisle and then forward to the teacher.

This system of checking and collecting the papers from quizzes, assignments and practice exercises may well have been developed specifically for a large class like this. The teacher is handling 7 classes, besides having an advisory class. At an average of 57 students per class, she would have a tremendous number of papers to mark notwithstanding that the long test has 50 items. Thus a system of checking and collecting papers was established. But this system of checking is not without risks. It limits the teacher's knowledge of what her students really do not know or cannot do. And the accuracy of the students' marking can only be assumed. The danger is that it is possible that the student gets the correct answer but the process that he used is wrong. Consider Roger's work. He got only 3 out of 5 points for he did not show his method of solution. If there were fewer pupils, she could have asked Roger about his solution. The problem was "The measures of 3 angles are in the ratio 1:2:3. What is the measure of each angle?" Using the numbers 2 and 3 in the ratio as divisors of 120, Roger got the following: 120/2 = 60 – measure of the 3rd angle, 120/3 = 40 – measure of the 2nd angle, 60 – 40 = 20 – measure of the 1st angle. 20 + 40 + 60 = 120. In a quiz with GANAS, Mary Jane correctly answered "none" to the question "How many lines can pass through 3 non-collinear points?" In interview, it emerged that her answer was prompted by the term "<u>non</u>-collinear." This raises the question as to the form of student understanding likely to develop in such a classroom setting.

On the last day of data collection, students took a long test covering the whole period of documented instruction. The printed test which covered 10-day lessons, had 10 of them multiple choice items, 10 true or false items, 10 completion/naming items, 5 illustration, and 2 word problems with all item types testing only knowledge, comprehension or computation. These are the typical components of such tests. Two days before the test, students had practice exercises and one day before they had graded recitation, both of which were announced. All items in the graded recitation were similar to those given in the long test, except for the changes on the given numbers or labels. Student scores ranged from 9 to 48 with a mean of 26.58 out of a perfect score of 50. Of the 20 students interviewed, 11 got scores above the mean and 1 did not take the test. Surprisingly, of the 19 interviewees who took the test, 17 claimed that they were at least average in mathematics.

Post-lesson interviews provided many examples to show that students' understanding of mathematics was superficial: devoid of visualization and care in considering the given task conditions. The students had their own methods of solving problems and though these might be related to the teacher's method, the students did not necessarily see the connections, leading to frequent student misunderstandings, which the teacher was unable to monitor, detect or correct.

4.4 Resources available to the learner

Twelve out of 20 students interviewed claimed that the best way to learn mathematics is by listening attentively to the teacher's lesson presentation. Rebeca said:" Because we need to listen to the teacher. We need to because for a child the reason that you come here (school) is to listen to the teachers. In order to understand her lesson especially in mathematics" This is consistent with the finding of Arellano (1997) that 5 out of 6 Filipino students are auditory learners. Thus, the predominant learning style is listening which matches a teaching strategy that is mainly exposition, which does not require much teaching/learning resource, and for which the large class size is not a problem.

There were those who said that besides focusing on what the teacher was saying it is necessary to copy the teacher's solution so that if you could not get it you can study it at home. Nolito said that he copied the notes on the board so that when they have a test he can review and somehow get a high mark in the test. Upon reaching home, he would read his notes, make an example on a paper and solve it. It was observed that students copied notes while listening to the teacher. So that they may have something to fall back on if they needed more time to think and reflect about the lesson and review for the examination. Students such as these are reflective learners (Dunn, Dunn, & Perrin, 1994) and it has been reported 3 out of 4 Filipino students are reflective learners (Arellano, 1997).

Some students also learn from their classmates. Donato took the initiative to teach his seatmate Karlo who did not get the correct answer. But he would also ask Karlo, whom he considers as good in mathematics, to teach him when he could not understand the lesson. Six out of the 20 interviewed students reported that they would ask two or three of their brighter classmates even when they are not their seatmates. When a student is absent from class, he would rely on his classmates to help him understand the lesson that he missed. Due to large class size and small classroom area, the closeness resulted in easier communications and the tendency for students to ask questions of each other and to help each other learn the lessons. The large class size might explain why students tended to ask the more substantive questions or seek the help of the brighter students since this required more time and attention which they could not get from the teacher.

There were 7 out of the 20 interviewed pupils who sought the help of family members such as their parents or older siblings. None of the students pay for tutors since this is probably not economically possible for them.

Another form of constraint arising from a large class size was the inhibition of asking questions because of the embarrassment it might cause if a student gave a wrong answer.

Interviewer:	Suppose you couldn't understand it right away, what do you have in mind?
Laurencio:	I would not tell my teacher about it, I'm embarrassed Interviewer: You're embarrassed to Ms. Santos? Has there ever been an instance when you had asked Ms. Santos questions? None yet? Why are you embarrassed?
Laurencio:	I have many classmates.
Interviewer:	Ahh, because there are many… hasn't there been even just one of your classmates who asked Ms. Santos?
Laurencio:	There is , sometimes. They are also embarrassed.
Interviewer:	Because there are many… so what if there are many students, why? Why is it embarrassing to ask questions?
Laurencio:	Because, Ma'am, if the answer is wrong they might laugh.

Students attach substantial importance to social acceptance. This could be the reason why in class despite getting the same answer which the teacher got by using her own method, Michaela, the most consulted and best student in mathematics, prefers to use the teacher's method in subsequent similar problems because she does not want to be different from what her classmates do.

4.5 Competence under constraint

Large classes and lack of materials have been consistently cited as reasons for teachers' inability to introduce innovative teaching strategies aimed at improving learner outcomes (HS Math Survey, 1995). Given the economic infeasibility of reducing class size and providing more resources in the Philippines, it appears that improvement may not be attainable. To the teacher described here, there seemed to be no alternative but expository teaching to the whole class. The crowded classroom conditions afforded students the opportunity to assist each other, and it appeared that many students made use of this opportunity.

It is possible that the real constraints to competence are not so much the physical or material aspects but the inadequacy of the teacher's repertoire of alternative teaching skills. However, it is clear that both teachers and students have developed forms of classroom practice as a consequence of the large class size and limited resources. International comparisons of classroom practice need to take such constraints (and affordances) into account.

5. THE LEARNER'S PERSPECTIVE ON CONTEXTUALISED MATHEMATICS (SOUTH AFRICA)

In South Africa, the LPS investigation took place in a context of curriculum fluidity, and offered some opportunity to observe the various ways in which teachers and learners respond to curriculum changes, and to describe the forms and substance of classroom mathematical cultures co-created. One of the three schools in which this investigation took place was Umhlanga High school, in the Kwazulu-Natal Province.

The new political dispensation in South Africa, marked by an election of a new ANC-led government, brought with it an opportunity to introduce a new education system. This system is based on a pedagogy of learner-centredness, integration, issues of relevance, Outcomes-Based Education, equality, equity and human rights. It is thus considered to be in contrast with the old apartheid education system which promoted separateness (DoE, 1997). Such attempts to reconstruct curricula to meet ideals of equity, and the consequent curricular fluidity, should be acknowledged in international research comparisons and not concealed in comparisons of national means or characterizations of typical practice.

During the LPS data collection period at Umhlanga High school (October 2001), the government-appointed Curriculum Review Committee had already released a Revised National Curriculum Statement for public comments. What still remains one of the common threads, though, between the initial National Curriculum for the new education system and the Revised National Curriculum is a commitment to an education system which is relevant to the lives of the learners (DoE, 1997:01 &. DoE, 2001:12). It is for this reason that this new curriculum, Curriculum 2005 (C2005), is termed by some as a 'boundary-bashing' curriculum (Muller and Taylor, 1995). In other words, it is seen as a curriculum that encourages the collapse of boundaries between different disciplines.

This discussion focuses on one teacher, Bulelwa, who used AIDS as a context for teaching number patterns. For C2005 Mathematics, an awareness of number patterns is considered one of the most important learning outcomes because the learners at this level are expected to be able to identify and analyse regularities in a given pattern (DoE, 2001:88). The relevance of a context such as AIDS can be argued for, both at provincial and national level. At a provincial level, Kwazulu-Natal, the province in which Bulelwa's school is located, is reported to have the highest infection rate in South Africa. In addition, a woman, Gugu Dlamini, was stoned to death following her public declaration that she was HIV-positive. This incident took place in a township that is situated about 60km from Bulelwa's school.

At a national level, the year 2001 saw a legal battle between the Government and the AIDS activists over the provision of a drug (Nevarapine) which, it is argued, reduces the mother-to-child transmission. Opposition parties (like the Democratic Alliance) have also voiced their disapproval of President Mbeki's stance on AIDS. The details and debates pertaining to this 'battle' are well beyond the scope of this paper. The point, though, is that AIDS, even as a context to advance mathematics, is not a play-reality or a 'benign' context. It is a sensitive and a realistic social concern, and one chosen because of its relevance to the particular community in which Bulelwa's mathematics class was situated.

Umhlanga High school is situated in one of the (apartheid-created African) townships in the Kwazulu Natal province. Its students are mainly Zulu-speaking Africans. The school is situated about 7 kilometres West of the main airport in the province, and 20 kilometres North East of the city centre (Durban). The school is a modern, double storey building with a total of about a thousand learners ranging from grade 8 to grade 12. For a township school, Umhlanga High school is relatively better resourced: there is an administration 'block' with a receptionist who 'mans' it. The school is electrified and there are computers and photocopiers in the administration section.

Bulelwa, a grade 8 mathematics teacher, is also the head of the school's Mathematics department. She is very positive about Curriculum 2005 (C2005) and what it means for mathematics. She has a Bachelor of Science degree with Mathematics and Statistics majors. The learners in Bulelwa's class are seated in groups of not more than six. She encourages the learners in her class to interact and discuss ideas. There is a chalkboard, which Bulelwa did not use extensively, at least during our data collection period, because she prepared worksheets for the learners.

The section that she was teaching, during our data collection period, was number patterns. Broadly speaking, this section entails an investigation of numeric patterns. The learners are expected to identify and analyse regularities and changes in these patterns in order to complete the patterns (as reflected in the table) as well as make predictions (i.e. form generalisations) about other numbers in the pattern. All the patterns we saw were presented in a table format. These tables were context-free, for example

Input	1	2	3	4	5
Output	3	5	8		

The current lesson was also based on number patterns. However, as previously hinted, the table that the teacher used for the worksheet was based on the context of AIDS sufferers and the world population. It took place during the second week of our visit and thus the learners were already familiar with the presence of 'strangers' in the classroom

ACTIVITY 7 (NUMBER PATTERNS IN NATURE)

Mathematicians have studied number patterns for many years. It was discovered that there are links between mathematics and our natural environment and sometimes events occurring in our societies. For this reason an understanding of algebra is central to using mathematics is setting up models of real life situations.

Study the tables given and answer the questions that follow.

Year	1960	2000	2040	2080	2120
World population growth	3 000 million	6 000 million	12 000 million		

Year	1997	1998	1999	2000	2001
World increase in the number of AIDS sufferers	16, 7 million	33,4 million	66,8 million		

From the tables, you can see that the AIDS and population figures follow trends, which can be seen, from the number patterns. These patterns allow researchers to predict what these figures will be for the future.

(a) Describe the pattern of population increase every 40 years as shown in the first table.

(b) Describe the pattern of the increasing number of AIDS sufferers as shown in the second table

(c) Fill in the missing numbers in each table

(d) Researchers believe the earth cannot support a population approaching 192,000 million people. If the population continues to double every 40 years, then in which year will it be 192,000 million? Explain how you worked out your answer.

(e) The world population in the year 2000 is said to be 6000 million . In which year will the number of AIDS sufferers be greater than 6000 million if the trend in the second table continues? Discuss what this means.

(f) How is HIV virus/AIDs transmitted?

(g) What can we do as a society to break the pattern of the increasing number of AIDS sufferers? (i.e. decrease the number of AIDS sufferers)

The introductory statement makes it explicit that the mathematical purpose of the task is to observe the number patterns in the two tables and describe what these would imply for real life settings. All the items, except for (f) and (g), are obtained from an OBE mathematics workbook i.e. text materials Bulelwa has as resource for planning her teaching for the new curriculum. Items (f) and (g) were added on by the teacher, and her intentions in this are discussed below.

At face value, the absence of letter symbols, the usage of the familiar context and the mode of expression in this worksheet first blur and then collapse the boundary between mathematics and the real world, familiar knowledge. More specifically, close examination of each of the question items reveals that mathematical demands (and this is Grade 8) do not extend beyond recognising and extending numerical patterns, doing the straightforward doubling and addition calculations required, and describing these mathematical actions in words. How this situation might be further mathematised is not part of this lesson (for example, worked on to produce mathematical models, be they equations or simple graphs, that are then used to manipulate possible scenarios and projections and in turn reflect these back on the real world).

In Bernstein's terms the worksheet has an element of a weakly classified text (Bernstein, 1996: 20), and in Skovsmose's terms (Skovsmose, 1994), we can see how the teacher draws in learners' backgrounds and foregrounds (for AIDS impacts significantly on their lived realities).

5.1 Contrasting perceptions of the lesson: the teacher

In this section, we focus briefly on the lesson and the teacher's reflection on this lesson on the basis of the interview conducted at the end of the data collection period.

<u>The Lesson</u>: The teacher started off by linking the current lesson with the previous ones on number patterns. She indicated that unlike the previous ones, the current lesson draws on the HIV-AIDS context. She then pointed out that the lesson will highlight the applicability of mathematics in addressing the escalation of HIV-AIDS.

Having set the scene for the lesson, the teacher goes through the items (a to g) in the worksheet to clarify briefly and broadly, what each item requires of the learners. She then follows this with a general discussion on AIDS. This discussion is opened up by a question to the learners: "When did you first hear about AIDS?" As some of the learners reflect and murmur the answer amongst themselves, she relates how she came to know about AIDS as a university student and how she and her friends used to corrupt the acronym AIDS as American Ideas for Discouraging Sex. She concludes this introductory part of the lesson by advising the learners on how careful they should be because AIDS is real.

After these introductory remarks, the teacher advises the learners to engage the worksheet. She, in the meantime, walks around from one group to another; monitoring progress and offering tips. Having made some observations from a few groups, the teacher advises the learners to inform their discussions on the basis of the number patterns. She expresses her concern that the learners seem to avoid items d and e in the worksheet, which require them to carry out calculations and reason about the basis of these calculations.

Reflections: There are two ways in which the role of AIDS is presented in this classroom. Firstly, as an epidemic that the learners should be careful of, secondly, as a context to study mathematics. In presenting AIDS as a real context, Bulelwa assumes the role of a concerned citizen whose responsibility it is to advise the learners about AIDS. She is all too aware that AIDS is not a benign but sensitive issue. Thus her classroom becomes some form of platform for the learners to reflect on and talk about AIDS, a point she articulates during the interview.

> I was concerned (with regard to the stigma that AIDS has) but the subject itself is a concern for debate. So even if there is somebody who has AIDS, I felt it would enlighten them more, it would make them feel that you don't have to ….it's a subject that we need to open up for debate. We need to discuss how it gets transmitted, for those who don't know about it yet. Because…In our community we still find people who are illiterate. Who feel there is no AIDS.

The second role that 'AIDS' played was that of a context to study mathematics. In this instance, we note another identity, that of Bulelwa as a mathematics teacher. In this respect, Bulelwa felt that the lesson was about mathematics (number patterns) and that mathematics was prioritised during the lesson. This is implicit in her utterances during the interview with Renuka (R). Below, a snapshot of that interview is provided.

R: What made you choose HIV-AIDS for your teaching?

B: Well, actually it was still number patterns. I wanted to choose something connected to real life. It's not that we learn mathematics in isolation. Just like when we started, we had an outbreak of cholera. I brought some statistics from the department; you know...the actual statistics from the department. So I taught them at the time how to get a table, a statistical table and analyse information. So it was learning mathematics, but with something that was happening at the time.

R: How did you feel about the whole issue of mathematics and context? Did you feel that there was one which you were prioritising?

B: I felt I was prioritising mathematics because most of the questions I asked were of a mathematics nature except the last two questions... "How it was transmitted" and "What can we do?" Because obviously if doing a lesson in class and the OBE context it need not just end up in a classroom situation. If you are dealing with the situation like this you need also to go out into the communities. So what I found out is that they (the learners) had more knowledge on AIDS...that they could handle most of the questions. That's why it was difficult for them to handle a question that was long.

Bulelwa is clearly conscious of the mathematical purpose of the lesson. In fact, she feels that she was prioritising the mathematics over the AIDS context since most of the questions in the worksheet were 'mathematical'. In this instance, Bulelwa presents herself more as a mathematics teacher advancing the mathematical purposes of the lesson.

In managing the blurring of the boundary between mathematics and real life knowledge (or in bringing in learners' backgrounds and foregrounds), we see Bulelwa enacting two different, perhaps competing identities (Setati, 2002) as a mathematics teacher of the new curriculum in South Africa. On the one hand, she is a mathematics teacher and is aware of her responsibilities in this regard. On the other hand, she is a responsible citizen who wishes to alert the learners to AIDS and its effects. Thus, at different stages, within the same lesson, the one role becomes foregrounded and the other backgrounded. In this way, she breaks or weakens the insulation between the mathematics and the everyday.

The question to ask and to be addressed in the next section is: How do learners respond to and experience this lesson? It is in this regard that the next section becomes relevant: The learners' experiences of the lesson.

5.2 Contrasting perceptions of the lesson: the student

We address the learners' experiences and response to the lesson by focusing on one group of learners. We are aware that a focus on one group is not a fair reflection of the classroom events as experienced by other learners in the classroom. However, from focusing on one group, we hope to be able to tease out that which may be masked as a result of a whole class observation. First, we give a brief discussion of the way in which a group of learners respond to this lesson and then their reflections of this lesson as espoused in their post-lesson interview.

<u>The learners' response to the lesson</u>: After Bulelwa had set the scene and discussed the worksheet with the learners, a six-member group, all of them boys, spent a considerable amount of time on items (f) and (g) in the worksheet. Their discussion on the transmission of HIV-virus in response to item (f) is mainly confined to sexual transmission. Afterwards they discuss some practical suggestions on how they can raise their communities' awareness of AIDS, this in response to item (g). There does not seem to be any intention from the learners to go through all the items in the worksheet. In particular, at no stage during the lesson, did we observe the learners carrying out the calculations, as would be expected, particularly with respect to items (a) to (e).

Towards the end of the lesson she highlights her awareness that most learners are trying to avoid item (e) and reminds them of the importance of completing the table. The focus group had not attended to item (e) of the worksheet as well. For example, when the answer 2040 was offered in response to item (e) by another group, they (the focus group) could not decide whether the answer was correct or not. Only one member of the group disputed this answer on the basis that 'it was a guess'. In sum, it is fair to suggest that a considerable amount of time was spent on items (f) and (g) and other 'off-the-worksheet' discussions.

<u>The reflections</u>: The discussion here is based on the interview by Godfrey (G) with three learners: Mandla (M), Nxunu (N) and Sthembiso (S). These learners were interviewed in a mixture of isiZulu (the learners' first language but the researcher's fourth language) and English (a second language to both the learners and the researcher). The first question required the learners' opinion on what they thought the lesson was about. Mandla was the first one to provide the answer.

M: (Softly) HIV…
G: Come again…?

M: HIV…the way one can contact it…
N: (takes over impulsively) Yes…the way people can get AIDS.

For these learners, what was visible in this mathematics lesson was the context – AIDS. Even though the teacher thought her lesson prioritised the mathematics, the learners' responses suggest it is the AIDS that was visible. The learners' did see the mathematics though, but they saw the mathematics as being supportive of the AIDS context. This is notable in the following part of the interview.

G: Now tell me, did you learn any mathematics today?
[S consults with M and N before responding]
S: Yes…we learnt about the year in which the AIDS became known.
M: (takes over). Yes, as well as how many people had it during particular years and how the numbers increased from year to year as well as the number of people who died.

Asked whether they saw any relationship between mathematics and AIDS, the Mandla and Nxunu suggested they did, and the interview continued.

G: Is it?
S: Because you have numbers which show how many people died of AIDS and how many are still alive as well as the percentages…So which means you needed to divide in order to get the percentages.
G: (looking at M)
M: Yes…I agree that Maths relates to AIDS because …everyday people die of AIDS and …
N: (takes over) We need a count of how many people die of Aids everyday.

The learners' comments on what the lesson was about and the relationship between mathematics and the context are centred around the context of AIDS. No reference, in the entire interview, was made in relation to the first table (which showed the world population growth).

Bernstein uses the term recognition rules to describe the students' orientation or awareness of the speciality of the context. The learners' failure to prioritize the mathematical intentions of the lesson may be regarded as a lack of recognition rules. In this respect, the way the learners engaged the worksheet, by attending to items (e) and (f), and not to other items, which demand them to calculate, may be seen as inappropriate for a mathematics classroom.

On the other hand, Skovsmose uses a different gaze to explain the learners' actions in class. He identifies two aspects that can support students' mathematical learning and interest, these are the students' foreground and background. Background refers to meanings that belong to the history of the person whilst the foreground refers to the learners' interpretation of his/her future (Skovsmose, 1994: 177; Skovsmose and Nielsen, 1996:1269). Given the cultural practice alluded to by the teacher, given the sensitivity and prevalence of AIDS in the province where the school is located; the context of AIDS may be regarded as being within the learners' social reality. For them, it may be argued, AIDS is not just a context to advance some mathematical intentions, it is an epidemic which they must be cautious of, and advise other members of the community of its dangers as well. If, however, such an instructional approach is to be evaluated on the basis solely of the mathematical understandings that students develop, then the aspirations of the curriculum are misrepresented. Further, if explanations of practice and effectiveness are sought on cultural grounds alone, the realities of politics, societal need, and cultural plurality are in danger of being omitted from consideration.

6. CONCLUSIONS

Experience in the Learner's Perspective Study suggests that one way to interpret the ICMI Study would be in relation to multiple claims of cultural affiliation, as these are experienced by teachers and students in countries such as Australia, Germany, the Philippines, South Africa and the USA. In these countries, students from a wide variety of cultural backgrounds participate in the same school systems (including some that might be considered 'East-Asian' or 'Western'). The problematics of a school system serving the needs of students from these different 'cultural traditions' is an important aspect of the ICMI Study that should not be overlooked. Teachers in Australia, Japan, The Philippines and South Africa face very different challenges with regard to cultural diversity of the communities they serve, class size and instructional resources, and societal and political priorities.

Our research must do more than document occurrence, whether it is of student achievement, curriculum content, teacher action, lesson structure, or teacher and student belief. Our research must also attend to the cultural homogeneity or heterogeneity of the student community, to the constraints that class size and resources put on the realization of curricular aspirations, and to the simultaneous need to address academic discipline-specific goals as well as political and social priorities. Our research must address the interrelationship of these things. From the studies that have been done, we have every reason to believe that it is in these interrelationships that the character and function of culture will emerge: in the teacher practice that mediates between curriculum content and the student, through the actions and the lesson structure that constitute the enactment of that curriculum in the classroom, together with the beliefs and expectations on which the student's participation is predicated, culminating in the learning of which student achievement is simply the most evident socially-constructed and culturally-mediated correlate. Culture is not outside these things. It is in the combination of these and other elements that culture itself is constituted. Nor, as has already been stated, is culture a synonym for nationality. As several studies have shown, the culture of the classroom can be constructed differently within a particular country or school system. There are, however, cultural values and beliefs that frame the educational endeavours of teachers, students and policy-makers within each country. These same endeavours are also afforded and constrained by economic, societal and political considerations. International comparative research must do more than document cultural differences, it must accommodate them by attending more closely to context and to voice – particularly the voice of the learner.

REFERENCES

Anderson, L.W., Ryan, D., and Shapiro, B., 1989, *The IEA Classroom Environment Study.* Pergamon, New York..

Anderson-Levitt, K.M., 2002, Teaching Culture as National and Transnational: A Response to Teachers' Work, *Educational Researcher*, **31**(3):19-21.

Arellano, E., 1997, Mathematics Learning Styles of Filipino High School Students: Towards a Theory of Learning. Doctoral dissertation, University of the Philippines Diliman.

Bernstein, B., 1996, *Pedagogy, Symbolic Control and Identity: Theory, Research and Critique*, Taylor and Francis, London.

Bracey, G.W., 1997, On Comparing the Incomparable: A Response to Baker and Stedman, *Educational Researcher*, **26**(4):19-26.

Clarke, D.J., 2001a, *Study Design: Learner's Perspective Study.* University of Melbourne (see website: www.edfac.unimelb.edu.au/DSME/research/lps).

Clarke, D.J., 2001b, *Perspectives on Practice and Meaning in Mathematics and Science Classrooms.* Kluwer Academic Publishers, Dordrecht, the Netherlands.

Department of Education (South Africa), 1997, Senior Phase (Grades 7 to 9) Policy Document.

Department of Education (South Africa), 2001, Revised National Curriculum Statement for Mathematics.

Dunn, R., Dunn, D., and Perrin, J., 1994, *Teaching Young Children through their Individual Learning Styles*, Allyn and Bacon, Boston.

Herrington, T., Sparrow, L., Herrington, J., and Oliver, R., 1997, *Investigating Assessment Strategies in Mathematics Classroom.* Edith Cown University, Perth: Mathematics, Science and Technology Education Centre.

Hess, R.D. and Azuma, H, 1991, Cultural Support for Schooling: Contrasts between Japan and the United States, *Educational Researcher*, **20**(9):2-8, 12.

High School Mathematics Education Group, 1995, *Needs Assessment Survey*, ISMED, Quezon City.

Hu, A., 2000, TIMSS: Arthur Hu's index. [On line] Available: http://www.leconsulting.com/arthurhu/index/timss.htm. (December 30, 2000). Cited in Wang (2001).

LeTendre, G., Baker, D., Akiba, M., Goesling, B., & Wiseman, A., 2001, Teachers' Work: Institutional Isomorphism and Cultural Variation in the US, Germany, and Japan. *Educational Researcher* **30**(6): 3-15.

Mariñas, B., 1999, *Philippines. An International Comparative Study of School Curriculum.* NIER, Tokyo.

Muller, J. and Taylor, N., 1995, Schooling and the Everyday Life: Knowledges Sacred and Profane, *Social Epistemology*, **9**(3):257-275.

Setati, M., 2002, Unpublished PHD dissertation, University of the Witwatersrand

Skovsmose, O., 1994, *Towards a Philosophy of Critical Mathematics Education*, Kluwer, Dordrecth.

Skovsmose, O. and Nielsen, L., 1996, Critical Mathematics Education, in: *International Handbook of Mathematics Education*, A.J. Bishop et al. eds., Kluwer Academic Publishers, Netherlands, pp.1257–1288.

Stigler, J.W., Patrick Gonzales, Takako Kawanaka, Steffen Knoll, and Ana Serrano, 1999, *The TIMSS Videotape Classroom Study: Methods and Findings from an Exploratory Research Project on Eighth-Grade Mathematics Instruction in Germany, Japan, and the United States.* U.S. Government Printing Office, Washington, DC.

Stigler, J.W. and Hiebert, J., 1999, *The Teaching Gap*, NY, Free Press, New York.

Thorsten, M., 2000, Once upon a TIMSS: American and Japanese Narrations of the Third International Mathematics and Science Study, *Education and Society*, **18**(3):45-76.

Wang, J., 2001, TIMSS Primary and Middle School Data: Some Technical Concerns, *Educational Researcher*, **30**(6):17-21.

Chapter 3-4

MATHEMATICS EDUCATION IN CHINA: FROM A CULTURAL PERSPECTIVE

ZHENG Yuxin
Nanjing University

1. There has been recently an increasing interest about East Asia in the circle of mathematics education in the world. In particular, some Western scholars, including people working in Hong Kong, have studied China's mathematics education through cultural comparisons.

In contrast to the traditional views, the above development represents a radical change in the West. While Western people were generally quite satisfied with mathematics education in their own countries, they at the same time used to take a more critical attitude to East Asia's mathematics education. This attitude for example was typically expressed as follows:

> They (i.e. mathematics education in East Asia) are very often content oriented and examination driven. The pedagogy is out of date and conservative, since the teachers seem to know nothing about the modern teaching methods and think it is enough to know the contents. Large class sizes are the norm and classroom teaching is usually conducted in a whole class setting. While the teacher directs the whole process, the students seldom play active roles in mathematics learning. Memorization of mathematical facts is stressed and students feel that their mathematics learning is mainly learning by rote. Students and teachers are subject to excessive pressure from highly competitive examinations, and the students do not seem to enjoy their mathematical learning. (Leung, 2000)

However, East Asian students consistently outperform their Western counterparts in international tests such as TIMSS. Therefore, in some Western people's eyes, it is really a paradox (*the Chinese learner's paradox*): how can such a poor mathematics education result in students' good performance? Furthermore, it seems to be a common conclusion of all the related researches that the traditional Western views about mathematics education in East Asia are in fact mainly misunderstandings. Some Western educators have also realized that there is some rationality in East Asia's mathematics education, and this rationality has resulted in students' good performance.

Generally speaking, the international studies of the mathematics education in the East Asian countries, which of course includes China, are still at a primary stage, and many Western researchers in the field aim at getting useful ideas to improve mathematics education in their own countries. But we Chinese mathematics educators on the other hand, should still confirm fully the contribution of such work to the further development of China's mathematics education, that is, the relevant international comparative studies can help us to realize better all the strengths and weaknesses of China's mathematics education. And it is just in this sense, we can say it is now a good time for us to make a serious summation of and reflection on China's mathematics education.

2. The related international studies have been concerned with various aspects of China's mathematics education, ranging from beliefs universally held by general Chinese people to teachers' ways of raising questions in the classroom. But, since the whole system of education in modern China, from the organization of schools, to the design of the curriculum and the teaching methods, has chiefly been introduced from abroad, and China on the other hand also has its special cultural tradition (which for example is often described as "Confucian-heritage culture"), then from the cultural perspective the following question is no doubt more important: *Are there are any cultural conflicts between the introduced system and the native tradition? Is the evolution of modern education in China primarily a process of assimilation or a process of alienation?*

Just as what has repeatedly happened in the long history of China, such as the final results of the invasion and ruling of some ethnic minorities over the Han nationality in history, I think what has happened here is also mainly a process of assimilation. That is to say, rather than being alienated by foreign factors, the latter were absorbed and assimilated into the Chinese culture. Some reasons for me to draw this conclusion can be found in the subsequent parts 3 and 4: they both focus on the features of China's mathematics education and thus show clearly how different it is from mathematics education in the West (it is also in this sense we can talk about "*the Chinese Mathematics education*"). But I would now only take "problem solving" as an example to analyze. As people all know, "problem solving" was the main slogan for mathematics education in the eighties of the 20th century, and it posed a direct challenge to the tradition in this field. Since mathematics educators in China have always attached great importance to the practice of solving problems for the learning of basic mathematical knowledge and skills, which, for example, is shown clearly by the dictum "practice makes perfect", the slogan "problem solving" was quickly and widely accepted in China even without any serious reluctance or resistance. But, rather than initiating any real change in mathematics education in China, the introduction of this slogan has only further strengthened the original tradition in this respect. For example, since repetitive practice and the grasp of the strategies of problem solving have been regarded in China as the prerequisites for the developing of the student's ability of problem solving, people now in mathematics education insist even more strongly on "over-loaded exercise", and instead of encouraging students to use their own ways to solve problems, more attention has been paid to the efficiency of problem solving, which in turn means that students should firstly study carefully all the strategies of problem solving. Thus, what happened here was really a process of assimilation: the slogan "problem solving" as an extraneous factor has been absorbed into the tradition of China's mathematics education and thus changed greatly to adapt to the prescriptive nature of China's education (cf. part 4).

Since the historical development of mathematics education in China is chiefly a process of assimilation through which all the foreign factors have been gradually absorbed into and changed by the original tradition, it should be taken as a basic principle for the comparative studies of China's mathematics education that all the concrete aspects of China's mathematics education should be studied in the context of the whole Chinese cultural tradition. In particular, although some ways of mathematics teaching and learning in China look the same as what are used in the West, such as the parallel between China's emphasis on repetitive memory and the rote memory in the West, and China's belief of "practice makes perfect" and the slogan "drill and practice" in the West, they may have different meanings and functions. Furthermore, what we should struggle for in this field is to develop a system of mathematics education which adapts both to the requirements of the time and China's reality

In fact, as many Western scholars working in this field have already noticed, "the cultural gap" is one of the most important origins of the above-mentioned Western misconceptions of East Asia's mathematics education. It is also from such a position, I think, while confirming fully the great importance of the related international studies on the search for identity of the Chinese mathematics education, China's mathematics educators should play a key role in this study, which of course does not exclude international cooperation on this point.

In what follows, I shall point out some main features of China's mathematics education in general.

3. The ideas of education are of course one of the most important aspects of the Chinese cultural tradition, which many Western scholars in studying China's mathematics education have already dealt with; but, from the cultural perspective, we at the same time should also consider the great influence of philosophical ideas in general, including ways of thinking, in particular, the primitive dialectical ideas, i.e., *the Yin-Yang theory of Taoism*. (Since this theory goes beyond the range of Confucianism, it is thus inappropriate to identify the Chinese traditional culture simply as a Confucian heritage culture.)

As a matter of fact, I think it should be regarded as a basic feature of the Chinese mathematics education that, instead of going to the extremes, China's mathematics educators prefer to get balances between various oppositions in education.

It should be noted here that some Western scholars, including people working in Hong Kong, have already pointed out from various angles some related features of China's mathematics education. For example, Dr. Leung from the school of education, Hong Kong University once pointed out:

> East Asians believe that their Western counterparts have gone too far towards the process extreme. They are re-affirming the importance of the content of mathematics in the process of learning mathematics.
>
> Educators in the West treasure intrinsic motivation in learning mathematics, and consider extrinsic motivation such as that derived from examination pressure as harmful to learning. Yet an optimal level of pressure is thought to be healthy in East Asian countries. ... East Asian thinks that both intrinsic and extrinsic motivation should be utilized in promoting students' learning of mathematics. (2000)

These words are clearly direct confirmation of the above-mentioned basic feature of China's mathematics education; but what we should do here is to extend our analysis to the various aspects of mathematics education, such as process and results, understanding and memorization, cooperation and independence in study, intrinsic and extrinsic motivation, effort and innate ability, and so on. Furthermore, it should also be regarded as the key to achieving success in the reform of mathematics education to get a good balance between the following oppositions:

The development of students' feelings, attitudes, values, and capacity in general, and the learning of basic knowledge and basic skills of mathematics;

Mathematics for all, and universal high-standards in mathematics education;

Students' active construction and teachers' directive role;

The application of mathematics and the formal nature of mathematics;

The inheritance of the established knowledge and innovation in learning;

Mathematical problem solving and the logical connections of mathematical knowledge;

The individual differences of students and large-class teaching which is the norm in China.

Finally, as far as the reflection on Chinese mathematics education is concerned, I think the above analysis also shows that our work should not be limited to the current practice, but should aim at a higher theoretical level. In doing so, we of course should use actively all the results of the Western research in this field; however, we at the same time should also notice their limitation as theoretical frameworks, i.e. their limitation in accommodating the dialectics of the oppositions.

4. Let's now go back to ideas of education.

Corresponding with the international studies, some Chinese scholars have also studied the differences between the Chinese and Western ideas of education. For example, Prof. D. Zhang from East China Normal University once pointed out that the differences between the Oriental and Western ideas of education can be summed up along the following lines: high vs. low pressure of examination, teacher-centredness vs. student-centredness, emphasizing exercise vs. emphasizing understanding, over-loaded vs. less homework, formal deduction vs. informal reasoning, stressing imitation vs. stressing innovation, working hard for reducing individual differences vs. polarization, and so on.

This analysis is illuminating. But, along with all the details, I think what is more important is the basic difference between the Chinese and Western ideas of education, which can be summed up as follows: education in China is chiefly of a prescriptive nature, in other words, it is mainly social-oriented, while in the Western countries, people pay more attention to the personal development of the students, and thus education there is chiefly individual-oriented.

What should be emphasized here is that the prescriptive nature of China's education does not exclude the personal development of students, but rather represents different ideas about the appropriate order of development. Just as J. Biggs says:

> In the West, we believe in exploring first, then in the development of skill; the Chinese believe in skill development first, ... after which there is something to be creative with. (In Watkins and Biggs, 1996, p. 55)

Thus, according to the Chinese tradition, obeying the rules (which is to say, conforming oneself to the society) is in fact a prerequisite or precondition for the following higher stage of innovation.

Furthermore, I think the Chinese pedagogy of mathematics as a whole could also be understood from such a position. For example, just as H. Stevenson and J. Stigler pointed out in their book THE LEARNING GAP:

> In Asia, the ideal teacher is a skilled performer. As with actor or musician, the substance of the curriculum becomes the script or the score; the goal is to perform the role or piece as effectively and creatively as possible (rather than to create a wholly new script or the score directly). (p.166-167)

What should be emphasized again here is that the above model of ideal teacher does not mean teaching in China is not a creative work, but rather, it represents a different understanding of creativity. It was also a point made clear by H. Stevenson and J. Stigler:

> It is hard for us in the West to appreciate that innovation does not require that the presentation be totally new, but can come from thoughtful additions, new interpretations, and skillful modification. (p.168)

And just along this line I think we can find the most valued aspects of the Chinese pedagogy of mathematics. Here, I would like to mention some:

- Highly efficient classroom teaching

In China, every lesson of mathematics has a definite target, which usually focuses on the learning of some concrete mathematical knowledge or mathematical skill, and the whole lesson is well designed around the target. As a matter of fact, mathematics lessons in China usually consist of the following five parts: review, introduction, teaching, exercises, and summation, and all the details of these parts, such as the division of the time and the design of hand writing on the blackboard, are results of deliberation aimed at making the teaching more efficient towards the fixed target – teachers in China always spend a lot of time and effort on preparing every lesson they teach, and efficient classroom teaching is also the goal for teachers' lesson study in group studies of the teachers. (In this sense, they are really "researchers in action"). For example, this feature of the Chinese pedagogy of mathematics can be seen clearly in the ways of introduction of mathematics lessons in China. That is, rather than spending a lot of time on creating some special situation, mathematics teachers in China prefer to stress the inner relationship of mathematical knowledge, because in this way, not only the new subject can be introduced much more easily, but the whole course will show clearly the character of coherence as well.

- Seeking for deeper understanding

It is a basic tenet, accepted by most mathematics teachers in China, that teachers should help students not only to know what and how, but also why. And it is only from such a perspective that some traditions in China's mathematics teaching, such as the ideas of "repetition creates new insight" and "practice makes perfect", can be understood. That is, what is recommended here is not rote memory but the dialectical relationship between memorizing and understanding: understanding helps memorizing and memorizing deepens understanding; and what people are seeking for here is not only the correctness and high-speed of mathematical operations but chiefly deeper understanding by repetitive practice. As a matter of fact, China's mathematics teachers have accumulated a wealth of experience in this aspect. For example, closely connected to the idea of "repetition creates new insight", the method "teaching through variation" is widely used in China, and one of its basic ideas is in fact to help students to learn better about the essence of mathematical knowledge by varying the related backgrounds or contexts, including ways of presentation. Besides, developing deeper understanding is also the core of the teaching method "concise teaching and abundant practice"; and therefore, great efforts have been made in the design and arrangement of the exercises so that there is a gradual development from simple to complex and from single to comprehensive.

- The Heuristic nature of teaching

Teaching in China is not regarded as a process of conveying well-developed knowledge. On the contrary, by paying more attention to the process of creation or discovery, and by making their own re-creation of the content, teachers in China are doing their best to make the content of their lessons really understandable to the students, and to disclose the underlying ways of thought. As a result, the students can not only grasp the related concrete mathematical knowledge or skills, but also learn to think mathematically.

In this connection, we should mention the study of the methodology of mathematics in China and its implication for mathematics teaching, because the heuristic nature of the Chinese pedagogy of mathematics could be summed up as "taking the methodology of mathematics as the guiding principle for the teaching of concrete mathematical knowledge and skills". (cf. Zheng, 1991)

Besides, the great emphasis on students' participation should also be regarded as one important implication of the heuristic nature of mathematics teaching in China. That is to say, contrary to the authoritative style, teachers in China all agree that good teachers in their teaching should do their best to get their students actively involved. This, for example, is why great importance has been attached to the ways of problem posing by the teacher in the classroom in China: such as questions should focus on the "process" rather than the "results", and should urge the students to think more actively.

Finally, in order to get a better understanding of this feature of the Chinese pedagogy of mathematics, we should also mention the reality of large-class teaching in China: in such a situation, strengthening the heuristic nature of teaching would make teaching beneficial to all the students rather than some of them. (In this sense, heuristic teaching is obviously closely connected with the above mentioned first feature of the Chinese pedagogy of mathematics, i.e. the highly efficient nature of classroom teaching, and should in fact be regarded as an important method to accomplish this.)

The above list of the valued aspects of the Chinese pedagogy of mathematics is, of course, not complete. It is also not based on some finished work, but shows some directions along which further research should be carried out.

5. From the above discussion it can be seen clearly that we should fully recognize the features and advantages of the Chinese mathematics education; but at the same time we should also see all its disadvantages or weaknesses.

Firstly, the ideal should not be identified with the reality. That is to say, under the high pressure of examination, the ideal of mathematics education has always been greatly distorted in reality. Besides, it is also a serious defect of China's mathematics education that teaching work frequently relies solely on teachers' own experience and fails to be developed to the level of theory, and therefore, can easily deviate from the right course. It is also from such a perspective, the following disadvantages or weaknesses of the Chinese mathematics education should be noted, because they are closely connected with the latter's main features mentioned above, and therefore, if the teacher involved is not careful enough, they are very likely to make the following mistakes:

- Long-term goals neglected

Because of the focus on the learning of some concrete mathematical knowledge or mathematical skill, the long-term goals of mathematics education, such as the development of students' feelings, attitudes, values, and capacity in general, can very easily be neglected. More generally, this neglect of long-term goals actually represents a limitation of the model of "skilled performer".

- Little space for students' innovation

Because of the prescriptive nature of education, what easily happens in China's classrooms is that mathematics lessons are completely controlled by the teachers and students are not left with enough space for innovation. For example, teachers in China always do their best to ensure the lessons are carried out strictly according to the plans designed beforehand and all the students are expected to grasp the relevant mathematical knowledge or skills along the recommended lines of thinking, so that there is really little space for students' creativity (and for students' discussion and interactions). Worse still, if the teacher were not aware of it, it would be very likely that the students' creativities are wholly stifled. This can happen, for example, when the teaching strategy "moving in small steps" is carried out to the extreme in teaching.

- Inadequate awareness of the application of mathematics

In comparison with their emphasis on the inner connections of mathematical knowledge, China's teachers pay less attention to the application of mathematics. As a matter of fact, in this respect we can also see the influence of cultural tradition. Since most Chinese students take studying hard as a norm, teachers in China do not feel strongly the necessity of motivating students to study hard for the sake of the application of mathematics.

- Students' individual differences de-emphasized

Since large class size is the norm and classroom teaching is usually conducted in a whole-class setting in China, teachers tend to pay less attention to the individual differences of students. What should be noted here is that this de-emphasis in fact concerns one of the basic principles of education. That is, what should be taken as the basis for education: the innate individual differences or the reduction of the possible distance among students? Furthermore, should we insist on high standards in mathematics education for every student? And as many people already noted, there is really a big difference on this point between the West and East Asia.

Therefore, the task we are facing is really not only the searching for the identity of Chinese mathematics education, but also its further development. In fulfilling it, we of course should be grounded in the practice of mathematics teaching and learning in China; but at the same time we should also work hard to go beyond the level of experience and develop the corresponding theories, so that our teaching will become a fully conscious work under the guide of theories, and further, through more practice some new theories may be developed, which can be more suitable to the requirements of the time and the reality of China.

REFERENCES

Leung, F., 2000, In Search of an East Asian Identity in Mathematics Education – the Legacy of an Old Culture and the Impact of Modern Technology, ICME-9, Japan

Ma, L., 1999, *Knowing and Teaching Elementary Mathematics*, Lawrence Erlbaum Associates.

Stevenson, H. and Stigler, J., 1992, *The Learning Gap – Why our Schools are Failing and What We can Learn from Japanese and Chinese Education*, Simon & Schuster.

Stigler J. and Hilbert, J., 1999, *The Teaching Gap – Best Ideas from the World's Teachers for Improving Education in the Classroom*, The Free Press

Watkins, D. and Biggs, J., eds., 1996, *The Chinese Learner: Cultural, Psychological and Contextual Influence*, Comparative Education Research Centre, Hong Kong, and Austrlian Council for Educational Resrearch, Australia.

Zheng, Y., 1991, *The Methodology of Mathematics*, Guangxi Educational Publishing House, China. (in Chinese)

Chapter 3-5

MATHEMATICS EDUCATION AND INFORMATION AND COMMUNICATION TECHNOLOGIES

An Introduction

Klaus-Dieter GRAF
Freie Universitaet Berlin

The discussion document for the ICMI Comparative Study refers to the role of ICT in Mathematics Education as an important issue under several aspects.

The Preamble notes the influences on the teacher-students relationship and asks that new roles for both the teachers and learners must be defined. Section IV.8, where ICT is embedded in the topic of methodology and media, notes that strong changes in teachers' and learners' activities in class are reported from many countries. Control of teachers on activities of students is diminishing. Students rely on outside sources or on information from their peers.

Some of these observations originate from the increasing role of the didactic component 'media' in class in general. Compared to classical media like blackboard and chalk or the OHP, modern information technology based media, like multimedia computers, have a stronger touch of educational intelligence. They can furnish students with information as well as with some advice for learning and about the correctness of findings. In addition, communication technology like networks can open up the classroom for geographically and – this is more important – culturally distant information, giving rise to activities like 'distance learning' in a virtual classroom.

Different reactions can be expected in different traditions.

Another general aspect in many societies, related to McLuhan's assumption "the medium is the message", plays its role in the context of ICT and math education as well: these societies have seen a 'metacognitive shift' going on at all levels of information and communication processes. There is less interest in the content of these processes and growing interest in the medium carrying the content.

Which different ways – if any – can uphold the intentions of mathematics education as far as content and methods are concerned?

The statements made so far refer mostly to developments that are not restricted to mathematics education alone. They give an idea, however, that ICT will and must have really fundamental influences on the structure and scope of this education, through changes in roles, artificially intelligent media and virtual classrooms.

Other potentialities from ICT more directly and concretely support traditional as well as new methods in mathematics education, which will allow deeper understanding of mathematical content and applications. Examples are visualisation of functions e.g. through powerful computer graphics, dynamic tools in synthetic geometry, numerical experiments. New mathematical content arises from ICT and, of course, teaching and learning tools like CBT – computer based training. These potentialities mean an extreme challenge to the content, methodology and intentions of mathematics education. They also bring a new quality in accessible applications and in problem solving.

Surprisingly to the members of the International Programme Committee there were not too many papers submitted referring to the aspect of technology or media in mathematics education. This is really surprising, since on one side there is much research going on about existing tools from ICT, the actual use of these tools, findings on their effectiveness and pedagogical and educational issues from the use of ICT. On the other hand, possibly not much attention has been given so far to the question of different reactions to the tools in different educational traditions. This may be due to the fact that – in the beginning at least – they were developed in the Western world and introduced without adaptation in East Asian schools.

Several papers were submitted from the field of intercultural distance learning in mathematics with classrooms in Australia, China, Germany and Japan engaged in pairs. This is a rather special field of ICT application in mathematics education, but a rather interesting and informative one as well. The reason is that distance learning covers and touches many of the aspects of the role of ICT in mathematics education mentioned above. Immediate influences on content and methods become visible. At the same time distance learning experiments promote rich experiences with intentions of general education within mathematics education, including examples of its openness to a variety of applications and methods of problem solving from different cultural traditions. Last but not least the papers turned out to be a rich source of information on different content and methods on the same topic as well as different reactions of students and teachers to the proceedings.

So the IPC decided to integrate elaborations of three papers in the Study Volume that had been presented at the Study Conference in a Symposium on Mathematics Education and ICT or in the Teaching and Learning Study Group.

a) M. Isoda, B. McCrae and K. Stacey:
Cultural Awareness Arising from Internet Communication between Japanese and Australian Classrooms.

b) Xu Fei and Gao Xiaofeng:
International Distance Learning Activities of the High School Attached to The People's University of China.

c) K.-D. Graf and S. Moriya:
Distance learning between Japanese and German Classrooms.

Most of the findings compiled in these papers relate clearly to the categories identified above. The students evaluated in project a) were meeting in a virtual "discussion room" which basically is a chat room in the web for collaborative mathematical problem solving, the chat being performed through writing in English language. Results, including comparative observation, were deduced from careful evaluation of the protocols of the discussions as well as from pre- and post-questionnaires.

So, for example, the Australian and Japanese students in project a) played teaching roles when carefully explaining their views and arguments on a mathematical problem in a more hermeneutic than deductive way. They applied a letter style comparable to scholarly exchanges 300 years ago instead of immediate interaction. They discovered English language as the adequate means to communicate about mathematical problems in their virtual classrooms. This may have had more relevance for the Japanese students. But even the Australian Western students were impressed that they could get access to their peers' different views and ways of thinking.

The students found support in their deeper understanding of what is mathematics when they encountered that you can approach the same mathematical problem in different ways, using different algebraic representations for example, or using pictures. Again, this was more striking for Japanese students, since their society is more mono-cultural than the Australian one.

Japanese and German students from project c) were meeting in a 'virtual classroom', composed of two real classrooms, which were tied up electronically by internet and with videoconference equipment. So the students could see each other and hear each other in real time, as in a real classroom. Teaching on mathematical topics took place in different modes, with one teacher teaching to both classes, students presenting results of the problems they solved, students teaching in small groups, demonstrating what they had achieved in problem solving and application of fundamental mathematical knowledge, or student-teacher interaction and student-student interaction.

Results were found from observers of the video-conference lessons, life and videotaping and from some questionnaires. Primary students experienced a new role in the teaching-learning process when they were confronted with the task to assist their teachers in forwarding what they had learnt and achieved to their foreign classmates. Secondary students found themselves in a real students-teach-students situation, prepared by internet exploration and teacher advice. All students profited from using the technical intelligence provided by websites, email, camera projectors, video cameras and large screens. Their virtual classroom, extended over a large geographical and cultural distance, allowed them to find a variety of motifs for their patterns and of their applications on fans, tiles, place mats or t-shirts. This environment, seen as a powerful new medium for the messages of mathematics education, created a feeling of really intimate closeness with their partners, which did strongly reinforce the message. More support to abilities like creativity, scholarship or methodology in mathematics came from impulses to find patterns in their environments, from learning about the importance of group tables or from learning about different tools used in geometry. Differences showed up in the behaviour of students, rather cool and oriented towards scholarly achievements in Germany, more enthusiastic about communication in Japan. Attention, zeal and endurance, on the other hand, were equally evident.

Paper b) refers to a set of projects carried out with schools in China, Canada and Japan as well as with US American schools in Canton. The setting in the project with students from Beijing and Osaka was equivalent to project c), with a ‘virtual classroom’ allowing real time visual and audio communication, supported by camera projectors and video recorders. Similarly to the other projects, interactivity between teaching and learning is one of the most remarkable examples of findings reported. Also, the experiment was considered to be a model for life-long-learning.

English as the working language in writing and speaking turned out to be an efficient tool for communication of non-native English speakers. Special support in mathematics education arose for the Chinese students in particular from the topic of the project: “Painting and Geometry”. They had to study modern geometry beyond Euclidean to be able to understand and handle perspective from a mathematical point of view. Using electronic and software tools (like powerpoint for example) provided them with additional support for their learning of mathematics. Finally they were really surprised to experience the interdependence between mathematics and art for the first time.

More related to general intentions of education it was noted that in all projects students expressed their appreciation to meet peers with different cultural background and to experience that there were common motives and ways for learning as well as different ones – which were well worth being understood. Students from both cultures were sensitive enough to note differences in the teaching style, be it from teachers or students: more exploring in Germany and more guiding in Japan. Many students admitted that they had false beliefs about their foreign partners and their environment, which could be discovered and corrected. The importance of such an attitude cannot be overestimated, thinking for example of the still existing tense relations between China and Japan because of hostile historical events

As mentioned before, findings like these were not restricted to students. In the course of the projects teachers as well as teacher educators and administrators learnt about their own conscious and unconscious educational traditions, theories and styles and about the existence of variations and alternatives.

The findings from the projects provide us with strong evidence that such activities fulfil a double purpose:

- they contribute to pursue this very important intention of general education in our so multifaceted networked world: preparing students for cooperative problem solving in intercultural and international teams, and
- they improve the quality of mathematics education through exploiting the motivating and supportive power of cultural distance learning.

Chapter 3-5a

CULTURAL AWARENESS ARISING FROM INTERNET COMMUNICATION BETWEEN JAPANESE AND AUSTRALIAN CLASSROOMS

Masami ISODA[1], Barry McCRAE[2] and Kaye STACEY[3]
[1]*University of Tsukuba;* [2]*Australian Council for Educational Research;*
[3]*University of Melbourne*

1. INTRODUCTION

How can we recognize cultural differences in mathematics education between East and West and how can we use them for mathematics education? One first becomes aware of one's own culture when a different culture is encountered (Lerman, 1994). This collaborative study illustrates that students from the East and the West became aware of their different cultures in mathematics and developed a new letter-style communication norm which considered their partners' perspectives.

The study established a collaborative problem-solving environment between East and West, Japanese and Australian students, on the internet using a Bulletin Board System (BBS) and illustrated what kinds of differences were recognized by the students. The study has significance related to developing a culture that can be a bridge between East and West differences. The first significant change in students' activities was that they became aware of the differences in their mathematics. Students encountered cultural differences about the mathematics that they knew and the ways of explaining mathematics, and then realized and developed alternative perspectives on mathematics and mathematics communication. The differences in mathematics caused them to develop a new view of mathematics. The second significant outcome through the activities was that a way of mathematical communication emerged which is similar to the letter style of communication among mathematicians. Students engaged in a hermeneutic effort (Jahnke, 2000), trying to understand the perspective of their communication partner and replying in an understandable representation. The letter style communication was both competitive and sympathetic, mindful of their partners' perspective and attitude (Isoda, 2002).

2. METHODOLOGY

The study created BBS sites named 'Discussion Rooms' as is shown in figure 1 for mathematics communication between classrooms in each country. In Japan, a secondary school was selected in which students have the ability to communicate in English and in Australia a school with a similar academic standing was selected. In this project, forty students, aged 15-16, participated from Japan and twenty-six students, aged 14-15, participated from Australia. Both boys and girls participated in Japan, but the Australian school was only for boys. The communication was organised between eight groups on each side. Because of differing class sizes, each Japanese group had five members and each Australian group had three or four members. Problem 1 (the left side of figure 1) for groups 1, 2 and 3, was about expressing numbers as the sum of consecutive whole numbers. Problem 2, for groups 4, 5 and 6, was about expressing numbers as the sum of powers, and Problem 3, for groups 7 and 8, was about a pantograph linkage. Each problem was divided into three or four parts (open questions) to structure the discussions. In both countries, the teachers asked groups not to proceed to the next part of a problem until they had successfully communicated a solution of the current part to the partner groups. Both teachers told students to cooperate with the other country's students and encouraged students to continue a social exchange whilst the mathematical discussion developed. The project took place over 3 weeks and included five exchanges of messages. At the beginning, students in each group introduced themselves to their partners. Thereafter, exchanges between partner groups continued with each group commenting on the other group's ideas and adding their own ideas. Because the communication between Japan and Australia was restricted on the BBS to a fixed number of opportunities for sending the messages, the students interpreted the messages very carefully.

To get quantitative data about changing students' beliefs, questionnaires were constructed based on the questionnaires of TIMSS. All students completed a pre-questionnaire prior to commencing the project and a post-questionnaire at the end of project. To get qualitative data, Japanese classroom activities were observed and video-taped by the Japanese researcher and Australian activities were observed and some were video-taped by an Australian researcher. Students' comments were gathered after the end of the project. Japanese and Australian researchers used a mailing list for sharing observations of the students' activity every day. For planning the project, analyzing the data and sharing the interpretation of the results, members of the team of researchers and teachers visited the other country before and after the experiment.

3. RESULTS

The results of the experiment are summarized as follows. Japanese students, the Far East students, were able to communicate mathematically with Australian students in English. Until this project, although they learned English, the participating Japanese students had never learned mathematics in English and never used English for communicating with people from other countries. At the beginning of the project, Japanese students needed time for translation from English to Japanese and from Japanese to English, for interpreting and sending messages very carefully. At the end of the project, they read English and directly wrote in English. The Australian, Western, students who use English in school were especially concerned with ways of communication. Through the communication, students from both countries encountered differences in the mathematics that they knew. For the collaborative problem solving, they tried to synchronize their communication and they developed a special style of communication that is similar to communication between mathematicians by letter. Through the project, students encountered differences between the mathematics that they learned and the ways of learning it in each country, compared their mathematical ideas, and considered good ways of explanation. The mathematics of the other students acted like a mirror which enabled students to see their own mathematics and made them consider better ways of mathematical communication between unknown people. They learned about the importance of a hermeneutic attitude for communication, to strive to understand the message from the viewpoint of the students in the other country. They built up rapport with the other students via their communication.

Next, we illustrate the student's growing cultural awareness arising from the differences in their responses, and then we describe the communication style that they developed.

4. OBSERVATIONS OF CULTURAL DIFFERENCES

Through the discussion, students encountered differences in mathematics between Japan and Australia, East and West. Figure 2, which is an extract of the mathematical part of the communication from the discussion of group 3, illustrates the differences in algebraic representation and explanation. At *(A)*, *(D)* and *(E)*, part of the social interaction is shown. At *(B)*, Australian students use only the representation x, y and z to stand for three consecutive numbers. They express other sequences in terms of x, y and z (e.g. (x+1, y+1, z+1)) but they do not express the relationship between x and y and z. At *(F)* and *(G)*, four days later, the Japanese students, who felt uncomfortable about the Australian representation, used the formal algebraic way to express x, y and z in terms of x and hence formally showed that the sum was a multiple of 3. Australian students made use of this Japanese representation for the first time at *(N)*, when they also corrected a mistake of the Japanese students, which revealed an interesting cultural difference. At *(J)*, the Japanese students had correctly obtained the expression 2(2x+3) for the sum of four consecutive numbers, beginning at x. They incorrectly concluded that "when x increases by 1, the answer increases by 2", probably by analogy with the earlier result that when x increases by 1, 3(x+1) increases by 3. The Australian students at *(N)* used the expression 4(x+1.5) instead of 2(2x+3) to remind Japanese students that 'the answer increases 4, not 2'. Although it caused no uneasiness for the Australian students, the use of the non-integer term '+1.5' did not fit comfortably with the Japanese custom for explaining integer problems algebraically. This is a cultural difference. The Japanese students accepted it at *(O)* because the Australian students thoughtfully used the idea of the Japanese students, 'As with your solution for part (a)' at *(N)*. At the omitted part in *(O)*, Japanese students introduced a pictorial representation of the sum of consecutive numbers such as shown at *(P)*. At *(Q)*, Australian students acknowledged its value but immediately brought up the negative case, which was not so easy to represent pictorially. In the Japanese reply to *(Q)* which is omitted from Figure 2, the Japanese students began to consider the negative case and used fractional representations such as (6*6/2+6/2)-(3*3/2+3/2) which they had not used before for integer problems. Finally students on both sides were well synchronized at this moment as they each were developing their ideas based on reviewing their partner group's work.

In this excerpt, both groups of students encountered differences in ways of using algebraic representation and differences in the way of explaining mathematics such as using pictures. They both experienced the difficulty and importance of grounding activity[1] (Baker, 1999) as they aimed to share their ideas. An Australian student in group 3 commented as follows: "I learned that it is easier to explain problems using very simple mathematics. [I recommend that you] explain things perfectly and do not say more than is necessary. Use Basic Mathematics. (Australia)" Although his group valued the Japanese picture representation at *(Q)*, he mentioned that they were uncomfortable with the Japanese representation such as at *(G)* from the viewpoint of sharing ideas and recommended that the Japanese students should use 'basic mathematics'. Because what is considered basic mathematics is different in the two countries, he was referring to what is basic for him. Through the project, he encountered cultural differences in mathematics and they functioned as a mirror on his own mathematics culture.

The explicit differences between Japan and Australia which surprised students on both sides were related to mediational means such as tools, representations, and the ways of explanation which they had learned before and used everyday. In problem 2, a spreadsheet was offered for student use. One Australian group sent their Japanese partner group a similar program for a graphing calculator, which the Japanese students had never had any experience of using. In problem 3, Japanese students made the physical model of the pantograph and sent the pictures to the Australian students. The Australian students did not have the physical model but, in return, sent the Japanese students a dynamic geometry Cabri file of the pantograph. The Japanese students were astonished to see it.

Through these kinds of experiences, it was expected that students from both countries would change their beliefs. Changes of their beliefs were evaluated by comparing the difference in responses to the pre- and post-questionnaires. Most of the comments after the project indicated strong positive experiences overall with the project and appreciation that the students had learned ways to communicate mathematics. However, many beliefs about mathematics measured by the TIMSS questionnaire did not significantly change. For example, there was no change evident on the question: 'How much do you like mathematics?' Only four questions (item nos. 3.1, 3.2, 3.3 and 10 in figure 3) had significantly different responses before and after the project, and this was only in the case of the Japanese students.

[1] Michael Baker et al (1999) defined grounding as the process for reaching common ground of mutual understanding, knowledge, beliefs, assumptions, presuppositions, and so on that were claimed to be necessary for many aspects of communication and collaboration. Isoda et al (2002) discussed the difficulties of sharing mathematical images by the communication using palmtop computers.

☐ Internet Project name :P P [1999/10/26,08:51:49]

(A) We have four members in our group. They are P P , E C , R t R and D d T . We are all between 14 and 15. We go to S in Melbourne, Australia. We all play *(OMISSION)*

For part (a):

(B) Let x be the first number, y be the second number and z be the third number in the consecutive sequence. The first sequence is 1,2,3 which adds to 6. The second sequence os 2,3,4 which adds to 9. The third possible values of x,y,z are 3,4,5 which add to 12. This is because the first sequence (1,2,3) can be represented by x,y,z. The second sequence will be (x+1),(y+1),(z+1) which is the same as x+y+z+3. The third sequence is (x+2),(y+2),(z+2), adding to x+y+z+3+3. The next equals x+y+z+3+3+3, and so on, adding 3 each time. So, starting at 6, they all go up in multiples of 3 when represented by consectutive numbers. The same method applies for adding with 5 consecutive numbers – it goes up in multiples of 5.

(C) Please reply to our suggestion for the answer of (a) and suggest spmething for part (b).

☐ [FIGURE] Group 3 -(58 KB)

(D)

[SUBJECT] **We are Japanese**
[NAME] R K [1999/10/26,14:06:06]

[MESSAGE] Hello. Our group has 5people. *(OMISSION)*

☐ Internet Project name :P P [1999/10/27,08:26:45]

P P is the Vice president of the *(OMISSION)*

(E) be a lawyer or a business man and R want to be rich and possibly a scientist or an engineer. Please reply to our suggestion for part (a) in our last letter, we look forward to hearing from you. :) :) :) :)

☐ Reacton and our idea to solve (a) ,(b) name :Y K [1999/10/30,22:20:51]

(F) #As you know, we are not so good at writing English. So please let us know if you don't understand.

(OMISSION)

We read your message. The answer is same as ours. But we solve it in a differnt way. I think this way is easier than yours. You used three letters. But to use only X is easier. I'll show you our way.

---- part (a) ------

(G) #In this problem, we have to think 3 consecutive numbers and 5 consecutive numbers separately.

< 3 consecutive numbers >

Let the first number be X. As three numbers are consecutive, the next number must be (X+1).
In the same way, the last number must be (X+2). So the sum of these 3 numbers is...
X+(X+1)+(X+2)=3X+3 =3(X+1)
X will be natural number. (It can be taken for only integer which includes negative numbers.)

(OMISSION)

(H) Gentlemen and Gentlemen! (You are only boys) I hope you will understand the meaning of this expression.
Actually, when X=2 3(X+1)=9, when *(OMISSION)* Anyway the answer is <u>multiple of 3 bigger than or equal to 6</u>

< 5 consecetive numbers >

(OMISSION)

(I) #Question from us (1)
About "A" consecetive numbers. When "A" is an odd number, you can express the sum as multiple of "A".
When "A" is an even number, you can't express the sum as multiple of "A". Can you tell us why?

---- part (b) -------

#We considered part (b) in the same way. I'll show you waiting for your pointing out our mistakes.

< 4 consecetive numbers >

(OMISSION)

(J) The expression 2(2X+3) means that when X increases 1, the answer increases 2.
The answer is <u>multiple of 2 bigger than or equal to 12</u>

< 6 consecetive numbers >

(OMISSION)

#Question from us (2)

(K) We considered this problem over an basic condition. It is that the "numbers" means natural numbers.
But as I discribed before, "numbers" can be taken for integer which includes negative numbers.
If "numbers" means integer, how does the answer change?

---- Message -----

(L) Are you happy? Be happy! *(OMISSION)*

Figure 3-5a-1. Excerpt from the Discussion of Group 3 on Problem 1

(To be continued to the next page)

(Continued from the previous page)

☐ Internet Project name :P P [1999/11/03,07:30:13]

(M) If negative numbers were included then the answer would be the same, but include all the answers as a negative as well as the positive.

part (b)

(N) The lowest number is 10 , this is because the numbers can be represented as (x, x+1, x+2, and x+3.) This is for the addition of 4 consectutive numbers.This works out as 4(x+1.5) As with your solution for part (a), It goes up in multiples of the amount of adding consecutive numbers, in this case, 4. This meanns the values are 10. 14. 18 22. 26, etc...

For 6 numbers..

(OMISSION)

☐ Let's think about part (c)! name :Y K [1999/11/05,21.05:34]

(O) We read your letter. Your answer of <u>#Question from us (2)</u> was perfect!

If negative numbers are included, there is no minimum value. I think we discussed enough about (a) and (b).

But, have you discussed on <u>#Question from us (1)</u> in your group?

I will tell you the answer of it in the next letter. Please think about it again before the next letter comes.

Anyway, we want to go to part(c).

(OMISSION)

☐ Internet Project name :P P [1999/11/08,

Discussion for part (c)

(Q) We like your idea, but we have another idea.

For this problem we will just focus on positive numbers, as you can get any numbers using negatives, eg (-3)+(-2)+(-1)+(0)+(1)+(2)+(3)+(4)=8

(OMISSION)

(P)

This chart means 1+2+3+4+5+6

0
0 0
0 0 0
0 0 0 0
0 0 0 0 0
0 0 0 0 0 0

It is similar to a right angled isosceles triangle.

<u>Instead of counting all the points, calculate its area.</u>

Figure 3-5a-2. Students' Evaluations of the Project (showing significant changes)

No.	Question [1]		JAPAN Mean	JAPAN Gain	AUSTRALIA Mean	AUSTRALIA Gain
3.1	We work on mathematics projects.	Pre	2.13		2.58	
		Post	2.51	0.39*	2.54	-0.04
3.2	We use computers.	Pre	1.85		1.88	
		Post	2.23	0.38 **	1.92	0.04
3.3	We work together in pairs or small groups.	Pre	1.98		2.75	
		Post	2.25	0.28 *	2.79	0.04
10	Mathematics is different in different countries.	Pre	2.48		3.17	
		Post	3.15	0.68**	3.04	-0.12
14	I enjoyed communicating with the students in another country.	Post	4.05		4.38	
15	I enjoyed working on the project.	Post	3.98		4.17	
16	During the project my ability to explain mathematical ideas improved.	Post	2.70		3.04	16

1) Used Likert items; Strong Agree (5), Agree (4), (Mid 3), Disagree (2), Strong Disagree (1)

* Significant Difference by t-value, one sided at 10% level, ** 2.5% level of significance.

Item no.10 for the Japanese students changed positively with the largest t-value. At the beginning of the research, we expected that the project would enhance students' belief that 'mathematics is common throughout the world'. However, the trend was to produce the opposite effect for the Japanese students, though some of them suggested that mathematics is common. For example, one Japanese student commented: "I learned that we could communicate in the mathematical world beyond language and country difference" Even though both groups of students had to deal with differences like those shown in the excerpt in figure 2, the project produced a positive change of the Japanese mean value only. There are two discussion points: the first point is why is there a significant positive change in the Japanese mean value, and the second point is why only by the Japanese students?

Apparently, Japanese students were affected more than Australian students. Indeed, 55% of Japanese students changed their response category (positive or negative). However, it is important to note that 35% of Australian students also changed their response category. Even if Australians' selections (agree/disagree etc) did not change, their comments did sometimes change – such as from "Mathematics should be common but teaching standards may be different" to "People use different ways to solve problems and base the learning around different mathematics theorems" (from a student who selected Agree both before and after). This student's description changed revealing that he experienced differences. Thus, we could say that students from both sides experienced differences, but that Japanese students had held a more mono-cultural view of mathematics until the project.

We could conclude that Japanese students recognized cultural differences more than Australian students in this project. From our observations and the students' comments, we identified four possible reasons for this difference. Firstly, Japan is a mono-cultural country but Australia is multi-cultural. This was reflected in some of the Australian students noting in the pre-questionnaire that, depending on the country, the teaching methods for mathematics are different. In many Australian schools, there are students who learned mathematics in another country before coming to Australia. Secondly, Japanese students had to communicate in a second language. Explaining mathematics in English is itself a cross-cultural activity. The following comment was typical of the Japanese students: "Explaining my idea in English required looking beyond mathematics and was hard to do but I could communicate mathematical ideas with our partners in English." Thirdly, Japanese students learn mathematics in a more isolated way than do Australian students. Indeed, the responses to item nos. 3.1 and 3.2 in figure 2 show that the collaborative methods of the Internet project were new ways of learning for the Japanese students. One Japanese student described it as follows: "I learned about the fun of collaborative problem solving in groups and via the Internet. I recognized that it is another mathematical interest which is different from solving problems alone." We did not see comments of this kind from the Australian students. Fourthly, although students from both countries recognized differences of mediational means, some methods used by the Australian students, such as the graphic calculator programs, were extremely far from the Japanese students' mathematical experiences before the project. The change in responses to 3.2 in figure 2 reflects this.

From these results, we conclude that Japanese students recognized two points of difference. The first point related to mediational means of using tools, representations and ways of explanation. Indeed, Japanese students rarely have a chance to use computers for mathematics, whereas they are commonly used at the Australian school. However, they are skilled in formal algebraic representation. The second point is the way of learning mathematics; Japanese students realized that their everyday learning style is solitary. The questionnaire results (3.2 and 3.3) support this. By contrast, Australian students recognized the difference in the way of explanation but their everyday learning style is not so far from the project, so there were fewer differences in the questionnaire results.

5. DEVELOPING A COMMUNICATION STYLE

Comments from students from both countries illustrate that they learned how to communicate across cultural and mathematical differences. For example, Australian students commented as follows:

> "I have learned how to communicate the mathematics problem over the internet";
>
> "I recommend trying to explain what you think of the answer and why you think that is the answer, instead of just writing the answer";
>
> "I learned that I had to explain my answers more clearly than normal. I also learned how to work with people who aren't normally my friends."

Japanese students also commented on this and additionally commented on their successful use of English. Indeed, at each stage of communication, students engaged in their hermeneutic effort, carefully interpreting the message, trying to uncover the idea communicated. They were trying to synchronize their mathematics with the message whilst also developing their mathematics beyond the message.

Students developed a letter style of communication that has conventionally been done from the age of the ancients (e.g. Archimedes

'Method'; the letters between Pascal and Fermat). It includes both a sympathetic and a competitive attitude that are not seen in a formal deductive mathematics textbook. The sympathetic attitude aimed at sharing ideas with empathy and continuing communication is evident at *(C)*, *(E)*, *(F)*, *(H)*, *(L)* and *(O)* of figure 2. The competitive attitude, which aims to go further than the other side or to develop stronger mathematics, is evident at *(H)*, *(I)*, *(K)*, *(M)*, *(N)*, *(O)* and *(Q)* of figure 2. From the pre-questionnaire, we knew that all students had experience of mathematics projects and writing reports. But their usual reports did not involve this kind of collaborative communication. The letter style emerged via the collaboration on the internet between countries with different mathematics cultures.

In figure 2, the synchronization of the letter style communication can be clearly seen to be developing from *(N)* when Australian students began to use the Japanese students' notation and pointed out the Japanese students' mistake at *(J)*. Indeed, earlier at *(F)*, the Japanese students had acknowledged the Australian reply but just began to describe their own answer: 'We read your message. The answer is same as ours. But we solve it in a different way.' On the other hand, Australian students began to reply to the substance of the Japanese message at *(N)*:'This works out as 4(x+1.5) As with your solution for part (a), it goes up in multiples of the amount of adding consecutive numbers, in this case, 4.' This method of persuading others to reach the truth based on their own ideas such as 'as with your solution for part (a)' has been a very important technique of mathematical and dialectic communication from the age of the Eleatic School. When the Japanese students replied to *(Q)* (omitted in figure 2), this time they used the idea of the Australian students. This way of using the partner's idea clearly demonstrates the students' hermeneutic attitude which considers the partner's perspective at the same time as explaining their new ideas based on their partner's idea. This is an integration of sympathetic and competitive attitudes.

Overall the total experience was very significant for the students. An Australian student in group 3 commented as follows: "I thoroughly enjoyed the project and found it interesting and fun." A Japanese student in group 3 described it as follows: "I never met them and I never heard their voice. But they exist far beyond the Pacific, and they explore the same problem. Only imaging this situation, this project is surprising." Historically, mathematicians used to communicate by letter even though they never met. Internet communication enabled this new situation and students engaged in dramatic and enjoyable communication based on a letter style with mutual hermeneutic effort of interpretation.

6. FINAL REMARK

Our experiments produced a cultural impact on students in both countries. The communication and its impact awakened students' cultural perspectives in mathematics and developed their hermeneutic attitude for collaboration and letter style explanation involving both sympathetic and competitive attitudes. Many stories about communication in mathematics, such as the dialectic between proof and refutation by Lakatos (1976), focus on the competitive process. On the other hand, the study illustrated the importance of a sympathetic attitude as well as a competitive attitude. Mathematicians usually communicate with each other by letter even they do not know each other. They have to write and read the letter imaging the mind of the reader or writer. Because communication between countries lacks much common ground, people have to make an effort to continue communication. We expected that communication in mathematics would be very far from everyday communication because we use shared formal language in mathematics. But the study illustrated that effort for grounding is necessary even in mathematics. Consequently, the study illustrated that Information and Communication Technology can help to develop cultural understanding in the mathematics community between East and West countries.

ACKNOWLEDGEMENT

The project was carried out with the help of the teachers and graduate students. The researchers want to say special thanks for the contributions of Brian Woolacott, Kyla Marston, Nobuaki Kawasaki and Yoshinori Kobayashi and to the students from both schools.

REFERENCES

Baker, M., Hansen, T., Joiner, R., Traum, D., 1999, The Role of Grounding in Collaborative Learning Tasks, in: *Collaborative Learning*, Dillenbourg, E. ed. Pergamon, Oxford, pp.31-63.

Lakatos, I., 1976, *Proofs and Refutations*, Cambridge University Press, Cambridge.

Lerman, S., 1994, Changing Focus in the Mathematics Classroom, in: *Cultural Perspectives on the Mathematics Classroom*, Lerman, S. ed., Kluwer, Dordrecht, pp.191-213.

Isoda, M., 2000, Awaken of Cultural Perspectives in Mathematics through Cultural Differences, *Tsukuba Journal of Educational Study in Mathematics*, **20**:39-48.

Isoda, M., Lee, Y., Aoyama, K., and Ishizuka, M., 2002. The Study of Mathematics Communication on Internet with Palmtop Computer. Ignatios Vakalis et al. *Proceedings of* *the 2nd International Conference on the Teaching of Mathematics (at the undergraduate level)*, Wiley Publishers, http://www.math.uoc.gr/~ictm2/Proceedings/pap308.pdf.

Jahnke, H. et. al., 2000, The Use of Original Sources in the Mathematics Classroom, in: *History in Mathematics Education*, Fauvel, J., and Maanen, J., eds. Kluwer, Dordrecht, pp.291-328.

Chapter 3-5b

THE INTERNATIONAL DISTANCE LEARNING ACTIVITIES OF *HSARUC*

XU Fei
The High School Affiliated to RENMIN University of China

Information education is one of the important trends of thought concerning today's education reform as well as the requirement of the age. The revolution of science and technology not only calls for an adaptation and innovation of what to learn, but also directly affects the means and way of how to teach and learn. Modern education technology is exerting a more and more profound influence on the way of teaching. Such up-to-date means as computer-aided instruction, distance learning and the learning and research on-line are booming.

Distance learning is a way of education aiming at the development of high-quality education by drawing on modern information technology. Compared with other forms of non-classroom education, the remarkable interactivity of distance learning between teaching and learning brings the education quality to the equivalent of classroom teaching.

Modern distance learning fortifies the traditional teaching with flexibility in time, openness of education, and optimization in content and resource pool. Unbinding the restriction of time, space and capacity, it makes teaching and learning more equalized, humanized and individualized. On the other hand, it lays a foundation for multiform and multilevel education and prepares updated resorts for the structure and perfection of life-long learning systems as well.

The High School Affiliated to RENMIN University of China[1] remains the leader of Chinese high schools in the field of distance learning.

In April of 1998, HSARUC conducted a pilot project of multiple media online distance learning with Reed High School of Canada. Both Chinese and Canadian students enjoyed a lively atmosphere communicating in

English via net meetings. As the first in China, this activity claimed the attention of the educational organizations and media of both countries.

HSARUC conducted another multimedia distance course of Conversational English with Canton College of Technology of SUNY Canton and Canton Central Schools of the United States in April of 2000. Via the visual conference system and ISDN, the students of both countries had the lessons by an American instructor. Discussions and question-and-answer could be performed between the two sides with the help of camera lens and all-directional microphone. After class, the students from both sides would be regularly involved in on-line assignments, including virtual discussion, paper writing, quiz and e-mail communications. So far, HSARUC has experienced three sessions of distance learning.

December of 2000 witnessed a realization of the Schools Connection Project sponsored by the Ministry of Education of China, which was firstly undertaken by HSARUC. As the central station, HSARUC launched a one-to-ten interactive distance course with ten schools of Beijing's remote counties on the basis of television, sound and IP data radio technology.

In 2001 a modern interactive distance-learning network with the digital satellite broad band of 54m was successfully installed in HSARUC. Based on positive conditions HSARUC was the first to undertake the Schools Connection Project of the Ministry of Education—a Construction for the Channel of the Secondary Education on the Basis of Multimedia Broadband Transmission.

The information reception system of sky net occurred in a distance teaching room where the instructor could present his lecture by the aid of multimedia and electronic white board. The communication went as smoothly as in classroom teaching but comparatively this was much more flexible, timesaving, and visual.

The advantage of distance learning will show itself far more evidently as the distance between the instructor and the students becomes greater. It can cope with individual interactive learning with a particular instructional design as well as large-scale teaching to form a resource pool. In so doing, students scattered in different areas of Beijing will be able to gain high-quality education as their counterparts will in the centre of the city.

In October of 2001, a distance course of mathematics between 20 students from HSARUC and 15 students from the Tennoji High School Affiliated to Osaka Education University started.

This distance course is a big breakthrough in the history of our distance learning as well as an unprecedented co-operation. The breakthrough is embodied by the change of instructor role and the content of the course. Specifically, instead of teachers, students of both sides are playing the role of instructors through the whole activity. On the other hand, rather than

English as the content, mathematics and painting is the topic discussed by the students.

This activity consisted of three teaching units:

1. Introduction of the usage of projection in Japanese ancient paintings from the perspective of analytic geometry by students from the Tennoji High School Affiliated to Osaka Education University
2. Introduction of the usage of perspective in Chinese ancient paintings from the perspective of Euclidean geometry by students from the High School Affiliated to RENMIN University
3. Communication between students from both China and Japan

English was used as the working language through the whole activity. Besides the instructional sessions, teachers and educators from both countries presented an evaluation on the student's performance and creativity and conducted a hot discussion about how to foster the students' creativity.

The theme of the distance teaching activity, "Painting and Geometry", is obviously a great challenge to Chinese students, for they had to confront their ignorance of scientific knowledge and shortage of Chinese references. When they were making efforts to search on the Internet for the relevant information they found that almost no one was even involved in this field. Although there were some limited Japanese materials, the language barrier became a huge obstacle to hinder them from going ahead. As is known to all, what they have learnt of geometry is Euclidean geometry, but geometry involved in this topic belongs to the field of Non-Euclidean geometry. What is more, lack of basic painting skills is also a serious problem. To solve all these difficulties, they divided themselves into several teams such as math team, computer team, painting team and information gathering team, with different teams for different tasks. For example, the math team studied the book *Modern Geometry*. The computer team and the painting team collected information about paintings and art, downloading a number of famous paintings, some of them were of the Renaissance while others were of the Chinese Ming and Qing dynasty. Some of the students even studied the introductory course of Art History. Each team drafted its own research program and completed its own presentation materials. Each team seemed individual, however, they communicated a lot with each other and helped each other.

All the preparatory work involved, from making PowerPoint, the program of the three-dimensional flash, to the accumulation of the materials, were designed and completed by the students. The entire activity was filled with an animated atmosphere. All the students showed such great passion that every one of the three teaching units exceeded the time limit.

With the development of science and technology, students' ability and structure of knowledge have to some extent over-run their teachers. For this reason we educators are confronting a challenge, that is, how should we fully tap students' potentialities in secondary education.

Through the distance teaching activity, the students, playing the main parts, were the protagonists. They acted as the chief roles while teachers just provided the backstage support for them. During the interaction with the Japanese students, all of them were not only students, but also teachers. The whole preparations for the presentations, such as searching information, making teaching-aids, composing presentation plan, were totally done by the students themselves. Through this process they learnt methods of research and experienced the segments of teaching. Moreover, they tasted the hardships that their teachers face every day. More importantly, they experienced the inner relationship between painting and geometry by making their own paintings. This distance teaching activity was a great success. The students enjoyed high ratings by Japanese friends. They attributed these achievements to TEAMWORK, which is the quintessence of Chinese collectivism.

The successful performance of this Sino-Japanese distance learning activity allowed us confidence to tap educational potential. We obtained illumination in the following aspects.

1. This distance learning activity has not only granted the students knowledge but also helped them understand each other. It is a significant practice for making a world peace as well as for making the international academic exchange.
2. A different idea of education is embodied through this activity. The traditional role of teacher and student has greatly changed. Once as a controller, the teacher now acts as a supervisor. In the meantime, the students, taking the main role of the entire activity, have fully brought their potentialities and aptitude into play.
3. Through this activity, the content of mathematics has been enriched. Moreover its vast function has been applied. Besides the lectures, the students made some analyses about some noted art works like the Monument of the People's Heroes and Tian An Men Gate Tower with math principles and synthesized them via high technology.
4. This innovative education attaches importance to student-development-oriented education. During this activity, the students made an exploration with an open mind, enjoyed the fruit of the creativity in completion of the open tasks and displayed their own achievements in an open class.

One of the participants of this activity, Ren Yuan, spoke out his thought in the follow-up questionnaire:

With the rapid progress of information technology and the apparent tendency of globalization, the world has entered its brand new era, which we call "knowledge-sharing" era. This will not only be a good opportunity for us teenagers, but a great challenge. It is the Chinese students' choice to accept this challenge, and to face it with an extraordinarily active attitude.

Before this activity we even never thought of relating painting with math, for they used to appear as two totally different subjects. We were studying math and art isolated assuming that math is math while art is art.

"Painting and Geometry", an aspect involving mathematics and art, is a newly developed category which perfectly integrates the logical principles of geometry and the aesthetic value of art. We improved our abstract thought of math, our imaginative thought of art, as well as creativity and the aesthetic appreciation. The training and exploration enabled us to fuse mathematical thought, painting admiration and artistic creation together as an "alloy". And this will definitely be significant for us.

NOTE

[1]**A brief introduction to the High School affiliated to RENMIN University of China**

HSARUC, founded in 1950, is among the first model high schools approved by Beijing Municipal Government. With Madame Liu Pengzhi as the head, the school is striving to become superior nation-wide, first – class worldwide.

HSARUC set up a fiber Ethernet of 100m with ten servers in 1998, via which all the offices, classrooms and laboratories of the school can access to the Internet. In order to promote communications between various depart-ments, a huge information pool was built involving banks of teaching aids, subject resources, audio-visual resources, programs, CDs and books. The information capacity of the school amounts to 700GB. In March of 2001, a modern interactive distance-learning network with the digital satellite broad band of 54m was successfully installed in HSARUC. Based on these positive conditions HSARUC was the first to undertake the Schools Connection Project sponsored by the Ministry of Education of China – a Construction of the Channel for Secondary Education on the Basis of Multimedia Broadband Transmission.

Chapter 3-5c

DISTANCE LEARNING BETWEEN JAPANESE AND GERMAN CLASSROOMS

Klaus-D. GRAF[1] and Seiji MORIYA[2]
[1]Freie Universitaet Berlin; [2]Kyoto University of Education

1. BACKGROUND FOR THE PROJECT

There were different observations about new objectives of education at the beginning of the work on our joint projects of distance learning between remote classrooms: the research group of Kiyoshi Yokochi in Japan was concerned with the growing demand for an improvement of the quality of mathematics education (Yokochi 1995), whereas Klaus-D. Graf was more concerned with reflecting general intentions of education like inter-disciplinary and intercultural context, which should be integrated in all subjects' education, and had sketched models for this purpose (Graf 1995, referring to Ruiz 1993). Yokochi's and Seiji Moriya's activities with classes from different Japanese districts had shown that this method of remote distance learning provided considerable profit for the intentions mentioned.

When exchanging these observations, we agreed that in Japan and in Germany, as well as in other countries, there are challenges to societies by universal problems such as the world economy or environment, which require immense international interaction. Any success is dependent on mutual understanding and acknowledgement of different traditions and attitudes related to such problems. Readiness and good will have to be developed in young people growing into these societies.

So we decided to start a series of international and intercultural distance learning experiments in mathematics education with these aims:

(1) Initiating mathematical originality and creativity of pupils and students
(2) Learning comprehensive uses of mathematics with other subjects through solving real problems
(3) Improving the mathematical scholarship of students
(4) Appreciating and acknowledging the mathematical cultural characteristics of each district or country in problem solving
(5) Cultural exchanges related to science and technology between students of two classes
(6) Interaction with peers from different districts or countries in mathematical problem solving, complicated by constraints like different languages, for example.

2. BASIC STRUCTURE OF THE LEARNING AND TEACHING EXPERIMENTS

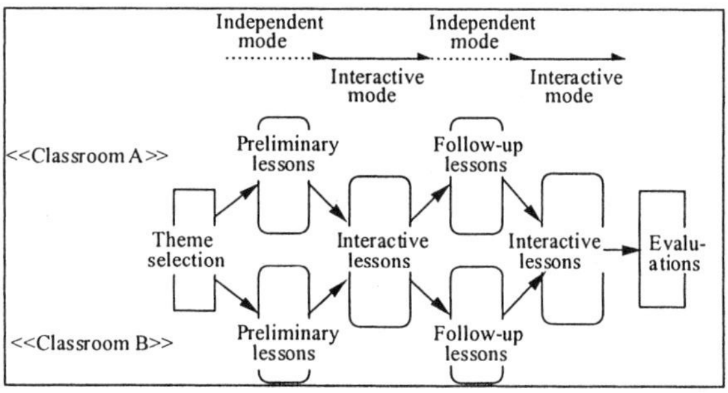

Each experiment consisted of different activities of two partner school classes and their teachers, together with researchers in didactics at universities, extending over six to twelve weeks. The activities started with preparation of a project by researchers and teachers, contacts running via mutual visits in Germany and Japan or via e-mail, Internet homepages or airmail. After this, teachers in the two countries started working with their classes concurrently on the same topic, mostly taken from mathematics, science, social science (environmental problems) or arts.

Besides learning about the topic in different ways the students, in more or less cooperation with their teachers, started to prepare demonstrations for their peers about the problems given to them, their findings and results. These demonstrations, followed by interaction between the students, were executed in several video conferences integrated into the total activity of 6 – 12 weeks. Material was also forwarded by e-mail or fax or put in the respective homepages on the Internet.

In the primary schools, some videoconferences consisted of teaching by the foreign teachers about the mathematical background of some products generated (for example, tiles with patterns) to the other class, allowing questions and answers.

In secondary schools, the teaching was done by the students themselves, followed by demonstrations and discussions.

In addition to concentrating on special subject matter, which was essential in our design of the learning units, there were passages in the activities, per e-mail or in the video conferences, where the students tried to find out more personal information about each other, concerning pets, grading systems, views on the other countries or people, including the students themselves.

At the end of each conference there was a detailed discussion of teachers and teacher educators, as well as administrators, about their impressions and observations.

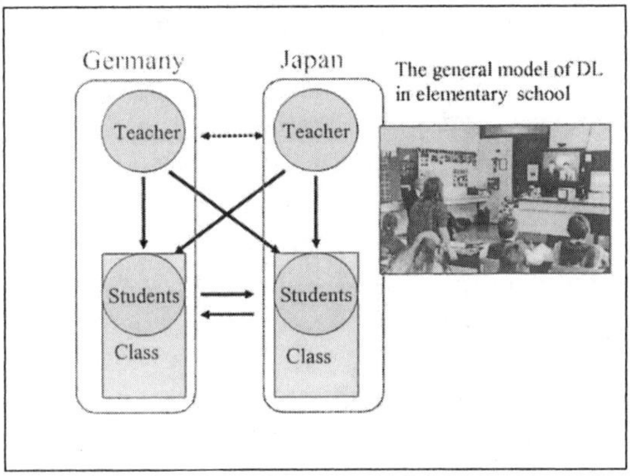

The general model

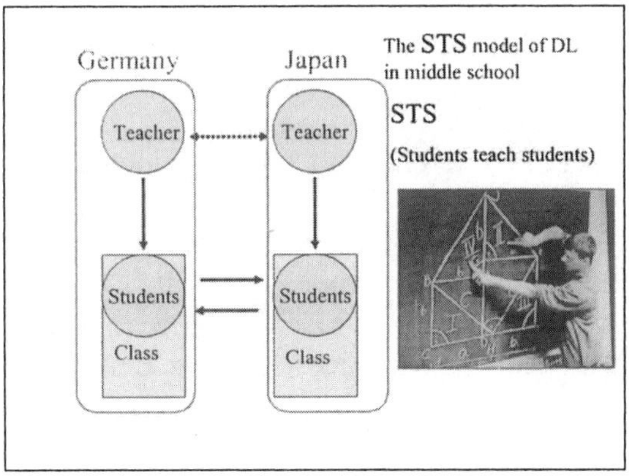

The STS model

3. AGE LEVELS AND CONTENT OF EXPERIMENT

The following partnerships and learning units between German and Japanese classes or Japanese and Chinese classes have been realised so far:

a) Peter-Witte-Primary-School in Berlin and a Primary School attached to Yamanashi University in Kofu. The students came from grades 5 and 6; there were 20 in Berlin and 40 in Kofu. Their subject matter was stripe patterns and rectangular patterns, seen under aspects of mathematics and art. There were four video conferences of one hour each included, two dominated by students' presentations and discussions, two by teaching of the teachers to remote classes. Two German teachers performed team-teaching for some periods.

b) Hildegard-Wegscheider-Secondary-School in Berlin and Irihirose High School in Niigata. The students came from grade 10 in Berlin and grades 8 and 9 in Japan. Subject matter was the Theorem of Pythagoras. Groups of German students explained proofs at different levels: experimental work, visualisation by graphics and logical proof based on constructions at the blackboard. There was one video conference of about 75 minutes.

c) Hildegard-Wegscheider-Secondary-School in Berlin and a Secondary School attached to Yamagata University in Yamagata. The students came from grade 10 in Berlin and grade 9 in Japan. Subject matter was the sundial and its mathematical background from the geometry of the globe. As in b) groups of Japanese students taught the German students using models and logical explanations at the blackboard. They put forward excellent questions to the German students and they showed great excitement about good answers.

There was one video conference of about 75 minutes.

Three more experiments took place between students from 3 different upper secondary schools in Tokyo and 3 in Berlin, centred around environmental problems, which shall not be discussed here in detail. In addition, we executed a couple of video conferences between Berlin teacher students in computer science education and Kyoto teacher students in mathematics education, with interaction about fascinating applications from history and real environments in classes.

4. METHODS OF TEACHING, LEARNING AND INTERACTION

Different methods were applied and experienced. In the primary school experiments both sides followed a rather conservative mode of teachers on both sides controlling the performance of the classes and asking most of the questions. The pupils were given opportunities to present their work with prepared statements. Students and teachers spoke in their mother tongues; they were translated by interpreters in both classrooms. All students showed great patience when waiting for translations for the other side.

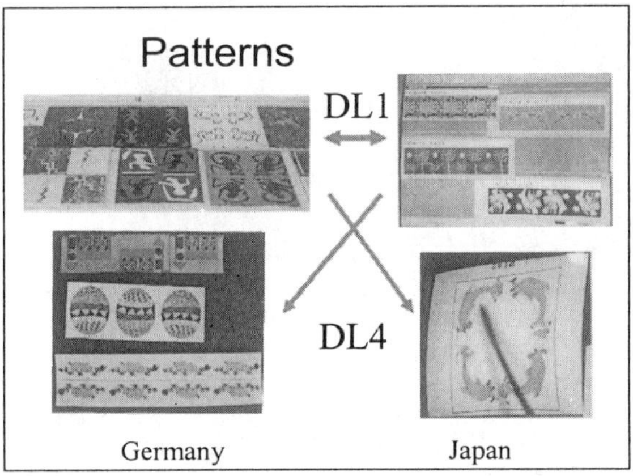

The change of motives (4)

Product table (3)

The Pythagoras experiment was characterised by independent teaching of German groups of students to the other side. They had explored the topic on the Internet and evaluated the results together with their teacher. The effect of learning was considerable on both sides, since there was a high motivation towards good teaching and understanding. This experiment was also characterised by both classes speaking English. This caused many misunderstandings and repetitions. Using figures and visualisations helped relieve this situation.

proving Pythagoras theorem (1)

explaining the principle of the sundial (1)

The sundial experiment was also conducted in English with the Japanese students doing most of the teaching in small groups. This was an outstanding event in a Japanese classroom. The Japanese students had prepared excellent material for the visualisation of shadows depending on the sun-earth constellations.

In the Berlin-Tokyo experiments there was a special situation in language, since all of the German students had taken Japanese language courses in their schools. So they had prepared their statements in Japanese, including a popular Japanese song, which was happily accepted by their peers in Japan. Japanese students gave their contributions in their own language and the Germans tried to understand. Interpreters in both classrooms translated, if necessary.

We would like to add a general remark here about the role of teachers in our special learning setting. We started our experiments with teachers organising most of the planning and the execution of work in classes. The students' roles were mostly to demonstrate what they had learned. Elder students then took over some of the teaching. This was still organised or "filtered" by the teachers. With growing experience, the students' demand grew to erase the filters totally, including the activities of preparing to teach and so to replace their own learning by preparing to teach. This of course, causes a gain in active learning, but at the same time a loss in efficient learning.

5. FINDINGS FOR GENERAL AND SUBJECT MATTER EDUCATION

It should be mentioned that our experiments did not only affect the learning and understanding of the students. At the same time the researchers from teacher education as well as the teachers active in preparation, execution, and evaluation of the distance learning experiments profited a lot. They had to reflect deeply on the educational theories they were relying on, so that they would be understood by the partners in the other country. Considerable feedback had to be given to make sure that the others' intentions had been understood. It became clear that understanding went far beyond the language problems. Still, using English to explain activities based on German or Japanese language was a problem in itself. The teachers had extra work to invest since they had to explain their project in their environment, for example to colleagues, school administrators and parents. They experienced that both sides were very ambitious to 'perform' good presentations and they had to control these tendencies for the sake of allowing genuine learning processes.

Comparing the teachers' styles, one point especially became apparent: There was a strong tendency with Japanese teachers to prepare each video conference very carefully, leaving no space for unexpected or unsolvable situations and thus guaranteeing an effective and efficient learning success. They were very interested to arrange with German teachers for a common "scenario" before the video conferences. German teachers, on the other hand, were ready for a more open proceeding, allowing mistakes by students for example, and thus a less efficient course of the conferences.

Japanese teachers experienced that for the experiments they needed knowledge from astronomy, history, art and language for this kind of mathematics education. In their teacher education at universities they had not learnt such subject matter, whereas German teachers have education in at least two subjects.

It is, of course, not possible to verify precisely to what extent the aims mentioned in section 1 could be fulfilled. Many of the observations we made relate to several aims, others bring in new aspects. Here are some of our findings.

1. Before the experiments in Peter-Witte-Elementary School, German students as well as teachers were not aware of the abundance of geometrical patterns around them. In the course of the video conferences real mathematical originality and creativity was displayed by all students when finding such examples in their environments for the different kinds of patterns they had to deal with, and even more when they applied patterns to decorate real objects like tiles or T-shirts. German students had no problems in finding these examples in their city environment, whereas Japanese students tended to look for traditional patterns in books or similar sources. Japanese secondary students did remarkable work when creating adequate models for sundials and for demonstrating sun-earth constellations, while their German peers were very capable in finding intuitive methods to exemplify Pythagoras' theorem before turning to a precise logical proof. After the experiments many students confirmed the following attitude as adequate: "If it is difficult to solve a problem in the usual way, then I think of another way".

2. Related to the Theorem of Pythagoras, Japanese teachers stated that their German colleagues think the Theorem's history and value as important as the pure formula, whereas they themselves turn to application very soon and let the students do many practical exercises. Comprehensive uses of mathematics were successfully realised in relation to arts, when classifying stripe or rectangular patterns. After a formal geometry lesson on symmetries in a video conference Japanese students expressed their feeling now that mathematics is a language which can be used all over the world. The application oriented kind of problems posed generated high motivation in all students.

3. The experiment offered many opportunities to further the students' mathematical scholarship, when tools like group tables were introduced or elements from spherical trigonometry were taught. Evidence for this became clear when they could easily find all possible patterns in squares or in equilateral triangles and even completed the corresponding group tables. The problem "draw a figure formed by the symmetry on a slanting (diagonal) line" was solved by 29% of Japanese students before the experiment and 71% after.

4. Appreciation and acknowledgement of cultural characteristics could be well observed. German students were really fascinated by the Japanese students' abilities in decorative art, which led them to mathematical questions; Japanese students were impressed, and some of them even intimidated by the German way of dealing with mathematical problems that were 'pure' or abstract, without a 'real', concrete problem. When looking for basic elements for their rectangular patterns German students in the beginning chose abstract motifs, which they had dealt with in art education (Paul Klee, for example). Japanese students selected rather traditional and real motifs like kites, fans, cherry trees, fireworks, etc for their stripe patterns. After demonstrating their works to each other, and after having learnt from the teachers how to generate these patterns, they changed the type of motifs. In the following demonstration Japanese students presented rectangular patterns with abstract motifs on T-shirts and German students showed up with stripe patterns on place mats, with motifs like a church or the Brandenburg Gate. Interesting discussions occurred about the reasons for selecting special combinations of colour, or when the students discovered that they were using different kinds of geometric tools.

5. Numerous and interesting observations could be made on cultural exchanges. Fr example, a comparison of grading systems was performed with deep interest on both sides. A rather surprising reaction happened when, in a conversation about pets, a Japanese girl really shocked the German students by showing her squirrel in a cage. This is absolutely unusual in Germany. In one experiment students were asking each other, which were their impressions and ideas about the other country. It turned out that there still exist many cut and dried opinions, which often seem incredible to the other side. One example: a Japanese student thinks Germany is a very clean country and those who litter will be put in jail.

6. Interaction with peers again allowed many fascinating and encouraging observations. There was a continuous zeal to report things to the foreign

peers as well as to listen and to understand them. This was accompanied by a remarkable patience, when they had to wait for an interpretation or when they had to ask again because of problems with English language. There was a clear atmosphere of competing, but in a friendly way. Secondary students in particular did not hesitate to express their appreciation for good explanations or demonstrations. Germans did this with a 'cool' attitude, while Japanese students expressed their acknowledgement more enthusiastically and cheered after good answers to questions they put. In all experiments students and teachers, as well as observers, expressed their feeling of an intimate closeness in the course of the video conferences. Due to large screens on both sides, the two classrooms seemed really close together.

7. To end this section a few more general observations shall underline the very positive atmosphere created through the experiments and appreciated by all participants. The students' great interest in each project was sustained for the whole period, despite the extra efforts they had to take. There were no drop outs at all. The students were planning with great zeal how to explain their results to their peers in the other country. They developed a very general interest in the other environment and country. They discussed their project with other students and teachers from their schools, and with their parents and families. Without problems in using the new technology they put direct questions to their peers in the other countries, forgetting about the distance.

6. CONCLUSIONS AND VISIONS

Neither the cost of the equipment and communication nor the extra amount of work related to the new media for distance learning should discourage us from integrating these media into the teaching and learning processes. This situation is comparable to the days when computers first entered schools and appeared uneconomical and unusable for education.

Teleconferencing systems, or what will be developed from them, will be a standard medium in schools within a few years. They will not only be used for communication with peers, which we consider a very fruitful application, but also for communication with any point on the globe and anybody on the globe who can contribute to the demands of students to get information in an interactive way. Teachers will have to become advisors and supervisors for this learning by exploring. They have to prevent students from wasting too much time by unstructured exploring.

Distance learning has originated from the problem of physical distance in time and/or space between a class and a teacher, or, more generally, between learners and knowledge bases including intelligence to interact – real or artificial. The more this intelligence and its means to interact (through multimedia) grow, the less the physical distance will matter. Instead, cultural distance as we used it in our experiments will become important for the learning process between learners and an intelligent source of knowledge. This distance can cause the learners to get more stimulated to learn from a distant source, be it a group of peers or a knowledge base, to learn more through competing, or just through imitation and variation.

Putting individuals into school classes for learning may have economical reasons. But – each student coming from a different environment (the culture of a family, for example) – it also means to initiate some kind of distance learning. This idea can be extended to the family of man.

REFERENCES

Downes, T. and Zammit, K., 2000, New Literacies for Connected Learning in Global Classrooms, in: *The Bookmark of the School of the Future*, Taylor, H. and Hogenbirk, P., eds., Kluwer Academic Publishers, Boston/Dordrecht/London.

Graf, K.-D., 1995, Promoting Interdisciplinary and Intercultural Intentions through the History of Informatics, in: *Integrating Information Technology into Education*, D. Watson and D. Tinsley, eds., Chapman& Hall, London, pp.139-150.

Graf, K.-D. and Yokochi, K., 1998, Educational Experiments of Distance Learning and Reorganisation of Mathematics Education, in *Information and Communication Technologies in School Mathematics*, D. Tinsley and D.C. Johnson, eds., Chapman & Hall, London, pp.265-274.

Koizumi, K., Dasai, T. and Moriya, S., 1997, Proposal of Interactive Distance Learning with CCV Educational System and Verification Experiments, in: *Proceedings of the Int. Conference for Computer Communications (ICCC'97)*, pp.107-114.

Koizumi, H., Dasai, T., Yokochi, K., Graf, K.-D., and Moriya, S., 1999, An Interactive Distance Learning System and Verification Experiments between Japan and Germany, in: *Communications and Networking in Education*, D. Watson and T. Downes, eds., Kluwer Academic Publishers, Boston/Dordrecht/London, pp.39-50.

Ruiz i Tarrago, F., 1993, Integration of Information Technology into Secondary Education: Main Issues and Perspectives, in: *Guidelines for Good Practice*, Van Weert, ed., IFIP Working Group 3.1, Geneva.

Wibe, J. et. al, 1999, Projects in Networked Learning: European Experiences, in: *Communications and Networking in Education*, Watson, D. and Downes, T., eds., Kluwer Academic Publishers, Boston/ Dordrecht/London, pp.101-126.

Section 4

VALUES AND BELIEFS
Introduction

Alan BISHOP
Monash University, Australia

Two of the significant areas of research in mathematics education in relation to this study concern beliefs and values. The reasons are many, but chiefly relate to three factors:

1. the awareness that problems of teaching and learning mathematics are much more deep-seated than commonly realised,
2. the recognition that teachers' attention to 'surface' explanations of students' errors and misconceptions are failing to address the real problems, and
3. affective and emotional aspects of learning mathematics have as significant an effect on learning as do cognitive aspects.

Beliefs and values are in focus in particular in this study because of the markedly different educational and cultural traditions of the 'East' and the 'West', and the opportunities this situation provides for research and development. As the Discussion Document says: "This study presupposes that the impact of cultural tradition is highly relevant to mathematics learning. Cultural traditions encompass a broad range of topics. It includes the perceived values of the individual and the society, as well as social structures such as the relationship between parents and children, or the relationship between teachers and students. There are clear differences in all these areas between Asian and Western traditions...The literature shows that there is a high correlation between the attitudes and beliefs of teachers on the one hand, and their instructional practices on the other, and it will be both important and interesting to see whether there are cultural differences in teacher attitudes and beliefs."

Indeed the chapters in this section amply demonstrate that important differences do exist in teachers' beliefs and values between the two main cultural traditions. Although the studies reported here are interesting in relation to these main traditions, such comparative studies and their analyses more importantly help us to understand more about the constructs of teachers' beliefs and values themselves. Indeed one of the main justifications for undertaking such comparative studies could be precisely to learn more about constructs such as these, and about their influences on mathematics learning. In that sense then these studies are also taking the opportunity presented by the comparative context as an important research site for studying these constructs themselves, and for raising important researchable questions. This is significant for both constructs but perhaps more so for the area of values, where the research base has not really been established and where the research itself is that much more recent than research on teachers' beliefs. So in this introduction to Section 4, we will be raising some of these questions ourselves.

The first contribution, Chapter 4-1 by Perry, Wong and Howard, is a report of their study comparing teachers' beliefs in Hong Kong and New South Wales (NSW), the most populous state of Australia. The precise context is important because Australian states have autonomy over education. The choice of research sites is always important in any study and in dealing with the 'East' and the 'West' this is undoubtedly so. For where or what is the 'East' and the 'West'? In general, one can ask: which countries/states/regions are sensible choices for comparing the influences of cultural traditions?

The authors point out that the Hong Kong community is deeply rooted in Chinese culture but, largely because of British colonisation, it has experienced many influences from the west. On the other hand, the dominant cultural influence in Australia results from the many years of white British/Anglo-Saxon control, modified by the recent migration from many countries including East Asia. So both research sites have experienced British influences but with other influences now strongly in play.

Perry et al. point to important educational differences in terms of class sizes, the selective system in Hong Kong compared with those in NSW, and the differences in mathematics performances in the TIMSS study. They base their comparison of teachers' beliefs, theoretically, on the distinction between the *transmission* view of education and the *child-centredness* view, and through questionnaires and interviews they relate the results of their comparisons to these two underlying views. In relation to doing this type of comparative research, the authors were also keen to point out that, in their view, the questionnaire used was equally relevant in the Australian and (in its translated form) in the Hong Kong context, a non-trivial issue for any researchers.

They found that teachers in both Hong Kong (HK) and New South Wales (NSW) reported beliefs across both the transmission and child-centredness factors. There were however, significant differences between the groups' loadings on these two factors, with HK primary teachers seeing themselves as more transmission-oriented than child-centred while the other three cohorts – HK secondary, NSW primary and NSW secondary teachers – saw themselves as more child-centred than transmission- oriented. The data are interestingly discussed, especially the differences between the two groups of Hong Kong teachers.

Chapter 4-2 reports another East/West comparative study carried out by An, Kulm, Wu, Ma and Wang, but in different country contexts from the previous chapter, namely USA and China. One of the challenges for the readers is to judge how well, and in what ways, the different research sites in these two studies 'represent' their relevant cultural contexts. Indeed one can legitimately ask to what extent the relatively small samples of teachers in such a study as this can adequately 'represent' their two countries' cohorts of mathematics teachers, let alone their two cultural contexts. In such huge countries there could well be as much, if not more, internal variation in beliefs and values as there is between the two traditions, although I suspect, but could be wrong, that that is less true of China than of the USA. However, this study, sensibly in my view, takes a qualitative approach to the research, as one can easily raise valid objections to the idea of studying constructs as deep-seated as beliefs and values through a relatively shallow instrument such as a questionnaire, even if that instrument does allow for assessing large numbers of respondents.

In their chapter, An et al. focus on four main aspects of teachers' beliefs: goals of education, primary focus on teaching mathematics, importance of teachers' knowledge, and planning for instruction. Once again, this study points to important differences between the two groups of teachers, but the authors here also show how these differences in beliefs translate into unique teaching practices, thus helping to secure that theorised relationship through this important comparison.

The next chapter, by Cai, moves us further towards the teaching practices of teachers in the same two cultural traditions. The focus here is on what Cai calls 'pedagogical representations' which are the representations teachers and students use in their classroom, as carriers of knowledge and as thinking tools to explain a concept, a relationship, a connection, or a problem-solving process. Cai claims that there is no doubt that teachers' conceptions of what mathematics is affects their conceptions of how it should be presented.

Eleven U.S. and 9 Chinese teachers participated in the study. The U.S. teachers were from Pennsylvania, North Carolina, and Wisconsin, while the Chinese teachers were from Guiyang, Guizhou. Interestingly both sets of teachers were selected on the recommendations of mathematics educators and were considered "distinguished" mathematics teachers in their respective regions according to local criteria...Thus Cai argues, "it is reasonable to assume that the teachers are recognized as distinguished in their respective regions because their teaching embodies the culturally accepted values of effective mathematics instruction."

The method used in this study is also interesting in that the teachers were presented with 28 student responses to a series of seven problems (Grade 6 level) which they had to score on a five point scale about the extent of students' understanding shown. Each student response had a correct answer (or a reasonable estimate for the answer) and an appropriate strategy that yielded the correct answer (or estimate), but the representations and solution strategies in these responses were different.

Generally speaking, the analysis of interview transcripts showed that the Chinese teachers focused their scoring on "what is missing", while the U.S. teachers focused their scoring on "what is there." Cai concludes that overall, the U.S. teachers were much more lenient than the Chinese teachers in their scoring. However, at a deeper level, the differences in expectations may reflect the differences in cultural beliefs about mathematics and the learning of mathematics. Although both groups of teachers agreed that mathematics has wide applications in the real world, the true beauty of mathematics for the Chinese teachers was its purity, generality, and logic. Thus, a solution strategy that lacks generality (e.g., a visual approach) should be discouraged. In contrast, argues Cai, the U.S. teachers heavily emphasized the pragmatic nature of mathematics: as long as it works, students can choose whatever strategies they like.

Of course, any teacher's teaching is also influenced by the teaching materials available, and the next study by Cao, Seah and Bishop takes as its focus the different textbooks used in the two cultural traditions of Australia and China, and explores the mathematical values represented therein. Besides presenting interesting data in itself, this chapter also raises the question of whether some of the previous research on teachers' beliefs may well have been looking at their values instead. The distinction between these two constructs is not clear cut, particularly when one is undertaking empirical research, but as we saw in the previous chapter, where the teacher has to make judgements or decisions, and where the teacher is faced with choices, is where values get revealed. And it is also where one sees that those values rest on certain beliefs, which in this ICMI study are argued to be rooted in the two different cultural traditions.

So in this chapter, the focus is clearly on the representation and influence of values, and the chief finding concerns the values emphasis given by the textbooks. Certain values are emphasised more than others, which means that teaching which relies on textbooks alone will almost certainly reflect those value preferences also. For a teacher who wishes to emphasise other sets of values the task is made that much harder. The other important result from this study is that the differences between the two cultures in terms of value emphasis were rather small, suggesting as the authors say, that "the image of mathematics taught in schools in the two cultures is basically the same." One can't help wondering whether this is true world-wide.

In Chapter 4-5 our attention turns to the students. We have seen how cultures influence teachers' and textbooks' beliefs and values, but how do these influence in their turn the students' beliefs and values? Does the fact that mathematics teachers so rarely focus overtly, directly or explicitly on different beliefs and values mean that they exert their influences covertly, indirectly and implicitly? And what does this say about how students form their own beliefs and values? Do they just have to accept what the teacher is implicitly 'giving' them?

So Keitel and Jablonka analyse data from Hong Kong and Germany, and present various classroom episodes of interactions between teacher and student(s) which demonstrate vividly how this "study of classroom interaction can reveal the implicit values operating in a classroom." Moreover they point out that "The values become more explicit if a conflict of values emerges. This is not necessarily visible as a straightforward conflict referred to by the teacher or a student but also in a more hidden way when the smooth flow of interaction is interrupted, when students or the teacher laugh or utter an ironic comment."

They also discuss the different ways in which the different cultural values impinge on the options the teacher chooses at such interruption moments, thereby giving rise to a demonstration of the different values at play. "Very tentatively it can be said that the Hong Kong teachers have a tendency to repair the pattern by switching more to the presentation mode, while the German teachers tend either to open it by introducing normal questions instead of directing questions or switch to a presentation."

This chapter also points out many of the difficulties of researching values in practice. For example: "Such research (into values) is predicated on the assumption that there is something stable behind the actions of people and that there are in fact some shared values. Because much of what constitutes a system of values that influences the behaviour of the participants of a distinct social practice is what is rarely questioned – that's just the way things are done – values are seldom made explicit in the form of evaluative, prescriptive or normative statements. As to mathematics teaching/learning this is usually the case only in curriculum documents. Most of the values remain implicit, and in addition, it is not even clear whether the participants of a social practice can easily articulate their values when asked."

The last two chapters in this section return to the topic of teachers' values, and illustrate in the first case how a change of teaching context presents the teacher with cultural conflicts which must in some way be resolved. In terms of the ideas in the last chapter, the chapter by Seah and Bishop explores the values-revealing interruptions of the normal situation, by studying the conflicts experienced by two teachers who emigrated from East Asia into Australia. They face a very different teaching/learning situation from the one 'back home', so what does this conflict situation reveal to us about their values?

By 'cultural value differences' the authors refer to "the immigrant teachers' perception of differences in which their respective home cultures and the Australian culture place importance over mathematical knowledge, school mathematics curriculum, or mathematics teaching/learning." They present findings that illustrate rather than prove the kinds of value differences which can be perceived by teachers in the immigrant situation and which give rise to significant value conflicts for them.

If these so-called 'cultural value differences' are the kinds of value conflicts which these immigrant teachers have to resolve, how does this help us to understand the values development of any teachers? In the final chapter in this section, Chin reports on a study which contrasted the values espoused by Taiwanese teachers with those demonstrated in an earlier study in Australia, and in analysing the data from observations and interviews, he introduces us to a new concept relating to values development, that of teacher identity.

Building on Wenger's (1998) description of *learning as a process of becoming* Chin argues that "All teachers are in the process of developing their pedagogical identities through which they learn to see themselves as becoming the teachers that they value most." Moreover he sees the teacher progressing through various stages of development starting with the student teacher stage, through *the probationary teacher* stage and the *novice* stage, and then becoming *an experienced teacher* and finally *an expert teacher*, who is viewed as *a master* in the teaching community. Chin explains that "this five-stage developmental sequence fits well a possible path of teacher's identity transition, within which a mathematics teacher commits her or himself to the compatible values of their pedagogical identities." He then uses this stage sequence to interpret the differences of values between two groups of teachers in Taiwan, one which was more experienced, ('expert' in his terms) and the other which was formed from probationary or student teachers.

So in this section we present to the reader a collection of fascinating studies, all of which explore in some way the differences which the cultural divide between East and West can reveal through mathematics teaching and learning, but here focusing on the more affective constructs of beliefs and values. As well as demonstrating important differences, the papers also reveal some of the specific conceptual and methodological issues which the student of beliefs and values faces in cross-cultural research contexts. As has been found in other new research sites, the 'pluses' offered by the richness and novelty of the new situation are matched by the 'minuses' of the conflicts and challenges presented by that new research situation. This is as it has always been with significant research developments: we venture into the new context optimistically but with the realisation that there will be new problems to overcome. The good news is that if we hadn't tried to venture down the new path we would not have been made so starkly aware of the new problems waiting to be tackled. I very much hope that more cross-cultural studies of values and beliefs can be undertaken in order to help all of us develop our understandings of these deep-seated and fundamental aspects of mathematics education.

Chapter 4-1

COMPARING PRIMARY AND SECONDARY MATHEMATICS TEACHERS' BELIEFS ABOUT MATHEMATICS, MATHEMATICS LEARNING AND MATHEMATICS TEACHING IN HONG KONG AND AUSTRALIA[1]

Bob PERRY[1], WONG Ngai-Ying[2] and Peter HOWARD[3]
[1]University of Western Sydney; [2]The Chinese University of Hong Kong; [3]Australian Catholic University

1. TEACHERS' BELIEFS ABOUT MATHEMATICS, AND ITS LEARNING AND TEACHING

In both Eastern and Western cultures, teachers' beliefs about mathematics, mathematics learning and mathematics teaching play a critical role in determining how teachers help their students learn mathematics (Ma, 1999; Pajares, 1992; van Zoest, Jones & Thornton, 1994). It is recognised that a student's prime, but by no means only, source of mathematical experiences is the classroom (National Council of Teachers of Mathematics, 2000) and what occurs in the mathematics classroom influences student beliefs (Relich, 1995). The teacher and, in particular, the beliefs of the teacher, are critical to

[1] In New South Wales, Australia, primary teachers are generalist teachers working with students in the age range of approximately 4 years 6 months to 12 years 6 months. Secondary mathematics teachers are specialist teachers working with students in the age range of approximately 11 years 6 months to 18 years. In Hong Kong, primary teachers are generalist teachers working with students in the age range of approximately 6 to 12 years. Secondary mathematics teachers are specialist teachers working with students in the age range of approximately 12 to 17 years.

the classroom implementation of mathematics learning and teaching (McLeod, 1992, Wong, Lam, & Wong, 1998) although exactly how this occurs and to what extent one influences the other has been questioned (Beswick, 2002; Sosniak, Ethington, & Varelas, 1991). All teachers of mathematics hold beliefs about mathematics learning and teaching. These beliefs are developed within the cultural contexts, community and student expectations in which teachers are taught as children, participate in teacher education and continue their professional development. Over this time span, cultural values and community expectations combine to influence teachers in the development of their personal beliefs about mathematics, mathematics learning and mathematics teaching. These beliefs influence and guide teachers in their decision making and implementation of teaching strategies. Indeed, it has been suggested that the investigation of beliefs about learning and teaching may well be the most critical factor in educational research (Pajares, 1992).

2. COMPARATIVE STUDY OF BELIEFS

In their book *The Teaching Gap*, Stigler and Hiebert (1999, pp. 87-88) make the following point about the importance of teacher beliefs and the cultures in which they are embedded:

> Cultural activities, such as teaching, are not invented full-blown but rather evolve over long periods of time in ways that are consistent with the stable web of beliefs and assumptions that are part of the culture. The scripts for teaching in each country appear to rest on a relatively small and tacit set of core beliefs about the nature of the subject, about how students learn, and about the role that a teacher should play in the classroom. These beliefs, often implicit, serve to maintain the stability of cultural systems over time. ... these systems of teaching, because they are cultural, must be understood in relation to the cultural beliefs and assumptions that surround them.

It seems clear that, in order to gain an appreciation of differences and similarities in the learning and teaching of mathematics across cultures, it is essential that the beliefs of both students and teachers be studied. The focus within this chapter compares the beliefs of both primary and secondary mathematics teachers from Hong Kong and New South Wales, Australia about mathematics, mathematics learning and mathematics teaching and seeks to explain similarities and differences in terms of the dominant cultures within the communities and the mathematics curriculum in schools.

This chapter has grown from the long term commitment of the authors – both nationally and internationally – to the study of the impact of teachers' beliefs on their teaching and their students' learning of mathematics (see, for example: Conroy & Perry, 1997; Howard, Perry, & Fong, 2000; Perry, Howard, & Tracey, 1999; Perry, Tracey, & Howard, 1998; Perry, Vistro-Yu, Howard, Wong, & Fong, 2002; Wong, Lam, Leung, Mok, & Wong, 1999). It continues this commitment through an in-depth bilateral comparison of these beliefs in Hong Kong and New South Wales, Australia. Data from both qualitative and quantitative instruments are analysed in order that possible links among the beliefs espoused by the four groups of teachers can be discussed in terms of the different educational and cultural contexts in which the teachers work.

3. BRIEF BACKGROUND

There are many differences and some similarities between the two research sites selected for this study. The Hong Kong community is deeply rooted in Chinese culture, but has experienced great influences from the west, particularly through the many years of British colonisation which ended in 1997 (Wong et al., 1999). The dominant cultural influence in Australia results from more than 200 years of white Anglo-Saxon control following the British invasion of the country in 1788. However, this has been gradually modified through a steady but consistent move to multiculturalism, influenced over the last 40 years by much immigration from Asian countries. So, both research sites have been British colonies. This is reflected in their school systems, although in both cases local changes have been made.

3.1 Hong Kong

A child growing up in Hong Kong receives nine years' compulsory primary and junior secondary education (from Primary 1 to Secondary 3), with secondary school allocation monitored by the Academic Aptitude Test in between. Although senior secondary education (Secondary 4 & 5) is not compulsory, about 90% of junior secondary school students are eligible for promotion to Secondary 4. These students sit the Hong Kong Certificate of Education Examination at the end of Secondary 5. The promotion rate from Secondary 5 to Secondary 6 in recent years is around 40%. To gain entry into university, sixth-formers sit the Hong Kong Advanced Level and Advanced Supplementary Level Examinations after two years' study.

Similar to many Asian countries, classes in Hong Kong are large, averaging 40 students per class. This is one of the highest among participating countries at the Third International Mathematics and Science Study (Wong et al., 1999). Hong Kong scored the lowest among Asian countries in the Third International Mathematics and Science Study (Beaton et al., 1996).

3.2 New South Wales, Australia

School education is a state responsibility in Australia so the following statement is specific to the state of New South Wales and is not true of all states of Australia. Children in New South Wales receive ten years' compulsory school education (Kindergarten to Year 6 in primary and Years 7 to 10 in secondary). In general, children progress through these years as part of their age cohort – repetition of classes is unusual. There is no qualifying examination for entry into secondary school. Approximately 85% of students across the state completing Year 10 move into Year 11, again without a qualifying examination. Almost all of these students undertake the Higher School Certificate examination at the end of Year 12 which provides both a school completion qualification and, for those students wishing to attend university, an entry qualification.

Class sizes in New South Wales schools vary from 20 in the early years of primary school and in secondary schools to about 30 in the senior years of primary school. Australia scored in the second major band in The Third International Mathematics and Science Study, significantly higher than the USA, significantly lower than the leading Asian countries and lower than Hong Kong. (Scores for New South Wales students were not significantly different from those for the total Australian sample.) The relevant comparative mathematics scores for eighth grade students in the two countries on TIMSS and TIMSS-R are given in Table 1 (adapted from National Center for Educational Statistics, 2003).

Table 4-1-1. Average scores on the 1995 TIMSS (rescaled) and 1999 TIMSS-R mathematics assessments in Australia and Hong Kong

Country	**TIMSS mathematics score**		**TIMSS-R mathematics score**	
	Average	Standard error	Average	Standard error
Hong Kong	569	6.1	582	4.3
Australia	519	3.8	525	4.8

4. METHOD

Data for this study were gathered using a researcher-designed questionnaire containing 20 items dealing with teacher beliefs about mathematics and its learning and teaching. The questionnaire was constructed from numerous sources (Australian Education Council, 1991; Barnett & Sather, 1992; Wood, Cobb & Yackel, 1992), trialled extensively with both primary and secondary teachers and used in numerous earlier studies (Howard, Perry, & Lindsay, 1997; Perry, et al., 1998, 1999; Tracey, Perry, & Howard, 1998) including cross-cultural studies in Australia, Indonesia, the Philippines and Singapore (Howard et al., 2000; Perry & Howard, 1999; Perry et al., 2002).

In Australia, the questionnaire was administered in English and in Hong Kong, it was translated into Chinese. Respondents completed the questionnaire by indicating on a three point Likert scale – Disagree, Undecided, Agree – to what extent they agreed with each statement. The sample for this implementation of the questionnaire were primary (n=377) and secondary teachers (n=179) in Hong Kong (HK) and primary (n=252) and secondary teachers (n=249) in New South Wales, Australia (NSW) – a total of 1027 teachers.

After the results for the questionnaire had been analysed, focus group interviews with small groups of primary and secondary teachers were held in both Hong Kong and New South Wales. In these interviews, teachers were asked to comment on the results from four specific statements in the questionnaire as well as to provide general comments on the results. The data derived from these interviews are used in this paper to explicate the quantitative findings from the questionnaire as well as to provide a general overview of the orientations of groups of teachers in each jurisdiction.

5. RESULTS AND ANALYSIS

An important initial result is the confirmation that the questionnaire used in this study is not only relevant in the Australian context (Perry et al., 1999) but also, in its translated version, in the Hong Kong context. In earlier studies (Howard, et al., 1997, 2000; Perry & Howard, 1999; Perry, et al., 1998, 1999; Tracey, et al., 1998), two categories of beliefs – transmission oriented and child-centredness – have been clearly defined. They can be described in the following ways:

transmission: the traditional view of mathematics as a static discipline which is taught and learned through the transmission of mathematical skills and knowledge from the teacher to the learner and where "mathematics [is seen] as a rigid system of externally dictated rules governed by standards of accuracy, speed and memory" (National Research Council, 1989, p.44);

child-centredness: students are actively involved with mathematics through "constructing their own meaning as they are confronted with learning experiences which build on and challenge existing knowledge" (Anderson, 1996, p.31).

Teachers in both Hong Kong and New South Wales reported beliefs across both the transmission and child-centredness factors, and, through factor analysis, measures of how each respondent sees her- or himself as a mixture of a 'transmission' and a 'child-centred' teacher were calculated. For each group of teachers, there were significant differences between the groups' loadings on these two factors with HK primary teachers seeing themselves as more transmission oriented than child-centred while the other three cohorts – HK secondary, NSW primary and NSW secondary teachers – saw themselves as more child-centred than transmission oriented.

By applying independent sample t-tests to the factor scores on the transmission and child-centredness factors for each pair of teacher groups, statistically significant differences in these factor scores across the four different cohorts of teachers were found. Table 2 shows the results of such an analysis for the transmission factor and Table 3 for the child-centredness factor.

Table 4-1-2. Analysis of independent samples t-test using transmission factor scores across teacher cohorts

Teacher cohort	HK Secondary		NSW Primary		NSW Secondary	
HK Primary	t=20.67	p<0.001	t=50.45	p<0.001	t=50.87	p<0.001
HK Secondary			t=62.93	p<0.001	t=64.04	p<0.001
NSW Primary					t=0.55	NS

The rank order of groups on the transmission factor was HK Secondary teachers (most positive), then HK Primary teachers and the two NSW cohorts which are not significantly different.

Table 4-1-3. Analysis of independent samples t-tests using child-centred factor scores across teacher cohorts

Teacher cohort	HK Secondary		NSW Primary		NSW Secondary	
HK Primary	t=59.59	p<0.001	t=27.16	p<0.001	t=27.18	p<0.001
HK Secondary			t=25.09	p<0.001	t=24.34	p<0.001
NSW Primary					t=0.34	NS

The rank order of groups on the child-centred factor was HK Secondary teachers (most positive), followed by the two NSW cohorts – which are not significantly different – and the HK Primary teachers.

The fact that the HK secondary teachers ranked most positively on both the transmission and child-centred factors begs further explanation which is provided later in this chapter.

The four statements highlighted in Tables 4 to 7 have been chosen for inclusion here because they show a clear distinction between the responses of the HK and NSW teachers and because they can represent the two previously identified factors. Statements 6 and 9 load onto the transmission factor and Statements 14 and 15 load onto the child-centred factor.

Table 4-1-4. Questionnaire date for Statement 6: Percentages of each cohort reporting Disagree (D), Undecided (U), Agree (A) for each statement

Statement	Teacher group	D	U	A
6. Right answers are much more important in mathematics than the ways in which you get them	HK Primary	9	57	35
	HK Secondary	0	5	95
	NSW Primary	87	8	6
	NSW Secondary	88	6	6

Table 4-1-5. Questionnaire data for Statement 9: Percentages of each cohort reporting Disagree (D), Undecided (U), Agree (A) for each statement

Statement	Teacher group	D	U	A
9. Mathematics learning is being able to get the right answers quickly	HK Primary	2	52	46
	HK Secondary	6	71	24
	NSW Primary	84	8	8
	NSW Secondary	82	11	7

Table 4-1-6. Questionnaire data for Statement 14: Percentages of each cohort reporting Disagree (D), Undecided (U), Agree (A) for each statement

Statement	Teacher group	D	U	A
14. Mathematics learning is enhanced by challenge within a supportive environment	HK Primary	15	52	34
	HK Secondary	21	64	15
	NSW Primary	0	2	97
	NSW Secondary	1	5	94

Table 4-1-7. Questionnaire data for Statement 15: Percentages of each cohort reporting Disagree (D), Undecided (U), Agree (A) for each statement

Statement	Teacher group	D	U	A
15. Teachers should provide instructional activities which result in problematic situations for learners	HK Primary	48	42	10
	HK Secondary	15	73	12
	NSW Primary	4	9	87
	NSW Secondary	2	15	82

Teachers in the interview groups were asked to provide some comment on the results for each of these statements. Representative responses are given in Tables 8 and 9.

Table 4-1-8. Teacher comments on the results from Statements 6 and 9

Teacher group	Comments
HK Primary	• Exams mainly count the 'answers'. Even in the aptitude test, students get points if their answer is correct. • The pace of living is faster in Hong Kong … 'faster is better'. • Teachers often assign students to do more math because they want to train their speed.
HK Secondary	• The need to get the right answers quickly might be due to public exams. • Teachers might use ways they learned when they were in school. The need to get the right answers quickly might pass from one generation to the other. • There are two groups of teachers. One asks students to do the questions quickly, so that they have enough time to check the answers. The other group asks students to go slowly and think thoroughly in order to gain accuracy.
NSW Primary	• There has to be a balance of both because if they have the process right but are not getting the right answer, there has to be something wrong. You have to have the right answer and the right process. • You have to identify individual needs but at the same time it comes down to accuracy and memory and learning things off by rote, particularly times tables. There is no other way. It has to work together.
NSW Secondary	• The process is important and questions can have a number of solutions. Maths is about logic and you can't learn that by just getting it right or wrong.

Table 4-1-9. Teacher comments on the results from Statements 14 and 15

Teacher group	Comments
HK Primary	• Hong Kong teachers do not have sufficient time. • I think Westerners are more willing to take risks, while we are more inclined to play safe, and avoid making mistakes and challenges. Obedience and conformity are also important.
HK Secondary	• The teaching schedule in Hong Kong is very tight, especially the classes that have to deal with public exams. There is not enough time for teachers to organise activities. • The ratio of students and teachers is an important factor.
NSW Primary	• I didn't really understand a lot of mathematical concepts until I started to teach them. I was taught through rote and memorisation and I didn't have a good understanding of what I was doing. That's why you have to vary your activities and build upon understanding. • As soon as you get bored with something you lose interest. There has to be challenge.
NSW Secondary	• Because of the way that the syllabus is written, you have to incorporate those types of things into assessment tasks. It is important to build upon students' prior knowledge and to link maths to real life as well. • The teacher education program emphasises constructivism rather than a behaviourist approach. The syllabus is structured on these principles and you have to teach the syllabus.

6. DISCUSSION

The HK primary teachers were less child-centred and less transmission oriented than their secondary colleagues. This might be explained in terms of the importance placed on giving students a sound foundation in the basics in the primary years in Hong Kong (see Table 8). Perhaps the primary teachers feel less empowered than their secondary colleagues to move beyond the obligatory syllabus or the set textbook and so do not feel in a position to form strong opinions of their own about mathematics and its teaching and learning. The looming importance of the Academic Aptitude Test seems to weigh heavily on the HK primary teachers (Wong et al., 1999). When HK primary teachers were asked about this result in the interviews, comments included:

- mathematics teaching in primary schools is less child-centred because that is too time consuming;

- other factors include the requirements of the curriculum and the expectations of high academic results from parents.

On the other hand, HK secondary teachers suggested that:

- the age of the students is a major factor – teachers may respect an older student as an individual;
- there are public exams in secondary school.

The NSW primary teachers were more child-centred and less transmission oriented in their beliefs about mathematics and its learning and teaching than their Hong Kong counterparts. These results could be a direct consequence of the relative freedom which is afforded primary teachers in NSW to interact with the mathematics syllabus in flexible ways such as integration with other subjects, variability in timetabling, use of teaching based constructivist approaches to learning and only a slight reliance on system-wide testing – although this is currently increasing in frequency (Table 9). There is a cultural expectation that Hong Kong students will reach and achieve particularly high standards of mathematical achievement in the primary years (Wong et al., 1999). This community expectation may well place demands upon teachers to emphasise the transmission of knowledge in their teaching strategies more so than for NSW teachers. When arriving in Australia, Hong Kong parents often state that their children are at a much higher level of mathematics, often meaning the ability to work with numbers, than children of the same age taught in NSW schools. There is a high cultural value placed on HK primary children's ability to calculate with numbers. The emphasis on numeracy skills may lead to an emphasis on a greater transmission orientation in the beliefs of HK primary teachers compared to NSW primary teachers. The HK primary teachers interviewed listed the following pressures on their approaches to mathematics teaching:

- parents' expectations;
- examination results;
- curriculum too tight;
- large classes; and
- amount of homework.

These findings reiterate HK teachers' concerns about the mathematics curriculum which were highlighted by the recent holistic review of the mathematics curriculum (Ad hoc Committee on Holistic Review of the Mathematics Curriculum, 2000). In particular, "most mathematics teachers reflected that it [the mathematics curriculum] was too bulky, lacked flexibility, and was unable to cater for individual differences and to provoke thinking" while "Almost all teachers pointed out that the existing mathematics curriculum was too packed, too boring, impractical and unrelated to real life" (Wong et al., 1999, p. 78). In contrast, the mathematics curriculum used by NSW primary teachers at the time of data collection (NSW Department of School Education, 1989) has been widely recognised by teachers and academics as one of the most usable, practical and teacher-friendly ever devised.

Secondary mathematics teachers in Hong Kong are both more child-centred and more transmission oriented than their counterparts in NSW. This apparent conflict is reflected in the high proportion of 'Undecided' scores given by HK secondary teachers in Statements 9, 14, 15 (see Tables 5, 6, 7). These teachers seem to be expressing beliefs about trying to meet the pragmatic requirements of teaching older – and, therefore, more respected - students and meeting their cultural and educational responsibilities in Hong Kong society. It is not unusual for teachers to express a mixture of beliefs, depending on the circumstances in which they perceive themselves (Beswick, 2002) and it seems that this might be the case with the HK secondary teachers.

HK secondary mathematics teachers experience high levels of community expectation for students to achieve well in mathematics, particularly in an examination-driven climate. The greater transmission orientation among the HK secondary teachers may be related to teachers having to ensure that the mathematics is taught and learnt to meet these expectations which are different from those in NSW (Table 8). At the same time, there is a climate of mutual respect between HK teachers and students which interview data suggest is higher than that in NSW. This may result in the HK secondary teachers focussing more on their students understanding their mathematics and therefore espousing a more child-centred orientation than their NSW counterparts. One NSW secondary teacher explained the dilemma in the following way:

> Sometimes it is easier and quicker to use a transmission approach to get through the maths content, particularly in Years 11 and 12. The amount of content affects the style that we use to teach. You have to have planning – time and effort – to use child-centred approaches while you cover a lot more and it is a lot easier to use transmission approaches but the long-term results are not as good.

7. CONCLUSION

This study has demonstrated a number of important aspects about the need for the consideration of teachers' beliefs about mathematics, mathematics learning and mathematics teaching. Foremost among these is that, as measured on the beliefs instrument used here, teachers in Hong Kong, at both primary and secondary levels, espouse significantly different beliefs from those espoused by their NSW counterparts. If we believe that teachers' beliefs affect their teaching and that teaching affects student outcomes, then it is possible that the differences in beliefs which are highlighted by this study could help explain some of the differences in student achievement which have been reported in international studies.

Teacher beliefs are rooted in, and constrained by, the culture of the society in which the teachers are living and working, in the culture of the education systems and traditions of the society and in their own experiences as school students, teacher education students and members of school communities. As well, teachers must respond to parental, societal and student pressures in terms of examinations and other assessment and pedagogical challenges. These pressures are also rooted in the various cultures of which the education systems are part. Hence, the expectation that cultural norms and patterns might affect teacher beliefs and practices is not surprising. In this paper, these aspects of Hong Kong and New South Wales, Australia cultures have been explored and related to the different responses of these societies to the mathematical education of their children. Of course, beliefs are not the entire story, but they are an essential part of cross-cultural comparisons in mathematics education which deserve greater recognition and research.

REFERENCES

Ad hoc Committee on Holistic Review of the Mathematics Curriculum, 2000, *Report on Holistic Review of the Mathematics Curriculum*. Curriculum Development Council, Hong Kong, retrieved June 6, 2003 from http://cd.ed.gov.hk/maths/document/Adhoc.htm.

Anderson, J., 1996, Some Beliefs and Perceptions of Problem Solving, in: *Technology in Mathematics Education*, P.C. Clarkson, ed., Deakin University Press, Melbourne, pp.30-37.

Australian Education Council, 1991, *National Statement on Mathematics for Australian Schools*. Curriculum Corporation, Melbourne.

Barnett, C., and Sather, S., 1992, Using Case Discussions to Promote Changes in Beliefs among Mathematics Teachers. Paper presented at the American Educational Research Association Conference, May.

Beaton, A.E., Mullis, I.V.S., Martin, M.O., Gonzalez, E.J., Kelly, D.L., and Smith, T.A., 1996, *Mathematics Achievement in the Middle School Years: IEA Third International Mathematics and Science Study*, TIMSS International Study Center, Chestnut Hill, MA.

Beswick, K., 2002,. Teacher Beliefs: Probing the Complexities, in: *Mathematics Education in the South Pacific*, B. Barton, K.C. Irwin, M. Pfannkuch and M.O.J. Thomas, eds.,

Proceedings of the twenty-fifth annual conference of the Mathematics Education Research Group of Australasia, **1**:139-146, Auckland: MERGA.

Conroy, J., and Perry, B., 1997, Australian and Indonesian Student Teacher Beliefs about Mathematics and Performance on a Classic Ratio Task, in: *Proceedings of the Twenty-first Conference of the Psychology of Mathematics Education Group*, E. Pehkonen, ed., PME, Lahti, Finland, pp.177-184.

Howard, P., Perry, B., and Fong, H.K., 2000, Beliefs about Mathematics, Mathematics Learning and Mathematics Teaching: Views of Some Primary Teacher Education Students in Singapore and Australia, in: *Mathematics Education beyond 2000*, J. Bana and A. Chapman, eds., MERGA, Perth, pp.331-338.

Howard, P., Perry, B., and Lindsay, M., 1997, Secondary Mathematics Teachers Beliefs about the Learning and Teaching of Mathematics, in *People in Mathematics Education*, F. Biddulph and K. Carr, eds., MERGA, Rotorua, NZ, pp.231-238.

Ma, L., 1999, *Knowing and Teaching Elementary Mathematics*, Lawrence Erlbaum, Mahwah, NJ.

McLeod, D.B., 1992, Research on Affect in Mathematics Education: A Reconceptualization, in: *Handbook of Research on Mathematics Teaching and Learning*, D.A. Grouws, ed., Macmillan, New York, pp.575-596.

National Center for Education Statistics, 2003, *Teaching Mathematics in Seven Countries: Results from the TIMSS 1999 Video Study*. US Government Printing Office, Washington, DC.

National Council of Teachers of Mathematics, 2000, *Principles and Standards for School Mathematics*, Author, Reston, VA.

National Research Council, 1989, *Everybody Counts: A Report to the Nation on the Future of Mathematics Education*, National Academy Press, Washington, DC.

NSW Department of School Education, 1989, *Mathematics syllabus K-6*, Author, Sydney.

Pajares, M.F., 1992, Teachers' Beliefs and Educational Research: Cleaning up a Messy Construct, *Review of Educational Research*, **62**(3):307-322.

Perry, B., and Howard, P., 1999, Beliefs about Learning and Teaching of Mathematics: Views from Australia, Singapore and Indonesia, in: *Proceedings of Eighth South East Asian Conference on Mathematics Education*, Ateneo de Manila University, Manila, pp.311-318.

Perry, B., Howard, P., and Tracey, D., 1999, Head Mathematics Teachers' Beliefs about the Learning and Teaching of Mathematics, *Mathematics Education Research Journal*, **11**(1): 39-53.

Perry, B., Tracey, D., and Howard, P., 1998, Elementary School Teacher Beliefs about the Learning and Teaching of Mathematics, in: *Proceedings of the ICMI - EARCOME 1, Vol II*, H.S. Park, Y.H. Choe, H. Shin and S.H. Kim, eds., Korea Society of Mathematical Education, Chungbuk, Korea, pp.485-498.

Perry, B., Vistro-Yu, C., Howard, P., Wong, N.Y., and Fong, H.K., 2002, Beliefs of Primary Teachers about Mathematics and its Teaching and Learning: Views form Singapore, Philippines, Mainland China, Hong Kong, Taiwan and Australia, in: *Mathematics Education in the South Pacific: Proceedings of the Twenty-Fifth Annual Conference of the Mathematics Education Research Group of Australasia*, B. Barton, K.C. Irwin, M. Pfannkuch, and M.O.J. Thomas, eds., MERGA, Auckland, NZ, pp.551-558.

Relich, J., 1995. Gender, Self-concept and Teachers of Mathematics: Effects of Attitudes to Teaching and Learning, *Educational Studies in Mathematics*, **30**(2):179-197.

Sosniak, L.A., Ethington, C.A., and Varelas, M., 1991, Teaching Mathematics without a Coherent Point of View: Findings from the IEA Second International Mathematics Study, *Journal of Curriculum Studies*, **23**(2):119-131.

Stigler, J., and Hiebert, J., 1999, *The Teaching Gap: Best Ideas from the World's Teachers for Improving Education in the Classroom*, The Free Press, New York.

Tracey, D., Perry, B., and Howard, P., 1998, Teacher Beliefs about the Learning and Teaching of Mathematics: Some Comparisons, in: *Teaching Mathematics in New Times*, C. Kanes, M. Goos and E. Warren, eds., MERGA, Gold Coast, pp.613-620.

Van Zoest, L.R., Jones, G.A., and Thornton, C.A., 1994, Beliefs about Mathematics Teaching held by Pre-service Teachers involved in a First Grade Mentorship Program, *Mathematics Education Research Journal*, **6**(1):37-55.

Wong, N.Y., Lam, C.C., Leung, F.K.S., Mok, I.A.C., and Wong, K.M.P., 1999, Holistic Reform of the Mathematics Curriculum – The Hong Kong Experience, *Journal of the Korea Society of Mathematical Education, Series D: Research in Mathematical Education,* **3**(2):69-88.

Wong, N., Lam, C., and Wong, K.P., 1998, Students' and Teachers' Conceptions of Mathematics Learning: A Hong Kong Study, in: *Proceedings of the ICMI-East Asian Regional Conference on Mathematical Education, Volume 2*, H.S. Park, Y.H. Choe, H. Shin, and S.H. Kim, eds., Korea Society of Mathematical Education, Chungbuk, Korea, pp.375-404.

Wood, T., Cobb, P., and Yackel, E., 1992, Change in Learning Mathematics: Change in Teaching Mathematics, in: *Redefining Student Learning: Roots of Educational change*, H. Marshal, ed., Ablex, Norwood, NJ, pp.177-205.

Chapter 4-2

THE IMPACT OF CULTURAL DIFFERENCES ON MIDDLE SCHOOL MATHEMATICS TEACHERS' BELIEFS IN THE U.S. AND CHINA

AN Shuhua[1], Gerald KULM[2], WU Zhonghe[3], MA Fu[4] and WANG Lin[5]
[1]California State University; [2]Texas A&M University; [3]University of Missouri; [4]Nanjing Normal University, China; [5]Mathematics Research Office in Jiansu Province, China

1. INTRODUCTION

Cultures and societies have varying philosophies and beliefs regarding the teaching and learning of mathematics. These variations of the beliefs and values concerning mathematics education result in different educational systems. However, the relationship between beliefs and teaching in the social context is complex because embedded in this context are the values, beliefs, and philosophical leanings of the educational system at large (Thompson, 1992).

Mathematics education in countries is strongly influenced by the cultural and social factors that build goals, beliefs, expectations, and teaching methods. Furthermore, many countries have different views about the impact of teachers' beliefs on teaching mathematics (Ernest, 1989). Cross-cultural comparisons can lead researchers and educators to a more explicit understanding of their own implicit theories about how children learn mathematics (Stigler & Perry, 1988). To understand teaching from teacher's perspectives, there is a need to understand teachers' beliefs about education with which they define their work (Nespor, 1987).

Numerous studies have focused on the changes in mathematics teachers' beliefs and practices (Cooney & Shealy, 1997; Franke, Fennema, & Carpenter, 1997). Teaching practices embody teachers' beliefs, which are a reflection of their own values, experiences, and cultural background (Cabello & Burstein 1995). Beliefs about teaching act as filters to reflect teachers' views of teaching (Kagan, 1992). However, important questions remain such as, "How do we cultivate a congruence relationship between teachers' beliefs and their teaching practice?" "How do teachers internalize their beliefs?" "How and to what degree do cultural factors influence teachers' beliefs?" "What is the relationship between teachers' knowledge and beliefs?" In order to address these questions, it is necessary to study beliefs and their impact in depth in international and cultural perspectives and to compare the differences in teachers' beliefs and their impacts.

This study attempted to investigate this complex relationship between teachers' beliefs and their teaching practices by comparing the differences between mathematics teachers' beliefs in American and Chinese middle schools. Specifically, this study examined how these relationships were impacted by different cultural and social contexts, and how these differences were revealed in teachers' knowledge, teaching methods, knowledge of learners' cognition, and planning for instruction. The ultimate goal of this study was to focus on the importance of teachers' beliefs and their impact on the practice of teaching, and to provide data and recommendations, which may be used to identify problems in mathematics in the schools of the U.S. and China.

The research questions are: What are the differences and similarities between the middle school mathematics teachers' beliefs about education in the U.S. and China? How are these differences revealed in their pedagogical content knowledge, planning for instruction, teaching method, and knowledge of learners' cognition? How are these differences influenced by culture differences?

2. THEORETICAL FRAMEWORK AND OBJECTIVE

2.1 Historical and philosophical influence on teachers' beliefs

For more than two thousand years the Chinese have been following Confucius (551-479 BC) as the father of Chinese education (Ashmore & Cao, 1997). One feature of beliefs of Confucianism in learning is that to acquire knowledge one needs to study, constantly to ask questions, and to review the basics continuously. The classic work *Arithmetic of Nine Chapters* in the Tang dynasty shows the characteristics of a sequencing of questions, answers, and principles throughout the whole book. The center of this instructional model is the questions, and the emphasis is on the computations, which had a significant influence on mathematics education in China.

In contrast, European philosophers and mathematicians such as Plato and Aristotle originally influenced U.S. mathematics education. In the early 20th century, Dewey's belief in "learning from doing" had a vital impact on mathematics education. NCTM (1989 & 2000) has supported the notion of mathematics as a dynamic process and structure, rather than product, since the 1980s.

The discrepant views in history and philosophy on mathematics education between the U.S. and China reflected the divergent thinking about mathematics in different cultural contexts, and produced distinct belief systems about mathematics in both countries.

2.2 The relationship between teachers' knowledge and beliefs

Teachers and teaching has been found to be one of the major factors relating to students' mathematics achievement in TIMSS and other studies. To address the important role of teaching, Stigler and Heibert (1999) state, "teaching is the next frontier in the continuing struggle to improve schools" (p.2). According to *Principles and Standards for School Mathematics* (NCTM, 2000) "Effective teaching requires knowing and understanding mathematics, students as learners, and pedagogical strategies" (p.17). In the domain of mathematics teachers' knowledge, pedagogical content knowledge has been recognized as the most important element for teaching mathematics (Shulman, 1987; Pinar et al., 1995). Pedagogical content knowledge is defined as the ability of the teacher to transform the content knowledge into forms that are "pedagogically powerful and yet adaptive to the variations in ability and background presented by the students" (Shulman, 1987, p.15).

Pedagogical content knowledge not only addresses how to teach mathematics content successfully but also focuses on how to understand students' cognition, including taking into consideration both students' cultural background and students' preference for various teaching and learning styles. One of the core components of pedagogical content knowledge is the knowledge of teaching. Knowledge of teaching in this study consists of knowing students' thinking, preparing instruction, and mastering effective teaching approaches (See Figure 1).

Figure 1 shows the interactive relationship between teachers' knowledge and their beliefs. Teachers' beliefs have a strong impact on teachers' pedagogical content knowledge. Pedagogical content knowledge, in turn, constantly promotes teachers' conceptual changes in beliefs. The process of reflecting on and developing pedagogical content knowledge facilitates the internalization of teachers' beliefs. To measure and analyze this internalization of teachers' beliefs, four major teaching components should be fully taken into consideration: pedagogical content knowledge, preparation for instruction, methods of teaching, and knowledge of student thinking.

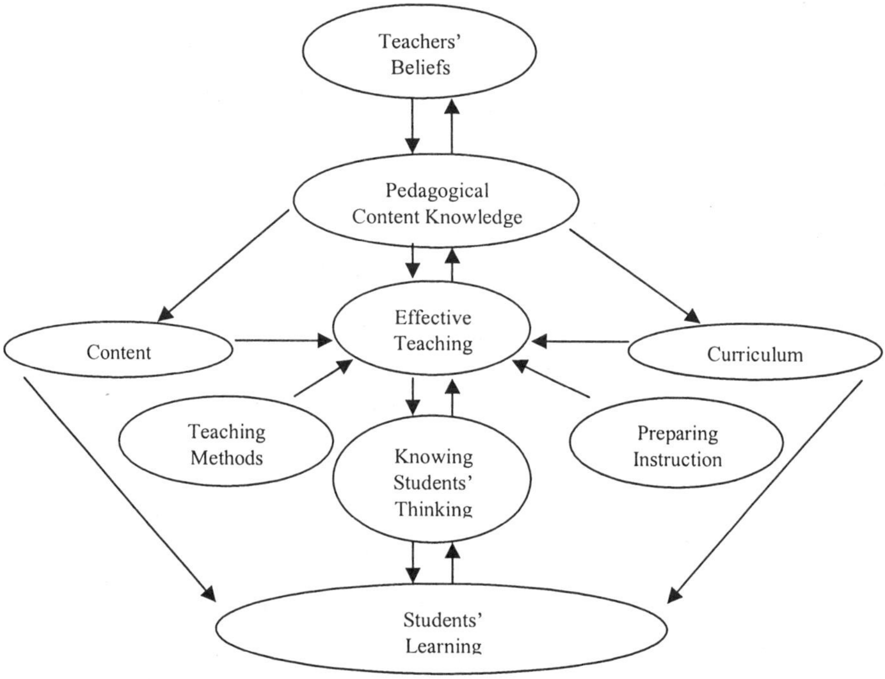

Figure 4-2-1. The Impact of Beliefs on Teachers' Knowledge and Practices

2.3 The four aspects of teachers' beliefs

In this study, we categorized teachers' beliefs into four main aspects: goals of education, primary focus on teaching mathematics, importance of teachers' knowledge, and planning for instruction. It is important to note how these four aspects of teachers' beliefs relate to each other. Table 1 addresses the essential components of each aspect of teachers' beliefs and corresponds with eight questions about teachers' beliefs. With a set of goals of teaching in mind, teachers understand their primary focuses in teaching, design and use various approaches in classrooms, and try to find an effective teaching method in their teaching in order to meet individual students' needs and help all students to learn successfully. To teach effectively at a continuous base, a teacher should enhance not only content knowledge, but also pedagogical content knowledge, including knowledge of students' thinking. Understanding students' thinking can be achieved through many approaches. One of the approaches is to know students' thinking through grading students' homework, in which the teacher can fully assess students' weaknesses and strengths and plan for further instruction according to students' needs.

Table 4-2-1. Categories for Describing Four Aspects of Teachers' Beliefs about Mathematics Teaching and Learning Questions

Four Aspects of Teachers' Beliefs	**Essential Components**
Goals of Education	Goal of Education in General
	Goal of Math Education
Primary Focus on Teaching	Conceptual Understanding vs. Procedural Development
	Effective Teaching Methods
	Meet Individual Students' Needs
Importance of Teachers' Knowledge	Knowledge of Students' Thinking
	Promoting Students' Ability to Thinking
	Pedagogical Content Knowledge
	Enhance Knowledge of Math Teaching
Planning for Instruction	Approach of Planning for Math Instruction
	Time Spend on Planning Instruction
	Students Homework
	Importance of Grading Homework
	Approach of Grading Homework

3. METHODS

3.1 Data source and procedure

The study was designed with a combination of qualitative and quantitative research methods to examine mathematics teachers' beliefs and their practice in the U.S. and China. Employing qualitative methodology (Guba & Linclon, 1996), data was collected via questionnaires from 28 mathematics teachers in 12 middle schools in U.S. and 33 mathematics teachers in 22 schools in China in 2000. A total of ten teachers, five teachers from each country, were observed and interviewed to confirm and clarify responses in the questionnaires. There were two forms of questionnaires. A "Mathematics Teaching Questionnaire" consisted of four problems designed to examine teachers' pedagogical content knowledge on the topic of fractions, ratios, and proportions, and a questionnaire "Questions on Teachers' Beliefs about Mathematics Teaching and Learning" included eight questions designed to investigate the teachers' beliefs about mathematics education. An in-depth treatment of the results on the teachers' pedagogical content knowledge is reported in a separate paper (An, Kulm & Wu, 2004). A comprehensive analysis and discussion of all of the results can be found in the dissertation by the first author (An, 2000).

3.2 Data analysis

Transcripts were made for these observations and interviews by examining field notes, and reviewing the responses to the questionnaire and interviews to gather and assess the data. A constant comparative data analysis method (Lincoln & Guba, 1985) was used. The responses from teachers were categorized into groups and assigned a descriptive code. In all, more than 30 different categories were identified which included the responses to the eight problems. These 30 categories were sorted into four groups according to the framework of beliefs in Table 1. Researchers used the resulting codes to analyze the responses and calculated the percentages for each response.

4. RESULTS

The results of the study indicate that mathematics teachers' beliefs in the two countries differ markedly, which has a deep impact on teaching practice in each country. According to the four aspects of teachers' beliefs, the findings are discussed below.

4.1 Goals of education

Goal of education in general. In this study, the U.S. and Chinese teachers differ in their goals for education. The U.S teachers provided various responses about the goal of education: 25% of the U.S. teachers believed that the primary goal of education is to teach students knowledge and skills; 25% of teachers thought that it is to help students to succeed in society; 15% of teachers reported that the goal is to cultivate students to be productive, responsible, and educated citizens; 11% of teachers stated that the goal is to support students to be lifelong learners. Notably, 11% of U.S. teachers claimed that education should provide equal opportunities to all children and that every child can learn compared to 0% of responses from Chinese teachers. Also, 28% of U.S. teachers believed that the goal of education is to increase students' confidence and self esteem compared to 0% of responses from Chinese teachers.

In contrast, for the same question, Chinese teachers responded in limited dimensions. They focused on three areas that are similar to U.S. teachers: 28% of responses from Chinese teachers indicated that the goal of education is to produce productive citizens, 34% of teachers believed that the goal of education is to help students to succeed in society, and 31% of Chinese teachers believed that the goal of education is to teach students knowledge and skills.

It is not surprising that U.S. teachers have varied beliefs about the goals of education. The U.S. teachers' beliefs are influenced by a democratic educational system. In turn, these beliefs impact the teachers' pedagogical content knowledge. This study found that the U.S. teachers try very hard to explore numerous teaching approaches to reach all children. For instance, teachers try to connect learning to concrete models and to students' experience in their daily life. Although there is no single focus on specific teaching approaches, U.S. teachers do respect the individual differences among students and attempt to provide equal opportunity to all children. These beliefs are supported by the NCTM (2000) philosophy that "excellence in mathematics education requires equity, high expectations, and strong support for all students" (p.12).

Similar to the U.S. teachers, the majority of Chinese teachers in this study believed that the goal of education is to teach students knowledge and to help students succeed in society. However, in the Chinese teachers' view, to be successful in society means to become productive human resources for the nation and to enhance the quality of the nation as a whole. In order to achieve that, the goal of education is to help students fully develop in four areas: ideology, morality, schooling, and discipline. Under this nationalistic education, the purpose of learning is to become a useful person and to be able to contribute to the country rather than to pursue personal interests. Consequently, the development of students' personalities and individual values are neglected. The rigorous examination system is another feature of the centralization of education. In order to enter high schools and universities, students must pass national entrance examinations. Students are sorted according to their grades so that they may enter schools at different levels. This results in more than half of students being weeded out of high school every year. Their only choice is to go to vocational schools, in which students have little chance to compete in entrance examinations for universities compared to those students who are in regular high schools. Under the pressure of the examination system, teachers believe that the goal of education is not only to teach students knowledge and skills, but also to help students to be successful in society, which is partly measured by passing or failing in national entrance examinations. This belief directs teachers' pedagogical content knowledge into a direction that focuses on intensity and proficiency in skills. Teachers work very hard to prepare students fully in basic concepts and skills in order to meet the needs of the examinations. Spring (1998) observed that the examination system determines students' chances in life and places tremendous pressure on students and teachers. Additionally, teachers were forced to teach to the test. Spring (1998) also pointed out that the examination system reproduces social class. However, in recent years, China has been trying to decentralize education by allowing the publication of different versions of textbooks, by abolishing the entrance examination for the middle school, by reforming the entrance examination for the university, and by learning educational experience from other countries, especially the United States.

Goals of math education. There are differences in viewing the goals of mathematics education between the two groups of teachers. U.S. teachers have various responses: 30% of the U.S. teachers agreed that mathematics should be taught in real life situations and that mathematics should be connected to concrete models compared to 0% of responses from Chinese teachers. In addition, 22% of U.S. teachers indicated that the goal of mathematics is to teach students how to solve problems in the real world. Under the influence of the notion that knowledge is socially constructed and cultural processes have an important impact on students' learning (Boaler, 2000), the U.S. teachers strongly believed that teaching mathematics in real life situations and connecting it to concrete models is an important approach in mathematics teaching. It makes learning meaningful, visible, and applicable. This type of learning is thought to take root more easily in students' minds.

However, unlike the U.S. teachers, Chinese teachers' responses focused on three areas: Gain math knowledge and skills (72%), enhance critical thinking and logical reasoning ability (69%), and solve and apply math in the real world (63%). It is interesting to note that 28% of Chinese responses indicated the importance of teaching "learning methods;" (No U.S. teachers mentioned this goal). However, no Chinese teachers considered teaching math in real life situations and connecting math to concrete models (30% of responses from U.S. teachers). These differences indicated that Chinese teachers understand the importance of "learning method" but ignore the importance of concrete models, while U.S. teachers like to use concrete models to teach mathematics but did not pay much attention to teach students "learning methods."

The belief of enhancing the ability of logical and critical thinking as the key component of mathematics teaching has a great impact on Chinese teachers' pedagogical content knowledge. In the interviews with Chinese teachers, most of them considered teaching a particular way of thinking as a main focus. In the observation of their classes, teachers posed different levels of questions to promote not only student thinking but also how to think. In the teachers' lesson plans, questions are layered and problems are designed to enhance students' ability to master the methods of thinking. Chinese teachers believed that to teach methods of thinking is an important way of improving students' learning. The 5th to 8th grades are critical stages of transition in mathematics learning. Students should be further trained in thinking logically and critically so they can more easily adjust and succeed in learning algebra. NCTM (1989) supports the view that middle school students must progress through using reasoning to making conjectures and they must apply both inductive and deductive reasoning.

4.2 Primary focus on teaching

Conceptual understanding vs. procedural development. The results of this study indicated that 41% of U.S. teachers focused on both conceptual understanding and procedural (skill) development in mathematics teaching. However, 33% of U.S. teachers emphasized conceptual understanding, and 22% of U.S. teachers considered procedural development as the primary focus in teaching mathematics. In contrast, Chinese teachers' responses were distributed evenly among three areas: 38% of them focused on procedural development, 31% of them concentrated on conceptual understanding, and 31% of them used both ways to teach mathematics.

This study found that the U.S. teachers try to focus on both conceptual understanding and procedural development in teaching by connecting learning to concrete models and to students' daily life experience, but they tend to separate conceptual understanding and procedural development into two disjoint dimensions. To U.S. teachers, to teach "procedure" is simply to teach steps only.

In contrast, Chinese teachers focused more on procedural development in teaching mathematics. However, they recognize the interplay between two areas and believe that concepts could be abstracted from procedural development and the main goal of conceptual understanding and procedural development is to develop students' thinking abilities. It is important to note that Chinese teachers believe that the concept of "procedure" is more than "a series of steps followed in a regular definite order" (Merrian Webster's Dictionary). To Chinese teachers, "procedure" is a "learning process" in which students engage in comparing, analyzing, applying, and synthesizing learning, and students' conceptual understanding and thinking abilities can be fully developed through the procedural development.

Effective teaching methods. The U.S. teachers were split on which method they most used: One group (44%) preferred student-centered instruction, while another group (40%) liked to use both student-centered and teacher-centered instruction. In the observation of classrooms, most of the U.S. teachers were able to apply student-centered or both methods.

In contrast, 81% of Chinese teachers identified student-centered inquiry as the most effective method of teaching mathematics. Their responses indicated that they were all aware of students being the main focus of learning, and knew that the student-centered approach is the process in which students play a role and experience learning. However, in the observation of classroom teaching, few Chinese teachers used a student-centered approach to teaching. The reason the teacher-centered method was used may be due to the large classes in China and the cultural influence on teaching. In China, teacher-centered instruction is considered to be a heuristic method, in which a teacher inspires and promotes students to think deeply and to learn actively.

Meeting individual students' needs. Both the U.S. and Chinese teachers used various ways to approach students' individual needs. However, Chinese teachers tend to design different assignments to meet the differences among students. In this study, 93% of U.S. teachers have less than 35 students in their classes, while 97% of Chinese teachers have more than 50 students in their classes. The differences in the number of students in classrooms in the two countries produced different ways of dealing with students' individual differences. The U.S. teachers would deal with students' individual differences using a variety of approaches: 63% of responses indicated that teachers would deal with students' needs using various teaching methods; 19% of responses liked to give different assignments; 15% of responses indicated they tutor students; and 7% of responses indicated the importance of motivating students. The diverse approaches used by U.S. teachers reflected their attention on students' needs

In contrast, Chinese teachers deal with students' individual differences mainly in one way: 75% of responses indicated that teachers would deal with students' individual differences by giving different assignments. Since most Chinese teachers have more than 50 students in each class, to deal with students' individual differences, Chinese teachers would design different levels of problems for students and ask different types of questions during the class. In addition, 44% of responses indicated that teachers would deal with students' individual differences by motivating them in different ways. Only 22% of responses indicated that teachers would deal with students' individual differences using various teaching methods; and 9% of responses indicated that teachers would deal with students' individual differences by tutoring students.

4.3 Importance of teachers' knowledge

Knowledge of students' thinking. All teachers agreed that it is important for teachers to understand students' thinking and understanding in learning mathematics. There are differences in the ways teachers gauge an understanding of students' thinking: 56% of U.S. teachers evaluated students' thinking from students' explanations and discussion; and 41% of U.S. teachers knew students' thinking by asking students questions. Although 72% of Chinese teachers understood students' thinking from students' explanations and discussion, 63% of Chinese teachers knew students' thinking by checking students' homework and conversation with students, while only 19% of U.S. teachers refer to homework.

Promoting students' ability to think. The U.S. teachers promoted students' ability to think in several ways: 22% of teachers promoted students' ability to think through cooperative learning; 15% of them asked different questions; 11% of them engaged students in various activities; 7% of teachers used inquiry and creativity; and 4% of teachers used problem solving strategies.

Although Chinese teachers promoted students' ability to think from different aspects, 44% of Chinese teachers promoted students' ability to think using different levels of practice compared to 7% of the U.S. teachers who used the same approach.

Pedagogical content knowledge. Both groups of teachers agreed that with deep pedagogical content knowledge, teachers would know a variety of methods, understand students' thinking, and know how to teach students at different levels. 41% of the U.S. teachers' responses indicated that a teacher with an in-depth understanding of both mathematics content and teaching methods is critical for effectively teaching compared to 84% of Chinese teachers with the similar view. 22% of the U.S. teachers believed that a teacher with an in-depth understanding of mathematics content and teaching would know students' thinking and understanding compared to 16% of Chinese teachers with the same view. In addition, 13% of Chinese teachers indicated the understanding of mathematics and teaching will help them better understand the curriculum and textbooks.

All the Chinese teachers in this study believed that it is very important for a teacher to have an in-depth understanding of the mathematics being taught and to have a deep understanding of mathematics teaching. This study found that most Chinese teachers' beliefs in the pursuit of more mathematical knowledge were influenced by Chinese culture. For example, as the reason for continuing learning, 15 Chinese teachers quoted the Chinese saying, "If you want to give the students one cup of water, you should have one bucket of water of your own." Chinese teachers believe that with deep knowledge, teachers would be able to teach mathematics at a higher level and explain mathematics in simpler ways. It also helps teachers to understand textbooks and grasp key points. Importantly, teachers should have the ability to link knowledge and connect other subjects with mathematics teaching.

Enhancing knowledge of math teaching. Both U.S. and Chinese teachers enhanced their knowledge using various approaches. 56% of the U.S. teachers enhanced their mathematics and teaching knowledge through in-service study and workshops. 33% of teachers developed their knowledge from independent study; 11% of teachers enhanced their knowledge from college study; and 11% of teachers gained more knowledge from sharing with colleagues.

In contrast, although Chinese teachers used numerous approaches to develop their knowledge, 88% of responses indicated that they enhanced their knowledge from independent study; 44% of responses indicated they enhanced their knowledge from continuing education in college; and 28% of responses indicated they improved their knowledge by sharing with colleagues and observing each other's classes.

The major different approaches in pursuing knowledge between the two groups of teachers shows that U.S. teachers relied on in service study and workshops, while most Chinese teachers enhanced their knowledge from self-study. By attending professional development meetings, U.S. teachers were able to learn new ideas and techniques for their teaching.

4.4 Planning for instruction

Approach of planning for math instruction. The results of this study showed that the U.S. teachers used a variety of ways to plan for instruction: 48% with a team; 19% by using textbooks; 11% by using the curriculum; and 7% according to the students' needs. However, Chinese teachers plan for mathematics instruction by focusing on two aspects: textbook and students' needs: 81% by using textbooks, 75% by gauging students' ability level, 34% according to the curriculum, and only 3% in a team. Among the responses, 59% of Chinese teachers planned for instruction according to a combination of their students, textbooks, and curriculum.

Although Chinese teachers often discuss their lesson plans with colleagues, most Chinese teachers would plan instruction according to textbooks and students' needs. Chinese teachers considered textbooks as the basis of planning, and students' needs are the main source of planning. Chinese teachers in this study frequently quote the very famous Chinese education philosophy: Teaching according to students' background and needs, and teaching according to the textbook.

Furthermore, Chinese teachers often observe colleagues' teaching in their schools. Most school districts also have open classes for teachers to observe once a month. Those open classes provide models for teachers and help them gain insights to teach. It is worth noting that the U.S. teachers do not write a detailed lesson plan, writing only a simplified outline for a lesson, while Chinese teachers' lesson plans are detailed teaching notes, which include the objectives, materials, teaching methods, the types of questions asked, the examples given, alternative ways of problem solving, summary, etc. In Chinese teachers' lesson plans, the sequential layered questions and problems are designed by teachers to enhance students' ability to master the methods of thinking. Usually, it takes at least two to five pages for one lesson plan for Chinese teachers, while U.S. teacher's lesson plan is a form for a week or outlines for several weeks.

Time spent on planning instruction. The results of this study shows that most U.S. teachers (74%) used 30 minutes to one hour for daily planning, while most Chinese teachers (84%) used one hour to two hours for daily planning. The different amounts of time in planning between the U.S. and China may be due to the educational system. However, with sufficient planning time, Chinese teachers are able to fully study and understand the subjects and their requirements, to write a detailed lesson plan, and to design an effective way of teaching mathematics.

Importance of assigning homework. The U.S teachers assign homework for different purposes: 44% of the U.S. teachers believed that the purpose of assigning homework is to review and practice; 34% of teachers thought it is to reinforce knowledge; and 19% of teachers stated it is to check understanding. In addition, 74% of the U.S. teachers believed that most of their students typically do their homework. In contrast, 94% of Chinese teachers thought that the purpose of homework is to reinforce and understand knowledge; 22% of Chinese teachers believed it is to check understanding; 13% of Chinese teachers believed it is to review and practice concepts. 94% of Chinese teachers believed that most of their students typically do their homework. An almost equal percentage of teachers in both countries agreed that assigning homework is for checking for understanding.

However, to most of the U.S. teachers, assigning homework is for review and practice, while to most Chinese teachers, knowledge and understanding could be reinforced from doing homework.

Approach to grading homework. Since it is known that U.S. teachers do not usually grade students' homework one problem by one, questions like "How do you grade your students homework? Do you grade each problem on each students' homework everyday?" were not asked in the survey; instead, these questions were asked during the interviews with teachers.

The results indicated that the U.S. teachers used various approaches to grade students' homework. However, they do not grade every problem. The results showed that 75% of U.S. teachers grade homework in class, that is they give answers to students and have students' check their own work; 25% of U.S. teachers grade homework by completion; 13% of teachers grade homework by effort, and only 25% of teachers actually grade students' homework one by one. Chinese teachers' approaches are totally different from U.S. teachers. In this study, 97% of Chinese teachers always grade students' homework one by one, analyze homework errors, and have students make corrections for errors. Besides grading students' homework, 33% of Chinese teachers also grade homework "face to face" with the student.

In the observation of the U.S. teachers' classes, most teachers grade students' homework during the class by calling the answers out to students or simply giving completion grades. This way of checking homework does not help students to progress in learning and neither does it help the teacher to know students' thinking. Consequently, the teacher later spends additional time to mend a big "hole" in students' understanding and skill. However, Chinese teachers not only grade each problem on a students' homework, but also correct students' misconceptions. Chinese teachers believed that to nip the "blind point" in the bud will help students realize and eliminate errors in the early stage.

5. EDUCATIONAL IMPORTANCE OF THE STUDY

This study included comparisons and contrasts of teachers' beliefs, not in a vacuum, but situated within the cultural milieu of each country. A new perspective on the two-way relationship between teachers' beliefs and teaching practice is developed in this study. The results of the study show that the mathematics teachers from different cultures have their own beliefs, which they translate into unique teaching approaches. The Chinese teaching system emphasizes gaining correct conceptual knowledge by reliance on traditional, more rigid procedural development and layered practices, which have proven their value for teaching content. The United States system emphasizes a variety of newer but less well-tested activities designed to promote creativity and independent thinking over concept mastery. Both approaches have benefits and limitations. The practices of each country may be partially adapted to help overcome deficiencies in the other, but wholesale transplantation of pedagogies without regard to the cultural environment and cultural tradition is not applicable and even harmful to each other's mathematics education.

REFERENCES

An, S., 2000, A Comparative Study of Mathematics Programs in the U.S. and China: The Pedagogical Content Knowledge of Middle School Mathematics Teachers in the U.S. and China. Doctoral dissertation, Texas A&M University, College Station, TX.

An, S., Kulm, G. and Wu. Z., 2004, The Pedagogical Content Knowledge of Middle School Mathematics Teachers in China and the U.S., *Journal of Mathematics Teacher Education*.

Ashmore, R.A. and Zhen Cao., 1997, *Teacher Education in the People's Republic of China*. Phi Delta Kappa Educational Foundation, Bloomington, IN.

Boaler, J., 2000, Exploring Situated Insights into Research and Learning, *Journal for Research in Mathematics*, **31**:113-119.

Cabello, B., and N.D. Burstein, 1995, Examining Teachers' Beliefs about Teaching in Culturally Diverse Classrooms, *Journal of Teacher Education*, **46**(4): 285-294.

Cooney, T.J., and Shealy, B.E., 1997, On Understanding the Structure of Teachers' Beliefs and their Relationship to Change, in: *Mathematics Teachers in Transition*, E. Fennema & B.S. Nelson, eds., Lawrence Erlbaum Association Publishers, Mahwah, NJ.

Ernest, P., 1989, The Impact of Beliefs on the Teaching of Mathematics, in: *Mathematics Teaching: The State of the Art*, The Falmer Press, New York.

Franke, M., Fennema, E., and Carperter, T.P., 1997, Changing Teachers: Interactions between Beliefs and Classroom Practice, in: *Mathematics Teachers in Transition*, E. Fennema and B.S. Nelson, eds., Lawrence Erlbaum Association Publishers, Mahwah, NJ.

Guba, E.G. and Lincoln, Y.S., 1994, Competing Paradigms in Qualitative Research, in: *Handbook of Qualitative Research*, N.K. Denzin and Y.S. Lincoln, eds., Thousand Oaks, CA: Sage.

Kagan, D.M., 1992, Implications of Research on Teacher Beliefs, *Educational Psychologist*, **27**:65-90.

Lincoln, Y.S., and Guba, E.G., 1985, *Naturalistic Inquiry*, SAGE Publications, Inc, Newbury Park, CA.

National Council of Teachers of Mathematics (NCTM), 1989, *The Curriculum and Evaluation Standards for School Mathematics*. Author, Reston, VA.

National Council of Teachers of Mathematics (NCTM), 2000, *Principles and Standards for School Mathematics*, Author, Reston, VA.

Nespor, J., 1987, The Role of Beliefs in the Practice of Teaching, *Journal of Curriculum Studies*, **19**:317-328.

Pinar, W.F., Reynolds, W.M., Slattery, P., and Taubman, P.M., 1995, *Understanding Curriculum*, Peter Lang, New York.

Shulman, L., 1987, Knowledge and Teaching: Foundations of the New Reform, *Harvard Educational Review*, **57**(1):1-22.

Spring, J., 1998, *Education and the Risk of the Global Economy*, Lawrence Erlbaum Associates, Mahwah, NJ.

Stigler, J.W., and Hiebert, J., 1999, *The Teaching Gap*, The Free Press, New York.

Stigler, J.W., and Perry, M., 1988, Cross-cultural Studies of Mathematics Teaching and Learning: Recent Finding and New Directions, in: *Effective Mathematics Teaching Directions*, Grouws, D., and Cooney, T., eds., National Council of Teachers of Mathematics, Reston, VA, pp.194-223.

Thompson, A.G., 1992, Teachers' Beliefs and Conceptions: A Synthesis of the Research, in: *Handbook of Research on Mathematics Teaching and Learning*, Grouws, D., ed., Macmillan Publishing Company, New York.

Chapter 4-3

U.S. AND CHINESE TEACHERS' CULTURAL VALUES OF REPRESENTATIONS IN MATHEMATICS EDUCATION

CAI Jinfa
University of Delaware

1. INTRODUCTION

A major goal of educational research is to improve learning opportunities for all students. Cross-national studies provide unique insights into issues in the teaching and learning of mathematics as well as diagnostic and decision-making information about how to improve students' learning. During the past several years, an attempt has been made to explore the impact of curriculum and instruction on U.S. and Chinese students' mathematical thinking. Previous studies have revealed remarkable differences between U.S. and Chinese students' mathematical thinking and reasoning (e.g., Cai, 2000; Cai & Hwang, 2002). By investigating factors that may contribute to these types of cross-national performance differences, we are just now beginning to understand these cross-national differences.

Although there is no universal agreement as to whether mathematics is a culturally bound subject, no one questions the idea that the teaching and learning of mathematics is a cultural activity (Bishop, 1988). Since teachers have an important effect on the ways students learn and think about mathematics, one may hypothesize that the differences between U.S. and Chinese students' thinking are related to the differences in their teachers' beliefs about mathematics and their conceptions about teaching mathematics. This paper is a progress report of a larger research project that examines the impact on students' thinking of U.S. and Chinese teachers' conceptions and their constructions of pedagogical representations. In particular, the purpose of this paper is to analyze U.S. and Chinese teachers' evaluations of a set of student responses in an attempt to understand U.S. and Chinese teachers' cultural values of representations and strategies in mathematics education.

2. THEORETICAL BASIS OF THE STUDY

In the field of educational research, there has been an increased interest in investigating how teaching and learning are connected (e.g., Fennema et al., 1996; Hiebert & Wearne, 1993; Stein & Lane, 1998). Rather than studying teaching and learning separately, educational researchers have started to study the mechanisms by which teaching and learning are related as well as the processes by which students construct meaning from classroom instruction. Since classroom instruction is a complex enterprise, researchers have attempted to identify important features of classroom instruction to investigate how teaching and learning are related. "Representation" is an important construct in the research about the teaching and learning of mathematics because it is both an inherent part of mathematics and an instructional aid for making sense of mathematics (Ball, 1993; NCTM, 2000; Perkins & Unger, 1994). In mathematics, a representation must necessarily be used to express any mathematical object, statement, concept, or theorem (Dreyfus & Eisenberg, 1996). "Virtually all of mathematics concerns the representation of ideas, structures, or information in ways that permit powerful problem solving and manipulation of information" (Putnam, Lampert, & Peterson, 1990, pp. 68).

2.1 Previous research on solution representations

Solution representations are both thinking and representational tools in problem solving. After a solver solves a mathematical problem, she/he communicates the thinking involved in the solution with certain representtations that express the solution processes. In other words, solution representations are the visible records generated by a solver to communicate his or her thinking about the solution processes. Clearly, solution processes can be recorded using different representations.

In previous studies related to this project, open-ended tasks were used to examine thinking and reasoning involved in Chinese and U.S. students' mathematical problem solving and problem posing (e.g., Cai, 2000; Cai & Hwang, 2002). These studies consistently revealed that Chinese students tended to use symbolic representations (e.g., arithmetic or algebraic symbols), U.S. students, on the other hand, tended to use visual representations (e.g., pictures). For example, when students were asked to find the number of blocks needed to build 20-step and 100-step staircases, over 20% of the U.S. 6th graders attempted to draw a 20-step staircase to arrive at an answer and nearly 10% of the U.S. 8th grade students tried to draw a 100-step staircase. By contrast, only a few of the Chinese 4th, 5th, and 6th graders answered the 20-step question this way and none of the Chinese students tried to draw 100-step staircase. When the U.S. and Chinese students were asked to generate mathematical problems based on a similar staircase situation, a considerable number of the U.S. students generated problems with pictures, but few Chinese students did.

Later, Cai and Hwang (2002) examined the nature of U.S. and Chinese students' generalized and generative thinking in mathematical problem solving and problem posing. Across the tasks, the Chinese students had higher rates of success than did the U.S. students. The disparities in the U.S. and Chinese students' problem-solving success rates were related to their use of different strategies and representations. Chinese students tended to choose abstract strategies and symbolic representations, while U.S. students tended to choose concrete strategies and drawing representations. If the analysis is limited to those U.S. and Chinese students who used concrete strategies, the success rates between the two samples become very similar. Therefore, the Chinese students' preference for abstract strategies seems to help them outperform the U.S. students on problems amenable to abstract strategies.

In these previous studies, the U.S. 6th grade students had not formally been taught algebraic concepts, but the Chinese 6th graders had received about 20 lessons on the topic. A recent study examined the extent to which U.S. and Chinese students' mathematical thinking is related to their opportunity to learn algebra (Cai, in press). The findings from the study showed that the Chinese 6th grade students' opportunity to learn algebra did not explain why they were less likely to use concrete visual representations than U.S. students. Even among the U.S. students who formally learned algebraic topics, a considerable number still used visual representations. In fact, U.S. 8th graders who had been taught formal algebra were more likely than Chinese 4th graders were to use concrete representations. These results suggest that we need to look beyond what was taught to understand the differences between U.S. and Chinese students' selection of strategies and representations.

2.2 Pedagogical representations

Adequate pedagogical representations play an important role in the way students learn and understand mathematics (Bransford et al., 2000; Hiebert & Carpenter, 1992). Pedagogical representations are the representtations teachers and students use in their classroom as carriers of knowledge and as thinking tools to explain a concept, a relationship, a connection, or a problem-solving process. As Dreyfus and Eisenberg (1996) indicated, "[a]ny representation will express some but not all of the information, stress some aspects and hide others" (pp. 267-268). In mathematics instruction, some representations might be more adequate than others as carriers of knowledge and thinking tools to explain a problem-solving process. Furthermore, pedagogical representations are effective in classroom instruction if they are known by students or are easily knowable. Indicators that students understand mathematics include their ability to use representations to express mathematical ideas and problems and their ability to move fluently within and between representations (Hiebert & Carpenter, 1992).

Although there is no universal agreement about what constitutes "good pedagogical representation" in mathematics teaching, no one questions the notion that teachers' beliefs, conceptions, and knowledge influence their selection of desirable pedagogical representations. There is no doubt that teachers' conceptions of what mathematics is affects their conceptions of how it should be presented (Thompson, 1992). A teacher's manner of presenting mathematics is both influenced by and indicative of what he/she believes to be most essential in it, thereby influencing the ways students understand and learn mathematics. However, we know very little about U.S. and Chinese teachers' conceptions and constructions of pedagogical representations in mathematics instruction. It is plausible that U.S. and Chinese students' use of different representations and strategies in problem solving reflects their teachers' differing views about various representations.

Although the Chinese 6th grade students were more likely than the U.S. 6th grade students to construct mathematical expressions and use symbols in their solutions, still a considerable number of them did not construct mathematical expressions or algebraic equations in their solutions. Nor did they choose to use a concrete, visual strategy. Why did these Chinese students not use concrete, visual approaches to solve the problems as the U.S. students did, since a concrete, visual strategy may provide entry-level, easily accessible tools for solving the problems? For example, about 80% of the Chinese 4th graders were unable to correctly conclude and justify that each boy gets more pizza than each girl if 2 pizzas were equally shared by 8 girls, and 1 pizza was equally shared by 3 boys. In fact, only 4% of the Chinese 4th graders used visual drawings even though a visual strategy might have benefited those Chinese students who did not use mathematical expressions (Cai, in press). Is it possible that teachers in China do not encourage visual strategies? If so, what are the Chinese teachers' cultural beliefs, if any, that made them discourage their students from using concrete, visual strategies?

A research project is currently underway to address these questions through extensive interviews and analyses of U.S. and Chinese teachers' lessons. The analysis of their lessons and interview transcripts contributes information about U.S. and Chinese teachers' cultural values of various representations from three aspects: (1) generating pedagogical representations for classroom instruction, (2) knowing students' representations and strategies in problem solving, and (3) evaluating students' representations and solution strategies. This paper reports some preliminary findings from a study analyzing how 11 U.S. and 9 Chinese teachers scored a set of 28 student responses.

3. METHOD

3.1 Selection of teachers

Eleven U.S. and 9 Chinese teachers participated in the study. The U.S. teachers were from Pennsylvania, North Carolina, and Wisconsin. The Chinese teachers were from Guiyang, Guizhou. U.S. and Chinese teachers were selected on the recommendations of a group of U.S. mathematics educators and a group of Chinese mathematics educators, respectively. All the selected U.S. and Chinese teachers were considered "distinguished" mathematics teachers in their respective regions according to local criteria. In particular, all selected U.S. teachers have taken leadership roles in their schools and/or school districts. They have led workshops or made presentations at regional or national mathematics education conferences. All of the U.S. teachers received at least one teaching award, such as "teacher leader," "district teacher of the year," or "the Presidential Award for Teaching Excellence in Mathematics and Science." All Chinese teachers have ranks of "first class teacher" or "special class teacher," the top two ranks in China for ranking teachers.

Teachers are recognized as distinguished in their respective regions because their teaching embodies the culturally accepted values of effective mathematics instruction. Therefore, the inclusion of distinguished mathematics teachers may help us understand U.S. and Chinese teachers' cultural values in mathematics education. During the time of this study, 8 of the U.S. teachers and all of the Chinese teachers were teaching 6th grade mathematics; the remaining 3 U.S. teachers were teaching 7th grade math. These 3 U.S. teachers had taught 6th grade math the year before the study. Three of the U.S. teachers and 4 of the Chinese teachers were selected from schools that were involved in a previous study examining U.S. and Chinese students' mathematical thinking (e.g., Cai, 2000). The inclusion of these teachers allows the establishment of a link between the teachers' conceptions of mathematics and their students' mathematical thinking. All the other U.S. and Chinese teachers were from schools not involved in the previous studies.

3.2 Interview procedures

All the teachers were interviewed by asking them to score a set of 28 student responses using a general 5-point scoring rubric (0-4):

4 points - correct and complete understanding
3 points - correct and complete understanding, except for a minor error, omission, or ambiguity
2 points - partial understanding of the problem or related concept
1 point - a limited understanding of the problem or related concept
0 point - no understanding of the problem or related concept

These 28 responses consisted of various students' solutions to seven problems. Each student response had a correct answer (or a reasonable estimate for the answer) and an appropriate strategy that yielded the correct answer (or estimate), but representations and solution strategies in these responses were different. The teachers were asked to explain the reasons for their scoring. After they completed their scoring, they were asked to judge the sophistication of the representations and strategies used in the responses to each problem. It should be indicated that all the problems and student responses upon which the interview questions are based were from previous studies of U.S. and Chinese students' mathematical thinking. All interviews were videotaped.

3.3 Translation equivalence

In a cross-national study, it is absolutely essential to ensure the equivalence of the two language versions of the instruments. Although the 28 student responses were selected from students' actual work, both the Chinese and English versions of these responses were re-written by an educator to avoid possible biases and misinterpretations. To ensure the equivalency of the two versions, two people literate in both Chinese and English contributed to the translation of the student responses. One person first translated them from English into Chinese. The second person then compared the translated Chinese version with the originally-prepared Chinese version to ensure equivalence and consistency except for intentional changes involving culturally appropriate words like personal names, object names, contexts, and terminology. The presentations of the students' work and explanations were identical except that one was in Chinese and the other in English.

4. RESULTS

Table 1 below shows the means of the scores that U.S. and Chinese teachers assigned to the 28 student responses. The U.S. teachers assigned higher scores than did the Chinese teachers on a vast majority of the responses. In fact, the overall mean score for the 11 U.S. teachers is 3.47, while the overall mean score for the 9 Chinese teachers is 3.09. On 25 out of 28 student responses, the U.S. teachers gave higher scores than the group of Chinese teachers did. On 2 of the 28 student responses, the Chinese teachers gave higher scores than the U.S. teachers did, but the differences were very small. On the remaining response, the Chinese teachers (mean = 3.67) scored it much higher than the U.S. teachers did (mean = 2.73). This response involves a Number Theory Problem, which allows for multiple correct answers. The U.S. teachers scored it lower because four of them did not recognize the correctness of the answer in this response.

Table 4-3-1. Mean Scores Given by U.S. and Chinese Teachers

Response	China	US	Response	China	US
A	3.56	3.73	O	2.44	3.73
B	2.89	3.45	P	4.00	3.91
C	3.44	3.82	Q	2.44	3.82
D	3.78	3.55	R	3.44	3.73
E	2.33	2.55	S	2.89	3.27
F	3.89	3.91	T	3.78	3.91
G	3.44	3.73	U	3.00	3.82
H	1.33	1.82	V	3.78	3.82
I	2.56	3.64	W	3.89	3.91
J	3.67	3.91	X	3.56	3.64
K	1.56	2.00	Y	1.33	2.18
L	3.56	3.73	Z	3.44	3.91
M	2.67	3.36	AA	3.67	2.73
N	2.67	3.82	BB	3.56	3.73

Across the 28 responses, both U.S. and Chinese teachers showed very high internal consistency in their scoring. In fact, Cronbach's alpha is .7217 for the U.S. teachers and .7031 for the Chinese teachers. Although there is high internal consistency for both groups of teachers, the analysis of the interview transcripts reveals differences between the two groups' scoring of particular responses. In addition, the interview transcripts show there are different underlying reasons for their scoring. Generally speaking, the analysis of interview transcripts showed that the Chinese teachers focused their scoring on "what is missing", while the U.S. teachers focused their scoring on "what is there". Specific differences, categorized according to four themes, are described below.

4.1 Algebraic approach: It is valued highly, but should it be expected?

Algebraic approaches were used in 3 of the responses. For example, Response P to the Odd Number Pattern Problem, shown below, involves an algebraic approach. In Response P, the student found and used the general expression (2n – 1) to find the number of guests that entered on the nth ring. To answer part C of the problem, the student set 2n – 1 = 99 and solved for n, which is 50. All the U.S. and Chinese teachers, except 1 U.S. teacher, gave it 4 points.

> **Odd Number Pattern Problem**: Sally is having a party. The first time the doorbell rings, 1 guest enters. The second time the doorbell rings, 3 guests enter. The third time the doorbell rings, 5 guests enter. The fourth time the doorbell rings, 7 guests enter. Keep going in the same way. On the next ring a group enters that has 2 more persons than the group that entered on the previous ring.
>
> A. How many guests will enter on the 10th ring? Explain or show how you found your answer.
>
> B. In the space below, write a rule or describe in words how to find the number of guests that entered on each ring.
>
> C. 99 guests entered on one of the rings. What ring was it? Explain or show how you found your answer.

Almost the all U.S. and Chinese teachers scored the responses with algebraic approaches the highest when compared to other responses to the same problem. For example, there are 5 responses (see below) for the following Map Ratio Problem:

> **Map Ratio Problem**: The actual distance between Grantsville and Martinsburg is 54 miles. On the map, Grantsville and Martinsburg are 3 centimeters apart. On the map, Martinsburg and Rivertown are 12 centimeters apart. What is the actual distance between Martinsburg and Rivertown?

Response G: The student first found the many of miles that a centimeter on the map represents (54/3=18), then multiplied the result by 12 to get the actual distance that the 12 centimeters on the map represents (18 x 12=216).

Response H: The student used a finger to measure the distance between Martinsburg and Grantsville on the map, then used the measurement unit to measure the length between Martinsburg and Rivertown on the map and to find the number of the unit of the length. By multiplying the number of finger lengths by 54, the student found the distance between Martinsburg and Rivertown.

Response I: The student first multiplied 3 by 4 and got 12. Since 3 centimeters represent 54 miles, the actual distance represented by the 12 centimeters was 216 (4 x 54=216).

Response J: The student set up a formal proportional relationship to find the actual distance (i.e., $3/12 = 54/x$, $x = 216$ miles).

Response K: A student used a paper clip as a unit to measure the distance between Martinsburg and Grantsville on the map, then used the measurement unit to measure the length between Martinsburg and Rivertown on the map and to find the number of the unit of the length. By dividing 54 by the number of the measurement unit between Martinsburg and Grantsville, the student found the number of actual miles per measurement unit. By multiplying the number of the measurement unit between Martinsburg and Rivertown by the number of actual miles per paper-clip unit, the student found the number of actual miles between Martinsburg and Rivertown.

Both the U.S. and Chinese teachers scored Response J the highest. A few U.S. and Chinese teachers deducted 1 point because they thought the explanation in Response J was not complete. For example, U.S. Teacher 1 wanted to see the proportion labeled and more explanation given about how the equation was set up. Chinese Teachers 2 and 3 gave a response involving the algebraic approach only 3 points because, in the response, the students did not explain what "x" meant. If these responses had included a sentence like "Let x be ...," the responses would have been scored 4 points by the 2 Chinese teachers.

Although all the U.S. and Chinese teachers highly valued responses with algebraic approaches, the U.S. teachers seemed to have different expectations than the Chinese teachers did. All of the U.S. teachers, except for Teacher 6, believed that in general 6th grade students in the United States should not be expected to solve problems using algebraic approaches. For example, U.S. Teacher 9 said, "I wish my 6th graders could do this. But in our school, only 7th or 8th grade students are taught algebraic concepts, and 6th graders are only learning pre-algebra and are not expected to solve problems using this kind of approach involving x's. At this point, I am happy if they can do it no matter what they use." On the other hand, all the Chinese teachers expected their 6th graders to solve problems using algebraic approaches.

4.2 Visual or concrete approach: It works, but is it efficient?

Chinese teachers consistently took the nature of the solution strategies into account in their scoring. If a response involved a visual or concrete strategy, Chinese teachers usually gave a relatively lower score even though the strategy was appropriate for the correct answer. For example, in order to find the number of blocks needed for building 5-step and 20-step staircases, Response N contains correctly drawn pictures of 5-step and 20-step staircases. The Chinese teachers gave only 2 or 3 points for this response, but 9 of the 11 U.S. teachers gave it 4 points, and the remaining two awarded 3 points. Most U.S. teachers acknowledged that the drawing in Response N was not a sophisticated strategy, and it was very time-consuming to use this strategy. However, they recognized that the drawing in Response N was a viable approach that produced correct answers. Moreover, almost all U.S. teachers stated that these visual drawings clearly showed how students thought about the problems and how they solved them.

Response Q involves the Odd Number Pattern Problem mentioned before. In Response Q, tables were created to solve the problem. In particular, a long table from ring number 1 to ring number 50 was created to list the number of guests entering on each ring, and then to determine the ring number when 99 guests entered. Like Response N, Chinese teachers gave only 2 or 3 points for Response Q, but 9 of the 11 U.S. teachers gave 4 points and the other 2 gave 3 points. Chinese teachers gave lower scores to the responses with visual or concrete approaches because "It is difficult to solve for larger numbers" (Chinese Teacher 1), "The approach is not efficient" (Chinese Teacher 4), "For [Response] N, the construction of the staircases are accurate, but the reasoning process is not as good as that in other responses" (Chinese Teacher 2), or "It is really troublesome to draw and not to find regularities among numbers" (Chinese Teacher 9).

The Chinese teachers seem to have a clear goal: students should learn more efficient strategies. The following excerpt from Chinese Teacher 7 is just one of the examples showing that Chinese teachers have such a goal: "Being able to solve a problem is good, but just the first step. Through mathematics instruction, we want students to learn generalized problem-solving methods. They should be able to 'Ju Yi Fan San' and 'Chu Lei Pang Tong (i.e., make generalizations and transfer them to other problem situations)." However, there is no evidence from the interviews that U.S. teachers have the clear goal that students should learn efficient strategies.

Instead, the U.S. teachers' goal seems to be that students solve a problem no matter what strategies they use.

Perhaps because the Chinese teachers believed that students should learn more efficient strategies, they seemed to have less internal consistency than did the U.S. teachers on scoring the responses involving drawing or making a list. The U.S. and Chinese teachers' scoring of Response U is a good example. Response U involves a drawing strategy to solve the following Hats Average Problem:

> **Hats Average Problem**: Angela is selling hats for the Mathematics Club. She sold 9 hats in Week 1, 3 hats in Week 2, and 6 hats in Week 3. How many hats must Angela sell in Week 4 so that the <u>average</u> number of hats sold is 7?"

In Response U, the student used a drawing to show how to use the leveling-off processes to solve the problem. The student viewed the average (7) as a leveling basis to line up the numbers of hats sold in Weeks 1, 2, and 3. Since 9 hats were sold in week 1, the drawing for Week 1 shows two extra hats beyond the average level. Since 3 hats were sold in week 2, 4 additional hats in Week 2 are needed in order to line up with the average. Since 6 hats were sold in Week 3, it needs 1 additional hat to line up with the average. In order to have enough hats to rearrange so that each week lines up with the average number of hats sold over the 4 weeks, 10 hats should be sold in Week 4.

The majority of the U.S. teachers gave it 4 points, and no one gave it less than 2 points. However, equal numbers of Chinese teachers gave it 2, 3, or 4 points. Chinese teachers seem to hold two different views regarding a response like this. At least three of the nine Chinese teachers felt that the drawing approach to leveling would be difficult to use when solving similar problems involving larger numbers, so it should not be scored 3 or 4 points. However, some other Chinese teachers maintained this approach should be scored 4 points because it shows students' creativity as well as their understanding of the averaging process.

4.3 Estimate of an answer: It is reasonable, but is it enough?

Responses H, K, and Y provided only estimates of answers. Both U.S. and Chinese teachers not only scored these responses the lowest, but also they scored them with the biggest variations. For Response H, 5 Chinese teachers scored it 2 points, 2 scored it 1 point, and 2 scored it 0 point. The U.S. teachers' scores for Response H ranged from 0 to 3 points. All the Chinese and 6 of the U.S. teachers liked the thinking processes involved in Response H and realized that the thinking processes in Response H were similar to those in Response I. However, these Chinese and U.S. teachers commented that the estimate of an answer is not enough even if the estimate is very reasonable and good. As U.S. Teacher 7 pointed out, "They've got everything to figure out the problem, but they seem not to care and don't use them. Regardless how good the estimate is, it is just a wrong answer." Four U.S. teachers gave it 3 points, citing that the approach of measuring with the paper clip is acceptable. "They could do better because of the information given to them, but I guess without using all the information they've attacked the problem well. This student proved an understanding of ratios" (U.S. Teacher 11).

The variation for each group of teachers was even bigger for Response Y than for Response H or K. Response Y involves an estimate for the following Score Average Problem:

> **Score Average Problem**: The average of Ed's ten test scores is 87. The teacher throws out the top and bottom scores, which are 55 and 95. What is the average of the remaining set of scores? Show how you found your answer.

In Response Y, the student said that the average for the remaining set of scores is between 55 and 95. But 87 is closer to 95 than 55. So the average for the remaining set of scores must be about 90. Chinese teachers gave this response either 0 or 2 points, and U.S. teachers' scores ranged from 0 to 3 points for this response. The 3 Chinese teachers who gave it 0 point expressed a concern about guessing instead of using rigorous mathematical reasoning. Eight of the 9 Chinese teachers explicitly commented that "it is a bad habit to guess for solving a problem like this and deducting points may help students overcome such a habit." The two U.S. teachers who gave it 0 or 1 point commented that students should provide precise answers for a problem like this. On the other hand, the remaining nine U.S. teachers, who gave it 3 points, thought that the response showed students' understanding of some properties about arithmetic mean. Because of that, they felt the response deserved some points.

4.4 Focus on details: It is important, but what is the consideration behind?

When Chinese teachers scored the 28 students' responses, they consistently focused on details. Were units attached to the answers? Did the students write their responses in an appropriate format? For example, in Response I (3 x 4 = 12. 4 x 54 = 216. Therefore the answer should be 216), five Chinese teachers only gave it 2 points and only one Chinese teacher gave it 4 points. In contrast, for Response G (54/3 = 18. 12 x 18 = 216.

Therefore, the answer should be 216), all the Chinese teachers scored the response 3 or 4 points. The reason some Chinese teachers scored Response I lower was that students in the response used the number "4" (3 x 4 = 12) which was not given in the problem. The following excerpt is from the interview with Chinese Teacher 4.

Teacher:	It [Response I] is 2.
Interviewer:	Okay!
Teacher:	It has no units and 3 x 4 = 12 should be 12/3 = 4.
Interviewer:	Why should 3 x 4 = 12 be 12/3 = 4?
Teacher:	Four is not there [in the problem]. Where is the 4 from? Students must have guessed mentally about the number 4. It would be very hard to guess if the number is larger. In examinations, you will always have big numbers, and I don't think students should just guess. Otherwise, their scores will be deducted in their examinations. It is better to ask them to pay attention to little things in daily practice, so they will have good habits to write their solutions.
Interviewer:	What if the student has had 12/3 = 4?
Teacher:	It is going to be 3, but still not 4, because there are no units there.
Interviewer:	What do you mean about units?
Teacher:	4 x 54 = 216. What does 216 mean? What is the unit for 216? They need to understand the units.

The Chinese teachers' concerns about the details of the written format and the inclusion of units for answers might be related to the examination culture in China. In fact, every Chinese teacher mentioned the grading criteria in city-wide or region-wide common examinations at least once. In particular, Chinese Teacher 3 referred 11 times to writing requirements in examinations in her scoring of the 28 responses. The Chinese teachers' concerns about the details of the written format and the inclusion of units in answers are also related to their beliefs about understanding mathematics. Chinese teachers seem to believe that the use of an appropriate written format and the inclusion of units in problem solving can help students develop their abilities to think logically.

In contrast, U.S. teachers were not as concerned with details about written formats. For example, in Response O, the following expressions were included to find the number of blocks needed to build 20-step staircase: 1 + 2 = 3 + 3 = 6 + 4 = 10 + 5 = 15 + 6 = 21 + 7 = 28 + 8 = 36 + 9 = 45 + 10 = 55 + 11 = 66 + 12 = 78 + 13 = 91 + 14 = 105 + 15 = 120 + 16 = 136 + 17 = 153 + 18 = 171 + 19 = 190 + 20 = 210. Eight out of the 11 U.S. teachers gave it 4 points. Such imprecise writing did not seem to bother them at all. As U.S. Teacher 10 commented, "No, it does not bother me. The little fellow just put down what he thinks in his head. Isn't it the way we think?" However, no Chinese teachers gave it 4 points. The explanation provided by Chinese Teacher 5 was typical: "The result is correct, but there are some mistakes in the writing. Two sides of an equal sign should be equal."

5. DISCUSSION

By analyzing their scoring of 28 student responses, this paper shows how cultural values of U.S. and Chinese teachers affect their appraisal of solution representations and strategies. Overall, U.S. teachers are much more lenient than Chinese teachers are in their scoring. However, U.S. teachers' leniency cannot be detected in their evaluation of students' responses involving conventional approaches, such as using algebraic equations and other mathematical expressions. In fact, almost all the U.S. and Chinese teachers valued the responses with algebraic approaches the highest when compared to other responses in the same problem. Although all the U.S. and Chinese teachers highly valued responses with algebraic approaches, it is clear that U.S. and Chinese teachers hold different curricular expectations. Chinese teachers expect 6th graders to be able to use equations to solve problems, but for U.S. teachers this expectation only applies to 7th or 8th grade students.

The U.S. teachers' leniency was reflected on their rating of responses involving both visual strategies and estimates of answers. Chinese teachers consistently took the nature of the solution strategies into account in their scoring. If a response involved a visual or concrete strategy, Chinese teachers usually gave it a relatively low score even though the strategy could be used appropriately to arrive at a correct answer. While U.S. teachers acknowledged that the drawing strategy may not be a sophisticated strategy and may be very time consuming, they appreciated the fact that drawing is often a viable approach that produces correct answers. Therefore, U.S. teachers felt that a response with an appropriate concrete drawing strategy should not be penalized. The Chinese teachers seem to have a clear goal that students should learn more generalized strategies and they expect 6th grade students to use algebraic approaches. The U.S. teachers, on the other hand, seemed to be satisfied as long as their students were able to use an appropriate strategy to solve a problem. Furthermore, the U.S. teachers believed that in general 6th grade students in the United States should not be expected to solve problems using algebraic approaches.

Chinese teachers also gave lower scores for responses involving estimation than did U.S. teachers. For Chinese teachers, if a problem includes all of the information to provide an accurate answer, it is not desirable to simply estimate the answer. For U.S. teachers, if the process of solving a problem is sound and the process shows an understanding of the concept involved, the student response should receive a high score even though only an estimate of the answer is provided. In addition, Chinese teachers seem to be much more concerned about the details of the written format and the inclusion of units for answers than U.S. teachers are. Chinese teachers believe that the use of an appropriate written format and units in problem solving can help students develop their abilities to think logically. Such details are also required on Chinese examinations.

The fact that U.S. and Chinese teachers hold differing curricular expectations is not surprising since the curricula of the two countries are very different. However, on a deeper level, the differences in expectations may reflect the differences in cultural beliefs about mathematics and the learning of mathematics. Although both the U.S. and Chinese teachers agreed that mathematics has wide applications in the real world, the true beauty of mathematics for Chinese teachers was its purity, generality, and logic. Thus, a solution strategy that lacks generality (e.g., a visual approach) should be discouraged. In contrast, U.S. teachers heavily emphasized the pragmatic nature of mathematics: as long as it works, students can choose whatever strategies they like.

The differences in U.S. and Chinese teachers' scoring of responses involving visual approaches and estimates of answers appear to suggest the different cultural values of representations in mathematics education. Cultural beliefs do not dictate what teachers do. Nonetheless, teachers do draw upon their cultural beliefs as a normative framework of values and goals to guide their teaching (Bruner, 1996). Evaluation and scoring of student responses is a routine activity for both U.S. and Chinese teachers. This study indicates that U.S. and Chinese teachers may use such a routine activity to foster students' learning in very different ways. The U.S. teachers seemed to believe that as long as students can solve a problem using whatever viable strategies are available, the students should be encouraged by giving them full credit (positive reinforcement). On the other hand, Chinese teachers seemed to use negative reinforcement to help students form good habits by deducting points for less desirable solutions or written formats.

This paper includes only some preliminary results from a larger research project. Additional analyses are in progress to better understand U.S. and Chinese teachers' conception and construction of pedagogical representations. Nevertheless, the preliminary results not only demonstrate the U.S. and Chinese teachers' differential cultural values of representations in mathematics education, but the preliminary results also suggest the feasibility of using teachers' scoring of student responses as an effective way to examine teachers' values. In addition, findings from this study show the impact of U.S. and Chinese teachers' conceptions of representations on their students' thinking. It suggests that U.S. and Chinese students' use of differential solution strategies and representations may be due, at least in part, to their teachers' different cultural values of various representations.

ACKOWLEDGEMENT

The research discussed in this paper was supported by a grant from the Spencer Foundation. Any opinions expressed herein are those of the author and do not necessarily represent the views of the Spencer Foundation. I gratefully acknowledge the valuable assistance of Vic Cifarelli for collecting U.S. data and Chunghan Lu and Bingyi Wang for collecting Chinese data. Assistance provided by Erin Meyers and Xiujun Qi for transcribing interview data and analysis is greatly appreciated. Special thanks go to Alan Bishop and John Moyer who made valuable suggestions concerning earlier versions of this manuscript, thereby contributing to its clarity and improvement. Of course, any errors are solely the author's responsibility.

REFERENCES

Ball, D.L., 1993, Halves, Pieces, and Twoths: Constructing and Using Representational Contexts in Teaching Fractions, in: *Rational Numbers: An Integration of Research*, T.P. Carpenter, E. Fennema, and T.A. Romberg, eds., Erlbaum, Hillsdale, NJ, pp.328-375.

Bishop, A.J., 1988, *Mathematical Enculturation: A Cultural Perspective on Mathematics Education*, Kluwer Academic Publishers, Dordrecht.

Bransford, J.D., Brown, A.L., and Cocking, R.R., 2000, *How People Learn: Brain, Mind, Experience, and School* (expanded edition), National Research Council, Washington, D.C.

Bruner, J.S., 1996, *The Culture of Education*, Harvard University Press, Cambridge, MA

Cai, J., 2000, Mathematical Thinking involved in U.S. and Chinese Students' Solving Process-Constrained and Process-Open Problems, *Mathematical Thinking and Learning,* **2**:309-340.

Cai, J., in press, Why Do U.S. and Chinese Students Think Differently in Mathematical Problem Solving? The Impact of Early Algebra Learning and Teachers' Belief, *Journal of Mathematical Behavior.*

Cai, J., and Hwang, S., 2002, Generalized and Generative Thinking in U.S. and Chinese Students' Mathematical Problem Solving and Problem Posing, *Journal of Mathematical Behavior*, **21**:401-421.

Dreyfus, T., and Eisenberg, T., 1996, On Different Facets of Mathematical Thinking, in: *The Nature of Mathematical Thinking*, R.J. Sternberg and T. Ben-Zeev, eds., Erlbaum, Hillsdale, NJ, pp.253-284.

Fennema, E. et al., 1996, A Longitudinal Study of Learning to Use Children's Thinking in Mathematics Instruction, *Journal for Research in Mathematics Education*, **27**:403-434.

Hiebert, J., and Carpenter, 1992, Learning and Teaching with Understanding, in: *Handbook of Research on Mathematics Teaching and Learning*, D.A. Grouws, ed., Macmillan, New York, pp.65-97.s

Hiebert, J., and Wearne, D., 1993, Instructional Tasks, Classroom Discourse, and Students' Learning in Second Grade Arithmetic, *American Educational Research Journal*, **30**:393-425.

National Council of Teachers of Mathematics, 2000, *Principles and Standards for School Mathematics*, Author, Reston, VA.

Perkins, D.N. and Unger, C., 1994, A New Look in Representations for Mathematics and Science Leaming, *Instructional Science*, **2**:1-37.

Putnam P.T., Lampert, M., and Peterson, P.L., 1990, Alternative Perspectives on Knowing Mathematics in Elementary Schools, in: *Review of Research in Education*, C.B. Cazden, ed., American Educational Research Association, Washington, DC, pp.57-150.

Stein, M.K., and Lane, S., 1996, Instructional Tasks and the Development of Student Capacity to Think and Reason: An Analysis of the Relationship between Teaching and Learning in a Reform Mathematics Project, *Educational Research and Evaluation*, **2**:50-80.

Thompson, A.G., 1992, Teachers' Beliefs and Conceptions: A Synthesis of the Research, in: *Handbook of Research on Mathematics Teaching and Learning*, D.A. Grouws, ed., Macmillan, New York, pp.127-146.

Chapter 4-4

A COMPARISON OF MATHEMATICAL VALUES CONVEYED IN MATHEMATICS TEXTBOOKS IN CHINA AND AUSTRALIA

CAO Zhongjun, SEAH Wee Tiong, and Alan J. BISHOP
Monash University, Australia

1. BACKGROUND

Values are standards for making judgments on what is important (Thompson, 1995), and they occupy a more central place in our belief system compared with other affective qualities such as attitudes and beliefs (Oppenheim, 1966; Mandler, 1989). Probably due to the central place of value in our belief system, the value issues have been attracting more and more researchers' interest in the area of mathematics education. For instance, Buxton (1981) and Fasheh (1982) found values were closely related with mathematics teaching practices. Martin (1997) showed how values could enter into the mathematical modelling process. Lim and Ernest (1997) grouped values into epistemological values, social and cultural values and personal values. FitzSimons, Seah, Bishop and Clarkson (2000) documented Australian primary teachers' conceptions of values and pointed out that there is a lack of a common language for teachers in discussing value issues in mathematics education. Bishop (1988) classifies mathematical values as three complementary pairs: Rationalism and Objectism, Progress and Control, Openness and Mystery, and points out that in the current teaching of mathematics there is a significant emphasis on the values of Objectism, Control and Mystery over their complementary values of Rationalism, Progress and Openness respectively.

There are a variety of sources that can convey values explicitly or implicitly in mathematics education. Teachers' teaching is one of them. For example, when a teacher tells a student lagging behind in mathematics that he is able to catch up with his classmates so long as he puts in more effort, he is using the value *working hard* explicitly to encourage the student; when a teacher gives students a lot of exercises to do in the class and after class, this may underline the value principle that *practice makes perfect.* Curriculum is another source of values. Research has suggested that different points of view are emphasised in the aims of the mathematics curricula in China and in Australia (Cao, 2002).

The mathematics textbook, as part of mathematics curriculum, is clearly a significant component in the teaching of mathematics and it is definitely a transmitter of values (Dowling, 1996; McBride, 1994). It is also a fact that teachers in different countries not only use different mathematics textbooks to teach mathematics, but also use textbooks differently (Leung, 1992). In some countries it is sensible to think of the textbook as only helping to determine the implemented mathematics curriculum because the teachers' ideas are dominant and the textbook is merely an aid for the teacher. In other countries the textbook is almost the key determiner of the implemented curriculum and thus is, in reality, best considered to be part of the intended curriculum. Textbooks in Australia and textbooks in China seem to fall into these two categories respectively, as teachers in Australia have relatively more freedom to choose textbooks in their teaching, and teachers in China can use only the standard textbooks in the mathematics classroom. Accordingly it would be informative if we could compare the values transmitted in the mathematics textbooks in two countries with different cultural backgrounds. In particular, this paper focuses on the comparison of the mathematical values portrayed by mathematics textbooks in China and Australia.

2. THEORETICAL ASPECT OF THIS STUDY

As has been mentioned in the last section, Bishop (1988) identified three complementary pairs of features of mathematics, which he later (Bishop, 1996) called mathematical values: Rationalism and Objectism, Control and Progress, Openness and Mystery. Rationalism involves the separation of an idea from any associated object, utilising "deductive reasoning as the only true way of achieving explanations and conclusions" (Bishop, 1988, p.62), while its complementary part, Objectism, involves the abstracting of ideas and treating these ideas as if they are objects. Progress represents the idea that mathematics demonstrates its growth and development, while Control is demonstrated when mathematics is used (or perceived to be used) as a tool to control the environment and its people, or when mathematics gives the learners a feeling of being in control. Openness "concerns the fact that mathematical truths, propositions, and ideas generally, are open to examination by all" (Bishop, 1988, p.75), while Mystery is represented when people (including mathematicians) feel mystified about what mathematics ultimately is and where particular ideas came from.

Based on Bishop's work on values, and on van Dormolen's and Skemp's work (van Dormolen, 1986; Skemp, 1979) on the classification of mathematics education, Seah (1999) compared the mathematical values and mathematics educational values represented in mathematics textbooks in Singapore and Victoria, Australia. The results support Bishop's assumptions about the imbalance between each pair of complementary mathematical values in Mathematics: the values of Objectism, Control and Mystery are emphasised over their complementary values of Rationalism, Progress and Openness respectively. However, it is not clear if this conclusion is true for Chinese mathematics textbooks, as China is a socialist country and of eastern cultural tradition, while both Australia and Singapore are capitalist countries, with a Western culture and an Eastern culture respectively. Indeed, it is also not certain what different values are portrayed in Australian and Chinese mathematics textbooks.

Based on Bishop's (1988, 1996) conception of types of mathematical values, and on Seah's (1999) coding of these values and the results in Australian textbooks in his study, this paper will examine how similarly or differently these mathematical values are emphasised in the mathematics books in China and Australia.

3. METHODOLOGY

To facilitate direct comparison of data, this study will use the same coding protocol for the analysis of the mathematics textbooks in China. It will also choose the same topics as used by Seah (1999). Results gathered from the Chinese mathematics textbooks will then be compared against the results from Seah's analysis of the Australian textbooks in his previous study. The procedures are illustrated as follows:

3.1 The selected textbooks and topics

The Chinese mathematics textbooks adopted for analysis in this study were part of a standard series, *Mathematics* (The People's Educational Press, 1995a, 1995b, 1995c, 1995d, 1995e, 1995f, 1995g). As the characteristic of a centralised education system in China, all the schools there use the standard textbooks designated by the Education Ministry of China. Similarly, the Australian textbooks used for analysis by Seah (1999) were also a then popularly-used series: *Mathematics for Australian Schools: Year 7* (Ganderton & McLeod, 1996a), and *Mathematics for Australian Schools: Year 8* (Ganderton & McLeod, 1996b). In fact, 23 out of a random sample of 40 secondary schools in Victoria were using this series of textbooks. Compared with the textbooks in Australia, the size of the Chinese mathematics textbooks is smaller (the paper size is B5, while the paper size for the Australian textbooks is between A4 and B5) and much thinner, the number of pages in one volume range from only 140 to 180, while the number of pages of the two Australian textbooks is 517 and 519 respectively. The topics used for analysis are "rate, ratio and percentage", and "area, perimeter, volume". As mentioned above, these are the same topics analysed in Seah's study. These topics accounted for a total of 108 pages in the series of Chinese textbooks. On the other hand, the same topics occupy 680 pages in the Australian textbooks.

3.2 The coding protocol

The coding protocol followed that used by Seah (1999). In this coding system, the mathematical values of Rationalism, Objectism, Control, Progress and Mystery and Openness are embodied as concrete value signals. The value of Rationalism is seen to be portrayed by value signals such as "abstraction involved", "use of logic connectors", "introduction of theorem & formula through guided induction" etc. The value of Objectism is flagged by value signals such as "symbolisation adopted", "awareness of moral aspects to the solution" etc. For the value of Progress, examples include "example of mathematics contributing to societal progress", "historical developments related to present situation" etc. Examples of value signals associated with the value of Control are "control over the environment", "use of imperatives" etc. The value of Openness corresponds to value signals such as "introduction of theorem and formula through proof ", " introduction of theorem and formula through worked example" etc. The value of Mystery is implied by such value signals as "introduction of theorem and formula through statement", " use of passive voice", "use of imperatives" etc. Value signals corresponding to each of the six mathematical values are listed in a checklist, and the number of such signals used in the selected textbooks were counted. The extent to which a value is emphasised in a topic in a selected textbook is then assumed to correspond to the number of times this value is flagged by the frequency of use of the related value signals. The major value signals used for coding each category of the mathematical values are summarised in Table 1:

Table 4-4-1. Major value signals used for coding each category of mathematical values

Rationalism	Abstraction involved; Use of logic connectors; Introduction of theorem & formula through guided induction; Context-nil.
Objectism	Symbolisation adopted; Awareness of moral aspects to the solution; Attempt to objectivise.
Progress	Example of mathematics contributing to societal progress; Historical developments related to present situation.
Control	Control over natural environment; Control over people and objects around us; Use of imperatives.
Openness	Introduction of theorem and formula through proof; Introduction of theorem and formula through worked example; Tone---assumption of peer relationship between writer and pupils; Use of 'we', 'you' and related forms.
Mystery	Tone---disregard for pupil prior, personal knowledge; Introduction of theorem and formula through statement; Use of passive voice; Use of specialist vocabulary; Introduction of procedure through---listing of steps.

3.3 The coding of the data

Coding reliability was achieved firstly through trial coding exercises between the first two authors. The first chapter of the Chinese textbook analysed was also coded separately by these first two authors, and the individual codings then compared. One evolving issue regarded the sentences of hidden passive voices in Chinese, as some sentences in Chinese are not so obvious in using passive forms of verbs even though they are in passive voice. This can be illustrated by the following example:

> Li 2 (2) ruo guo ba ta(bian chang shi 1 mi de zheng fang xing) fen cheng biang chang shi 1 fen mi de xiao zheng fang xing, ke yi hua fen duo shao ge? Ta de mian ji shi duo shao ping fang fen mi? (*Mathematics*, *Volume 6*, 1995, p.131).

> Translated into English, the expression would be:

> Example 2 (2) If a square with the side length of 1 meter is divided into small squares with the side length of 1 decimeter, how many small squares can it be divided into? How many square decimeters is the area of the square?

The first sentence is not obviously in a passive voice in the Chinese expression (ruo guo ba……, ke yi hua fen wei duo shao ge?), however it actually implies a passive voice, therefore it should be treated as a passive voice and as a signal of the value of Mystery.

4. RESULTS

The number of value signals identified in the selected Australian and Chinese textbooks is listed in Table 2:

Table 4-4-2. Number of value signals identified and percentage of each category of value signals accounted for against total value signals in textbooks in China (CHN) and Australia (AUS)

	CHN	AUS
Rationalism	92 (10%)	486 (0.5%)
Objectism	261(28%)	86,043 (95%)
Progress	82 (9%)	30 (0.03%)
Control	197 (22%)	980 (1%)
Openness	75 (8%)	469 (0.5%)
Mystery	203 (23%)	2,431 (2.7%)
	910 (100%)	90,529 (100%)

It can be seen from Table 2 that in both countries' textbooks the value of Objectism is emphasised more than its complementary value of Rationalism. Similarly, for the other two pairs of complementary values, the value of Control is emphasised more than the value of Progress, and the value of Mystery is emphasised more than the value of Openness. Moreover, comparing each category of value signals between China and Australia, Objectism seems to be very much more emphasised in Australian textbooks than in the Chinese ones, while the value signals of the other categories such as Control, Mystery, Rationalism and Openness are all emphasised more in the Chinese textbooks.

In order to gain details of what sorts of value signals dominate in each category of values, the major value signals of each value were also identified by the times the signals appeared in the textbooks. The results are indicated in Table 3.

From Table 3, it can be seen that the major value signals are quite similar in most categories of values. However, there are also differences in a few categories. For instance, for the value of Mystery, one major signal evident in the Chinese textbooks analysed is the "introduction of procedure through listing of steps", while this value signal is not so strongly present in the Australian textbook. For the value of Openness, the value signal "Tone – assumption of peer relationship between writer and pupils" is one major value signal in the Chinese textbook, but it is not a significant feature of the value of Openness in the Chinese textbooks.

Table 4-4-3. Major value signals in Chinese (CHN) and Australian (AUS) textbooks

	CHN	AUS
Rationalism	Context-nil; Abstraction involved	Context-nil; Use of logical connectors
Objectism	Symbolisation	Symbolisation
Progress	Example of mathematics contributing to societal progress	Example of mathematics contributing to societal progress
Control	Use of imperatives; Control over the environment	Use of imperatives
Openness	Tone---assumption of peer relationship between writer and pupils; Use of 'we', 'you' and related forms	Use of 'we', 'you' and related forms
Mystery	Use of imperatives; Introduction of procedure; through---listing of steps; Use of specialist vocabulary	Use of specialist vocabulary; Use of passive voice; Use of imperatives

Apart from those major value signals, there are also other value signals that function differently in each country's textbooks. For example, the value signal "computer programming exercise" appears four times in the Australian textbooks, while this value signal doesn't appear at all in the Chinese textbooks. The value signal "Awareness of moral, ethical, etc aspects to solution" appears in the Australian textbooks eleven times, but it doesn't appear in the Chinese textbooks at all.

5. CONCLUSIONS AND DISCUSSION

The results further confirm Bishop's (1996) assumptions regarding the imbalance of mathematical values represented in school mathematics education. The values of Objectism, Control, and Mystery had been found amongst Australian and Singaporean textbooks to be emphasised more than their respective complementary values (Seah, 1999). This study notes a similar trend in the selected Chinese textbooks, even though there are slight differences in the degree of emphasis. Comparing across each of the six mathematical values, it was also evident that the Chinese textbooks demonstrate relatively less emphasis on the value of Objectism, and relatively more emphasis on all the other values. The results also reveal that generally the texts in both countries are more abundant with features like symbolisation, imperatives, and specialist vocabulary terms compared with other features such as use of logical connectors, examples of mathematics contributing to societal progress, and introduction of theorem and formula through proof, while the feature of the symbolisation seems to be more evident in the Australian textbooks than in the Chinese textbooks. In other words, in the context of similar topics in the selected two countries' textbooks, there were more incidents of the use of symbols to convey or to illustrate the relevant information.

The fact that the two countries' textbooks demonstrate similarities in the relative emphases of the three pairs of complementary mathematical value seems to indicate that the image of mathematics taught in schools in the two cultures is basically the same. Even though the content of the same year level may not be the same in the two countries (the selected topics were taught to Chinese students at an earlier level compared to their Australian counterparts), the spirit of the discipline of mathematics is the same. This may in fact indicate the influence on publishers and writers of mathematics textbooks by mathematicians, with regards to the nature of how mathematical content may be presented.

However, the above conclusion does not necessarily mean that the design of mathematics textbooks is not influenced by societal forces. On the contrary, the societal factors present an unavoidable influence on the formatting of mathematics textbooks. For example, the fact that the value signal "computer programming exercise" appeared in the Australian mathematics books a few times, but did not appear in the Chinese mathematics textbooks suggests that the Australian textbook designers have more awareness of the role of computer technology in mathematics education than their Chinese counterparts, which certainly reflected the popularity of technology use in the two societies at that time. When the standard Chinese mathematics textbooks were published in 1995, computers were not even popular in primary schools in most of the urban areas, not to say in schools of rural areas. In this context, it would not have been realistic for the textbook designers to put any computer programming exercises in the mathematics textbooks. On the other hand, many schools in Australia have had the opportunities to access the computers when the textbooks *Mathematics for Australian Schools* were published in 1996.

As is noted by Bishop (1988), the imbalance in emphasis amongst these values might be the cause that mathematics is losing its appeal amongst more and more learners. Learners usually feel lost in their ways of learning mathematics as they feel that mathematics is too far away from their daily life and do not know where or how the mathematics ideas come from. A solution for the problem is probably that more examples related to daily life, more details regarding the process of how some mathematics ideas are generated, and the historical development of mathematics related to the current situation should be added to the mathematics curriculum and teaching process to increase the emphasis on the values of Rationalism,

Openness and Progress. It is only when all these six mathematical values are better balanced in mathematics curricula and mathematics classrooms, that it is possible that learners' interest in mathematics can then be aroused and promoted.

REFERENCES

Australian Educational Council, 1991, *A National Statement on Mathematics for Australian Schools*. Australian Curriculum Corporation, Melbourne.

Bishop, A.J., 1988, *Mathematical Enculturation: A Cultural Perspective on Mathematics Education*, Kluwer, Dordrecht.

Bishop, A.J., 1991. Mathematical Values in the Teaching Process, in: *Mathematical Knowledge: Its Growth through Teaching*, A.J. Bishop et al., eds., Kluwer, Dordrecht, pp.195-214.

Bishop, A.J., 1996, How Should Mathematics Teaching In Modern Societies Relate to Cultural Values-Some Preliminary Questions, in: *Proceedings of Seventh Southeast Asian Conference on Mathematics Education*, Nguyen Dinh Tri et al., eds., Vietnamese Mathematical Society, Hanoi, pp.96-102.

Board of Studies Victoria, 2000, *Mathematics: Curriculum and Standards Framework II*. Board of Studies, Carlton, Australia.

Cao, Z, 2002, Values in Mathematics Education: Differences Reflected in the Intended Mathematics Curricula in China and Australia, *Reflections*, **27**(2):2-6.

Cao, Z., Ng, Q. and Bishop, A.J., 2002, Comparing Mathematics Teachers' Values in China and Vietnam, in: *Valuing Mathematics in the Society*, C. Vale, J. Horwood and J. Roumeliotis, eds., The Mathematical Association of Victoria, Melbourne, Australia, pp.30-40.

Clarkson, P.C., FitzSimons, G.E., and Seah, W.T., 2000, Values Relevant to Mathematics? I'd Like To See That!, *Reflections*, **25**(2):1-3.

Dormolen, J.van, 1986, Textual Analysis, in: *Perspectives on Mathematics Education*, B. Christiansen, A.G. Howson and M. Otte, eds., D.Reidel Publishing, Dordrecht, pp.141-171.

Dowling, P., 1996, A Sociological Analysis of School Mathematics Textbooks, *Educational Studies in Mathematics*, **31**:389-415.

Fasheh, M., 1982, Mathematics, Culture, and Authority, *For the Learning of Mathematics*, **3**(2):2-8.

FitzSimons, G.E., Seah, W.T., Bishop, A.J., and Clarkson, P.C., 2000, Conceptions of Values and Mathematics Education held by Australian Primary Teachers: Preliminary Findings from VAMP, in: *Proceedings of the HPM 2000 Conference, History in Mathematics Education: Challenges for the New Millennium* (Vol. II), W.-S. Horng and F.-L. Lin, eds., Taipei: National Taiwan Normal University, Taipei, pp.163-171.

Ganderton, G.D., and McLeod, J.K., 1996a, *Mathematics for Australian Schools: Year 7 (3ed)*, MacMillan Education, Melbourne.

Ganderton, G.D., and McLeod, J.K., 1996b, *Mathematics for Australian Schools: Year 8 (3ed)*, MacMillan Education, Melbourne.

Hsu, F.L.K., 1981, *Americans and Chinese: Passage to Differences*, University Press of Hawaii, Honolulu.

Krathwohl, D.R., Bloom, B.S., and Masia, B.B., 1964, *Taxonomy of Educational Objectives: The Classification of Educational Goals*, (Handbook II: Affective Domain), David McKay, New York.

Leung, F.K.S., 1992, *A Comparison of the Intended Mathematics Curriculum in China, Hong Kong and England and the Implementation in Beijing, Hong Kong and London.* Unpublished doctoral thesis. University of London, London, UK.

Leung, F.K.S., 2000, *In Search of an East Asian Identity in Mathematics Education–The Legacy of an Old Culture and Impact of Modern Technology.* Paper abstract presented in ICME9, Tokyo, Japan.

Lim, C.S., and Ernest, P., 1997, Values in Mathematics Education: What is Planned and What is Espoused?, in: *Proceedings of the Day Conference held at University of Nottingham*, British Society for Research into Learning Mathematics (BSRLM), Nottingham, Britain, pp.37-44.

Ling, L., Burman, E., and Cooper, M., 1998, The Australian study, in: *Values in Education*, J. Stephenson, L. Ling, E. Burman, and M. Cooper, eds., Routledge, London, pp35-60.

Martin, B., 1997, Mathematics and Social Interests, in: *Ethnomathematics: Challenging Eurocentrism in Mathematics Education*, A.B. Powell and M.Frankenstein, eds., State University of New York Press, New York, pp.155-171.

Mandler, G., 1989, Affect and Learning: Reflections and Prospects, in: *Affect and Mathematical Problem Solving: A New Perspective*, D.B. Mcleod and V.M. Admas, eds,. Springer-Verlag, New York.

McBride, M., 1994, The Theme of Individualism in Mathematics Education: An Examination of Mathematics Textbooks, *For the Learning of Mathematics*, **14**(3):36-42.

McLeod, D.B., 1992, Research on Affect in Mathematics Education: A Reconceptualisation, in: *Handbook of Research on Mathematics Teaching and Learning*, D.A. Grouws, ed., National Council of Teachers of Mathematics, Reston, VA, pp.575-596.

Oppenheim, A.N., 1966, *Questionnaire Design and Attitude Measurement*, Heinemann, London.

Rokeach, M., 1973, *The Nature of Human Values*, The Free Press, New York.

Robitaille, D., and Dirks, M., 1982, Models for Mathematics Curriculum, *For the Learning of Mathematics*, **2**(3): 3-21.

Saxe, B.G., 1990, *Culture and Cognitive Development: Studies in Mathematical Understanding.* Lawrence Erlbaum Associates, Hillsade.

Seah, W.T., 1999, *The Portrayal and Relative Emphasis of Mathematical and Mathematics Educational Values in Victoria and Singapore Lower Secondary Mathematics Textbooks: A Preliminary Study.* Unpublished master thesis, Monash University, Australia, Melbourne.

Seah, W.T., and Bishop, A.J., 1999, Realising a Mathematics Education for Nation-building in Southeast Asia in the New Millennium [CD-ROM], in: *Proceedings of MERA-ERA Joint Conference 1999*, Vol.3, S.P. Loo, ed., Malaysian Educational Research Association and Educational Research Association Singapore, Malaysia and Singapore, pp.1241-1249.

Swadener, M., and Soedjadi, R., 1988, Values, Mathematics Education and the Task of Developing Pupils' Personalities: An Indonesian Perspective, in: *Mathematics Education and Culture*, A.J. Bishop, ed., Kluwer, Dordrecht, pp.193-208.

Skemp,R.R., 1979, *Intelligence, Learning, and Action*, John Wiley& Sons, Chichester.

The People's Educational Press, 1995a, *Mathematics: Volume 6* (in Chinese), 1st ed, The People's Educational Press, Beijing.

The People's Educational Press, 1995b, *Mathematics: Volume 7* (in Chinese), 1st ed, The People's Educational Press, Beijing.

The People's Educational Press, 1995c, *Mathematics: Volume 8* (in Chinese),1st ed, The People's Educational Press, Beijing.

The People's Educational Press, 1995d, *Mathematics: Volume 9* (in Chincsc), 1st cd, Thc People's Educational Press, Beijing.

The People's Educational Press, 1995e, *Mathematics: Volume 10* (in Chinese),1st ed, The People's Educational Press, Beijing.

The People's Educational Press, 1995f, *Mathematics: Volume 11* (in Chinese),1st ed, The People's Educational Press, Beijing.

The People's Educational Press, 1995g, *Mathematics: Volume 12* (in Chinese), 1st ed, The People's Educational Press, Beijing.

Thompson, D., 1995, ed., *The Concise Oxford Dictionary of Current English*. Oxford University Press, Oxford.

Wilson, B.J., 1986, Values In Mathematics Education, in: *Values across the Curriculum*, P. Tomlinson and M. Quinton, eds., The Falmer Press, Lews, pp.94-108.

Chapter 4-5

VALUES AND CLASSROOM INTERACTION: STUDENTS' STRUGGLE FOR SENSE MAKING

Reasoning Discourse and Patterns of Student Participation in Classrooms from Germany and Hong Kong

Eva JABLONKA[1] and Christine KEITEL[2]
Freie Universitaet Berlin

1. INTRODUCTION

Values are commonly conceptualised as frames of references or as blueprints for perception and behaviour that operate as a generic background and provide meta-rules and norms (cf. the definitions quoted in the chapter by Chien Chin in this volume; Bishop, 1999). One important aspect of the concept of values is that they are shaped by cultural factors and life in the social realm. Thus they are different from belief systems and personal visions. Values are the principles, standards and qualities explicitly or implicitly considered worthwhile or desirable by the participants of a distinct social practice. This definition refers to the set of values a group of people share. Values are always both product and living condition; they are not *a priori* to interaction.

Values operate at all levels of mathematics education. Bishop (1988) identified values developed in the history of mathematics as a cultural practice. These values refer to ideology, attitude and ownership. They are not fixed, but have changed during history and are sensitive to context influences. Different communities of mathematical practice share different sets of values. Richards (1991) distinguishes at least four linguistic domains with mathematical content that are associated with different cultures concerning the expectations regarding the assumptions, the goals and underlying methodologies (see also Burton, 1999; Ernest, 1999; Love & Pimm, 1996).

In the process of recontextualisation of mathematical practices for pedagogic purposes, the mathematical discourse is transformed. FitzSimons (1999) points out that in classrooms, as a consequence of often conflicting influences from the macro-level (societal, institutional) and teachers' and students' goals, some of the public values of curriculum and pedagogy might be suppressed, whereas others might be more openly admitted. Political arguments about curriculum and pedagogy exemplify Bernstein's (1990) view that these are arguments between different conceptions of the social order and thus are profoundly value-driven. Bernstein (1996) shows that pedagogic discourse cannot be identified with any of the discourses it has recontextualised. Consequently imaginary subjects are created (p.46 ff.). However, these subjects might resemble different variations of the recontextualised practice.

In acknowledging the socio-cultural nature of mathematics and mathematics education, learning mathematics has to be conceived as initiation in mathematical culture (van Oers, 2002) or as the process of becoming mathematical, that is of being able to think and speak mathematics (Lerman, 2002). Teaching is the mediation of institutional culture by local personnel (Lerman & Tsatsaroni, 1998). From this perspective it seems obvious that values operate at the community of practice in the classroom (cf. Seah & Bishop, [ICMI book]). The public values initiate students into what counts as mathematical practice, including discursive and non-discursive elements of it, in that particular classroom. The "mathematical perspective" and its accepted values (e.g. relevance of dealing with numbers, non-contradiction, systematicity, accuracy) emerges neither by itself nor do all students come to share it. In mathematics classrooms values relate to both social behaviour and the learning and teaching of mathematics. When mathematical content is enacted in a classroom, these different sets of values merge.

The values of social behaviour that define what to do and what to refrain from are usually not explicitly warranted, though in an educational practice they are likely to be substantiated when being introduced or contravened. For some situations, giving a reason may be culturally sanctioned; a case in point is excuses. In classrooms, we find many situations in which a teacher feels obliged to offer a reason after a student has violated a norm (such as not to wear a base-cap or a headscarf during a lesson, not to give an answer without being asked etc.). The warrant might refer to an authority or might contain an explanation, depending on the general educational values in operation. In an academic mathematical discourse – if described as an ideal – giving a reason is a value in itself. One shared value linked to mathematical reasoning, is that it should not draw on any authority but the human mind and the stock of scrutinised results. And these results are neither empirical facts nor rules, but logical or theoretical relationships between the meanings of mathematical objects. The style of the reasoning discourse established in a classroom is most likely to exhibit the values in operation and to show how more general cultural values productively intersect with the local norms of that particular classroom.

The research reported in this paper aims to identify ways in which role-related asymmetries and culturally sanctioned ways of interaction serve as an orientation for the participants in mathematical reasoning discourses in classrooms. Episodes from classroom discourse and student interviews are interpreted in the course of a contrastive analysis. One goal of this ongoing study is to identify links between structure and students' agency.

2. COMMONALITIES AND DIFFERENCES IN MATHEMATICS CLASSROOMS

Germany and Hong Kong are different in structure and policy of their educational systems, but both are providing a highly advanced education in mathematics, focussing on pure rather than applied mathematics, to an increasingly large group of the school population. One of the most significant differences is the fact that Hong Kong schools are examination driven. Several tests have to be passed in order to proceed to higher levels of school and to university. Although teachers are encouraged to adopt student-centred, activity-based and hands-on approaches to organise student learning, the atmosphere of examinations and relatively strong competition suggests that teachers and schools consider the attainment of good exam results as a priority, and students have to be exposed to constant drill on skills in order to secure their abilities in written exams. Mathematics is considered a core subject and is taught every day. There is no graduation without an exam in mathematics, and therefore it is expected that Hong Kong students spend more time out of school doing extra homework, extra lessons or studies in mathematics, than in other countries (Lam 2002).

Furthermore, teaching and learning must be focused on the public exam syllabus and on students' proficiency to work out problems of this syllabus. They perceive mathematics often as a terminology only and doing mathematics as applying a set of rules rather than thinking. (Lam et al. 1999). In international comparisons Hong Kong students perform well in mathematics tests, but do not necessarily show positive attitudes towards mathematics. (Mullis et al.1997)

In contrast, in Germany even the final exam of the most advanced level of schooling, at the university-bound high school, follows mostly the teacher-based assessment mode. The syllabus, similar to the Hong Kong syllabus, is rather focusing on abstract and technical aspects (e.g. of algebra in our examples), on strict use of terminology and on following clearly stated formal rules, though the focus varies across the federal states. It is to be interpreted as a recommendation rather than a prescription. However, a common understanding of what mathematics teaching should be about among teachers is constructed and secured in the practical phase of teacher training, the pre-requisite to the entrance into the profession. Some see the most prominent and new challenges in Germany in the large groups of immigrants from different social, ethnic and cultural groups. This has increased the diversity of the students in many areas, but the university-bound Gymnasium is not yet very much affected.

The values assigned to teaching or learning mathematics in the intended curriculum can hardly be separated from the general educational values. If argumentation and critical thinking are educational values then mathematics is more likely to be seen as the outcome of deliberate intentions and valued as a means for analysing socially relevant problems. If functioning in the workplace is an important aim, then mathematics is more likely to be seen as a kind of generic tool-box (cf. Jablonka, 2003a).

In mathematics classrooms, mathematical practices are re-contextualised in a specific manner for the purposes of enculturation. The higher the value attributed to academic mathematics in an education system, the stronger we expect the orientation of the classroom practice to be towards the "quintessence" of this practice, such as mathematical argumentation, reasoning and proving. As to the intended curriculum, this applies to Germany in the university-bound strand of schooling, and particularly to Hong Kong, where mathematics is the core subject. In the mathematics classrooms under study, an observer would expect to find a high proportion of reasoning discourse, the stylistic characteristics varying according to the general educational values in operation: When a teacher explicates and elucidates the meaning of a new concept or demonstrates a mathematical proof, when students try to solve a problem in group work, or when in a "whole class" discussion mathematical meaning is negotiated. Consequently, reasoning discourse can be a sensible focus of analysis.

In classroom discourse "reasoning events" were identified on the level of utterances as parts of a discourse in which a person offers a reason for something because s/he interprets something as being not evident, doubtful or disputable, or is asked by another person to give a reason. The attempt to increase evidence or acceptance has to be visible for the other participants in the discourse. Some patterns in classroom discourse are already visible in small units containing just a few turns (e.g. the initiation-response-evaluation sequence), others can only be interpreted as parts of larger patterns. A focus of the analysis is to identify the distortions and systematic transformations of mathematical reasoning – as compared to reasoning in situations in which knowledge and power are more equally distributed than in classrooms - with respect to commonalities and differences in the cultural environments of the classrooms under study.

2.1 Ideal and practice: distortions

The teacher – whose role is to mediate between the intended curriculum and the students – as well as the students, bring their own values to the classroom. Consequently, classrooms are not homogeneous in terms of values. But if there were not at least some shared values, interaction would not be possible. The students and the teacher find their ways of resolving the tension of competing obligations and constraints, which derive from their affiliations with various groups competing for their allegiance (cf. Clarke, 2001). In addition, the affiliations that an individual sees for herself may be quite different from that attributed to her by others.

In classrooms, subversive practices are likely to occur due to the role-related asymmetries. Students might choose to conform to a desirable behaviour without, at the same time, sharing the values (cf. Drew and Heritage, 1992).

The observation that social behaviour providing a warrant is more likely in a situation of breakdown (or when such a situation is anticipated) can also be made with respect to the mathematics, at least in the classrooms under study. The teachers are much more likely to provide a reason when something is considered as wrong. This seems natural because skilful participation – as reflected in obtaining a correct result when solving a task – is taken as a sign of mathematical understanding.

A study of classroom interaction can reveal the implicit values operating in a classroom. The values become more explicit if a conflict of values emerges. This is not necessarily only visible as a straightforward conflict referred to by the teacher or a student, but also in a more hidden way when the smooth flow of interaction is interrupted, when students or the teacher laugh or utter an ironic comment. Students more likely reveal their values in an interview with a researcher who is not directly collaborating with the teacher, while the teacher might be more cautious to make his or her values public especially if s/he conceives of these values as not desirable or in accord with the intended; these values are visible only in their consequences for classroom practice. Consequently, the interviews with the students do give easier access to students values not coinciding with those of the teacher, while analysis of classroom practice of both, teacher and students, reveal values of the teacher as well as those demonstrated by students as desirable behaviour.

This points to a general methodological problem of researching values. Such a research is predicated on the assumption that there is something stable behind the actions of people and that there are in fact some shared values. Because much of what constitutes a system of values that influences the behaviour of the participants of a distinct social practice is what is rarely questioned – that's just the way things are done – values are seldom made explicit in the form of evaluative, prescriptive or normative statements. As to mathematics teaching/learning this is usually the case only in curriculum documents. Most of the values remain implicit, and in addition, it is not even clear whether the participants of a social practice can easily articulate their values when being asked. There is still another level at which values are involved, that is that the interpretation of classroom episodes or interviews itself is value-driven. However, members of a research community may become more aware of their values in a setting in which these values are drawn from different cultural backgrounds.

2.2 Patterns of whole class interaction

In the German as well as in the Hong Kong classrooms there are a couple of episodes of discourse that could – albeit only from the viewpoint of logic - be interpreted as a collective process of reasoning in order to obtain a previously unknown solution of a mathematical problem. The teacher and the students collectively provide a chain of (minor) premises ("reasons") and inferences, though the discourse does not necessarily contain any single utterance that can be interpreted as a request for, or a provision of, a reason. This is when the teacher exhaustively pre-structures the discourse by breaking down the chain of inferences into a series of closed questions, which – if answered correctly – in sum warrant the resulting conclusion, that is the solution of the problem. Typically, in such a discourse, the students' turns consist of providing short, mostly single word or single number sentences as the answers to the teacher's questions (cf. Voigt, 1995). In these episodes it is doubtful whether the students perceive the process as a collective process of argumentation as identified by Krummheuer & Brandt (2001) because in none of the single steps does anything seem to be not evident, doubtful or disputable.

Unlike normal questions, which aim at retrieving knowledge from the addressee that the speaker does not have, the questions of the teacher aim at conducting the students' thinking. The questions exactly specify the domain in which to answer. In normal questions this is the knowledge that the questioner thinks is shared by the addressee.

Example 1 (Hong Kong)

04:43	**T**	Class, please read the question. What does it remind you of?
04:59	**T**	What does it remind you of, Patrick?
05:06	**Patrick**	Addition and subtraction.
05:08	**T**	It reminds you of adding and subtracting numbers. Thank you. Why do you think about addition and subtraction?
05:17	**Patrick**	Simplify complicated things.
05:26	**T**	Who's more imaginative? What comes to your mind when you're reading this question, factorizing polynomials? What do you know and what do you want to know?
05:41	**Mark**	Learning factors.
05:42	**T**	Mark...huh?
05:43	**Mark**	Learning factors.
05:44	**T**	Learning factors. Factors...Thank you. We've learnt factors in primary school, like common factors, largest common factors etc. In this lesson, we'll talk about factorizing polynomials. This is a new topic. You haven't learnt it in form one or form two. First, we look at a simple example. [T writing m times brackets a plus b is equal to m times a plus m times b]

This episode can be interpreted as a strictly planned teacher presentation with allocated parts (cf. Ehlich and Rehbein, 1986). It can be reconstructed as a consistent chain of assertions uttered by one speaker. However, the teacher's questions aim at still maintaining a minimum of students' participation. In order to guarantee that it goes off smoothly, this pattern of interaction needs some skills and experience in formulating the directing questions. The decision about what exactly is stated as the known and as the unknown in the question is based on the expected mental operations performed by the students and aims at influencing these according to the direction in which the teacher wants the discourse to continue (cf. Jablonka, 2003b).

The reasoning involved in this pattern is not intended to be self-initiated by students. The teacher wants Patrick to *explain* his assertion; the answer does not fit the teacher's propositional plan. Consequently, a new directing question follows which is less ambivalent.

This pattern is common in all the classrooms under study. However, there are differences in the ways the teachers deal with unexpected comments and answers, that is with ruptures. Very tentatively it can be said that the Hong

Kong teachers have a tendency to repair the pattern by switching more to the presentation mode, while the German teachers tend either to open it by introducing normal questions instead of directing questions or switch to a presentation. In an arbitrarily chosen Hong Kong lesson, an episode of teacher talk in the course of a demonstration of a problem solution takes the form of a "dialogue" in which the teacher introduces the reasons he provides by rhetorical questions or rhetorical contradictions. In the classical European tradition of rhetoric this would be interpreted as a stylistic means used in speeches that aim at convincing the audience.

2.3 Barring empirical arguments

The following two episodes, one from a Hong Kong and one from a German classroom, are similar in that they contain self-initiated comments of students who introduce empirical arguments. In both episodes the students are asked to solve a word problem and both problems refer to the context of farming. The rejection of the empirical referents is not warranted. However, if the speaker decides not to utter a reason, the lack of understanding remains unproductive. It is likely to lead to frustration or disapproval on the side of the listener. This might be the case in the two episodes. In both cases the teacher decides not to give a reason for the fact that the points made by the students are not to be argued about or are not of interest (see example 2, 11:20; example 3, 33.15 – 33:30).

Example 2 (Hong Kong)

Some minutes before this, the teacher stated a problem to work on: A farmer has some rabbits and some chickens. He does not know the exact number of rabbits and chickens...but in total there are ten heads... remember...ten heads...and there are twenty-six legs. The teacher shows a strategy to find the answer by trial and error: for example start with the assumption that half of them are chickens, then there are ten chicken legs …

[S wants to say something]
11:13 **Teacher** Okay...please say.
11:16 **S** Rabbits have only two legs as the other two are hands.
[whole class laughs]
11:18 **Keith** [laughs] He is impolite.
11:20 **Teacher** We usually say they are legs instead of hands. Don't argue about this point...okay?

It can be argued that the laughter of the other students shows that in this classroom the practice of solving a typical textbook problem is well established. Most students know that the chicken is to be taken as a substitute for an animal with two legs and the rabbit for one with four, and that the farmer, who knows the sum of his animals' legs, is non-essential. In a previous conversation, in which Louis and Kelvin try to solve the problem, and in the interview, one of them talks about pigs instead of rabbits without noticing. Thus the student's self-initiated comment can only be taken as sarcastic by some. Interestingly Keith labels the student as being "impolite", which is exactly the attribute a person can expect when not conforming to the implicit rules of a distinct type of a routinised discourse. In another context, the introduction of an alternative assumption can be a legitimate move in a mathematical reasoning discourse. In the given example it would lead to a contradictory problem statement and could have been refused on these grounds. However, the teacher might have interpreted the comment as a subversive act that aims to distract participants from the proper theme of their conversation.

Example 3 (Germany)

31:01	**Teacher**	Well, now we want to address ourselves to the following problem.
31:17	**Teacher**	[draws a square on the board] For that purpose we look at this square, please. This is supposed to represent a square plot of land owned by a farmer. Now you have to tell the rest, would you please tell the story?
31:39	**Researcher**	The neighbour approaches the farmer and says 'It would be very advantageous for my planning if you could give me a strip of one meter of your land from the cross side [Teacher points to the upper horizontal edge of the square], I then would give you in return a strip of one meter from my property on the other edge' [Teacher points to the right vertical edge of the square] Would you agree?
32.20	**Teacher**	Okay.
32:26	**Anton**	No...it isn't the same...hm...hm.
32:28	**S**	Unfair.
32:41	**S**	No...well...because he then has less (...) You can...you can already see that from the drawing that you would then get less from that plot.
32:47	**S**	Well...somehow...well I would refuse because this is (...)//
32:48	**Teacher**	//Wait...I didn't get that...this is such a mumbling//

32:49	**S**	//I would refuse because somehow...ehm...you have a piece...you havn't such a corner (...) plot there around.
32:58	**Teacher**	No you have...ehm...what's the shape of the new plot...it's not a square anymore...but it's still at least a rectangle. So you would refuse because you'd prefer a square over a rectangle.
[Laughter]		
33:15	**S**	I mean...ehm...want to take away on one side and put on the other side...where...oh where is then the neighbour's garden...on the left side or there below?
33:22	**Martin**	Thus it is around.
33:23	**Teacher**	It is around.
33 24	**S**	But that's a bit illogical...isn't it?
33:27	**S**	Why?
33:29	**Teacher**	That's...that is...ehm//
33:30	**Martin**	//Man//
33:30	**Teacher**	//uninteresting. Our question is...ehm...whether the one concerned...well...whether he should exchange... whether this is favourable...or whether...ehm...it doesn't matter or...it is just this what it is about... whether for this very farmer who now is up to mischief in there...it is...ehm...yes a problem.
33:56	**Kerstin**	Well I would say...I would assume...ehm...if I...in this piece he wants to...I don't know...steal from me...if I hadn't anything there...I mean a shed and stuff...then I could as well take the meter down there because the area is the same.
34:15	**Martin**	Ey?
34:16	**Teacher**	Ah...that was...is the area the same?//
34:16	**Felix**	[putting up his hand] Yeah//
34:17	**Teacher**	//That is the good question.
34:20	**S**	No...isn't the same...because...ehm...before...well now...he is missing exactly one piece...he has the full side and now he has from his shorter plot also...ehm...only the shorter half from the bottom... and so he misses exactly one square meter.
34:33	**Teacher**	Hm.

In this episode some students immediately give the answer, one of them refers to the drawing. Perhaps because of lack of evaluation of these comments, the students then start to introduce different reasons and the discourse moves from visual evidence to empirical arguments and eventually (not reproduced here) to an algebraic description of the drawing, which was intended by the teacher. The teacher explicitly keeps away the empirical argument of the student who is wondering about the shape of the neighbour's property, similar to the teacher in the Hong Kong classroom. He gets some help from Martin in doing this. But the comment is not interpreted as a diversionary tactic. However the student expresses her dissatisfaction with the answer, again self-initiated. This is one of the rare instances in this classroom in which a student expresses disagreement with a statement of the teacher.

2.4 Ambivalent reactions and tactical behaviour

Students and the teacher bring to the classroom their knowledge of the interactional pattern of reasoning in everyday contexts. This knowledge becomes explicit only in situations in which the co-operation of the interlocutors is at stake or breaks down. If a person utters a reason, this aims at transforming the knowledge or values of the addressee in a way that he or she accepts or understands an action or an assertion of the speaker. In an everyday context this is only necessary if the speaker has a reason to think that otherwise the listener would not understand her action or assertion. If the speaker wants to prevent a breakdown of co-operation, which would lead to a breakdown of the system of actions in which the participants are engaged, she decides to utter a reason. Such an attempt can be successful or unsuccessful. Success is assumed when the listener utters or shows a sign of appreciation and understanding. If unsuccessful, several cycles may follow. Success is only possible if the participants share a system of knowledge and values. If this were not the case, the reasoning would theoretically lead to an infinite recursion. In everyday contexts the breakdown usually happens after a few unsuccessful cycles (Ehlich & Rehbein, 1986).

In the classrooms under study, the pattern of reasoning is commonly missing the step in which the listeners utter or show a sign of appreciation and understanding when the teacher is addressing the whole class. It is usually not the case that all students get a turn to express their understanding. In the Hong Kong classroom from the third school the teacher frequently addresses the whole class by asking something like "Did you understand, class?" and is then confronted with no reaction. A non-reaction is ambivalent; it can be taken as both a sign of understanding or as a lack of understanding. It is likely that the students even use a non-reaction to pretend understanding in case of a lack of understanding. This helps them to circumvent or to undermine the formal and compulsive character of classroom interaction. This means the students are using their knowledge of the functioning of the reasoning pattern in order to pretend approval and co-operation. The teacher might use the same reaction, that is neither showing a sign of understanding nor of not doing so after a student has uttered a reason, to provoke further reasoning on the side of the student. This might be the case in example 4. Such a use of the knowledge about patterns of interaction can be called *tactical behaviour*. It even takes the form of showing a reaction that expresses the opposite of the mental state (e.g. approval in the case of disapproval) instead of only performing the ambivalent action of not reacting.

In the following example, the ambivalent reaction is explicitly expressed by the student.

Example 4 (Hong Kong)

36:07	**S**	[to S] How did you get that? [in Chinese]
37:10	**S**	(...)
38:27	**T**	Ida, Understand what I'm talking about?
42:19	**IDA**	Maybe.
43:27	**T**	Maybe?
44:27	**IDA**	Yeah.
45:17	**T**	Where don't you understand?
46:13	**S**	That simple? [in Chinese]
47:13	**S**	This was what he said. (He) didn't do the calculation himself. [in Chinese]
47:19	**IDA**	I don't know.
49:05	**S**	[Class giggles]

The following short conversation is typical of the first German classroom from the LPS. The teacher offers an explanation because the student has got an incorrect result and the student acts like being convinced, though by analysing his written productions and from the interview it can be said that this is tactical. In the interviews some other ways of acting as a "professional student" were discussed.

Example 5 (Germany)

09:20	**Teacher**	No that's not right...this negative sign here in the brackets...you have to...that means we have to envisage that it's multiplied by minus one...which means you write three minus z plus one half minus two.
09:38	**Günther**	Oh right...so I have to.

More interestingly, tactical behaviour also appeared when students discussed a problem in group work. It was only the interview that gave information about Otto's tactical approval and Norbert's strategic behaviour – both of them could not make a lot of sense of the task:

Example 6 (Germany)

31:53	**Norbert**	Yeah we've got that. I don't, but I don't know how we can – that's the measurement of the area. Oh we probably have to uh explain how we've arrived at *a* minus *b* in brackets squared. Here I mean.
32:15	**Tom**	It says here we only have to make a picture.
32:17	**Albert**	Finished?
32:17	**Norbert**	No wait, give me another sheet of paper [to Albert]. I haven't got one. I'm poor. No, I need that, I don't want that spoiled.
32:20	**Otto**	Haven't got a sheet.
32:24	**Norbert**	Sheet, sheet...Right look now we've got the square.
32:33	**Tom**	Yeah.
32:37	**Norbert**	That's the size here.
32:39	**Otto**	Yeah do it, get it done.
32:41	**Norbert**	Right and this is *a* squared, this is *b* squared, this is minus *ab* and minus *ab* right?
32:47	**Otto**	Yeah
32:48	**Norbert**	So, because that's, and because that's minus, because that's now minus, um this has, this has to go.
32:58	**Otto**	Yeah, yeah that's clear.
32:59	**Norbert**	This here then stays.
33:02	**Otto**	*a* squared?
33:03	**Norbert**	*a* squared then, this thing here and *b* squared.
33:10	**Norbert**	Yeah, yeah like that and then well like that, if you were to do that as a line, this here would be *a* and this bit *b*, then you take it away, then this piece here is *a* minus *b*.
33:25	**Otto**	Yeah.
33:26	**Norbert**	Yeah? Now when you have this bit *a* minus *b* and this bit *a* minus *b* here, and then the square of *a* minus *b* in brackets, then that must be the area.
33:45	**Otto**	Ah.
33:46	**Norbert**	Now look, *a* times *a* must be the area or *b* times *b*.
33:52	**Norbert**	[dismissive gesture in the direction of Otto and Seppi] Don't understand any of it, do you.
33:59	**Norbert**	But it's logical, look.
34:06	**Tom**	The microphone's not right.
34:09	**Norbert**	a minus b is this distance here, <u>this here</u> [emphasized]
34:14	**Otto**	And this as well (points to it on the sheet)

Consequently, when interpreting classroom discourse – be it as a participant or as a researcher – assuming sincerity can be problematic. Especially in a situation of unequal distribution of power and control over interaction, tactical behaviour is likely to occur. In mathematics classrooms this is often the case when the disapproval of a reason provided by a student can be taken as a lack of understanding, which, in turn, is evaluated in terms of achievement. This may establish a dynamic: the more tactical approval a student shows, the fewer reasons he/she is offered as explanations, the less chance he/she gets to make sense.

3. VALUES AND LEARNING MATHEMATICS

3.1 High achievers do (not) know mathematics better?

The German students in this selection of interviews attend their second year in a (university-bound) secondary school for the better achieving students, an inner city German Gymnasium, and belong to the upper third of the school population within the highly selective school system. Concerning their personal achievement in mathematics, their mathematics teacher considers them as a bit better than the average. Typically, their parents belong in majority to the middle class, among them quite a few academics, in particular teachers and lawyers. They value education very much; and they strongly support their children's schooling. Only very few students in the class come from an immigrant background. The 8th grade lessons in algebra are about transformations of algebraic terms.

It has to be emphasised here that students who have reached the Gymnasium, at the age of 10 or 12, are privileged in many aspects: They normally have a well-educated teacher specialised in mathematics (this is not regularly the case in the other secondary school types), and are by their teachers – and socially – considered as the high-achievers and possible future elite. In the TIMSS-results, students from this school type achieved much better results compared to students that have been placed into the German school types for low-achievers. They generally show much self-esteem and do not attribute problems of understanding mathematics or disliking it to themselves as a personal problem, they know that they can do well in other school subjects and are generally successful in school. For them it is just the mathematics that can cause trouble (if not the teacher). Given, that among these privileged students many utter their lack of sense-making, an aversion towards, or a frustration in mathematics, it can be assumed that mathematics must be more threatening as a selection means for students in other school types, who feel that they have already lost their chances for prestigious professional careers or jobs, because they were placed in a school for low achievers. In the following, statements from German students are contrasted with those from the Hong Kong school.

3.2 Searching for meaning, importance and significance

The biggest struggle in mathematics is about meaning and significance, which is not revealed by the teaching practice. Why is mathematics to be learnt and taught? Why is mathematics applicable? These questions go beyond ordinary daily lesson plans and are rarely touched on by the teacher.

3.2.1 Why do we learn mathematics and what is important? What is math about?

Students learn to abandon the question 'Why' and 'For what' and have nearly no ideas about where to apply math beyond the shopping mall; their knowledge about application and applicability is very limited, and mostly wrong.

Interviewer: What does this sort of math mean to you essentially? What's it good for, what can you possibly do with it?
Steffi: It's not clear to me at all.
Sharon: Well, it just belongs to basic knowledge. You just have to be able to do it, and then you're allowed to forget it.
Interviewer: But what's the use of knowing it?
Sharon: Well, I don't know if//
Steffi: //Perhaps if you want to become a mathematician
Sharon: Or working in a bank. (Germany)
Interviewer: Uh, you like this lesson?
Peggy: Uh, a little bit. I can, I mean, I can discuss with my classmates.
Interviewer: Uh, what do you think is the most important thing to learn in this mathematics lesson?
Peggy: The most important? Calculation.
Interviewer: Calculation is the most important?
Peggy: Yep.
Interviewer: What have you learnt?
Peggy: I've learnt…(learnt). I don't know. (Hong Kong)

In the German interviews, the answer to the question if they like or dislike mathematics or are interested in doing mathematics has often nothing to do with mathematics as a subject – which they feel they do not know about – but with mathematics as a school subject and the way it is taught. This can be also seen in the Hong Kong interviews: The mathematics teachers might be nicer than other teachers, they can chat with their classmates during the lesson, the atmosphere is more acceptable. Sometimes quite surprising values are assigned to mathematics; on the other hand, aversions against mathematics depend on the teaching style, their own lack of understanding and the helplessness they experience in the lesson.

Interviewer: Um... Do you like this lesson?
Rachel: I like it. Uh, I don't know why, but I am interested in mathematics lessons. All teachers, no matter now or before, are quite good.

...

Interviewer: Do you like Mathematics lessons?
Rachel: Yes.
Interviewer: Why?
Rachel: Well, because, I do not know how to tell you. But I think that mathematics can make me think faster.
Interviewer: Yep.
Rachel: That is, I need not to use my brain to think in a lesson usually. That is, I only need to memorize all the information in books. Also (in mathematics lessons), it can- that is I can find the answer when I see the question, so there is no need to find the information in books. (Hong Kong)

3.3 Enjoyment and math

'Doing math' offers extra practice and some advantage in comparison to other students. However, enjoyment or fun is rarely connected to intrinsic features of doing mathematics, in contrast to other school subjects. The fun might be rather limited, and does not offer sense making either, but a kind of substitute for a frustration that is going alongside with mathematics. To feel already some kind of enjoyment in being quicker than other students when solving some tasks, and then being able to recapitulate possible mistakes, is alarmingly modest. However, solving tasks or problems correctly, i.e. avoid mistakes or errors, is considered a very necessary action for the assessment procedure. Errors are not allowed, there is nothing to learn from them, but one has to hide them in front of the teacher.

Interviewer: Um, do you like studying mathematics or having mathematics lessons?

Peggy: Having mathematics lessons? Mm, I like the atmosphere of the lesson.
Interviewer: What atmosphere?
Peggy: That is very relaxed. He would not bind you inside the classroom and not allow you to say anything.
Interviewer: What about learning Mathematics?
Peggy: Learning Mathematics? Mm, I do not do that. I// seem to be, on average.
Interviewer: // Do you like that?
Peggy: Mm, if- if it is not too difficult, I am interested in it.
Interviewer: You are - are interested in (...)
Peggy: Yep, I like to learn new things, but when I learn the new thing, I learn in a slower speed. If he teaches too fast and I cannot catch up, I will not like this lesson lastly. (Hong Kong)

Interviewer: Do you like this lesson?
Polly: Yes. //We calculate in every mathematics lesson so I am not bored.
Interviewer: //Why... Why you are not bored?
Polly: I have to think during calculation. I can use my brain to think.
Interviewer: Others are boring? It cannot be recorded by nodding your head.
Interviewer: What do you think is important to learn?
Polly: This lesson or the normal lessons?
Interviewer: Normal lessons.
Polly: Uh, what is important to learn?
Interviewer: Yep. What do you think is important to learn?
Polly: My- my brain can think faster.
Interviewer: Yep, then this lesson?
Polly: Also the same.
Interviewer: Have you learnt something that makes you think faster?
Polly: Yep.
Interviewer: What is it?
Polly: For example, that parenthesis, I forget to write it and I could think of it. (Hong Kong)

....

Interviewer: Do you like mathematics or mathematics lesson?
Polly: I like.
Interviewer: Why? (laughing) It is difficult to answer.
Polly: Yep.

Interviewer: You can say if you do not.
Polly: Because this lesson is the one which let us think a lot. (Hong Kong)

Osbert: I didn't figure that out
Interviewer: You didn't figure that out. But why do you think it's important?
Osbert: Because I didn't understand.
Interviewer: It's important because you didn't understand?
Osbert: Yes.
Interviewer: What did you understand? You've said that you understand those methods at the beginning. Do you think it's important after you learnt those?
Osbert: I don't have to learn about it or think about it after I learnt it. (Hong Kong)

Interviewer: You enjoy doing (math)?
Friedrich: Most of the time, yeah.
Interviewer: Do you think you have a talent for it?
Friedrich: I don't think I am excessively talented. Because I enjoy doing it, I have the feeling that I get more practice, I don't know. Take these daily exercises that we always do, I usually get through them quite quickly and have, well I work out pretty quickly and then I've always enough time to check them through again. That is I can go through each problem twice and can usually spot most of the mistakes.
Interviewer: Is speed then the criterion so to speak, the quickest is the best.
Friedrich: Not necessarily, with me it's always much better if I finish quickly, as I've then still got time to go through it all again, because I've usually got one or two mistakes and then I've got more time for them. (Germany)

The Hong Kong teacher supports extra homework to foster students' proficiencies, but only few students can do so:

Interviewer: How often would your students do mathematics exercise at home?
Teacher: ...Usually, they will do homework. For the extras...they will do extra mathematics in advance. I've emphasized that students could do the following exercise if they could. For example, if they still have a lot of time after finishing their class work and homework, I allow them to do other

mathematics exercise. Some students have already finished half of the exercise in trigonometry although I haven't taught it.

Interviewer: Does it happen frequently? Does it happen frequently? Would a lot of students do exercise by themselves?

Teacher: Huh?

Interviewer: Would a lot of students do exercise by themselves?

Teacher: I…we call it beyond syllabus.

Interviewer: Beyond syllabus means…

Teacher: It means you go beyond the syllabus. About four or five students will do that. (Hong Kong 1, TI2)

3.3.1 Math is fun? An ironic remark

The ironic turning of "fun" in mathematics into magicians' or conductors' work shows the sad state of school mathematics: you do things correctly in following the rules, but it is just magic and does not make sense to you. And at the same time again: 'math has nothing to do' with the students, it is not their math, but is someone else's math, a strange or foreign propriety you have to master, but have no access in understanding.

Martin: Math is one of my favourite subjects. It is fun.

Interviewer: What is fun about it?

Martin: No idea. Numbers that nobody understands. You try to understand what you are supposed to do with all the x's and that's fun, juggling with numbers.

Interviewer: Like a sorcerer?

Martin: Yeah, to do something magical, you get a result, and then someone says that it is right. Yeah, but sometimes you get nothing out, just something you can't understand like why x equals y or the like. But it's fun anyway. Well...if the maestro says that it's so...then so be it....I really have to believe what he says. (Germany)

3.4 It is the assessment that counts: The only important goal is performance

Mathematics is considered even by the majority of these privileged German students at the Gymnasium as a compulsory enterprise without significance for them; it is for selection, for checking the mind, does not make sense, is not understandable, an empty set of rules, and can become terrifying when turned into assessment tasks.

3.4.1 Dislike of mathematics as selection means: math is just terrible

Sabine: Classroom tests written classwork in math always drive me crazy in advance. These are the only ones which make me so nervous, in other school subjects I do not mind at all, but in math it is just....uaaah!

Interviewer: What is it that makes math so terrifying for you?

Sabine: No idea, it is just terrible. I do accept all other subjects, but math ... it's fully stupid. And I do not know why! (Germany)

Interviewer: Yep. Uh, do you like learning mathematics or having mathematics lessons?

Olivia: I do not like it very much.

Interviewer: You do not like it very much.

Olivia: I like Mr. Ng's lesson as I can chat. However, because I... I did not like mathematics since I was a child, I did not like listening to the lessons and my results were not good, so I did not like having the lesson. Also, mathematics is not like other subjects. You do not listen in the lesson of other subjects, then you pick up the book. There are some words in the book. If you revise it, you will pass the tests. However, mathematics is different. You... even if I pick up the book, but without the explanation of Mr. Ng and others who teach me: It changes like this and this changes like this and it substitutes like that. I cannot see that. Therefore... therefore, what have you asked?

Interviewer: Do you like having mathematics lesson or learning mathematics? (laughing)

Olivia: //Yes, I do not like it, I do not like it. (laughing) (Hong Kong)

3.4.2 Good performance and active participation in the classroom: one has to care for being recognised

In German schools, assessment of oral participation is an important aspect of evaluating performance in mathematics. Consequently one has to show that one has done correctly, and especially because of the teacher based assessment modes, one has to constantly care for being recognised as belonging to those who can do so. To avoid to be recognised as somebody who did not do correctly is frequently shaping the pattern of classroom participation. It can be argued that the teacher-based assessment mode forces students and teachers to exhibit an ambivalent behaviour due to a conflict of values: On the one side the teacher is the facilitator of learning, maybe a friend who cares, on the other side, s/he is the (sole) responsible for assessment results, who can easily become the enemy of the students. Even if they like the teacher, by the very system, they might keep a distance.

Interviewer: Martin, I noticed you went to ask Mr. Reimer if your solution of the given task was correct. But you were completely sure already before that you had a correct solution. Why did you ask Mr. Reimer then?

Martin: I have to make sure that he knows and notes that I did it correctly, that I have well done. You have to care for that. (Germany)

To be recognised as a good and an actively participating student is considered as important in Hong Kong schools as well, but there are different perceptions of what might be honoured by teachers and valued by classmates:

Michael: Well, here. … I think we were not energetic enough.

Interviewer: Energetic means...

Michael: You were not answering the question actively. That means not - actually, it is not necessary to - not necessary to reach the level of active...the situation was that there was no one put up his or her hand. We only stood up to answer the question when teacher called us. This was not so good.

Interviewer: You think the whole class...

Michael: Yep, to see the whole class as a whole. Also including me. (laughing)

Interviewer: (laughing) How about you, why don't you put up your hand?

Michael: Ha?

Interviewer: You were shy?

Michael: No, think...I think that others would put up their hand to answer. Also, anybody else liked to answer- answer the questions. I don't know why.

Interviewer: Some students like answering the questions?

Michael: Yep, there is one. He would answer, so I gave the chance to him to answer. (laughing) He would like to be recorded. So. (laughing) (Hong Kong)

Interviewer: After watching the whole tape, do you have any other thing you would like to say, as a supplement or about the school, no... what the classmates and the teacher did and said that you think it is important?
Nina: In- I only think that there- there is one person speaks a lot. That means he always put up his hand.
Interviewer: (laughing)
Nina: Mm, there are only one or two people like this.
Interviewer: //(...) What?
Nina: //Not much
Nina: I don't know. He is also like this normally. He always likes...how to explain...want to be focused. (laughing)
Interviewer: (laughing) I see, I see.
Nina: Yep.
Interviewer: He put up his hand, why? //Answering the question?
Nina: //Answering the questions... always. Yep, sometimes he answers it wrong and he loses face.
Interviewer: (laughing) You think that this is special?
Nina: Yep. (Hong Kong)

Interviewer: What were you doing at that time? Were you afraid that Mr. Ng would ask you?
Osbert: No. He seldom asks me. Usually, he asks...the ones who often answer his questions
Interviewer: Huh? Whom does he ask usually?
Osbert: The students who often answer his questions (Hong Kong)
Interviewer: You didn't understand these?
Interviewer: Did you raise your hand?
Osbert: No. Not many students raised their hands.
Interviewer: Not many students raised their hands
Osbert: Not many students understood. (Hong Kong)

3.5 Meaning and understanding in learning and doing mathematics

Some of the German students recognise that they might need alternative ways of approaching a mathematical topic, but in most cases are only offered one, which might be appropriate only for some, but not for the majority. If they do not know why they do not understand, they learn by heart.

3.5.1 Understanding why or learning by heart

Most of the German students exhibit in the interview that they have given up in trying to constructively participate, they follow the teacher's guidance and hide if they fail to understand the mathematics taught: It does not seem worthwhile to try to understand why mathematics is like it is, because you never find out by yourself.

Interviewer: What do you both prefer? Like this (a given discovery problem)? To reflect about, to draw something and to try to find out why? Or to calculate those other tasks.

Cordula: My, yeah, ... to find out why it is so, is more fun.

Interviewer: This is more fun for you?

Cordula: Yes.

Sabine: For me, I prefer tasks, it is too stressful for me to find out why, why it is just like this and not otherwise – alone, I would never have found out at all! So for me, it is enough to do tasks of which one just knows how to do them. It is too stressful to think about why – you never find it by yourself anyway. It is ok, when the teacher says it is ok.

Cordula: Yeah, but I would appreciate more when we now solve problems and also learn to know why this is like it is or not, the case that I just do problems and think yeah, it will be ok, this is not my favourite case. It makes much more fun to think about...why something does work or not. More than just do some algebraic transformations correctly.

Sabine: But it does not help me at all if there is a drawing or the like, I might have understood, but I would not be able to imagine, I am more such a type who can learn by heart.

Cordula: But when I have seen why, for example by a drawing, and understood why it is like this, it is much easier, I can reconstruct what we have done.

Sabine: But I cannot cope with those drawings, maybe there is another way that works for me – I don't know. If there is another way to work on this problem on your own and to find out, maybe that it seems to be more logical to me or the like, then I could probably also cope better and prefer it, but like this I cannot get a meaning. ... I have always questioned why to calculate with letters and who invented something as strange as math! (Germany)

3.5.2 Collaboration and discussion might be a very good means in the struggle for understanding and meaning

The German teacher usually did not encourage collaboration, as 'assessment has to be individually' done anyway. He therefore also does not usually organise collaborative work besides some very few exceptions. Students more or less have to hide if they prefer to collaborate, even if they have developed collaboration into a very successful working pattern and found out some very good reasons for collaboration:

Interviewer: Do you work together on these problems?
Steffi: Yes, we do always. We talk about them as well. We help each other.
Sharon: Each one says a bit and then we agree on a happy medium.
Steffi: We work through together to see where a mistake is when we have different results or we just share work.
Sharon: Because I can actually multiply quite quickly on paper or something, you know, and then she does, she can write it down quickly, so we really share the work.... – But you can't do in a test. Yeah, really very impractical.
Steffi: It really helps to talk about because you recognize the mistakes before, well before it is done on the board; if you can go through the problems well with someone and find the mistakes, or where the mistakes have been made, it is really better. You put together what you know, this makes you secure and you feel competent, even at the black board! (Germany)

The Hong Kong teacher is referred to by students as encouraging or allowing student discussion in groups as an explicit value.

Michael: I think that this part is quite important. That is when students did not understand, they should er...discuss.
Interviewer: What was happening?
Michael: Er...her- and I, she asked me a question.//...
Interviewer: //Who asked you a question?
Michael: Natalie asked me.
Interviewer: Ha, what she asked you? Do you remember?

Michael:	She asked me how to calculate that question and I said that she is stupid.(laughing)
Interviewer:	Which question? (laugh)
Michael:	I forgot, but she asked the one at the back and she was doing, she ...had not done. I said that she disturbed my rest.//...
Interviewer:	//Disturbing your rest.(laugh)
Michael:	Yep. I was closing my eyes and having a rest, then I said that you were stupid. It was like this. I have made her a new nickname. Then- then we called her by this nickname. I think that this is better because there is a gap between teacher and us. With a gap, we cannot ask so happily- cannot ask so easily. Asking our classmates can be very natural and easy and learn- I think we can learn faster. (Hong Kong)

REFERENCES

Bernstein, B., 1990, *The Structuring of Pedagogic Discourse*, Routledge, London.

Bishop, A., 1999, Mathematics Teaching and Values Education – An Intersection in Need for Research, Part 2, *Zentralblatt für Didaktik der Mathematik*, **99**(1):1-4.

Bishop, A.J., 1988, *Matematical Enculturation*, Kluwer Academic Publishers, Dordrecht, The Netherlands.

Burton, L., 1999, The Practices of Mathematicians: What Do They Tell Us about Coming to Know Mathematics?, *Educational Studies in Mathematics*, **37**:121-143.

Clarke, D., 2002a, The Learner's Perspective Study: Exploiting the Potential for Complementary Analyses. Paper presented as part of the symposium *Primary Research, Secondary Research and Research Synthesis: Theory, Value and Linkage* at the Annual Conference of the American Educational Research Association, April 2-5, New Orleans.

Clarke, D. et al, 2000b, Mathematics Classroom Practice: The Learners' Perspective. in: *Proceedings of the 10th Annual Conference of SAARMSTE* (Southern African Association for Research in Mathematics, Science and Technology Education), C. Lubisi and C. Malcolm, eds., University of Natal, Durban.

Clarke, D.J., ed., 2001, *Perspectives on Meaning in Mathematics and Science Classrooms*. Kluwer, Dordrecht.

Drew, P., and Heritage, J., eds., 1992, *Talk at Work. Interaction in Institutional Settings*, Cambridge University Press, Cambridge.

Ehlich, K., and Rehbein, J., 1986, *Muster und Institution. Untersuchungen zur schulischen Kommunikation*, Gunter Narr Verlag, Tübingen.

Ernest, P., 1999, Forms of Knowledge in Mathematics and Mathematics Education: Philosophical and Rhetorical Perspectives, *Educational Studies in Mathematics*, **38**(1-3):67-83.

FitzSimons, G.E., 1999, Values, Vocational Education and Mathematics: Linking Research with Practice, in: Changing Practice through Research: Changing Research through Practice, *Proceedings of the 7th Annual International Conference on Post-Compulsory*

Education and Training (Vol. 3), Centre for Learning and Work Research, Griffith University, Brisbane, pp.11-21.

Gellert, U., Jablonka, E., and Keitel, C., 2001, Mathematical Literacy and Common Sense in Mathematics Education, in: *Sociocultural Aspects of Mathematics Education*, B. Atweh et al., eds., Lawrence Erlbaum, New York, pp.57-73.

Jablonka, E., 2003a, Mathematical Literacy, in: *Second International Handbook of Mathematics Education*, A. Bishop, M.A. Clements, C. Keitel, J. Kilpatrick and F. Leung, eds., Kluwer Academic Publishers, Dordrecht , pp.75-102.

Jablonka, E., 2003b, Constraints and Affordances of Reasoning Discourses in Mathematics Classrooms: Examples from Germany, Hong Kong and the United States. Paper presented as part of a symposium on social interaction and learning in mathematics classrooms at the 10th European Conference for Research on Learning and Instruction, Padova, Italy, Auguste 26-30, 2003.

Keitel, C., 2000, Cultural Diversity, Internationalisation and Globalisation: Challenges and Perils for Mathematics Education, in: *Cultural Diversity in Mathematics Education*, A. Ahmed, J.M. Kraemer, and H. Williams, eds., Ellis& Horwood, Chichester, pp.41-60.

Keitel, C., and Kilpatrick, J., 1999, Rationality and Irrationality of International Comparative Studies, in: *International Comparative Studies in Mathematics Education*, Kaiser, G., Huntley, I. and Luna, E., eds., Falmer, London, pp.241-257.

Krummheuer, G., and Brandt, B., 2001, *Paraphrase und Traduktion*, Beltz, Weinheim.

Lam, C.C., Wong, N.Y., and Wong, K.M.P., 1999, Students' Conception of Mathematics Learning: A Hong Kong Study, *Curriculum and Teaching*, **14**(2):27-49.

Lam, L., 2003, *Mathematics Education Reform in Hong Kong*, Department of Mathematics, Hong Kong Institute of Education, Hong Kong.

Lerman, S., 2002, Cultural, Discursive Psychology: A Sociocultural Approach to Studying the Teaching and Learning of Mathematics, *Educational Studies of Mathematics*, **46**(1-3): 87-113.

Lerman, S., and Tstsaroni, A., 1998, Why Children fail and What Mathematics Education Studies can do about it: The Role of Sociology, in: P. Gates, ed., *Proceedings of the First International Conference on Mathematics, Education and Society*, Centre for the Study of Mathematics Education, University of Nottingham, pp.26-33.

Love, E., Pimm, D., 1996, This is So: A Text on Texts, in: *International Handbook of Mathematics Education*, A. Bishop et al., eds., Kluwer Academic Publishers, Dordrecht, pp.371-410.

LPS, 2002, The Learners' Perspectives Study – An International Research Collaboration on Mathematics Classroom Practice, http://www.edfac.unimelb.edu.au/DSME/lps/.

Mullis, I.V.S., Martin, M.O., Beaton, A.E., Gonzales, E.J., Kelly, D.L., and Smith, T.A., 1997, *Mathematics Achievement in the Primary School Years: IEA's Third International Mathematics and Science Study (TIMSS)*. Boston College, Chestnut Hill.

Richards, J., 1991, Mathematics Discussions, in: *Radical Con-structivism in Mathematics Education*, E. von Glasersfeld, ed., Kluwer Academic Publishers, Dordrecht, pp.13-51.

Richards, J., 1991, Mathematics Discussions, in: *Radical Constreuctivism in Mathematics Education*, E. von Glasersfeld, ed., Kluwer Academic Publishers, Dordrecht, pp.13-51.

Seah and Bishop, [ICMI book] Trip for the Body, Expedition for the Soul: An Exploratory Survey of Tweo Asian Teachers of Mathematics in Australia.

van Oers, B., 2002, Educational Forms of Initiation in Mathematical Culture, *Educational Studies of Mathematics*, **46**(1-3):59-85.

Voigt, J., 1995, Thematic Patterns of Interaction and Sociomathematical Norms, in: *The Emergence of Mathematical Meaning: Interaction in Classroom Cultures*, P. Cobb and H. Bauersfeld, eds., Lauwrence Erlbaum, Hillsdale, N.J, pp.163-201.

Transcript protocol:

S Unidentified student

// Overlapping talk

... Pause of three or less seconds, respectively where it would be in the English version of the original

(...) Indecipherable words

Chapter 4-6

TRIP FOR THE BODY, EXPEDITION FOR THE SOUL: AN EXPLORATORY SURVEY OF TWO EAST ASIAN TEACHERS OF MATHEMATICS IN AUSTRALIA

SEAH Wee Tiong and Alan J. BISHOP
Monash University

1. INTRODUCTION

Much of the (mathematics) educational research community's knowledge, understanding and comparison of different cultural and educational traditions has come about through studies which compare institutional policies, curriculum statements, textbooks, learning environments, student work and/or teacher/student affect in their respective cultural settings, a notable example being the series of international mathematics and science studies. The teacher, too, is an important focus in any study of mathematics education in different traditions (ICMI, 2000). While this chapter aims to contribute understanding and knowledge in this specific area involving teachers, their values and their subsequent decisions and actions, it also seeks to do so through using a different approach, i.e. through investigating the professional experience and practice of individual teachers of mathematics across different East Asia-Western cultural settings.

Specifically, this chapter represents the voices of two East Asian teachers of mathematics who had migrated to Australia over different periods in time. Although it is multi-ethnic in its demographic makeup, Australia's historically colonial links with Britain and a period of 'learning from the United States' in major cities such as Melbourne have meant that the dominant societal culture in Australia is Anglo-Saxon. Through the value differences these teachers encountered in the Australian mathematics classroom, we can "identify similarities and differences, and ... interpret and explain the similarities and differences identified" (ICMI, 2000) in the mathematics education systems in East Asia and in Australia. The chapter also reveals that as East Asians in the West, these immigrant teachers may have transported themselves physically into the Australian classroom, but the corresponding journeys of their souls have often been ones with numerous obstacles, dilemmas and uncertainties. Lastly, but no less importantly, the ways in which these immigrant teachers negotiate the value differences in their mathematics classrooms point to possibilities of how these differences may lead to an optimisation of school mathematics instruction in a Western culture such as Australia's, and also in the East Asian context.

2. SOCIO-CULTURAL NATURE OF MATHEMATICS AND MATHEMATICS EDUCATION

Some of the underlying factors behind different mathematics education systems in East Asia and in the West are due to the socio-cultural nature of both mathematical knowledge and mathematics education. The birth and development of mathematical systems are linked to particular needs of respective societies (Ascher & D'Ambrosio, 1994; Bishop, 1991). For example, it is reasonable to assume that 'Western' mathematics had not developed from the same impetus accounting for the beginning of the Chinese mathematical system in the early Chinese civilisation. Of course, it may be difficult to define 'Western' mathematics because it has certainly evolved through direct influences from many cultures itself, including the Chinese culture (Bishop, 1990). In fact, that the two traditions have developed differently reinforces the socio-cultural gap existing between (East) Asia and the West.

Also, school mathematics education is a socially-referenced activity (Cobb, 1996; Keitel, Damerow, Bishop, & Gerdes, 1989). Since countries in East Asia and in the West are in different stages of national development, the mathematics curricula in different nations are designed to serve the respective nations' current and unique developmental, societal and economic needs. Also, what appear to be similar mathematics curricula may be taught through different perspectives and approaches. Even similar teaching aids may be used to achieve different pedagogical aims (see Brenner, Herman, Ho, & Zimmer, 1999, for example, for an illustration of the different purposes of using manipulatives in Asian and American classrooms).

3. VALUES RELATED TO MATHEMATICS AND MATHEMATICS EDUCATION

Cultures are intricately related to values. Den Brok, Levy, Rodriguez and Wubbels (2002) define culture as

> the ever-changing values, traditions, social and political relationships, and worldview created and shared by a group of people bound together by a combination of factors (which can include a common history, geographic location, language, social class, and/or religion), and how these are transformed by those who share them. (p.448)

To McConatha and Schnell (1995), culture is "an organised system of values which are transmitted to its members both formally and informally" (p.81). Thus, differences in the mathematics education systems in East Asia and in the West can possibly be understood in terms of differences in values. As a cultural construct, values may be seen to have been internalised from affective qualities such as beliefs and attitudes (Krathwohl, Bloom, & Masia, 1964; Raths, Harmin, & Simon, 1987).

Values in the context of mathematics education represent one's internalisation, 'cognitisation' and decontextualisation of affective variables (such as beliefs and attitudes) in one's socio-cultural context. They are inculcated through the nature of mathematics and through one's experiences in one's socio-cultural environment and in the mathematics classroom. These values form part of one's ongoing and developing personal value system, which equips one with cognitive and affective lenses to shape and modify one's ways of perceiving and interpreting the world, and to guide one's choices of course of action. In general, values in mathematics education may be categorised as mathematical, mathematics educational, and general educational (Bishop, 1996), and are collectively part of broader categories of personal, institutional, epistemological and societal values (Seah & Bishop, 2000). According to Bishop (1988), 'western' mathematics may be regarded to value mathematical qualities like 'rationality', 'objectism', 'control', 'progress', 'mystery' and 'openness'. Teacher professional practice and mathematics education norms also transmit mathematics educational values such as 'technology', 'reasonableness' and 'applications', although the extent to which each is emphasised (or not) can differ amongst cultures. Certainly, interactions within a micro-version of society such as the mathematics classroom sees the portrayal of general educational values such as 'honesty', 'organisation' and 'respect' as well.

The data shown in this chapter is part of a bigger-scale questionnaire survey. The two immigrant East Asian teachers filled out the questionnaire in a place and at a time which each felt most comfortable and least restrictive.

The 45-item, 12-page questionnaire consisted of open- and closed-ended questions which encouraged the teacher participants to explore and examine the mathematics education traditions, and their pedagogical practices, in their respective home cultures and in Australia.

4. VALUE DIFFERENCES BETWEEN MATHEMATICS CLASSROOMS IN EAST ASIA AND IN THE WEST

Cultural value differences in this chapter refer to the immigrant teachers' perception of differences in which their respective home cultures and the Australian culture place importance over mathematical knowledge, school mathematics curriculum, or mathematics teaching/learning. What constitutes the Australian culture in this chapter would be the cultural order which each of the immigrant teachers perceives of Australia. In this sense, different notions of Australian culture can be associated with different immigrant teachers, but in many ways it can be argued that the Australian culture is indeed a construction within the lived experience of the individual involved. Importantly, it is the reference against which the immigrant teacher interprets the social world around him in his Australian mathematics classroom, and against which value differences are perceived.

The two immigrant teachers identified seven main value differences (see Table 1). Corresponding to each of these value differences is the respective teacher's identification of the factor(s) underlying the observed differences. These opinions generally correspond to each of the teacher's professional identity, a glimpse of which was discerned from the teacher's responses to the inter-related questionnaire items. It is perhaps expected that these value differences experienced in the mathematics classrooms cut across the intended, implemented and attained curricula (the curriculum comparative model adopted by the IEA's International Mathematics and Science Studies). More useful in our ongoing attempt to better understand the differences across cultures perhaps, is to relate these value differences to the five universal dimensions of cultural variability proposed by Geert Hofstede (1997). These dimensions are, namely, power distance, relationship of self to community, achievement orientation, uncertainty avoidance, and life orientation. They were conceptualised as a result of a survey of more than 100,000 employees of a multinational firm in more than 50 countries in the 1970s. Based on the scores calculated for the participating countries, each country occupies a position along each of the dimension continua, so that comparisons may be made between/among countries along any of these dimensions. Alternatively, each country may be seen to be uniquely identified by its cultural characteristics in five-dimensional space. The variety of relations between people in different countries, categorised into four co-existing but conflicting 'cultural projects' by Douglas (1996), may indeed be seen to be one way through which such five-dimensional cultural characteristics find expression.

Table 4-6-1. Value differences experienced by immigrant teachers of mathematics in Australia

Value difference (Australian culture / home culture)	*Underlying factor(s)*	*Dimension of cultural variability*
Teacher co-constructs knowledge / teacher dispenses knowledge	Authority of teacher in society	Power distance (small — big)
Mathematics as object of fear / mathematics as object of beauty	Difference in emphasis on contributions of mathematics to human civilisation	Self and community (individualism — collectivism)
Less content / more content	Differing relative emphasis on knowledge and skills	Self and community (individualism — collectivism)
Many mathematics subjects / few mathematics subjects	Differing relative emphasis on knowledge and skills	Achievement orientation (relationship — task)
Realistic question context / artificial question context	Differing relative emphasis on knowledge and skills	Uncertainty avoidance (weak — strong)
Specific aims / holistic aims	(Not identified by teacher)	Life orientation (short-term — long-term)
Surface understanding / deep understanding	Nature of aims of mathematics education	Life orientation (short-term — long-term)

The power distance difference between cultures may be used to explain one of the value differences experienced by Xiaoming, an immigrant teacher from China. According to him, a major part of the typical teacher training process in China is concerned with the mastery of mathematical knowledge. As such, the mathematics teacher is a respected dispenser of academic knowledge in the Chinese classroom. However, the nature of the secondary mathematics teacher training program in Australia may not be significantly different. That Xiaoming found himself playing the role of co-constructor of mathematical knowledge in his Australian classroom may actually be attributed to Australia being a relatively low power distance country (compared to China). In contrast, Hofstede points out: "in the large power distance situation …. the educational process is highly personalized … what is transferred is not seen as an impersonal 'truth', but as the personal wisdom of the teacher …. In such a system the quality of one's learning is virtually exclusively dependent on the excellence of one's teachers" (Hofstede, 1997, p.34), thus underlining the significance of the teacher's authority in China.

Xiaoming also noted that "in China, mathematics as a discipline is perceived as an object of beauty; [whereas] in Victoria, people don't seem to see this side of mathematics, and most of them express fear of it [and perceive it] as a discipline belonging to a select few". He attributed this cultural difference to the fact that there was a greater inculcation of student awareness in China to the contributions Chinese mathematicians had made to the world and to mathematical knowledge. There is a sense that the Chinese contributions constitute a significant part of the Chinese civilisation, that it has been an achievement of the 'we' group, from which one's identity is derived. Thus, this value difference concerning students' perceived nature of mathematics can be explained by the relatively more collectivistic Chinese society and a correspondingly more individualistic Australian society.

Thomas, an immigrant teacher from Singapore, was of the opinion that the relatively bigger number of pre-tertiary mathematics subjects on offer in Australian secondary colleges can be understood in the perspective of Australia's emphasis on skills over knowledge, and of a relatively lesser Singapore emphasis in this aspect. The bigger range of mathematics subjects in Australia allows a student to take up a mathematics unit which is more related to his/her intended vocation. This difference in orientation between the two educational systems may be related to Hofstede's (1997) distinction between masculine (task orientation) and feminine societies (relationship orientation). In particular, he says: "organizations in masculine societies [in this case, Singapore] ... reward ... on the basis of equity, i.e. to everyone according to performance .... and organizations in ... [feminine] societies ... reward people on the basis of equality ..., i.e. to everyone according to need" (Hofstede, 1997, p. 93).

Thomas also noted that "assessment questions at all levels of school mathematics in Victoria tend to be more realistically contextually-based, i.e. not standard, 'designed' contexts", the latter of which tended to be found (in his view) in Singapore assessments. Here, it is supposed that 'designed' contexts refer to those where given conditions are necessarily utilised in the solution process, and where the contexts can be easily interchanged without affecting the solution strategy expected of students. In other words, context is superficially (though appropriately) crafted into the mathematics assessment items. As with some of his other observations, Thomas attributed this difference to the two countries' difference in relative emphasis over student attainment of knowledge and skills. What underpins this cultural value difference in this case, however, may well be the countries' perceived difference towards uncertainties in their common desire for content relevance to daily life. In this context, Australia is a relatively weak uncertainty avoidance society, where the nature of the mathematics assessment question context reflects open-ended learning situations, and where multiple solutions/answers may exist.

The last cultural dimension to be considered here, that of long- versus short-term life orientations, is exemplified by Xiaoming's observation that the aims of mathematics education are (at least explicitly) different in China and in Australia. According to Xiaoming, school mathematics in China is seen as a subject which provides the necessary exercise to both hemispheres of the brain; in Australia, school mathematics promotes student reasoning and problem-solving skills. While this view highlighted particular emphases of the Chinese and Australian (mathematics) educational goals only, it does reflect the general differences in the two systems (see, for examples, Board of Studies, 2000; People's Republic of China National Education Committee, 1992; Tian, 2002). Although the desired outcomes in both countries may be largely similar, i.e. to equip students with the necessary capabilities to function in and contribute to the society, this value difference concerning aims exemplifies Xiaoming's perception of the countries' respective societal orientations. Australia's identification of more specific aims is in line with an expectation of quicker results, and may be interpreted to be expressing a concern with identifying and possessing the 'Truth' (e.g. the way to apply mathematical knowledge). On the other hand, the Chinese 'mathematics exercises the brain' approach signals a perseverance towards slower results, and while not identifying particular 'Truth', is concerned with student attainment of a 'Virtue'. These values are also reflected in the nature of the Eastern and Western religions (Hofstede, 1997), as well as in the nature of Chinese and Western medicinal approaches. Herein is a difference in terms of long-term (China) and short-term (Australia) societal orientations. In fact, China was the most long-term oriented country in Michael Bond's survey of 23 countries (including Australia at 15th place) along this dimension (Hofstede, 1997). Interestingly, in that survey, China was followed by four other East Asian nations, namely, Hong Kong, Taiwan, Japan, and South Korea.

5. NEGOTIATING VALUE DIFFERENCES ACROSS CULTURES

The value differences perceived by Xiaoming and Thomas in their own Australian mathematics classrooms have sharpened their awareness of their own cultural values relating to secondary mathematics teaching and learning. These values were inculcated in the immigrant teachers through their personal experience as children, students, and teachers in their respective home countries. Such cultural capital accompanied the teachers' arrival in Australia. The responses from both teachers have been positive and constructive. Specifically, Xiaoming felt that both his content and pedagogy needed to reflect the Australian cultural values, and his teaching in Australia has reflected changes to his former practice in China. His ranking of teaching styles revealed an increased use of textbooks in Australia, although he did not clarify if he was referring to the school-appointed textbook or to the big range of commercially available textbooks in Australia. Xiaoming referred to his teaching 'the Australian style' several times in his questionnaire response, which included establishing a more equal relationship with his students. Yet, he also expressed frustration whenever he perceived as confrontational student responses to his admonition. This perhaps demonstrates that Xiaoming might not have fully internalised his own attempts at practising with a local perspective.

At the same time, Xiaoming felt the need to continue to help his students appreciate the beauty inherent in mathematics, and to encourage them to share his view that mathematical practice provides the necessary exercise for the whole brain. He desired to enrich the local teaching culture with these values. Clearly, Xiaoming believed that these home cultural values related to mathematics teaching/learning can potentially contribute to a more effective pedagogical practice in Australia.

Like Xiaoming, Thomas' experiences of value differences in the Australian mathematics classroom have caused him much cognitive dissonance. For example, Thomas expressed dismay that some Australian teachers' treatment of the school-appointed textbook as seemingly teacher-proof has led to many lost opportunities of enthusing students through classroom activities, discussions and alternative modes of lesson presentations. To him, the local mathematics classroom environment was a far cry in all respects from his classroom in Singapore. At the same time, Thomas was aware of more positive aspects of the Australian mathematics curriculum, such as the structuring of mathematical problems in realistic contexts, and he was glad to be able to incorporate these into his teaching repertoire. As he put it, "I can only try to bring together the best features of the two countries' systems".

The questionnaire response from Thomas and Xiaoming appears to indicate that these immigrant teachers were involved in culture blending (see Ninnes, 1994) approaches to negotiating perceived value differences. For these teachers, perceived differences, and potential dissonance arising from the ways the home and Australian cultures value aspects of school mathematics education, were resolved through combining parts of the two cultures. At least in their professional lives, it appears that the 'melting pot' analogy of the interaction of different cultures was a useful one with which to interpret the immigrant teachers' experience, even if their personal lives may have contributed to a 'mosaic' image of multiculturalism in an essentially Anglo-Saxon Australian society.

These teachers' responses to perceived value differences demonstrate that the mediating role of cultural (pedagogical) values needs to be taken into account in an understanding of the Piagetian processes related to equilibration. In a related study of immigrant teachers of mathematics in Australia, Seah and Bishop (2001) identified a range of teacher approaches to value differences encountered in the mathematics classroom, namely, culture-blind, assimilation, accommodation, amalgamation, appropriation (see also Bishop, 1994). In this context, the two East Asian immigrant teachers in this chapter have responded to perceived value differences in ways which generally correspond to the amalgamation and appropriation approaches. Based on the questionnaire returns alone, it is understandably not clear the extent to which each of these two approaches have been adopted by Thomas and Xiaoming in their negotiation of the value differences. Both these approaches represent the blending of cultural values, albeit in different ways. The amalgamation approach involves the immigrant teacher adding certain Australian values to his/her personal value schema, while values associated with the home country continue to shape the teacher's worldview and disposition. Thus, in the light of Xiaoming's comments discussed earlier, if his practice in the Australian classroom features distinct periods of demonstrating (the Chinese cultural valuing of) 'beauty' (e.g. through teaching proof) and instances of working with contextualised, applications-based problems, it is likely he is amalgamating practices related to the two values in his practice. His conception of mathematics teaching would still be one which emphasises the beauty inherent in the structure and organisation of the discipline; that he includes applications problems as part of his classwork/homework/assessment may simply be a reflection of his valuing of reflecting the Australian culture in his professional practice.

On the other hand, an appropriation approach to negotiating difference in values involves the modification of personal value schemas as a result of incorporating new values into the schemas. The teacher values both the home and host countries' values to an extent which sees both these being integrated; in a way, the valuing of one demonstrates the valuing of the other as well. Take the same value difference mentioned in the last paragraph, for example. If Xiaoming's responsive approach is appropriative, it is likely that he would teach for the applications of mathematical concepts, and within these applications problems Xiaoming would highlight any inherent beauty rather than treating them as questions which students have to learn to answer simply to satisfy external assessment requirements. Thus, an appropriation approach would see Xiaoming creating for himself a professional practice based on the Australian 'model', enriched by his Chinese cultural and pedagogical values.

While no generalisation may be made here with regards to any representative approaches East Asian immigrant teachers may tend to adopt, it is useful to note that these are examples of positive interaction of culturally-based values related to the mathematics education systems of East Asia and the West. That Xiaoming and Thomas remain relatively empowered in the Australian teaching service exemplifies the opportunities for constructive and meaningful interaction between aspects of the mathematics education systems in these two parts of the world.

At the same time, the sense of frustration and helplessness the two immigrant teachers experienced at times serves to remind us how these negotiation processes may be less than smooth sailing for the psyche/soul. Thomas' note that some students and their parents in Australia question him over 'inadequate' textbook use (because he did not rely on the textbook closely in his lessons) highlights the limitations confronting teachers beyond the immediate confines of their mathematics classrooms. In Thomas' words, "I may not be teaching these kids next year, but I hope that some of them would have captured what I feel to be important this year so that they bring it [sic] with them to their future mathematics study and future lives. It's hard to change the culture here, but hopefully this [students picking up his values] will be something more permanent".

Cultures transform and 'move on' with time. As collective socio-cultural values undergo inevitable changes in level of relative emphases, cultures evolve in form and expression. It is helpful to remind ourselves that Thomas and Xiaoming's description of the Singapore and Chinese cultures respectively were based on their personal experiences as students and teachers in the relevant culture. Thus, the current-day mathematics educational scenes in Singapore and China are likely to be different from what is represented through the teachers' feedback in this chapter. The teachers' personal value schemas were shaped by home cultural values, and it was the perspectives offered by these value schemas through which Thomas and Xiaoming perceived differences in cultural valuing. Indeed, there is evidence in the first author's ongoing thesis research study that this phenomenon of 'culture freeze' is not affected by any ongoing knowledge of cultural change in the teachers' respective home countries, such as through communications and personal visits. It is worth noting, too, that development in home country's school mathematics curriculum often shows a tendency towards the embracing of relatively 'western' values, such as a greater integrated use of ICT in mathematics teaching, or the use of student group discussions in the pedagogical repertoire. Yet, these immigrant teachers would continue to interpret perceived value differences from the perspectives afforded by their very own knowledge of home cultural values.

6. CONCLUSION

This exploratory survey of two East Asian immigrant teachers has identified several similarities and differences in the school mathematics education systems in East Asia and in Australia as seen by these teachers. At the same time, it has also highlighted several teacher perceived value differences which reinforce Hofstede"s (1997) five universal value dimensions. There is also indication that teacher readiness and ability to blend the perceived values in difference may lead to the evolution of more effective mathematics teaching practices which incorporate meaningful and relevant values of different cultures, and which lead to successful professional socialisation experiences of immigrant teachers of mathematics. The former outcome is clearly desirable as Australia, having achieved reasonably well in school mathematics in international comparative studies such as the Third International Mathematics and Science Study (TIMSS) (e.g. Lokan & Greenwood, 2001) and Programme for International Student Assessment (PISA) (Organisation for Economic Co-operation and Development, 2001), looks to further improve its school mathematics education system so that the Australian workforce is more able to compete with leading knowledge and innovative economies occupying top spots in these international comparative studies. The significance of the second outcome is best understood against a looming shortage of qualified teachers of mathematics in Australian schools, due to a decreasing pool of mathematics graduates and to an ageing teaching workforce (Australian Education Union, 2001). A systemic failure to support the professional socialisation experience of immigrant teachers more generally, and indeed all teachers in transition, represents costly loss/ wastage of human capital in the society in general, and in the teaching service in particular. While the transition of these teachers across geo-political borders may be a relatively easy trip at the physical level, the adaptation of their spiritual/affective soul to the new, Australian professional culture may be an expedition fraught with uncertainty and even hostility. A collective, systematic and informed approach to scaffolding this expedition for individual immigrant teachers will not only benefit these 'new' professionals individually, but will also be a worthwhile, long-sighted investment in the intellectual future of the society!

REFERENCES

Ascher, M., and D'Ambrosio, U., 1994, Ethnomathematics: A Dialogue, *For the Learning of Mathematics*, **14**(2):36-43.

Australian Education Union, 2001, 2001 October 19, *A National Teacher Shortage: A Solution from the Australian Education Union*. Retrieved November 26, 2001, from http://www.aeufederal.org.au/Campaigns/index2.html#TS.

Bishop, A.J., 1988, *Mathematical Enculturation: A Cultural Perspective on Mathematics Education*, Kluwer Academic Publishers, The Netherlands, Dordrecht.

Bishop, A.J., 1990, Western Mathematics: The Secret Weapon of Cultural Imperialism, *Race and Class*, **32**(2):51-65.

Bishop, A.J., 1991, Mathematical Values in the Teaching Process, in: *Mathematical Knowledge: Its Growth through Teaching*, A.J. Bishop, S. Mellin-Olsen and J.V. Dormolen, eds., Kluwer Academic Publishers, Dordrecht, The Netherlands.

Bishop, A.J., 1994, Cultural Conflicts in Mathematics Education: Developing a Research Agenda, *For the Learning of Mathematics*, **14**(2):15-18.

Bishop, A.J., 1996, *How Should Mathematics Teaching in Modern Societies Relate to Cultural Values-Some Preliminary Questions*. Paper presented at the Seventh Southeast Asian Conference on Mathematics Education, Hanoi, Vietnam, June 3-7.

Board of Studies, 2000, *Mathematics: Curriculum & Standards Framework II*. Board of Studies, Carlton, Australia.

Brenner, M.E., Herman, S., Ho, H.-Z., and Zimmer, J.M., 1999, Cross-national Comparison of Representational Competence, *Journal for Research in Mathematics Education*, **30**(5): 541-557.

Cobb, P., 1996, *Accounting for Mathematical Learning in the Social Context of the Classroom*. Paper presented at the 8th International Congress on Mathematical Education, Sevilla, Spain, July 14-21.

den Brok, P.J., Levy, J., Rodriguez, R., and Wubbels, T., 2002, Perceptions of Asian-American and Hispanic-American Teachers and their Students on Teacher Interpersonal Communication Style, *Teaching and Teacher Education*, **18**:447-467.

Douglas, M., 1996, *Thought Styles*, Sage Publications, London.

Hofstede, G., 1997, *Cultures and Organizations: Software of the Mind*, revised ed., McGraw-Hill, New York.

ICMI, 2000, 2001 August 24, *The Thirteenth ICMI Study on Mathematics Education in Different Cultural Traditions: A Comparative Study of East Asia and the West: Discussion Document*, [Online]. ICMI. Available: http://www.mathunion.org/ICMI/bulletin/49/ Comparative_East_West_Asia.html [2001, September 3].

Keitel, C., Damerow, P., Bishop, A.J., and Gerdes, P., 1989, *Mathematics, Education and Society*, (Science and Technology Education Document Series No.35). UNESCO, Paris, France.

Krathwohl, D.R., Bloom, B.S., and Masia, B.B., 1964, *Taxonomy of Educational Objectives: The Classification of Educational Goals (Handbook II: Affective Domain)*. David McKay, New York.

Lokan, J., and Greenwood, L., 2001, *Maths and Science on the Line: Australian Year 12 Students' Performance in the Third International Mathematics and Science Study*, Australian Council for Educational Research, Victoria, Australia.

McConatha, J.T., and Schnell, F., 1995, The Confluence of Values: Implications for Educational Research and Policy, *Educational Practice and Theory*, **17**(2):79-83.

Ninnes, P., 1994, *Values and Cultural Change amongst Vietnamese Students in Adelaide*, Paper presented at the International Conference for Cultural Democracy, Adelaide, Australia, May.

Organisation for Economic Co-operation and Development, 2001, *Knowledge and Skills for Life: First Results from the OECD Programme for International Student Assessment (PISA) 2000*, Organisation for Economic Co-operation and Development, Paris, France.

People's Republic of China National Education Committee, 1992, *Jiu nian yi wu jiao yu quan ri zhi chu ji zhong xue shu xue jiao xue da gang (shi yong) [Mathematics syllabus for the full-day junior high school in the nine-year obligatory education]*, People's Education Publisher, Beijing, China.

Raths, L.E., Harmin, M., and Simon, S.B., 1987, Selections from 'Values and Teaching', in: *Value Theory and Education*, J. P.F. Carbone, ed., Robert E. Krieger, Malabar, FL, pp.198-214.

Seah, W.T., and Bishop, A.J., 2000, *Values in Mathematics Textbooks: A View through Two Australasian Regions*. Paper presented at the 81st Annual Meeting of the American Educational Research Association, New Orleans, LA, April 24-28.

Seah, W.T., and Bishop, A.J., 2001, Crossing Cultural Borders: The Negotiation of Value Conflicts by Migrant Teachers of Mathematics in Australia, *Proceedings of the 2001 Annual Conference of the Australian Association for Research in Education*, [CD-Rom]. Australian Association for Research in Education..

Tian, Z.J., 2002, *Quan ri zhi pu tong gao ji zhong xue jiao ke shu (shi yan ben, bi xiu), shu xue:. 'shu xue' di yi ce (shang) jian jie [An introduction to the revised edition of volume 1A of 'mathematics' (full-day normal senior high school textbook series: mathematics)]*, [Online]. China Basic Education. Available: http://www.cbe21.com/subject/maths/article.php?article_id=1530 (2002, July 16).

Chapter 4-7

CONCEPTUALISING PEDAGOGICAL VALUES AND IDENTITIES IN TEACHER DEVELOPMENT

A Comparison of Taiwanese and Australian Mathematics Teachers

CHIN Chien
National Taiwan Normal University

1. INTRODUCTION

Values are an inherent, more likely implicit, part of all educational processes, and they are a crucial component of the cognitive, affective, and conative environment of mathematics teaching (FitzSimons, Seah, Bishop, & Clarkson, 2001; Fraenkel, 1977). Mathematics teachers, in this case, inevitably teach particular values, and students thus perceive implicitly the values through classroom teaching activities. The extent to which teachers are aware of or willing to teach their own values, and such values' selection process, is important, although not clear, for mathematics teaching and learning. And yet, there is little knowledge about values teaching in mathematics classrooms and values education of mathematics teachers (Bishop, FistzSimons, Seah, & Clarkson, 2000).

Parallel research on values in mathematics teaching from two countries Taiwan and Australia (Bishop, FitzSimons, Seah, & Clarkson, 2001; Chang, 2000; Chin & Lin, 2000a; Leu & Wu, 2000), where one is a typical Chinese and the other is an Anglo English society, shared a common belief that the quality of mathematics teaching can be improved if there is more understanding about values teaching and the values education of mathematics teachers. Their purposes were to investigate mathematics teachers' values, the extent to which teachers can clarify or gain control over their own values, and forms of classroom teacher-student values interaction.

Values represent one's internalisation and awareness of affective variables in the socio-cultural context (Krathwohl, Bloom, & Masia, 1964), including the cognitive, affective, and behavioral elements (Rokeach, 1973). They form the underlying standards of a teacher's choice and judgment of action, attitudes toward and knowledge about students and the ethos of the classroom. Values serve as abstract frames of references (McConatha & Schnell, 1995), are conceived and expressed as decontextualised words (Seah & Bishop, 2000), or underlying substances of (pedagogical identities about) personal testable propositions (Chin & Lin, 2000b; 2001a). Beliefs, in the light of this, are seen as personal testable propositions or the actualisation of values in contextualised situations (Chin & Lin, 2001a; Seah & Bishop, 2000). For example, a teacher's valuing of 'individual thinking' may be the result of his or her internalization of such belief propositions as 'there is no learning if students do not think' and 'in teaching, all I have to do is to initiate student thinking on their own'.

The research approaches adopted, although slightly different in its application and form for this cross-nation project, were basically questionnaire survey, classroom observation, and pre and post lesson interview. Several common issues emerged from a comparative analysis of the values research in Taiwan and Australia (Victoria) relating particularly to the professional development of mathematics teachers, and these will be discussed in this chapter. A model of incorporating teacher professional identities and pedagogical values will also be proposed.

2. CONTEXT OF MATHEMATICS TEACHING IN TAIWAN AND AUSTRALIA

Several differences in the context of classroom mathematics teaching are salient between Taiwan and Australia (see Chin & Lin, 2001a; Seah & Bishop, 2001). Teachers' social backgrounds are multi-cultural in Australia but not in Taiwan. In Victoria, mathematics teachers may come from very different cultural backgrounds, ranging from English, Chinese, Malaysian, to Indian. But, Taiwanese mathematics teachers are quite uniform in that over 90% of them are born and educated in the same Taiwanese (Chinese) culture. So values conflicts here are more about pedagogical issues rather than cultural issues as in Victoria. The student background in Victorian schools is normally multi-cultural, and is more uniform in Taiwan except for the areas near the aboriginal regions. In this case, values teaching and values education of mathematics teachers in Taiwan are likely to be didactically rather than culturally oriented.

The values portrayed, which underlie school mathematics curricula in the two societies, are varied. For example, Australian schools addressed the values of individual difference, cooperation, and listening to students (Seah & Bishop, 2001), while the values of uniformity, individual work, and listening to teachers were central in Taiwan (Chang, 2000). As informed by the specific Taiwanese examination culture, helping students pass all kinds of tests, getting higher scores in those tests for the students, and training them to be skillful in solving varying kinds of mathematical questions were the three top priorities for school teachers to teach mathematics (Lin & Tsao, 1999). As a result, the major purpose of teaching mathematics is to induce students to acquire content knowledge and procedural skills for solving school mathematical items given in the textbook or pseudo-text book (Leu & Wu, 2002; Lin & Tsao, 1999). In this chapter, the goals of learning and teaching mathematics for school teachers are driven by the values of getting higher student scores and passing the examination.

Therefore, the values portrayed by the mathematics curricula and taught by mathematics teachers in different cultures, may differ enormously. As values teaching often takes place implicitly and unconsciously, it is then important for us to compare the varying degree of values clarification and articulation, and the awareness of and willingness to teach certain values within and across these two societies.

3. VALUE ISSUES IN TEACHER PROFESSIONAL DEVELOPMENT

3.1 Clarification and articulation

It was through intensive reflection on the values taught that school mathematics teachers became clear about their value positions, although some of them did not choose the values that we expected them to have (Chin & Lin, 2001b; Change, 2000). This seems to be very different from Australian teachers, as most of them claimed that they acknowledged the important roles that values should play in mathematics teaching (FitzSimons et al., 2001). This also reveals the societal differences between the West and East Asia, in which values are often a generally recognised issue for the Western people to discuss but not so obvious for us to talk about. It seems to me that the issues of values clarification and articulation are natural for teachers in the Western cultures but not the Eastern. In this case, the concept of values is there for the Western mathematics teachers, and yet, it might be quite a difficult concept for the Eastern teachers to even think about in relation to classroom teaching of mathematics. Another essential difference is related to the orientation of clarification and articulation. For Australian teachers, more efforts should perhaps be used in clarifying value conflicts from different cultures, and the extent of values articulation to which they could gain control. The issues for Taiwanese teachers are concerned mainly with the extent of clarification that they can make within their own value systems and between different values of self and others. However, it seems to be a problem for both sets of teachers to recognise the varying degrees of values clarification within the intended and implemented phases. Some teachers might have learned about the issues but not others, as the high and low values clarification of Taiwanese teachers showed (see Chin, Leu, & Lin, 2001) compared with different degrees of values nomination and explicitness among Australian teachers (Bishop et al., 2001).

In short, there are cognitive issues concerning the need to clarify teacher values in terms of different cultures. For Australian teachers, the concept of values in relation to mathematics teaching may be relatively easy for them to think through but not an easy affair for Taiwanese teachers, as they are not so sure about the notion of values in mathematics education. There are also different aspects that teachers have to articulate and clarify. For Australian teachers, more difficulties exist concerning the different values that teacher and student share within a multicultural environment of mathematics teaching. However, they all have similar problems with value discrepancies between self and others, and with the inconsistencies of one's own intended and implemented values.

3.2 Awareness and willingness

Two affective aspects about teacher values have emerged. The first issue is about the extent of consciousness of imposing personal values. Being unaware of the reality of classroom values teaching, it took quite a long time for the 6 out of 7 Taiwanese teachers from primary to senior high schools to realize the roles that values play in their mathematics teaching (Chang, 2000; Leu & Wu, 2000). The one exception was a senior high school master teacher who was in a short period of time passing through the five stages of values clarification, moving from a value carrier to a value communicator, and who described the significant features of his values teaching to others in terms of his own classroom incidents (Chin & Lin, 2000b). This notion is also supported by the Australian research findings (FitzSimons et al., 2001), in which different values were taught by different teachers according to the degree of awareness of their own intentions to teach particular values. The second issue is about teachers' willingness or conation to teach particular values. Varying degrees of the explicitness of the values teaching were examined and discussed in the Australian research (Bishop et al., 2001; FitzSimons et al., 2001), in which the number of values explicitly taught by the teachers, either consciously or unconsciously, was greater than the implicitly taught values. However, in our researches in Taiwan (Chang, 2000; Chin & Lin, 2001b; Leu & Wu, 2000), not only were fewer teachers aware of the taught values, they were also unwilling to teach certain values and even refused to teach the values that they had taught before. As a result, it is clear that there are pedagogical and socio-educational tensions surrounding the extent of the explicitness, willingness, and awareness of teachers' classroom values teaching.

In short, there appear to be varying pedagogical and socio-educational tensions for mathematics teachers in different cultures. For Australian teachers, the tensions of values teaching were more about personal consciousness and the explicitness of engaging in some particular values. However, for the Taiwanese teachers these tensions seemed more to do with the conative and socio-educational aspects than to the individual awareness of values teaching. These observations are also evident in the light of Leung's (2002) six-dichotomy model of conceptualizing/contrasting the features of the East Asian and the Western countries' identities in mathematics education and their underlying values.

The above analyses led me to integrate the concept of values and (pedagogical) identities within the professional development of mathematics teachers. This idea takes the socio-cultural aspects of individual development into account, in which mathematics teachers from different cultures establish certain different (pedagogical) identities framed by their own socio-educational values. Different cultures, in turn, create different tensions both for the teaching of values and for the development of pedagogical identities for the mathematics teachers.

4. VALUE TRANSITION AND IDENTITY DEVELOPMENT OF MATHEMATICS TEACHERS

Three approaches are prevalent in the literature of teacher development. The first aspect is to conceive the process as professional knowledge growth/change, for example, the change or growth of teachers' pedagogical content knowledge or knowledge about students. Researchers such as Even, Tirosh, & Markovits (1996), imply a theoretical framework of necessary knowledge for teaching a specific mathematics topic in order to design activities in the teacher education programme for teachers to learn. Another aspect is to educate teachers through the construction of their beliefs, for instance, by identifying beliefs in terms of the role of "authority" and seeing it as a growth indicator to conceptualize the process by which a teacher becomes a reflective person. This research (e.g., Cooney, Sheatly, & Arvold, 1998) uses a theoretical framework of change in beliefs to set up activities in an educational program for individual teachers to articulate. The third aspect is to consider the process as an intertwined web of conceptual changes, including knowledge, beliefs, and context, using a situated learning model to monitor the professional growth of a group of prospective teachers. This approach (e.g., Llinares, 1996) combines knowledge and beliefs into the situated context, as a mean for promoting teachers' development.

Although these approaches have been productive, however, they have missed a crucial issue for mathematics teachers concerning the nature of *being a teacher*, as he or she is potentially in the process of learning different ways *to be a mathematics teacher*. In other words, all teachers are in the process of developing their pedagogical identities through which they learn to see themselves as becoming the teachers that they value most. Therefore, a fourth aspect of conceptualizing teacher professional development is to conceive the process as the development of (pedagogical) identities and the transition of values.

There have been similar ideas explicitly developed, concerning the role of identity and value in education and learning. For example, Chickering and Reisser (1993) identify a seven-vector model, in which the *establishing identity* vector addresses the development of a person from confusion about 'who I am' to clarification of self-concept through roles and lifestyles in the process of education. To maintain self-image as a teacher, Furlong and Maynard (1995) identified four interconnecting dimensions in establishing a professional relationship for student teachers, in which awareness of self to the notion of *me-as-a-teacher* is related to individual teacher identity. Another vector of Chickering and Reisser's model (1993) is *developing integrity* concerning the inculcation of humanizing and personalizing individuals' values through education. In this case, the value systems may properly reflect those features of the development of eight identity stages, from early youth to old age (Erikson, 1963). The unity of personality that one perceives and understands bears the imprint of the ego and is close to what the person thinks of him or herself (Loevinger, 1976), in which the conscientious, autonomous, and integrated stages are more aligned with individuals' values transition and identities development in the process of searching for one's standards, uniqueness, and integrity.

Moreover, Wenger's (1998) description of *learning as a process of becoming* echoes my idea for mathematics teacher professional development:

> Because learning transforms who we are and what we can do, it is an experience of identity. It is not just an accumulation of skills and information, but a process of becoming – to become a certain person or, conversely, to avoid becoming a certain person. Even the learning that we do entirely by ourselves eventually contributes to making us into a specific kind of person. We accumulate skills and information, not in the abstract as ends in themselves, but in the service of an identity. (p.215)

On the other hand, researchers in mathematics education have also called for attention to be paid to the intertwined relationship between identity development and values, in which the aspects of articulation, clarification, awareness, and conation were used to empower professional development (e.g., Bishop et al., 2000; FitzSimon, Seah, Bishop, & Clarkson, 2000). For example, Bishop (2001) suggested several ways of educating and clarifying student teachers' ideas about values in mathematics teaching, such as through the sensitizing of value issues, engaging in values clarification and values-related activities, and providing critical teaching incidents. Lerman (2001) in a review of research perspectives on mathematics teacher education, argued for a focus on the development and articulation of teacher identities; and Chin, Leu, and Lin (2001) described activities that can be used to educate student teachers about values in mathematics teaching through values clarification and values articulation. A similar argument was also made by Boaler (2002) in terms of the knowledge-practice-identity triad, in which *mathematical identity* is used to describe the resulting intrinsic relationship of the students and their mathematics, which represents the varied aspects of (mathematical) knowledge and its (classroom) practice. Most importantly, different relationships with mathematics might inculcate different mathematical identities which accompany different pedagogical beliefs and values.

As far as this scheme is concerned, five transitional stages of teachers' identities seem to be crucial in the process of learning to teach. At the student teacher stage, a teacher sees his or her identity as *a student* in which learning and thinking about a different ethos and ideas are of importance for *the learner*. In *the probationary teacher* stage, one might see oneself as *a classroom practitioner* and his or her jobs are to integrate theory into practice and to inform theory with practice. In the following stage, the teacher becomes *a novice* in the teaching community as *a newcomer*. Later, she or he is gradually reaching the stage of *an experienced teacher* seeing her or himself as *an old-timer*. Then the teacher might approach the final stage of teaching profession as *an expert teacher*, and being viewed as *a master* in the teaching community. This five-stage developmental sequence fits well a possible path of teacher's identity transition, within which a mathematics teacher commits her or himself to the compatible values of their pedagogical identities.

For example, the case of Ming (Chin & Lin, 1999), who taught mathematics over 20 years, recollected his own profession as being developed through three stages in the same school, from a textbook *follower* and own *style builder*, to a *value characteriser*. His pedagogical identity moved from being *a knowledge transmitter* and a *mathematics tutor*, to a *contextual knowledge initiator*. Each stage entailed specific pedagogical values, from being committed to mathematical structure and knowledge to reality and pleasure. For teacher Yi (Chin, 2002a), there was a higher developmental stage of *socialized follower*, in which he followed fully the textbook content and procedure in teaching but abiding by and rectifying the pedagogical values that he preferred most. This teacher acts as *a philosopher* who respects the ways of teaching that the community shares, however, incorporating his personal ethos in teaching that he thinks are better for his students to learn and understand mathematics. He has gone through three schools' ethos during which his pedagogical identity as *knowledge initiator* was stable and consistent with the value of thinking, reasoning, and communication. The other two senior high school teachers are also well situated within this developmental scheme (see Chin, 2001, 2002b). One of them valued the knowledge and structure of mathematics, and identified herself as *a transmitter* throughout 30 years of teaching; and the other committed to the values of knowledge and structure at the beginning years, and then a few years later he moved to the stage of including student mathematical reasoning and thinking. The model has been further examined on the basis of a group of 3 pre-service mathematics teachers at the student/learner stage (see Chin, 2001, 2002b).

The identity transition and the values entailed throughout the stages of professional growth that these four school teachers showed represents an interwoven relation between pedagogical identities and values evolving in the process of teacher professional development, in which different teacher identities embody different pedagogical identities and values. The transitional phase of (pedagogical) identities and values is thus incorporated into an intertwining path of the teacher professional development, within which different teacher identities entail different pedagogical identities and values.

This aspect of considering mathematics teacher development from the transition of identities and the articulation of values adds to that of Lerman's (2001) proposal and echo also what Llinares (1996) suggests, addressing the contextual, integrated, active, and practical aspects of teacher development.

REFERENCES

Bishop, A.J., 2001, Educating Student Teachers about Values in Mathematics Education, in: *Making Sense of Mathematics Teacher Education*, F.L. Lin, and T.J. Cooney, eds., Kluwer, Dordrecht, The Netherlands, pp.233-246.

Bishop, A.J., Clarkson, P., FitzSimons, G., and Seah, W.T., 2000, Why Study Values in Mathematics Teaching: Contextualising the VAMP Project?, in: *Proceedings of the HPM 2000 Conference on History in Mathematics Education*, Vol.2, W.S. Horng, and F.L. Lin, eds., National Taiwan Normal University, Taipei, pp.147-154.

Bishop, A.J., FitzSimons, G., Seah, W.T., and Clarkson, P., 2001, Do Teachers Implement their Intended Values in Mathematics Classroom?, in: Proceedings of *the 25th Conference of the International Group for the Psychology of Mathematics Education*, Vol.2, Marja van den Heuvel-Panhuizen, ed., Utrecht University, Utrecht, The Netherlands, pp.169-176.

Boaler, J., 2002, The Development of Disciplinary Relationships: Knowledge, Practice and Identity in Mathematics Classrooms, *For the Learning of Mathematics*, **22**(1):42-47.

Chang, C.K., 2000, Score-ism as their Pedagogical Value of Two Junior High Mathematics Teachers, in: *Proceedings of the International Conference on History in Mathematics Education*, Vol.2, W.S. Horng, and F.L. Lin, eds., NTNU, Taipei, pp.172-177.

Chang, C.K., 2001, The Taiwanese Teachers' Beliefs and Values in Mathematics Education, in: Proceedings of *the 25th Conference of the International Group for the Psychology of Mathematics Education*, Vol.1, Marja van den Heuvel-Panhuizen, ed., Utrecht University, Utrecht, The Netherlands, p.295.

Chickering, A.W., and Reisser, L., 1993, *Education and Identity*, Jossey-Bass, San Francisco, CA.

Chin, C., 2001, *Pedagogical Values in Mathematics Teacher Education Project: First Year Research Report*, a Taiwanese NSC funded research project, National Taiwan Normal University, Taipei. (In Chinese)

Chin, C., 2002a, Researching Mathematics Teachers' Pedagogical Values: A Synthesis of the Three Taiwanese Studies, in: *Proceedings of the 2nd East Asia Regional Conference on Mathematics Education and 9th Southeast Asian Conference on Mathematics Education*, Vol.1, D. Edge, and B.H. Yeap, eds., Nanyang Technological University, Singapore, pp.80-89.

Chin, C., 2002b, *Pedagogical Values in Mathematics Teacher Education Project: Second Year Research Report.* A NSC funded research project, National Taiwan Normal University, Taipei. (In Chinese)

Chin, C., and Lin, F.L., 1999. *A Value-oriented Mathematics Teaching*. Paper Presented at the Annual Conference of Mathematics Education Research Projects Funded by Taiwanese National Science Council. (In Chinese)

Chin, C., and Lin, F.L., 2000a, A Case Study of a Mathematics Teacher's Pedagogical Values: Use of a Methodological Framework of Interpretation and Reflection, *Proceedings of the National Science Council Part D: Mathematics, Science, and Technology Education*, **10**(2):90-101.

Chin, C., and Lin, F.L., 2000b, Values and Values Statement Emerged in Students' Preferences on Test Items: A Case Study from Mathematical Induction, in: *Proceedings of the International Conference on History in Mathematics Education*, W.S. Horng, and F.L. Lin, eds., National Taiwan Normal University, Taipei.

Chin, C., and Lin, F.L., 2000c, *Values in Mathematics Teaching Project: Third Year Research Report*, A Taiwanese NSC funded research project, National Taiwan Normal University, Taipei. (In Chinese)

Chin, C., Leu, Y.C., and Lin, F.L., 2001, Pedagogical Values, Mathematics Teaching and Teacher Education: Case Studies of Two Experienced Teachers, in: *Making Sense of Mathematics Teacher Education*, F.L. Lin, and T.J. Cooney, eds., Kluwer, Dordrecht, The Netherlands, pp.247-269.

Chin C., and Lin, F.L., 2001a, Value-loaded Activities in Mathematics Classroom, in: Proceedings of *the 25th Conference of the International Group for the Psychology of Mathematics Education*, Vol.2, Marja van den Heuvel-Panhuizen, ed., Utrecht University, Utrecht, The Netherlands, pp.249-256.

Chin, C., and Lin, F.L., 2001b, Mathematics Teacher's Pedagogical Value Clarification and its Relationship to Classroom Teaching, in: *Proceedings of the National Science Council Part D: Mathematics, Science, and Technology Education*, **11**(3):114-125.

Cooney, T.J., Shealy, B.E., and Arvold, B., 1998, Conceptualizing Belief Structures of Pre-service Secondary Mathematics Teachers, *Journal for Research in Mathematics Education*, **29**(3):306-333.

Erikson, E.H., 1963, *Childhood and Society*, Norton & Company, New York.

Even, R., Tirosh, D., and Markovits, Z., 1996, Teacher Subject Matter Knowledge and Pedagogical Content Knowledge: Research and Development, in: *Proceedings of the 20th Annual Conference of the International Group for the Psychology of Mathematics Education*, Vol.1, L. Puig, and A. Guitierrez, eds., University of Valencia, Valencia, Spain, pp.119-134.

FitzSimons, G., Seah, W.T., Bishop, A.J., and Clarkson, P., 2000, Conceptions of Values and Mathematics Education held by Australian Primary Teachers': Preliminary Findings from VAMP, in: *Proceedings of the HPM 2000 International Conference on History in Mathematics Education*, Vol.2, W.S. Horng, and F.L. Lin, eds., National Taiwan Normal University, Taipei, pp.163-171.

FitzSimons, G., Seah, W.T., Bishop, A.J., and Clarkson, P., 2001, Beyond Numeracy: Values in the Mathematics Classroom, in: *Proceedings of the 24th Annual Conference of the Mathematics Education Research Group of Australasia*, Vol.1, J. Bobis, B. Perry, and M. Michaelmore, eds., University of Sydney, Sydney, pp.202-209.

Fraenkel, J.R., 1977, *How to Teach about Values: An Analytic Approach*, Printice-Hall, Englewood Cliffs, NJ.

Furlong, J., and Maynard, T., 1995, *Mentoring Student Teachers*, Routledge, London.

Krathwohl, D.R., Bloom, B.S., and Masia, B.B., 1964, *Taxonomy of Educational Objectives, the Classification of Educational Goals: Handbook 2: Affective Domain*, Longmans, New York.

Lerman, S., 2001, A Review of Research Perspectives on Mathematics Teacher Education, in: *Making Sense of Mathematics Teacher Education*, F.L. Lin, and T.J. Cooney, eds., Kluwer, Dordrecht, The Netherlands, pp.33-52.

Leu, Y.C., and Wu, C.J., 2000, An Elementary Teacher's Pedagogical Values in Mathematics: Clarification and Change, in: *Proceedings of the International Conference on History in Mathematics Education*, Vol.2, W.S. Horng, and F.L. Lin, eds., National Taiwan Normal University, Taipei, pp.178-195.

Leu, Y.C., and Wu, C.J., 2002, Mathematics Pedagogical Value System Oriented Towards the Acquisition of Knowledge in Elementary School, in: Proceedings of *the 26th Conference of the International Group for the Psychology of Mathematics Education*, Vol.3, A. Cockburn, and E. Nardi, eds., University of East Anglia, Norwich, England, pp.297-304.

Lin, F.L., and Tsao, L.C., 1999, Exam Maths Re-examined, in: *Rethinking Mathematics Curriculum*, C. Hoyles, C. Morgan, and G. Woodhouse (Eds.), Falmer, London, pp.228-239.

Llinares, S., 1996, Improving Knowledge, Professional Growth and Monitoring the Development of Mathematics Teacher: A Necessary Integrating of Theoretical Frameworks, in: *Proceedings of the 20th conference of the International Group for the Psychology of Mathematics Education*, Addenda, L. Puig, and A. Guitierrez, eds., University of Valencia, Valencia, Spain, pp.23-31.

Loevinger, J., 1976, *Ego Development: Conceptions and Theories*, Jossey-Bass, San Francisco, CA.

McConatha, J.T., and Schnell, F., 1995, The Influence of Values: Implications for Educational Research and Policy, *Educational Practice and Theory*, **17**(2):79-83.

Rokeach, M., 1973, *The Structure of Human Values*, Free Press, New York.

Seah, W.T., and Bishop, A.J., 2000, Researching Values with Migrant/Expatriate Teachers of Mathematics, in: *Mathematics: Shaping the Future*, J. Wakefield, ed., The Mathematical Association Victoria, Melbourne, Australia, pp.348-355.

Seah, W.T., and Bishop, A.J., 2001, Teaching more than Numeracy: The Socialization Experience of a Migrant Teacher, in: *Proceedings of the 24th Annual Conference of the Mathematics Education Research Group of Australasia*, Vol.2, J. Bobis, B. Perry, and M. Michaelmore, eds., University of Sydney, Sydney, pp.442-450.

Wenger, E., 1998, *Communities of Practice: Learning, Meaning and Identity*, Cambridge University Press, Cambridge.

Section 5

OUTLOOK AND CONCLUSIONS
Introduction

LEUNG Koon Shing Frederick[1], Klaus-Dieter GRAF[2] and
Francis J. LOPEZ-REAL[1]
[1]*The University of Hong Kong;* [2]*Freie Universitaet Berlin*

1. METHODOLOGY

In the Discussion Document, we argued for cultural sensitivity in carrying out comparative studies. We pointed out that there are differences in methodology between cultural traditions, and there is a danger in "applying certain methodologies from one cultural tradition inappropriately in another cultural tradition". It is therefore fitting in this last Section of the Study Volume that we include a chapter by Winslow and Emori which not only interprets findings of comparative studies in mathematics education in terms of the subtle context of the underlying cultural traditions, but also addresses methodological differences between the East and the West.

Winslow and Emori analyze characteristics of the secondary mathematics classroom in Japan and relate them to the wider context of the semiotics of the Japanese society and culture. They show how Japanese mathematics education, in particular the mathematical way of viewing and thinking, is uniquely Japanese, despite Western influence and superficial similarity with Western mathematics education. The role of the teacher, or the *sensei*, as one who has preceded the students on this "way", is to guide students to discover the "way". This echoes well with the concept of "*do*" discussed by Hirabayashi in Chapter 1-1. Implicit in this "way" are the good way of thinking, good way of writing and good way of speaking. The individual student, and teacher, is a "loyal member of nested *iemotos*" rather than a "critical participant in a socially and historically situated dialectic process", and the *iemoto* characteristics of the mathematics classroom correspond to the ideology of "harmony" in the Japanese system. Winslow and Emori contrast these with the Western "way" in terms of the mathematical sign system, mathematical discourse, mathematical statement, agents, goal of learning and the "ideal state", and point out that although there are superficial similarities, there are much deeper differences.

2. OTHER COUNTRIES

Such differences of course do not only occur between cultural traditions in East Asia and the West. The chapter by Alekseev, Barton and Knijnik reminds us that while this Study Volume focuses on cultural traditions in East Asia and the West, there are other conventions or traditions that may seem to be less widespread or less dominant, but which are no less important.

The paper by Alekseev *et al* starts by describing some philosophical background to different mathematical world views and argues that indigenous knowledge can only be properly understood from within the world-view of that indigenous group. How mathematics education is embedded in indigenous knowledge is then illustrated by an ethnomathematical approach to mathematics and mathematics learning in a project at a settlement school in Brazil that involved the joint action of peasants, students, teachers, and agronomists in constructing pedagogical work in mathematics focused on the productive activities in the settlement. A second example describes an internal differentiation within mathematics education in Russia, and shows how features of mathematics education are determined both by national elements, such as geographical position and socio-cultural traditions, and political structures. Finally, the chapter contextualizes the discussion within the politics of knowledge, questioning whether less dominant world views are given due respect for their potential and actual contributions to mathematics and mathematics education.

3. FUTURE COLLABORATION

In the final chapter, Bishop asks the question: "what comes next?", and explores the collaborative possibilities following this Study. This study takes as its basic assumption the notion that difference is to be welcomed and celebrated, and Bishop argues that in order to gain maximum benefit from our comparative work we need to focus explicitly on issues of cultural difference and contrast. This should be conducted in a collaborative rather than a competitive spirit. The chapter then discusses four issues to be overcome if potentially fruitful collaborative activities are to be developed: research collaborations using cross-cultural approaches; school collaborations; university teaching collaborations; and building international/inter-cultural networks.

4. CONCLUSIONS

So what can we conclude from the Study and from the papers presented in this Volume?

The focus of this Study is on differences in educational traditions in East Asia and the West which have cultural and social implications for the environment and practice of mathematics education. It is a comparative study for understanding, not competition, and in that regard, we have understood quite a lot more about mathematics education in East Asia and the West from the Study than what we started off with. The themes covered in the Study and the papers in this Volume have helped us understand the phenomenon of difference and correspondence between cultures. The papers have also presented results of attempts to exploit that understanding and apply it to mathematics education. In particular, we have a greater understanding of the impact different cultural traditions exert on the four aspects of mathematics education presented in the four Sections of this Study Volume.

By contrasting the different cultural traditions in East Asia and the West, we have gained a deeper understanding of various contexts or the givens from which we organize and practice mathematics education. The papers in Section 1 have provided us with an insight into how cultural beliefs and values, history, social structures and social needs have shaped mathematics teaching and learning in East Asia and the West.

Different cultural contexts give rise to different curricula, different practices in teaching and learning, and different values and beliefs. The papers in Section 2 have illustrated how cultural traditions, ideologies, political and historical factors have shaped the mathematics curricula and mathematics textbooks in various countries. Different traditions have also given rise to different assessment practices, which in turn have affected student performance both within their own countries and in international studies such as TIMSS and PISA. The obvious lesson to learn from these findings is that any curriculum reform should always take the cultural values underlying curriculum practices into account, and importing curriculum reform measures from "more successful" countries to "less successful" ones, without regard to the complexities of the cultural differences among the countries concerned, is doomed to fail.

Although the intention of the Study was to focus on the cultural diversity of mathematics education between countries from East Asia and the West, the papers in Section 3 show that large variations of practice in teaching and learning can be observed from within the same culture, and even within the same classroom. Notwithstanding this variability, the comparatives studies presented in the Section suggest that the differences in classroom practice within a country or culture are less than the differences between countries or cultures, and all the papers share "the common perspective of the existence of a strong cultural determination of mathematics lessons". Studies on distance learning reveal how technology is able to bring students and teachers from different parts of the world together via the internet, and thus contribute to a better understanding of teaching and learning in different cultures, and to a better understanding of mathematics itself.

Finally, the papers in Section 4 have shown that important differences do exist in teachers' beliefs and values between the two main cultural traditions. These different beliefs and values are manifested in various curriculum media, such as the textbooks, and translate into unique teaching practices. The comparative studies reported in this section of the volume have helped us understand more about the constructs of teachers' beliefs and values in different cultures and their influences on mathematics teaching and learning.

For the participants of the Study and the authors of the Volume, it has been an inspiring process of learning and of exchange of experiences and expectations. Through being exposed to different traditions, we have had the opportunity of reflecting upon the way mathematics education is conducted in our own culture, which we often take for granted. We have been able to look at our usual practices and beliefs with a fresh eye, and through this we have gained a better understanding about our own traditions. In the process, we have also shared among us the latest developments and research in mathematics education in our part of the world, and we have learned from each other's successes and failures.

Our "outlook" into the future is a global mathematics education community that stresses collaboration rather than competition. We hope to collaborate in both research and development, learning the best practices from each other, yet preserving what is valued in mathematics education in each of our own cultures. Our belief is that it is only through a better understanding of each other's cultures that we will attain a richer mathematics education worldwide.

Chapter 5-1

ELEMENTS OF A SEMIOTIC ANALYSIS OF THE SECONDARY LEVEL CLASSROOM IN JAPAN

Carl WINSLØW[1] and Hideyo EMORI[2]
[1]*Danish University of Education, Denmark;* [2]*Utsunomiya University, Japan*

1. INTRODUCTION

It is well known that in recent major international surveys on mathematical ability, such as TIMSS and PISA, students from Japan and other East Asian countries are consistently outperforming their peers in Western industrialised countries (see e.g. Robitaille et al., 1994, and Beaton et al., 1996). This has led many to believe that Asian systems of schooling must be somehow more efficient, and that importing its methods would lead to similar efficiency in the West. Some proponents of this idea admit that the success of Asian schooling (as measured by comparative surveys) is *culturally contingent*, e.g. it depends on popular values and attitudes: *one cannot hope to learn how education systems can be improved...without understanding the actions, beliefs, and attitudes that exist within the culture* (Stevenson, 1998, cf. also Stevenson et al., 1992). Now, our basic question is: *can anything more specific be said about this in the context of mathematics?*

More precisely, we shall describe some central characteristics of the *secondary level mathematics classroom in Japan* (abbreviated SMCJ), their cultural contingency, and their relation to what is found in similar contexts in the West. In order to do so, many theoretical frameworks and methods could be chosen. Our approach is (to our best knowledge) entirely novel in this context. It is based on *semiotics*. This paper, then, is about the intersection of two semiotic contexts: that of formalised, secondary education in Japan, clearly situated and interpretable only in a wider context of 'Japan'; and that of mathematics (as taught in secondary school), or more precisely the corresponding class of semiotic registers and their current usage. In particular, we draw on studies of Barthes (on various segments of Japanese culture), Rotman, Ernest and Duval (on practices of mathematics and mathematics education). The main ideas and notions in this approach are introduced in Section 1. Using these ideas, we first describe the semiotic *context* (visible assets, agents, general codes) of the SMCJ in Section 2. Then, in Section 3, we try to describe and exemplify specific *codes* (rules and forces) governing the SMCJ. It is a main point of the paper that these appear to be profoundly rooted in socio-cultural structures that are specifically and uniquely Japanese; the last section (4) is a (somewhat tentative) attempt to formulate this point by contrasting the global assets of the codes identified with what is typically found in the West.

Even if we do not address the question of 'method import' alluded to above, it is clearly an implicit suggestion of this study that the internal mechanisms of the Japanese schooling system are much less akin to Western analogues (present or past) than its surface features may suggest. A comprehensive account of this more general point can be found in Shimahara (1979). In fact, one of the key points in the work of Shimahara is the impossibility to understand the deep orientations of Japanese education from its concrete instances, or even from its institutionally defined context as a whole, in isolation from the general cultural assets of Japanese society. A semiotic analysis, however, pretends to much less: the identification and interpretation of sign systems and their interaction.

2. BACKGROUND: NOESIS AND SEMIOSIS

Learning mathematics is inevitably linked to various very specific forms of *communication*. Mathematics textbooks and blackboards in mathematics classrooms immediately show the most striking characteristics of this communication: the abundant and highly structured usage of *symbols* and *diagrams* of various sorts. The oral communication in mathematics lessons will also, at least usually, refer to such symbolic or graphical inscriptions, typically produced on the spot (on blackboards, computer screens etc.). This particularity is not just a superficial, exterior matter; the material 'inscriptions' (that we shall refer to as *signifiers* or *semiotic representations*) are more than arbitrary ways of articulating mathematical 'meaning' (what we will call *signifieds*). Any experienced mathematics teacher would agree that teaching mathematics without symbolism or diagrams – using only natural language – is impossible. Most would probably also contend that they could not change the *usage* of signifiers dramatically without causing severe troubles both for the students and for themselves. Mathematical concepts are not independent from their representations. Just think of using the symbolic representation '$x(f) = f^2 + 1$' to define a quadratic function in a test for secondary school students!

The dialectic relation between conceptual structure and its semiotic representation is a fundamental issue in modern philosophy, including its affinities with education. Here, the notion of *representation* is in itself problematic because of the often tacit implication that the signifier is a somewhat secondary or even dispensable way of pointing to the autonomous mental object.

A basic rationale for semiotic analysis is that the relation between signifier and signified is much more complex. The coming into being of this relation – by production or perception of the sign as a whole – is called *semiosis* or *signification.* Correspondingly, cognitive acts such as understanding an inference or concept, is called *noesis.* Duval (1995) argues that although *noesis* is apparently possible without *semiosis* and commands the latter, the reality is just opposite: *there is no noesis without semiosis; it is the semiosis, which determines the conditions of possibility and exercise of noesis* (p. 4, translated from French). Much of Duval's evidence for this general conjecture comes, in fact, from the context of French secondary level mathematics education, where the *coordinated apprenticing* of distinct semiotic registers is demonstrated to be a key element of the genesis of mathematical competency. The mutually constitutive nature of mathematical signifier and signified is also central to Sfard's recent work on discursive aspects of mathematics learning (Sfard, 2000).

Another basic point of the main semiotic analyses of mathematics is the prevalence of 'self-reference', that is, of signifieds being themselves mathematical signs (Rotman, 1988, cf. also Winsløw, 1997). The fact that the sign system of mathematics is referentially closed makes it impossible for real agents to 'act' directly upon its signs. In a sense, they create their own reality: *The system becomes both the source of reality, it articulates what is real, and provides the means of 'describing' this reality as if it were some domain external and prior to itself; as if, that is, there were a timeless, 'objective' difference, a transcendental opposition, between presentation and representation* (Rotman, 1987, p. 28). In this 'virtual reality', one may say that *noesis* is embedded (and embodied) in *semiosis*, if one is not willing to entirely do away with this opposition. One may say that mathematics and its learning can be seen as a privileged object of semiotic research because of the highly (and explicitly) structured semiotic systems that are so central to mathematical activity. However, the vast majority of existing semiotic studies fall within the fields of literary and cultural studies. The latter area is, of course, quite as important for us.

Namely, in the cultural context of this paper – broadly speaking, 'Japan' – we find a similar insistence on the primacy of *semiosis* in the famous essay 'The Empire of Signs' by Roland Barthes (1970). In fact, Barthes also argues that in this cultural context, one finds the (to the Westerner) surprising phenomenon of the *pure signifier* that is simply not amenable to signification in the usual sense (logical, metaphorical etc.). According to Barthes, the assumption of secondary or implicit meaning is a main source of the Westerner's misconception – or perhaps, misperception – of the Japanese sign-world, as found in poetry or in the fine arts. That is, a main challenge for the Westerner is to overcome his tendency to confuse the emptiness of the sign with non-sense. The point seems to be the same as in the case of mathematics: *signification is a meaningful activity in itself.*

In general, semiotic analysis is not limited to signs and their relations. More important are the *codes* (or 'formal rules') and *functions* (mostly 'purposes', that is, intended outcomes) of signification; a semiotic system with a coherent code and a well-defined set of functions is called a *register*. A register could pertain to contexts as different as algebra, traffic regulation, formal dressing, or tea ceremony; in each case, there are definite rules of signification, and specific functions related to the context. But actual events of sign production will also depend on exterior and 'local' aspects, particularly *agents* (persons or institutions that produce, or react to, signs) and *media* (material environment for signification, e.g. a blackboard). In fact, it is only in this broader perspective – that is, when including codes, functions, agents and media – that it becomes possible and meaningful to relate the scientific context of mathematics with the socio-cultural context of Japanese secondary school.

Our subject here, then, is (to at least initiate) the semiotic analysis of the SMCJ. In a rough sense, existing analyses of the mathematical and of the cultural context (outlined above, but further developed in the sequel) will be combined to address this subject. In particular, the analysis will be based on notions of a Western origin and will (partially as a result of this) both explicitly and implicitly be held up against contrasting Western views and practices. The semiotic systems considered will be those of mathematics, but embedded in the cultural context of Japanese secondary education. We may now sharpen our question from the introduction as follows: *How are the forms and functions of mathematical signification embedded, adapted (or even changed) in SMCJ? How does this relate to the wider educational context? And to Western practices?*

3. THE SEMIOTIC CONTEXT OF THE CLASS: SITE, AGENTS AND CODES

The *classroom* is the fundamental site of signification in at least primary and secondary mathematics education. It can be described from three points of view: the exterior, physical aspects, the agents present therein, and the general codes governing their interaction.

The physical assets are, at first sight, deceivingly banal; typically, a somewhat worn-down room, in grey tones, with metal chairs and single tables lined up in tight rows in front of the teacher's desk, which is of equally modest style. Behind the teacher's desk, a blackboard and a clock on the wall. The classroom always has two doors, one in front and one in the rear of the class. This allows for possible latecomers to enter the class without disturbing the attention directed towards the front (teacher and blackboard). Apart from the remains of Japanese characters on the board and perhaps on posters or other material in the classroom, one finds no immediately 'Japanese' flavour. Nothing peculiar, in fact – except that the apparent non-descriptness is in fact *prescribed*: dimensions and inventory of public school classrooms are legally defined and thus (in principle) identical in all classes (cf. Schmidt et. al., 1996). The absence of reference to the particular school or class can thus be seen as a silent signifier of the ideal of *isomorphism between individual concerns and organizational goals* (Shimahara, 1979, p. 159).

The prescribed central location of the blackboard – a primary material *medium* of signification – is particularly important for the mathematics lesson. Here, it becomes the privileged place to launch and work on the 'mathematical realm' of signs. This, in itself, is a very rich semiotic realm: besides the three sets of characters (*kanji*, *hiragana* and *katakana*) which are always used in written Japanese – and hence in natural language elements of Japanese mathematics writing – we have Arabic numerals, special mathematical symbols (e.g. $\perp$), Latin and occasionally Greek letters (e.g. for algebraic symbols). Notice that Japanese characters are never used in algebraic or other mathematical symbolic expressions, giving the Western imports a special role which they do not have elsewhere. While the art of good writing (especially in the context of *kanji*) is taught in other lessons, almost all Japanese mathematics teachers at the secondary level are very systematic and careful in their writing on the blackboard, and so at least implicitly – and not seldom explicitly – teaching the 'way of writing' (*kakukata*, cf. Section 3) as an important part of the mathematical 'way' (described below).

The placement of the teacher between the blackboard and the class may be seen as a signifier of his role as 'mediator' between the class and the sign world launched on the blackboard. It is possible and indeed common for the teacher to invite students to participate in the material production of signs; a standard form of devolution (in the sense of Brousseau, 1981) in the Japanese classroom consists of a segment of independent student work on a more or less open problem, culminating with student writing within fields on the blackboard designated by the teacher.

As visible agents in the classroom, we have the pupils and the teacher, or rather the *sensei*, meaning literally 'the one who precedes'. The group of youngsters, labelled by the name and number of the class, becomes a 'mathematics class' only when preceded by the *sensei* of mathematics (who, by the way, teaches only this subject). In what sense does he 'precede'? A metaphor is immediately suggested by the general objectives of mathematics teaching in Junior High school, including that of helping the students *to appreciate the mathematical way of viewing and thinking* (Nagasaki et al., 1990[1]). The *sensei* has preceded the students on this 'way', and guides them in discovering it. Notice that he is not merely to 'show them' the way, but that he is to *help* them follow and appreciate it. Also, this concerns not only the single pupil, but foremost the group of pupils (in actual teaching, the class). We shall return to this point in Section 4.

One may say that the agent system of *sensei* and pupils, in the ideal[2] state, forms a kind of *iemoto*, defined by Hsu (1975, p. 62) as *a fixed and unalterable hierarchical arrangement voluntarily entered into among a group of human beings who follow a common code of behaviour under a common ideology for a set of common objectives*[3]. This system, of course, can not be fully understood except as part of a wider system of agents and constraints; ultimately, the need for learning mathematics in order to pass the college entrance examinations (and hence, the gate to social success) would

[1] From 2002, a new 'mathematics program' has been put into effect. The quoted formulation is essentially unaltered, but the programme as such – as the general course of study – has been oriented more towards the 'zest for living' (*ikiru chikara*) of the students. We find it premature to discuss the possible consequences in this paper. Some of the basic ideas are explained in (National Commision for Educational Reform, 2000).

[2] That is, when the rules of the agent system are followed. In fact, the nature and force of rules are sometimes best seen from observing the effects of breaking them: *We know that normal action is rule-following because we nearly always know when we have broken the rules* (Collins, 1995).

[3] Shu argues in his book that iemoto forms the basic structure in many apparently different social relationships of Japanese society: *Iemoto characteristics are to be found in all aspects of Japanese society, in religion, in business, in schools and universities, in workshops and offices* (ibid, p. 69). Note that the term iemoto has a more restricted meaning in ordinary use, and is used here in the more abstract sense defined by Hsu.

have to be taken into account[4]. At all levels of secondary school, students are 'measured' through frequent *written tests*, and indeed students take great interest in the results; at no moment is their eagerness greater than when they discuss and compare these. But for our aims here, the fact is more important than its causes, as the topic here is to analyse the nature of the *iemoto* characteristics of the mathematics classroom in its forms of semiotic interaction. And the main fact is that pupils and teachers are – at least in principle – not in a conflictual position with respect to the objectives of this interaction; the common 'objective' is locally that of pursuing the 'mathematical way' in a collective rather than competitive manner, orchestrated by the *sensei*; and more globally, it corresponds to the ideology of 'harmony' (*wa*, cf. Sekiguchi, 2000) in an agent system (ultimately, in society).

Deviation from the corresponding 'code of behaviour' is obviously found – in fact, Japanese classrooms are often surprisingly noisy, especially to the Westerner with different expectations – but it is not common for the *sensei* to address disciplinary matters directly. His main function is to direct the students' exploration of mathematics, and indeed, as one proceeds through Junior and Senior High school, the sense and signs of *iemoto* tend to grow[5]. This means above all that the codes ruling semiotic interaction become still less a matter of explicit attention, allowing for the students to 'appreciate' with still more confidence the modes and ways of the mathematical sign world. In fact, the sensei may typically try to direct the 'energy' of the younger students towards heated discussions on a particular problem, as illustrated in the following excerpt of a classroom protocol resume (from Junior High School, grade 1):

> 12 minutes into the lesson, the teacher turns to the PC [connected to a TV screen in front of the class] which displays three figures numbered 1, 2, 3 (illustrations of the first three square numbers: 1, 4, 9). Two boys are so excited about the computer that they go to stand behind the teacher to look at the computer screen instead of at the TV from their seats – the

[4] About these, Shimahara (1979, p. 125) writes: *...the CEE is the most critical rite of passage for adolescents. It requires intense experience, self-denial, durability, the ability to accept psychological mutilation, and, above all, the resilience to accept a certain cognitive and motivational orientation to the society. In other words, it demands the highest degree of individual motivation to excel in a culturally prescribed frame of reference. One of the primary functions of the CEE is the sorting and stratifying of individuals – a vital and final phase in the process of adolescent socialization.*

[5] For instance, during three weeks in the summer of 2000, CW observed four classes of Junior and Senior High school at grade levels 7, 9, 10 and 12. Here, the teachers' interventions to regulate of code of behaviour were frequent in grade 7, occasional at grade 9, virtually absent at grade 10 and entirely absent at grade 12.

teacher doesn't try to prevent this. Together, they count the number of triangles in the three figures (1, 4 and 9). Then the students are asked to find the next number. As the students work, the teacher walks around the class to check and help. Some of the noisiest boys are first to find the next number 16, and even to suggest that the next again should be 25. A table of the first five numbers is made on the blackboard. The computer is used to confirm the numbers by displaying the figures, which the students themselves have also made. The teacher then asks one boy what the 100th number should be, and he proudly says 10000. No attempt is made to check this, but the teacher says it's correct, and jokes that 'I guess you didn't make a drawing to find out that'.

The teacher then explains, pointing to the figures, that the numbers can also be found as 1, 1+3, 1+3+5 etc., and the table on the board is extended with the greatest numbers in each sum. The teacher then makes the student observe that the differences between successive square numbers are exactly these odd numbers, and they use concrete instances of the formula $(n+1)^2 = n^2 + 2n + 1$ to connect this observation with the previous identification of the numbers as square numbers. They return to the number 10000, and using their new insight, they find – still in an ongoing conversation controlled by the teacher – that there are 199 triangles in the bottom of the triangle figure corresponding to this number. Two minutes are spent on discussing how much time it would take to draw such a figure. The last five minutes are spent on discussing the general formula $1+3+5+\ldots+2n-1 = n^2$.

The working mode in Senior High School is typically much more disciplined and may in fact display almost 'ritual' forms of action code, with explanations (by the *sensei*) and exercise solving (by students or *sensei*) at the blackboard being the main activity. The sense of 'group harmony' has then reached perfection, even if many students have little personal affection for the subject. An entire lesson may, for instance, pass with 3 or 4 sequences of the type 'exercises posed by the *sensei*, students solving them individually, voluntary students writing their solutions simultaneously at the board, plenary evaluation by the *sensei* in interaction with students'.

4. SIGNIFICATION IN THE CLASSROOM: *KANGAEKATA, KAKUKATA, HANASUKATA*

Japanese society is often described by foreigners as dense with 'unwritten rules', especially in the context of immediate social interaction. In fact, the existence of *codes*, in the sense of semiotics, is a prerequisite for any systematic form of social interaction. It is only natural than an exterior observer may notice these more vividly. The following three 'mathematics specific types of codes' for semiotic interaction seem particularly important in the SMCJ:

- *Ii kangae-kata* (good way of thinking): refers to a particular strategy or technique for attacking a problem posed in class. Used explicitly by the teacher in oral evaluation of a suggestion from a pupil, but more widely representative of certain aspects of both common teaching practices and of the official aims of mathematics teaching at secondary level.
- *Ii kaku-kata* (good way of writing): refers to the form of students' writing on the blackboard when presenting their solutions of a problem posed in class. Used explicitly by the teaching in oral evaluation of such written work, and referring both to symbolic and verbal parts.
- *Ii hanasu-kata* (good way of speaking): refers to the form of student's oral explanation of a 'way of thinking' to the teacher, typically in front of the entire class. In contrast to the other two, not explicitly used by the teacher, but a necessary complement for the analysis.

In official documents, such as (Nagasaki, 1990, *opus cit.*), and indeed in much of modern Japanese research on mathematics education, only *kangae-kata* is explicitly mentioned (the aims being phrased in terms of students' understanding, thinking, viewing etc.). The main point of this section is to explain how *kaku-kata* and *hanasu-kata* (codes of semiosis in the usual sense) are in practice inclusive of the domain designated above as *kangae-kata*, namely that of *noesis* in the usual sense. This is not to be confused with a kind of behaviourist attitude or ideology, where learning is measured and conceived according to ideal patterns of response (which, according to Stevenson and Stigler (e.g. 1992) are more characteristic of the American style of teaching). It means that to achieve the state of *wa* in the classroom – and more globally, to develop the social setting of *iemoto* – what matters is not the individual's cognitive actions as such, but rather its socialised forms which may only appear through shared codes and instances of signification. As a consequence, *negotiation of meaning* (or, *hanashi-ai* as explained in Sekiguchi, 2000) is an important aim of semiotic activity in the class. This is illustrated, for instance, by the following example (from a Junior high school class, third grade):

> Arriving in class, the teacher goes to the board and says 'Let's begin'. The students settle down, but there is still some unrest for the first few minutes. She then says: 'Last time, we solved the following kinds of

equations by using square roots [*writes*]: $x^2 = q$, $(x + p)^2 = q$.' She reminds the class of an unsettled equation [discussed, but not solved, at the end of the previous lesson]: $x^2 + 6x - 1 = 0$. She then writes down the formula $(x + p)^2 = x^2 + 2px + p^2$, which the students studied some weeks before. Then, she poses the following problem: to fill out the missing parts of $(x+...)^2 = x^2 + 10x+...$; immediately, a student says that 5 and 25 are the missing numbers. The teacher insists that this must be *explained.* She develops, using very suggestive writing on the board (using systematically coloured circles and boxes around elements of the general formula as well as of the previous concrete instance) how the missing parts are filled out *first* by inserting half the right hand coefficient to *x* in the parenthesis on the left, then taking the square of this number as the missing element on the right. She then goes back to the problem $x^2 + 6x + 1 = 0$, and says: 'We will now see how this can be solved'. All students, including the usual sleepers, are attentive during the following explanation, centred on the transformations necessary to produce, successively: $x^2 + 6x - 1 = 0$, $x^2 + 6x = 1$, $x^2 + 6x + 9 = 1 + 9$ [*here, particularly, is written: half of 6 is 3, and the square of 3 is 9*], $(x + 3)^2 = 10$, $x + 3 = \pm\sqrt{10}$, $x + 3 = \sqrt{10}$ or $x + 3 = -\sqrt{10}$, $x = -3 + \sqrt{10}$ or $x = -3 - \sqrt{10}$. The teacher then turns back to the third transformation ('here, particularly...') and emphasizes the importance of this step. We are 10 minutes into the lesson, as she asks them to look at their notes for a moment to check if they have understood or if they have questions. Then, they are asked to explain that the obtained two numbers are really solutions. She then walks around in the class and talks to several students. As much as 13 minutes passes this way, before a student is called to the blackboard to explain the validity of the solutions. The student writes down the whole equation ($x^2 + 6x - 1 = 0$) with $-3 + \sqrt{10}$ in the place of *x*, and repeats the equation while calculating the left hand side until he has '0=0'. The teacher criticises the 'way of writing', as it apparently assumes what is to be shown; but also, after making the necessary corrections, lets the student understand that his calculation is essentially correct.

In this event, the key *kangae-kata* – essentially the idea of 'completion of the square' as a means to solve quadratic equations – is introduced by the teacher, and its importance and difficulty is signalled by the careful and structured *writing* that the students meticulously copy in their notebooks. The importance of *hanasu-kata* – whose function here is *reflective discourse* in the sense of *justifying steps of semiotic transformations* (processing) – is emphasised both implicitly and explicitly ('...this must be *explained*'). And, as is often the case in *sensei* evaluation of students' blackboard writing, the importance of *kaku-kata* is really the exterior manifestation (and check-point) of the good *kangae-kata*. A standard phrase used by the *sensei* as evaluation of an argument whose symbolic parts are given at the board (by himself or by students) is: *Kore wa ii kangae-kata to omoimasu* ('This is a good way of thinking, I think'). The fact that imprecise oral 'ideas' are less valued than formal argument supported by correct writing is, of course, not independent from the fact that all student evaluation (tests and exams) are *written*. And it means that the three 'codes' are in fact inseparable corner stones of 'the mathematical way' as a purely *semiotic* and *discursive* enterprise. The students are led to pay great attention to *kaku-kata* as a manifestation of *kangae-kata*, with *hanasu-kata* as a bridge between the two, which is important only in the learning situation of the classroom.

In the formation of the *iemoto* of the SMCJ, the assimilation and use of these codes are paramount. For the individual student, they are simply part of the socialisation into adult society.

5. THE MATHEMATICAL WAY: SHARED ACTS OF SEMIOSIS

Based on the previous sections and additional excerpts from the official documents regulating mathematics teaching at the secondary level, we now establish a more global picture of how the 'mathematical way' is conceived and actually presented in secondary level teaching. Mathematics as taught in secondary schools can be viewed as a semiotic system which was historically imported from the West, and which continues to carry unmistakably foreign elements (not least in the symbolic inventory). However its functions and codes in teaching are to a large extent illustrative of Shimahara's (by now) classical point: *Modernization in Japan has been superficially Western in a variety of aspects, but its substantive features of modernization are uniquely Japanese* (Shimahara, 1979, 31). How is mathematics education 'uniquely Japanese', despite its surface similarities with corresponding Western practices?

In the West, two rivalling functions of mathematics are cultivated: the 'philosophical side', in which mathematics is a source of truth about abstract phenomena, and the 'utilitarian side', where it is a tool to describe, regulate and control social and physical reality. Of course there are many competing Western views of the nature of mathematics, including the relations among these 'sides', but social constructivism (cf. Ernest, 1998) is a very pragmatic one with explicit links to educational practice[6]. The idea that mathematical meaning and truth is constructed and reconstructed social interaction may seem less shocking to the Japanese observer than to the Westerner. The primacy of semiosis over noesis (cf. Section 1, especially Barthes' view of Japan, and Section 3) may indeed be elaborated as a Japanese version of 'social constructivism' as a model of mathematics teaching and learning. Nothing is more social in nature than acts of signification and discourse. The difference arises exactly from what is counter-intuitive to the Westerner: that the 'rock solid truths' of mathematics should be grounded in something as shaky as social consensus. Most Western accounts of the social nature of mathematics are based on conceptions of the *genesis* of mathematics, in particular Lakatos' (1976) dialectic view of the history of the subject, where the occurrence of *refutations* of previously accepted 'truths' are highlighted. To 'demonstrate' the dialectically constructed nature of mathematics, Lakatos' imaginary teacher leads the students through a sequence of 'proofs' and 'refutations' of versions of the Euler polyhedron formula. Ultimately, mathematical truth, then, is not only socially but also historically bound. It derives meaning from the historical context of knowledge warranted by historically situated agents. These agents partake in a continuing discursive 'war' against gaps and mistakes in their warranting discourse. The educator trains students to make sense of this 'war' (cf. Sekiguchi, 2000).

This contrasts sharply with the timeless conception of mathematics found for example in Japanese Senior High school, with its official objective 'to help students deepen their understanding of the basic concepts, principles and laws of mathematics' (Nagasaki et al., 1990). We have already noted that socialisation into certain forms of interaction are also crucial to the enactment of this aim in SMCJ. The authority of the *sensei* consists, not in 'forcing to believe', but in 'helping to understand' the timeless reality spanned by mathematical signs. The *sensei* is responsible for introducing new semiotic representations, codes and relations to the students, according to the official curricula. The description of 'contents' of each grade ends with a section 'Terms/symbols', which, for instance for the section 'geometrical figures' in first grade of Junior High School reads: 'arc, chord, solid of revolution, π, //, ⊥, ∠, Δ'. He is further responsible for the coherent development of the class as a group (*iemoto*), which one teacher expressed as follows: *The school's basic goal [is] having students learn as members of a group* (Stevenson, 1998). As mathematical 'meaning' resides in its internal

[6] Whether social constructivism is a genuine (not to speak of correct) 'philosophy of mathematics', is not the issue here. But this view has the advantage in educational settings of paying close attention to practice, especially in education.

relations, which again are accessible only in semiotic representations, this aim of collective learning requires the students to 'willingly apply mathematical ways', hence the *sensei*'s primary role is not to explain or impose this way, but to engage students in it through guided, and above all shared, acts of semiosis. Warranting of individual semiotic acts by peers is just as important as the warranting of the *sensei*. The SMCJ is strictly stratified, however: lessons are carefully planned and executed by the *sensei*. Learning requires students to *voluntarily* enter this hierarchical arrangement (cf. Hsu, 1975). This does not necessarily mean that 'not entering' is an actual option for students, but that the respectful treatment of resistance is a crucial challenge for the teacher (especially in the low grades, cf. Section 2).

The above discussion can (tentatively) be summarised as in the following table:

	Social constructivism (Ernest, 1998)	**Japanese analogue**
Mathematical sign system	Historically and materially rooted tools for negotiating *meaning*	Free sign system, one among many 'ways of coping and viewing'
Mathematical discourse	Language game of 'proof and refutations', ruled by 'logic of language'; in principle conflictual	Language game of 'guided semiosis', ruled by transmitted codes; in principle non-conflictual
Mathematical statement	Actual state of dialectic process, warranted by institutionalised forms of control	Result of shared acts of semiosis, guided and warranted by superiors (e.g. *sensei*) and agreed upon with peers
Agents	Institutionally situated	Members of stratified *iemoto*
Goal of learning	Ability to produce and critically analyse 'new' discourse	Adaptation to the codes of semiosis that almost define the *iemoto*
Ideal state	Continuing dialectics of proofs and refutations	Harmony inside local *iemoto* and in society ('global *iemoto*')

In a superficial sense, the practices of Japanese secondary schools confirm many basic tenets of social constructivism: The discourse of school mathematics is the central component of the culture of school mathematics. Mastery of this discourse (…) is the main intended learning outcome (Ernest, 1998, 232). The main difference is in the role of the individual student, teacher and class: critical participant in a socially and historically situated dialectic process, or loyal member of nested *iemotos*.

REFERENCES

Beaton, A. et al., 1996, *Mathematics Achievement in the Middle School Years. IEA's Third International Mathematics and Science Study*, CSTEEP, Boston College.

Brousseau, G., 1981, Problèmes de didactique des décimaux (Problems on the teaching of decimals), Recherches en Didactique des Mathématiques 2 and 3, pp.37-127.

Collins, H., 1995, Humans, Machines, and the Structure of Knowledge, *Stanford Humanities Review*, **4**:2. http://www.stanford.edu/group/SHR/4-2/text/collins.html.

Duval, R., 1995, *Sémiosis et pensée humaine*. Registres sémiotiques et apprentissages intellectuels. Peter Lang, Bern.

Ernest, P., 1998, *Social Constructivism as a Philosophy of Mathematics*, SUNY Press, New York.

Hsu, F., 1975, Iemoto: *The Heart of Japan*, John Wiley & Sons, Inc., New York.

Lakatos, I., 1976, *Proofs and Refutations*, Cambridge University Press, Cambridge.

Nagasaki, E. et al., eds., 1990, *Mathematics Program in Japan*, Japan Society of Mathematics Education, Tokyo.

National Commission for Educational Reform, 2000, 17 Proposals for Changing Education, http://www.kantei.go.jp/foreign/education/report/report.html.

Robitaille, D. and Nicol, C., 1994, Comparative International Research in Mathematics Education, in: *Didactics of Mathematics as a Scientific Discipline*, R. Biehler et. al., eds., Kluwer, Dordrecht, pp.403-414.

Rotman, B., 1987, *Signifying Nothing: The Semiotics of Zero*, St. Martin's Press, New York.

Rotman, B., 1988, Toward a Semiotics of Mathematics, *Semiotica*, **72**:1-35.

Schmidt, W. et al., 1996, *Characterizing Pedagogical Flow*, Kluwer, Dordrecht.

Sekiguchi, Y., 2000, Mathematical Proof, Argumentation, and Classroom Communication: A Japanese Perspective, in: *Proof and Proving in Mathematics Education*, Boero, P. et al., eds., http://www-cabri.imag.fr/Preuve/ICME9TG12/index.html.

Sfard, A., 2000, Symbolizing Mathematical Reality into being: How Mathematical Discourse and Mathematical Objects Create Each Other, in: *Symbolizing and Communicating: Perspectives on Mathematical Discourse, Tools, and Instructional Design*, P. Cobb, K.E. Yackel, and K. McClain, eds., Erlbaum, Mahwah, NJ, pp.37-98.

Shimahara, N., 1979, *Adaptation and Education in Japan*, Praeger Scientific, New York.

Stevenson, H. and Stigler, J., 1992, *The Learning Gap*, Touchstone, New York.

Stevenson, H., 1998, A Study of Three Cultures: Germany, Japan and the United States. An Overview of the TIMSS Case Study Project, *Phi Delta Kappan*, **79**(9):1524-1529.

Winsløw, C., 1997, A Case Study of Mathematical Semiosis, *Philosophy of Mathematics Education Journal* 10 (http://www.ex.ac.uk/~PErnest/pome10/art10.htm).

Winsløw, C., 2001, *Quadratics in Japanese*, Preprint, submitted.

Chapter 5-2

OTHER CONVENTIONS IN MATHEMATICS AND MATHEMATICS EDUCATION

Valeriy ALEKSEEV[1], Bill BARTON[2], Gelsa KNIJNIK[3]
[1]Lebedev Physical Institute, Muscow; [2]The University of Auckland, New Zealand; [3]Universidade do Vale dos Sinos, Brazil

Mathematics education is about something, and usually we do not bother to discuss the question of what exactly it is we are talking about because mathematics as an academic discipline is conventionally agreed. There are, of course, disputes about what aspects of mathematics should be emphasised, or what approaches favoured, but the broad characterisation of mathematics is rarely challenged. The current ICMI Study Group is unusual in its focus on difference within the field. The Study Group implicitly acknowledges two major traditions in mathematics and mathematics education: an Eastern one centred in China, and a Western one centred in Europe. Neither of these major traditions are homogeneous, nor are they located in any one place.

The acknowledgement of two conventions, however broad, legitimises the idea that there can be other conventions. This chapter explores such legitimisation by first describing some of the philosophical background to different mathematical world views, then illustrating with two examples, one from The Philippines, and one from Brazil. It then notes the nature of difference within conventional mathematics with the example of Russian mathematics education. Finally, the chapter contextualises the discussion within the politics of knowledge, questioning whether less dominant world views are given due respect for their potential and actual contributions to mathematics and mathematics education.

1. BACKGROUND

Different mathematical approaches, such as those exemplified in Eastern and Western traditions, involve different philosophies and sociologies of knowledge (Ernest, 1994; Restivo, Van Bendegem & Fischer, 1993), and different cognitive and conceptual approaches. Within the worlds of mathematics and of mathematics education, these are masked because there has always, through the history of mathematics, been a tendency to focus on what is common, not on what is different or cannot easily be articulated across cultural boundaries.

Part of this tendency is due to the nature of the subject itself. For a subject that defines itself as pure human thought, and argues for absolute or a priori status for its results, the idea of cultural difference is anathema. A focus on commonalities establishes the illusion (at least) that there is something about the subject that transcends any subjectivity, something about the truths of mathematics that are implicitly known to all. The nature of the classical mathematical task (in the Greek tradition) is the reduction to essential axioms; the search for foundations of the early 19th century was the pursuit of certainty. These grand designs, and the turning to mathematicians in both Eastern and Western traditions as the final arbiters of truth and taxes, has established, in history, the status of acultural (or pan-cultural) knowledge.

It is a major turning point, therefore, to raise, in an international forum, the issue of different mathematical traditions, and the spectre of more fundamental differences within mathematics than those of approach, education, or relationship to society.

It is well acknowledged by anyone with even passing knowledge of the history of mathematics, that contributions to the mainstream of mathematical development have come from many different cultures. It is becoming increasingly recognised, however, that there are cultural components to these well-springs that have been either ignored or misunderstood in the process of assimilation into one "acultural" discipline. The writings of Joseph (2000) have highlighted the unique nature of mathematics in, for example, Indian culture and from different parts of Africa. (Both Indian and African traditions are important exemplars of this theme, but are too large to be adequately treated in this chapter. Readers are referred in the first instance to Joseph (2000) and AMUCHMA—the African Mathematics newsletter).

When mathematical world views are well-established and robust, as they are within Eastern and Western mathematical conventions, there is a possibility of equal and comparative debate about their nature and consequences. However, when the world view is held by a small group, especially one that has been colonised or is in interaction with a larger group, then comparative debate is likely to be difficult and invidious. The dominant paradigms become the measure against which other conventions are judged, and when some of the criteria are aspects like generality or width of application, then conventions based in smaller communities will not count. It should be noted that the effect of dominant conventions operates within mathematics as well as externally.

A more fundamental aspect of this same problem is that the very idea of mathematics is defined by the dominant paradigm – any other system will not be called 'mathematics' and therefore will not be acceptable to 'mathematicians'. A requirement, therefore, for considering the existence and nature of other mathematical world-views is a definition of 'mathematical' which is inclusive and not dependent on Western or Eastern conceptions. An attempt at resolving this problem is D'Ambrosio's "jargons and codes, which clearly encompass [mathematisation], that is the way [people] count, measure, relate and classify, and the way they infer" (D'Ambrosio, 1984). Another is Barton's "'QRS systems', that is, systems by which we make meaning of quantity, relationships or space" (Barton & Frank, 2001).

But the importance of acknowledging and understanding other conceptions and conventions of mathematics is very great. First of all there is the need to embody in our academic activity fundamental humanitarian characteristics such as respect for others, acknowledgement of difference, and basic equality of all humans of access to education and economic health. Every mathematician has a responsibility to help reverse the exploitative history of colonialism and the role played by mathematics in that process (Powell & Frankenstein, 1997).

In addition, there is the educational imperative for all people to increase their mathematical (and hence scientific and technological) literacy so that there is a chance of continued emancipation and increased ability to understand and control aspects of human development (D'Ambrosio, 1994).

Less well recognised, but also vitally important, is the potential of mathematics from other world views to contribute to the broad conventions of mathematics. The history of mathematics teaches us that many major mathematical breakthroughs came from specific cultural origins – indeed the Greek tradition of argumentation that gave rise to the axiomatic method itself is an example. The cultural roots of much mathematics are now not well known, but that does not mean that these origins were unimportant. The very size and integration of the near-universal, conventional form of mathematics of this age is a danger to itself because it increasingly denies the possibility of life-giving contributions from outside itself. For example, the increasing dominance of English as the language of international mathematics precludes ways of thinking that can only be expressed in other languages (Barton, 1999).

It needs to be recognised that cultural difference in mathematical thinking does not just refer to different number systems, or alternative units of measurement, or ways of expressing relationships (although all of these can signal significant mathematical differences, and usually signify significantly different sociological difference – see below). The work of Pinxten, van Dooren, and Soberon, (1987) alerted us to ways of seeing the world mathematically that were different from the Indo-European one. There are other examples of organised systems of mathematical thinking that are not reducible to the conventions of mainstream mathematics, for example Kolam drawings (Ascher, 2002) or Maori weaving patterns categorisation (Barton, 1995). Strong arguments have also been mounted for recognition of the fundamentally mathematical nature of other systems, for example Turnbull (1991) with respect to Pacific navigation systems (see also Gladwin, 1990; Kyselka, 1987; Lewis, 1975), or Cooke (1990) with respect to genealogical systems.

A consideration of other mathematical world views relates to the rise of awareness of the nature of indigenous knowledge. Indigenous knowledge refers to the unique knowledge held by indigenous people about their social, political, cultural and ecological life (Grenier, 1998). It is stored in memories, captured in songs, folklore, proverbs, dances, values, beliefs, rituals, and is expressed in community laws, language and specific practices. Some characteristics include (Alangui, 2002):

- it is embedded, tied to place, or locale;
- it has evolved over time, and experience, and is orally-transmitted;
- it is holistic, and may include beliefs contrary to conventional science;
- it is dynamic.

But indigenous knowledge does not bear categorisation in terms other than those of the indigenous group concerned. The moment a piece of indigenous knowledge is categorised as "biology", "gynaecology" or "mathematics", it has been shifted from its original world view, and becomes something less than it really is in its own cultural context. Indigenous knowledge can only be properly understood from within the world view of that indigenous group. Indigenous knowledge can, of course, be largely understood and used by people outside its originating group, however its proper understanding is inextricably linked to that world view.

In what sense can we talk about mathematical knowledge as arising from indigenous knowledge? Ethnomathematics is different from indigenous knowledge because it is seated in the relationship between a specific cultural group and the world-wide field of mathematics. Even ethnomathematicians who work within their own culture are involved in the mathematical interpretation of that culture. Such activity is still dependent in a theoretical way on some concept of mathematics – a concept that, in its international sense, is not internal to any one culture. The fundamental differences in the parties to this relationship highlight political and equity issues in mathematics education (Knijnik, 1999).

Indigenous knowledge therefore raises issues in the sociology of knowledge: questions of universality, rationality, relativism and reflexivity (Turnbull, 2000; Walkerdine, 1988). The relativist assumptions of both indigenous knowledge and ethnomathematics are at odds with accepted philosophies of mathematics.

The next two sections of this paper give examples of other mathematical traditions: the first is embedded in indigenous knowledge and exemplifies an ethnomathematical approach to mathematics and mathematics learning, the other describes an internal differentiation within mathematics. The final section returns to a consideration of politico-social issues.

2. THE BRAZILIAN LANDLESS MOVEMENT: AN ETHNOMATHEMATICAL PERSPECTIVE

This section presents work which was developed with the Landless Brazilian Movement some time ago (Knijnik, 1998). This national movement involves approximately two hundred thousand families of peasants. At the center of the Landless Movement struggle is the implementation of a Land Reform which will contribute to the democratisation of wealth in a country with the largest concentration of land in the world. One of the dimensions of this struggle is education, where the Landless Movement has made an original contribution, thanks mainly to the ideas of Paulo Freire from the 1960s. The struggle for land can be said to be so amalgamated with education that each reinforces the other. It was the problems that arose in the struggle for land that indicated the need to set education as one of the priorities. Thus the structural struggle for land reform, takes as one of its priorities the education of its members. The Education Sector of the Landless Movement is developing work on primary and secondary schools, the education of youths and adults (with priority to literacy and numeracy), infant education, and in-service and pre-service teacher education.

In this rural educational context marked by cultural difference, mathematics education plays an important role, as it has at its centre the analysis of the interrelations between popular, technical and academic knowledge. This means the assumption of an ethnomathematics approach, understood as (Knijnik, 1997):

> the investigation of the traditions, practices and mathematical concepts of a social group and the pedagogical work which is developed in order for the group to be able to interpret and decode its knowledge; to acquire the knowledge produced by academic Mathematics and to establish comparisons between its knowledge and academic knowledge, thus being able to analyze the power relations involved in the use of both these kinds of knowledge.

This concept opposes the ethnocentric view with which popular cultures have often been treated, and articulates relativistic and legitimising perspectives in examining the mathematical practices of socially subordinated groups. However, it avoids the relativism which would end up producing what Grignon (1992), called "ghetto-isation of the subordinated groups". In the case of Landless Movement, a social movement whose action is in permanent relationship with the dominant groups, this ghetto-isation process would occur if the pedagogical process were limited to the recovery of native knowledge and the glorification of this knowledge.

In the last few years, the young people have begun seeking work alternatives in the cities, thus moving away from the specific struggles of the Movement. This second generation pressure for new possibilities of work and leisure arises now that their material needs are fulfilled. Thus the Landless Movement has sought to implement new projects involving the settlement youths. These needs inspired a project which began with 7th graders of a settlement school, involving the joint action of peasants, students, teachers, and agronomists in constructing pedagogical work in mathematics focused on the productive activities in the settlement. These activities are organised by groups of peasants who carry out all stages of production, from planning to commercialisation.

Initially, the students analysed the bank loan contracts of each group of settlers, to configure the profile of each debt. This was the first opportunity these youths had had of looking at official documents, and it required an understanding of financial mathematics and previously unknown mathematical tools, such as compound interest. The debt profile was presented at meetings with each group of settlers. For the peasants – many of whom were illiterate and most of whom had at most 4 years schooling – this was the first time they had access not only to the final amount to be paid to the bank, but to the details that produced this result.

This first stage triggered further stages, each involving the problematicsation of the production of a specific crop. What follows is an example that began at a joint meeting of the agronomist, students, and teachers, with the settlers of the "Rice Group". This group consists of peasants (originating from a distant region, a region in which soybeans, maize, and bean crops predominate) and of former employees of the farm which, after expropriation by the State, gave rise to the settlement. The characteristics of the soil render it appropriate for rice. Thus, in the group there are women and men for whom rice production is part of their life trajectories, and also those for whom it is a foreign element with which they still have difficulty. As Seu Arnoldo explained: "I am from elsewhere, I have been here for ten years but I have not yet caught up with the pace". The agronomist seeks to understand this "pace", attempting to speed it up by technical qualification.

During the first meeting organised in order to plan the rice crop, one of the questions concerned the amount of land which would be planted. One settler suggested that 30 quadras could be planted, another showed the possibility of planting "up to one colônia". When they heard these terms, one of the students interrupted the discussion to ask how many quadras there were in a colônia. (The expression colônia is used with different meanings in Brazilian rural areas. In this situation the settler was using it to signify 2.5 hectares). The settler answered: "Look, I deal in quadras, they in colônia". The dialogue continued:

> Agronomist: One quadra is 1.7 ha, i.e. 17424 meters.
> Seu Helio (settler): That is saying it in meters, in braças it is 3600 braças.
> Márcio (student): What is braça, Seu Hélio?
> Seu Helio: One braça is 2 meters and 20...is a braça, see? Let's say: cuba here, cuba there... 60 braças like this, the four strips here: see, 60 there, here, 60,60,60,60, to see how it makes a quadra, we will have exactly the 3,600.

Seu Helio was saying that one quadra is the equivalent of a square of 60 braça on each side, that is, a square of 3600 "square braça". The use of measures such as braças in the Brazilian rural environment has been examined by authors like Abreu & Carraher (1989) and Oliveira (1997).

Initially, it appeared that the answers given by the agronomist and the settler were enough for the youths. But when they returned to the classroom the explanations proved unsatisfactory and required more detailed study. The discussion began with what the specialist said, explaining that in the rice plantation, peasants deal mainly with quadra, although there are also those who use colônia as a measure, but he does everything in hectares since the bank loan contracts are in hectares.

Several questions arose: What kind of translation occurs when quadra is expressed in hectares instead of braças? How is colônia translated into quadra? And how do both connect to hectare? How can one establish bridges and shifts between these understandings? What are the effects, in terms of power relations, of these translation processes amongst the Rice Group and in the community?

The pedagogical work sought to problematise these questions. It was not a matter of performing translations which would be limited to numerical equivalencies. This would reduce the study to the demonstration that if a braça is 2.2 meters, then 60 braças are 132 meters, and therefore a quadra is 17424 square meters = 1.7424 ha. An approach which limited itself to this kind of operation would precisely be reducing the work to the formal academic mathematics in which "the practice operates by means of suppression of all aspects of multiple signification" (Walkerdine 1988, p.96). The approach of the project coincides with "the position ... that the object world cannot be known outside the relations of signification in which objects are inscribed" (Walkerdine, 1988, p.119).

The use of specific surface measures produces meanings which are culturally constructed. The imposition of standard measures was not the result of a consensus based on arguments of precision, nor by arguments of universalisation. On the contrary, there are examples of popular revolts such as the "Kilo Revolt "which took place in Brazil in 1871 (Sotto Mayor, 1978). This revolt had as one of its causes the imposition in the country of the French metric system. This part of history is not usually mentioned in the school curriculum, but was part of the work developed in the project, and allowed the construction of bridges between the history of mathematics and mathematics education.

Past and present cultural practices were examined as part of the struggle to impose meaning. Examples of non-official knowledge, vocalised by peasants from different regions with different traditions, were recovered, and were confronted with dominant knowledge, vocalised by the agronomist. In this process the traditions behind quadra, braça, hectare and colônia were also translated.

This episode points to several questions that might be relevant in other social contexts. Peasants, students, teachers and a technician were experiencing the construction of an educational process in which local and more global knowledge interact, where native and technical knowledge are confronted and incorporated. The pedagogical work overflowed the school limits, producing the double movement of making community life penetrate the school at the same time as knowledge produced during this process emanated from the school space. This two-way movement created a pedagogy that did not reinforce the hegemonic ways of learning and teaching mathematics marked by the western, white, urban male culture (Knijnik, 1996).

The pedagogical approach focused on problems of practical and material needs, rather than symbolic control problems, indicating other possibilities in the field of mathematics education, especially in mathematics education that is carried out in different cultural settings, such as the Landless Movement.

3. INTERNAL DIFFERENTIATION—THE CASE OF RUSSIAN MATHEMATICS EDUCATION

Russia is a good example of differentiation within mathematical education. Its special features are determined both by its geographical position (on the border between Europe and Asia), its socio-cultural traditions, and its political structures. Mass mathematics education occurred in Russia in the 20th century and has developed as one of the best in the world. What have been the national features behind such development?

The objective reasons for the intense development of mathematics and mathematics education in Russia were, as in other countries, the necessities of economic and military development. However, being a huge country, Russia had especially strong economic and military motivations because the feudal economy of 19th century Russia needed to cope with the Russian-Japanese war (1905) and two World Wars.

The basis for advanced mathematics education in Russia was the elitist system that existed at the beginning of the 20th century. The elite taught the elite and this education naturally reached high levels. An example is the famous theoretician of space flights Tsiolkovskiy, who reached the pinnacle of physics although he was a schoolteacher in a small city.

Another base was the existence of good textbooks generated by the reforms of Peter Great. In 1725 he founded the Academy of Sciences in Russia and invited famous European mathematicians such as Euler to work there. This not only established a high level of mathematical research in Russia but also resulted in the publication of mathematical textbooks that were to be the basis for many others. From that time Russian mathematics textbooks changed very little. For example, a geometry textbook by Kurganov (based on Euler's own text) was used from 1765 till 1845. This was followed by a textbook by Busse (new version of Kurganov's text), which was in turn followed by one by Kiselev that was based on Busse's work. Kiselev's text was published in 1893 and was in use as the main textbook until 1976 (throughout the socialist revolution).

Mathematics education programmes evolved over more than a century and proved to be successful. Arithmetic calculation, text problems, Euclidean geometry (including 3D-geometry) became the basis of the curriculum. Much attention was paid to development of logic: along with the question "how", the main question was "why". This orientation corresponds to a distinctively Russian way of thinking.

One of the main features of the development of education in Russia is that it was conducted by a strong centralized power. Decisions were made at a very high level and were strongly controlled, particularly after the Socialist Revolution of 1917. In the 1930s, the government decided to implement fast industrialisation, so that a lot of new specialists were needed. The decision was made to intensify fundamental education, and socialist concepts of equality required the development of a common education, rather than an elitist model. Secondary education became available for all and later became obligatory. In mathematics education there was no simplification of programmes. In spite of their high level, the government obliged young people of all regions to try to master mathematics according to these programmes. Many new pedagogical institutes were opened and graduating students had to work for several years where directed by the government. The journal Mathematics in School played a very important role. Its articles were mostly devoted to mathematics itself rather than to methodology.

Assessment was another part of centralized control of knowledge. In the 1930s students took national examinations each year from the 4th grade. In the 1950s they took examinations only after 4th, 7th and 10th (last) grades. Now they take examinations after 8th and 11th (last) grades. The examinations in mathematics are in both written (algebra) and oral (geometry) forms. To enter universities or institutes they need to take further special examinations. A special feature of oral examinations at Moscow State University is that they include proofs not only in geometry but also in algebra.

Despite the socialist imperative of equality, mathematics education in Russia was also strongly oriented to the special education of gifted students. An important factor of Russian mathematics education was that high school professors took part in the work with school students. In the socialist period, social activity (without salary) was very much appreciated and even obligatory. For university staff and high level teachers a suitable form of the (necessary) social activity was teaching students extra mathematics in special evening lessons. Such groups were organised in many universities and institutes. To make these lessons interesting, professors either taught those topics that were not considered in secondary schools or dealt with interesting logical and geometry problems. All this contributed to a higher level of mathematics education for top students. Later these ideas transformed into special mathematics classes, and eventually into the development of special mathematics schools in many cities. In 1963 some special mathematical schools with dormitories were opened in universities. These catered for gifted students from the whole country, for example, about 200 students graduated from the special mathematical and physical school of

Moscow State University each year. Since 1963 the Moscow State University organized a Distance Mathematics school for students from the whole of Russia. Since 1970 a special mathematical and physical journal Quant for students has been published.

The idea of competition in production was very popular in the 1930s. The concept also took hold in mathematics. From the beginning many famous Russian mathematicians took part in their organisation. For example, Academician A. N. Kolmogorov was the chairman of several mathematical Olympiads. In 1961 the first All Russian Mathematical Olympiad was organized with several stages: school, region, state, with about 300 school students taking part in the finals in Moscow. Moscow and All Russian mathematical Olympiads have been organised every year since that time, resulting in good results in international competitions.

Since the 1970s several changes have been made in mathematics education curricula. Elements of mathematical analysis, linear algebra, analytical geometry, and geometrical transformations are included, while Euclidean geometry remains the basis. Both geometry and algebra include now more theory than before and less problem solving. School mathematics has become difficult for students. As a result of this and of the democratisation of society, there are different opinions about mathematics education. A vigorous debate is now developing around mathematics education in Russia, and future directions are far from clear.

It can be seen from this short case description that national features have had a considerable influence on the development of both mathematics and mathematics education in Russia. The high level formal mathematics emanating from that country has its genesis in a particular socio-political history. Current debates have developed partly in response to outside interactions. Thus developments internal to mathematics respond to similar circumstances as those of external mathematical knowledge systems.

4. MATHEMATICS EDUCATION AND THE POLITICS OF KNOWLEDGE

The question of cultural difference in mathematics is a question not only from the anthropological standpoint, but also a question about seeking to understand mathematical difference sociologically, including the way differences constitute inequalities.

Thus, in returning to a discussion about the relationship between hegemonic mathematical knowledge transmitted by schools and the mathematical knowledge that is part of the students' culture, we argue that the main issues are those of the politics of knowledge and the politics of identity. Although one of the main purposes of schooling is to assure everybody has access to the knowledge of conventional mathematics, the price to be paid for this is appears to be the erasure of other forms of mathematical knowledge that have been marginalised throughout history. Boaventura dos Santos, when referring to the destruction of the knowledge of a given social group, refers to this as epistemicide (Silva, 2003, p. 196):

The opposition to such suppression cannot depend upon benevolence. What is required is more than mere respect for other modes of dealing mathematically with the world. It is not a matter of the "return" of voices repressed by Eurocentric discourses (Grossberg 1993, p.91). We are interested in this return, but we are aware that it is pregnant with complexity. The contemporary debate in mathematics education has attempted to understand this complexity in order to avoid naïve positions that would lead to folklorising indigenous knowledge, and to benevolent attitudes that would include these forms of knowledge in the school curriculum as long as they stayed on the edges. McDowell (2003), mentions Edward Said's Orientalism, and argues that the author is:

> attempting to open time for listening to a multiplicity of previously silenced voices, voices drowned out by the controlling master narrative. This movement of giving a certain sight to those "blind to other histories" in itself, then, is a form of resistance. Consequently, resisting the discursive hegemony becomes, then, it should be added, a morally significant matter that is shaped by the construction of alternative visions or ways of telling the story that more comprehensively incorporate and retain the distinctiveness of these voices. (p.2)

The articulation of these voices, in the field of mathematics education as elsewhere, is directly connected to the production of social identities. Both the topics that are selected as the object of study, and the theoretical tools which we use with these topics, reinforce certain identities and weaken others. Such identities are neither fixed nor unique, but are subject to our social world. As Woodward (2000, p.14) showed, identity is relational and is connected to social and material conditions, such as the school curriculum. For example, at school we teach the meanings of mathematical reasoning and ways of communicating this reasoning, highlighting those that are "right" and which of them should be ignored because they are not sufficiently important. Thus, for instance, the specific ways in which Brazilian peasants calculate the area of their lands, or the way New Zealand Maoris categorise weaving patterns, are devalued, and are usually silenced in the schooling processes. This creates a dichotomy between "high" and "low" mathematics. This dichotomisation is aligned with the politics of dominant knowledge: it shapes specific identities in teaching other things besides mathematics content. That is, it positions the students in certain places in the social world and not in others. It ends up by excluding particular world views, and thereby reinforces social inequalities.

Ultimately, this is what we mean when we observe the master narrative of academic mathematics reigning alone over the school curriculum, without allowing the presence there of other, non-hegemonic, narratives. This is why it is important to consider other conventions, both within conventional mathematics and external to it.

REFERENCES

Abreu, G.M. and Carraher, D.W., 1989, The Mathematics of Brazilian Sugar Cane farmers, in: *Mathematics, Education and Society*, UNESCO, Paris, pp.60-70, (Document Series 35).

Alangui, W., 2002, *Stone Walls in Agawa and Gueday*. Unpublished paper, The University of Auckland

Ascher, M., 2002, The Kolam Tradition, *American Scientist*, **90**:57-63.

Barton, B., 1995, Making Sense of Ethnomathematics: Ethnomathematics is Making Sense, *Educational Studies in Mathematics*, **31**(1&2):201-233.

Barton, B., 1999, Une Archéologie des Concepts Mathématiques: Tamiser le langue pour des significations mathématiques (An Archaeology of Mathematical Concepts: Sifting Languages for Mathematical Meanings), in: *Proceedings 1999 Annual Meeting of the Canadian Mathematics Education Study Group*, John McLoughlin, ed., Memorial University of Newfoundland, St John's, Canada, pp.57-70.

Barton, B and Frank, R., 2001, Mathematical Ideas and Indigenous Languages: The Extent to which Culturally-specific Mathematical Thinking is carried through the Language in which it takes place, in: *Sociocultural Research in Mathematics Education: An International Perspective*, B. Atweh, H. Forgasz and B. Nebres, eds., Mahwah, Lawrence Erlbaum Associates, NJ, pp.135-149.

Cooke, M., 1990, *Seeing Yolngu, Seeing Mathematics*, Batchelor College, Northern Territory, Australia.

D'Ambrosio, U., 1984, Socio-Cultural Bases for Mathematical Education, in: *Proceedings of ICME-5*, Adelaide.

D'Ambrosio, U., 1994, Cultuiral Framing of Mathematics Teaching and Learning, in: *Didactics of Mathematics as a Scientific Discipline*, Biehler, R., Scholz, R., Strasser, R. and Winkelmann, B., eds., Kluwer Academic Publishers, Dordrecht, pp.443-456.

Ernest, P., ed., 1994, *Mathematics, Education and Philosophy: An International Perspective*, Studies in Mathematics Education Series No.3, The Falmer Press, London.

Gladwin, T., 1970, *East is a Big Bird: Navigation and Logic on Puluwat Atoll*. Harvard University Press, Cambridge, MA.

Grenier, L., 1998, *Working with Indigenous Knowledge: A Guide for Researchers*, International Development Research Centre, Ottawa.

Grignon, C., 1992, Teoria & Educação, *Porto Alegre*, n.5:.50-54.

Grossberg, L., 1993, Cultural Studies and/in New Worlds, in: *Race, Identity and Representation in Education*, ed., Routledge, New York, pp.89-108.

Joseph, G., 2000, *The Crest of the Peacock*, Revised edition, Penguin.

Knijnik, G., 1996, *Exclusão e Resistência: Educação Matemática e Legitimidade Cultural*, Artes Médicas, Porto Alegre.

Knijnik, G., 1997, Popular Knowledge and Academic Knowledge in the Brasilian Peasants'Struggle for Land, *Educational Action Research Journal*, **5**(3).

Knijnik, G., 1998, Ethnomathematics and Political Struggles, Zentralblatt für Didaktik der Mathematik, Jahrgang, 30(6) December.

Knijnik, G., 1999, Indigenous Knowledge and Ethnomathematics Approach in the Brazilian Landless People Education, in: *What Is Indigenous Knowledge? Voices from the Academy*, L. Semali and J. Kincheloe, eds., Falmer Press, New York.

Kyselka, W., 1987, *An Ocean in Mind*. University of Hawai'i Press, Honolulu.

Lewis, D., 1975, *We, the Navigators: The Ancient Art of Landfinding in the Pacific*, Australian National University Press, Canberra.

Oliveira, Lucas De H., 1997, Educação Rural e Etnomatemática. Monografia de Curso de Especialização, UFRGS. Texto Digitado.

McDowell, J., 2003, Edward Said on Orientalism: Knowledge as Power. www.geocities.com/ johnnymcdowell/papers/. Accessed 04.10.04.

Pinxten, R., van Dooren, I. and Soberon, E., 1987, *Towards a Navajo Indian Geometry*, K.K.I. Books, Gent.

Powell, Arthur B. and Frankenstein, M., 1997, Ethnomathematics: Challenging Eurocentrism in Mathematics Education, State University of New York Press, Albany.

Restivo, S., Van Bendegem, J. P. and Fischer, R., eds., 1993, *Math Worlds: Philosophical and Social Studies of Mathematics and Mathematics Education*, State University of New York Press, Albany, N.Y.

Silva, T.T., 2003, Currículo e identidade social: territórios contestados, in: Silva, T.T. (org.), Alienígenas na Sala de Aula. Uma introdução aos estudos culturais em educação, Vozes, Petrópolis.

Sotto Major, A., 1978, Quebra-quilos: lutas sociais no outono do império, Nacional, São Paulo.

Turnbull, D., 1991, *Mapping the World in Mind: An Investigation of the Unwritten Knowledge of the Micronesian Navigators*, Deakin University Press, Geelong.

Turnbull, D., 2000, Masons, Tricksters and Cartographers, in: *Comparative Studies in the Sociology of Scientific and Indigenous Knowledge*, Harwood Academic Publishers, Amsterdam.

Walkerdine, V., 1988, *The Mastery of Reason*, Routledge, London.

Woodward, K., 2000, Identidade e diferença: uma introdução teórica e conceitual, in: *Identidade e diferença*, SILVA, Tomaz Tadeu, Vozes, Petrópolis, pp.7-72.

Chapter 5-3

WHAT COMES AFTER THIS COMPARATIVE STUDY – MORE COMPETITIONS OR MORE COLLABORATIONS?

Alan J. BISHOP
Monash University, Australia

This conference and the whole study has offered a remarkable opportunity for scholars from very different cultures to explore the similarities and differences in the ways we carry out the various aspects of mathematics education. Crucially it takes as its basic assumption the notion that difference is to be welcomed and celebrated. There is no attempt being made to colonise one culture by another; there is no notion that one way is 'better' than another. There is an implicit and often explicit recognition of equality in the sense of 'equal' scholars struggling to understand similar questions and deal with similar issues. More than that, the scholars who formed this Study group shared a fundamental belief that in order to gain maximum benefit from our comparative work we needed to focus explicitly on issues of cultural difference and contrast.

Much of the background to this study, and much of its impetus, has come from a competitive era, where international surveys of mathematics achievement attained the status of international Olympiads on a grand scale, with countries, their governments, and other interested parties poring over test results to see who had 'won' and who had 'lost'. Clearly much has been learnt during the prosecution of studies such as the Second International Mathematics Study (see for example, Travers & Westbury, 1989), the Third International Mathematics and Science Study (see Beaton & Robitaille, 1999), and the OECD International Programme for Student Assessment (see www.pisa.oecd.org for an overview), but their main emphasis has been on developing comparative studies rather than specifically focussing on the cultural differences and what one can learn from them.

Collaboration and competition are however not in direct opposition, but they are complements of each other. One needs collaboration in order to compete, as has happened with the previous studies, but one does not need to compete in order to collaborate. In the act of collaboration there will inevitably be comparison, either explicitly or implicitly, but in this study the participants don't just want to compare in a simplistic way, but to understand difference and similarity at a deeper level than is normally possible. A good example of this from my own work has been the collaboration with colleagues in Taiwan and Malaysia over values research, where comparison becomes a form of research procedure where each others' systems and procedures act as surrogate experiments. The importance of such cross-cultural research activity should not be under-estimated.

It is clear that we all grow in our learning by contrast, but we must construct intentional, not just accidental, collaborative activities in order to nurture that growth. In the rest of this paper I will therefore outline what I see as the issues to be overcome if potentially fruitful collaborative activities are to be developed.

1. RESEARCH COLLABORATIONS USING CROSS-CULTURAL APPROACHES

If we seek to develop more cross-cultural research studies then the first major issue concerns which cultures should one choose? I believe it is important to seek important differences in relation to the research questions in the study, rather than just take any differences. In the values research study mentioned above, it was clear that Taiwan, with its Confucian heritage and Malaysia with its Muslim religion represented two very different cultures from the mainstream culture of Australia. Those differences would surely involve very different sets of values (see Bishop, Seah and Chin, 2003).

Another issue is that sometimes there are special grants and government interests which focus on certain cultural comparisons, and these may well drive the research study. It may also be a case of countries having common problems in which case it is usually the potential partners who decide. But we may not necessarily all face the same problems, yet still be interested in some kinds of comparison. TIMSS is a good example of a study where the different countries involved wanted to address different issues, but were all addressing the major problem of how to provide a worthwhile mathematics education for the majority of students.

A third issue is that often cross-cultural research studies can expose issues of equity, in terms of funding differences, participation differences, political and social differences etc., and these should not be ignored because they often reveal significant differences in educational policy and practice which affect any of the results and findings. A particular example, and one which is relevant for this study and in an ICME year, concerns conference and publication participations and contributions.

It is always important for those working on international research studies to present their work in international conferences and publications, and there are nowadays many possible opportunities for doing this. However it may not be possible for some of the partners in the cross-cultural comparisons to attend conferences, because of financial, geographical, or even political reasons. Equally it may not always be possible for all the partners to contribute to written reports or papers because of the restricted set of languages used in international publications. In all these cases both contributors and readers must always be aware of the equity issues raised: Whose voices are heard? Who does the 'talking'? Which countries/cultures are under-represented in any particular study? Which countries/cultures are always being under- or mis-represented? And how can this issue of under-representation be dealt with?

A further aspect of equity concerns the research procedures used in any study, and one can ask: are these procedures culturally appropriate for all countries in the research study? An excellent paper on this point is by Valero and Vithal (1999) who argue that most research procedures used in our field have come from what they call the 'North' – using the word in not just its geographical sense but as a euphemism for the developed/colonial/high economy countries, who have imposed not just their cultures but specifically their research procedures on their colleagues who work in less-developed, or colonised, or low economy countries. One strategy which is gaining international currency is to use a combination of procedures, for example combining large, quantitative surveys with several small-scale studies that focus on specific aspects highlighted by the quantitative study. Once again, though, this strategy may not be applicable or appropriate for those working in developing countries.

Research collaborations are certainly on the increase, and provided that the issues mentioned here are addressed, the possibilities for development are limitless.

2. SCHOOL COLLABORATIONS

Also worth developing are school collaborations, fuelled by two developments – the relative cheapness of international travel, and also the increasing use of ICT in school mathematics education. Regarding the first, where students from one school will travel to others in another country, the experiences of teachers and schools involved in language exchanges are salient, and raise issues which will need addressing. Certainly there are enormous benefits associated with such exchanges, but there are some crucial issues.

In particular, the language and cultural differences will present a challenge although, since the focus will not be on language learning *per se* but on mathematics learning, there is no need to focus attention just on exchanges with a different language-based country. Of more importance for mathematics education could be differences in economic and human resources, and this is where the benefits for both teachers and students will exist. For school students from Australia to experience mathematics learning in schools in its nearest neighbour, Papua New Guinea, would be both interesting and challenging. To see how teachers there use local materials as well as the local languages to develop similar mathematical concepts to those they study 'back home' would indeed broaden their knowledge and attitudes! And of course, developing an exchange program means that students from PNG should be assisted to visit their counterparts in Australia, and no doubt this would also be a broadening experience for all concerned.

As yet, collaborations of this kind are few and far between and are building on the experiences and contacts in foreign language learning, e.g. Japanese and Chinese exchanges with Australian schools. They are certainly proving to be an interesting experience for both students and teachers.

School collaborations of the second kind are more likely to develop in the near future through the increase in ICT use in schools everywhere, and not just because of the use of ICT in distance learning activities. Once again there are cost and equity issues to be borne in mind. The cost issues are about the relative costs for each of the partners, both in terms of setting up, and the running costs, much of which can be hidden in the form of infrastructure costs. Another issue here is that the costs will include the establishment of compatible hardware and software.

It is unlikely that the order of costs involved with both kinds of collaboration can be borne within normal school budgets, and therefore it is usual for special funding to be sought either in the form of special government support schemes or by sponsorship from industry. There are problems with both of these, in particular the limited time the support will usually be given for. Most collaborations of this type are occurring through various projects, and it is necessary to build into the costs of such projects appropriate evaluations, taking into account costs, curriculum benefits, students' learning outcomes, teachers' impressions, time commitments etc.

Also it is rare for the financial supporters not to have an agenda of their own, and it may well be important to remember here that computers, WWW, email, video-conferencing etc. were not created for educational use. They do allow certain kinds of collaborative activities to occur, but cannot replace face-to-face collaboration. Clearly there are enormous curricular implications and it is essential that the teachers are fully involved in any projects being planned. Although the overall mathematics curriculum can appear to be very similar from country to country, there are often important differences in terms of when particular topics are taught, the depth to which that teaching goes, the sequencing of topics, as well as the style of the pedagogy.

3. UNIVERSITY TEACHING COLLABORATIONS

At the university level, teaching collaborations are much easier to arrange. The curriculum is not so controlled by outside bodies, there is usually sufficient funding available for equipment, and there is often specialist technical assistance available for ICT developments. Also the students can be more independent in terms of the kinds of collaboration activities they do.

Teaching collaborations have often occurred where a university in a developing country has sought teaching assistance from a university in another, often more developed country. For example, there has been a very successful project between a Danish university and its staff supporting doctoral studies in South Africa, involving both student and staff visits to each others' institutions (see Skovsmose & Valero, 2002). Teaching collaborations have also frequently occurred where there has been a former colonial relationship between the two countries, or where two or more universities in a developing region have combined forces to offer jointly taught courses which neither could mount alone. These 'shared teaching' collaborations are now being fuelled by the ICT developments, where web-based teaching and learning materials can be easily used and supported, without the need for costly and time-consuming student or staff travel.

Typically in the mathematics education field, the collaborations have been created at the postgraduate or research student level, because of an understandable desire to maintain the initial teacher-training courses within the cultural values of the country concerned. Even at the postgraduate level there is a clear need for the hosting institution to be sensitive to the issues of language, cultural, and societal differences. There are also huge financial issues to be tackled, and typically governments and their agencies are involved in providing financial support, often as part of their 'aid' programs.

In these kinds of collaborations either the teaching staff, or more commonly the students, often need to travel to the other institution. There has been a long history of university staff spending their 'sabbatical years' at institutions in other countries in order to further their research experiences. At the student level, when the two universities are offering similar level courses, a regular student-exchange program can be created. This has long been a tradition in some countries, for example the USA, where the 'junior year abroad' schemes have helped many young people travel to other universities either for language experiences, or specialist cultural courses, or just to see the world. Governments do regulate such schemes in terms of 'exchange' study visas, and there are clearly cost implications for any university offering to award credit for such 'overseas' and 'part-time' students. Nevertheless the benefits of such schemes are enormous, not only for the travelling students. The hosting institution also benefits, and provided the university does not consider these overseas students as just of financial benefit, its courses and subjects can be enriched by their presence.

4. BUILDING INTERNATIONAL/INTERCULTURAL NETWORKS

Finally let us move to consider another level of collaboration, that concerning international educational networks. The chapter by Jacobsen (1996) in the first International Handbook on Mathematics Education is an excellent introduction to the many international networks that exist in our field, and he gives many examples of international collaboration which can arise under the umbrella of such networks. Of course this all helps to remind us that, generally speaking, any of the international collaborations above will rely on the existence of supporting networks offering finance, government help with visas etc. A good example of such a network is the SEAMEO organization (The Southeast Asian Ministers of Education Organization) which oversees several regional centres of educational activity such as the Regional Centre for Science and Mathematics located in Penang, Malaysia. These are important centres to support as they can facilitate many different kinds of educational collaboration.

In addition SEAMEO members can initiate regional educational activities which particularly interest them, and which can generate much collaborative work and study. Such was the project on Regional Cooperation on Quality and Equity in Education which was Thailand's initiative to the 37th Southeast Asian Ministers of Education Organization (SEAMEO) Council Conference in 2002. For this project a Regional Coalition of Schools on Quality and Equity in Education (RCS-QEE) was formed by selecting schools from each SEAMEO Member Country. The selected schools for the RCS-QEE served as pilot schools in the region, thus becoming deeply involved in a collaborative educational initiative

Furthermore, groups like SEAMEO can help by connecting with other international educational institutions such as UNESCO, a good example being the upcoming SEAMEO-UNESCO Congress and Expo to be held in Thailand in May 2004. This congress will bring together educational professionals from many countries, which will enable various contacts to be made having the potential of leading to activities of a collaborative nature.

Jacobsen (1996) however writes rather pessimistically at the end of his chapter when he says: "The institutions set up to provide world co-operation, the United Nations, and its education agency UNESCO, the World Bank, there are many, are being starved of funds and their activities have had to be curtailed." (p.1253) UNESCO should be a particularly important hub for networking, serving many regions of the world. However at the present time there is no full-time mathematics education officer in the whole of UNESCO so the science education staff take on mathematics education as part of their responsibility. UNESCO's educational mission is interestingly more about numeracy than mathematics, in line with their worldwide work in basic literacy and numeracy, and this example serves to remind us that all international networks have their own agendas, missions, priorities and exclusions.

In response to the decline above, Jacobsen offers this comment: "I come to the same conclusion as did Miguel de Guzman, President of ICMI, when he opened ICME-7 with a call for solidarity in mathematics education (de Guzman, 1994), that it was up to professional mathematics educators to work to improve mathematics teaching worldwide....National mathematics education associations should seek international co-operation, providing funds for participants from the third-world to join their conferences, and to provide free copies of their journals to at least other associations" (p. 1253)

These things are now happening, and there are now many regional conferences in mathematics education each year, held under the auspices of particular groups, such as the South-East Asian Conference on Mathematics Education (SEACME), or the more recent Conference of the European Society for Research in Mathematics Education (CERME). More are needed.

5. CONCLUSION

This current ICMI study is itself a fine example of mathematics educational professionals working together to see what developments can be generated by studying the intercultural similarities and differences between East and West. Moreover whilst this is the final chapter of the book, which in some way represents the end of this ICMI study, it is the hope of the IPC and of all involved, that this study is just the beginning of more and deeper collaborations. We have made great progress in our own understandings through this collaborative work and we look forward to helping develop future collaborations to benefit a much wider group of professionals.

REFERENCES

Bishop, A.J., Seah, W.T. and Chin, C, 2003, Values in Mathematics Teaching – the Hidden Persuaders?, in: *Second International Handbook of Mathematics Education*, A.J. Bishop, M.A. Clements, C. Keitel, J. Kilpatrick and F.K.S. Leung, eds, Kluwer, Dordrecht, pp.717-765.

Beaton, A.E. and Robitaille, D.F., 1999, An Overview of the Third Mathematics and Science Study, in: G. Kaiser, E. Luna and I. Huntly, eds., *International Comparisons in Mathematics Education*, Falmer Press, London, pp.19-29.

de Guzman, M., 1994, Presidential Address, in: *Proceedings of the 7th International Congress on Mathematics Education*, C. Gaulin et al., eds., University of Laval, Sante-Foy, Quebec, Canada.

Jacobsen, E., 1996, International Co-operation in Mathematics Education, in: *International Handbook of Mathematics Education*, A.J. Bishop, M.A. Clements, C. Keitel, J. Kilpatrick and C. Laborde, eds, Kluwer, Dordrecht, pp.1235-1256.

Skovsmose, O. and Valero, P., 2002, Mathematics Education in a World apart – Where We Are all Together, in: *Proceedings of the 3rd International Mathematics Education and Society Conference*, P.Valero and O.Skovsmose, eds., Danish University of Education, Copenhagen, pp.1-9.

Travers, K. and Westbury, I., 1989, *The IEA Study of Mathematics 1: Analysis of Mathematics Curricula*, Pergamon Press, New York.

Valero, P. and Vithal, R., 1999, Research Methods of the "North" Revisited from the "South", *Perspectives in Education*, **18**(2):5-12.

Index

Geographical Areas

NEW ICMI STUDY SERIES

1.	M. Niss (ed.), Cases of Assessment in Mathematics Education. An ICMI Study. 1992 (The 6th ICMI Study)	0-7923-2089-1 HB
2.	M. Niss (ed.), Investigations into Assessment in Mathematics Education. An ICMI Study. 1992 (The 6th ICMI Study)	0-7923-2095-6 HB
3.	G. Hanna (ed.), Towards Gender Equity in Mathematics Education. An ICMI Study. 1996 (The 7th ICMI Study)	0-7923-3921-5 HB 0-7923-3922-3 PB
4.	A. Sierpinska and J. Kilpatrick (eds.), Mathematics Education as a Research Domain: A Search for Identity. An ICMI Study. 1998 (The 8th ICMI Study)	0-7923-4599-1 HB 0-7923-4600-9 PB
5.	C. Mammana and V. Villani (eds.), Perspectives on the Teaching of Geometry for the 21st Century. An ICMI Study. 1998 (The 9th ICMI Study)	0-7923-4990-3 HB 0-7923-4991-1 PB
6.	J. Fauvel and J. van Maanen (eds.), History in Mathematics Education. The ICMI Study. 2000 (The 10th ICMI Study)	0-7923-6399-X HB 1-4020-0942-9 PB
7.	D. Holton (ed.), The Teaching and Learning of Mathematics at University Level. An ICMI Study. 2001 (The 11th ICMI Study)	0-7923-7191-7 HB 1-4020-0072-3 PB
8.	K. Stacey, H. Chick and M. Kendal (eds.) The Future of the Teaching and Learning of Algebra: The 12th ICMI Study. 2004	1-4020-8130-8 HB
9.	F.K.S. Leung, K.-D. Graf and F.J. Lopez-Real (eds.) Mathematics Education in Different Cultural Traditions- A Comparative Study of East Asia and the West: The 13th ICMI Study. 2006	0-387-29722-7 HB

Information on the ICMI Study program and on the resulting publications, including Studies 1 to 5, can be obtained at www.mathunion.org/ICMI/ or by contacting the ICMI Secretary-General, whose email address is available on the ICMI website.